Robert T. Mason   Michael P. LeMaster
Dietland Müller-Schwarze (Eds.)

# Chemical Signals in Vertebrates 10

With 115 Figures

Robert T. Mason  
Department of Zoology  
Oregon State University  
Corvallis, OR 97331  
USA  
masonr@science.oregonstate.edu

Michael P. LeMaster  
Western Oregon University  
Monmouth, OR 97361  
USA  
lemastm@wou.edu

Dietland Müller-Schwarze  
State University of  
New York, ESF  
1 Forestry Drive  
Syracuse, NY 13210  
USA  
dmschwarze@esf.edu

Library of Congress Control Number: 2005923605

Proceedings of the Tenth International Symposium on Chemical Signals in Vertebrates, held July 29–August 1, 2003, in Corvallis, Oregon, USA.

ISBN-10: 0-387-25159-6  
ISBN-13: 978-0387-25159-2  
e-ISBN: 0-387-25160-X

© 2005 Springer Science+Business Media, Inc.

All rights reserved. This work may not be translated or copied in whole or in part without the written permission of the publisher (Springer Science+Business Media, Inc., 233 Spring Street, New York, NY 10013, USA), except for brief excerpts in connection with reviews or scholarly analysis. Use in connection with any form of information storage and retrieval, electronic adaptation, computer software, or by similar or dissimilar methodology now known or hereafter developed is forbidden.

The use in this publication of trade names, trademarks, service marks and similar terms, even if they are not identified as such, is not to be taken as an expression of opinion as to whether or not they are subject to proprietary rights.

Printed in the United States of America.

9  8  7  6  5  4  3  2  1

springeronline.com

# PREFACE

The editors and contributors to this volume should be justifiably proud of their participation in the tenth triennial meeting of the Chemical Signals in Vertebrates International Symposium. This meeting was held 27 years after the initial gathering of participants in Saratoga Springs, New York from June 6$^{th}$ to 9$^{th}$, 1976. Subsequent meetings have been held every three years in Syracuse, New York; Sarasota, Florida; Laramie, Wyoming; Oxford, England; Philadelphia, Pennsylvania; Tübingen, Germany; Ithaca, New York; and Kraków, Poland. This tenth anniversary symposium was held from July 29$^{th}$ through August 1$^{st}$ in Corvallis, Oregon and was hosted by the Zoology Department and Biology Programs of Oregon State University. This book also represents the tenth in a series of books on chemical communication, chemical ecology, olfactory and vomeronasal research in vertebrate species. The species covered in the chapters herein range from fish to mammals including humans. By taxonomic breakdown the mammals are the most represented in number of species and chapter contributions. However, the hosts of the meeting endeavored to have some representative contributions covering all of the major vertebrate taxa.

As in past years, the meeting was well-represented with just over 100 participants from 13 different nations. Plenary talks focused on some of the non-mammalian groups that have tended to be less represented in these symposia. Thus, we had a very nice overview of comparisons and contrasts of invertebrate chemical communication to vertebrate systems. We heard about the myriad toxic compounds sequestered in the dart poison frogs, their chemical composition, their sequestration from certain insect prey species, their physiological effects in mammalian systems and finally their use in up and coming research as a whole new category of analgesic. In the neurobiological field we saw an elegant example of how the common garter snake's penchant for its earthworm prey could lead to a series of experiments that elucidated the biochemical nature of the attractant, its detection and perception by the vomeronasal system, the cellular cascade of second messenger systems that result in transduction of the signal to the brain and subsequent initiation of behavioral responses. We rediscovered that birds are important contributors and participants in the world of chemical ecology with an examination of the natural history, ecology and evolution, and chemical defenses of the Pitohui of New Guinea, the first group of toxic birds to be described.

Perhaps most important, to lead off the symposium, Dietland Müller-Schwarze gave a retrospective examination on how far the field has progressed since that first meeting that he co-hosted in Saratoga Springs. In his examination, he discussed the exciting new breakthroughs that weren't even imagined at that first meeting as well as areas where the research questions of that time seem to have been passed by or overlooked. It was an insightful look back by perhaps the only member of the current group qualified to do so. In honor of his founding of this series, his active participation in all of the last 10 Chemical Signals in Vertebrates Symposia, and the considerable influence he has had on almost all the books of this series as well as the field of chemical ecology itself, the other two co-editors of this volume (RTM and MPL) would like to dedicate this tenth anniversary symposium proceedings to Dietland Müller-Schwarze.

Finally, as this volume was being edited by the publishers, the world received the news that the 2004 Nobel Prize in Physiology or Medicine had been awarded to two scientists, Richard Axel and Linda Buck whose research careers have been dedicated to the elucidation of how the olfactory system codes and transduces the seemingly innumerable number of odorant molecules. Clearly the fields of chemical communication, olfaction and vomeronasal chemoreception are on the cutting edge of some of the most fascinating areas in science. The interdisciplinary nature of these fields is just as vibrant and evident as it was back in 1976. The future seems both bright and exciting for our field. We all look forward to our next meeting when we reconvene in Liverpool, England.

We thank the President of Oregon State University, The Dean of the College of Science, the Department of Zoology, and the Biology Program for their support of the symposium proceedings. Special thanks to Jamie LeGore of the LaSells Stewart Center who organized and managed the conference. Deborah Lutterschmidt helped to work out all the audio-visual technicalities without a hitch. We thank the owners and staff of the Tyee Winery and Valley Catering for the wonderful banquet and the staff of Squirrels for facilitating after hours discussions. Finally, many graduate students and undergraduate research students from the Department of Zoology helped in the behind the scenes chores that made this conference a total success.

Robert T. Mason  Corvallis, June 2004
Michael P. LeMaster
Dietland Müller-Schwarze

# CONTENTS

## PERSPECTIVES IN CHEMICAL ECOLOGY

1. Thirty years on the odor trail: From the first to the tenth international symposium on chemical signals in vertebrates........................... 1
   D. Müller-Schwarze

2. Pheromones: Convergence and contrasts in insects and vertebrates.............. 7
   T. D. Wyatt

## PART I: INTRASPECIFIC BEHAVIOR

## COMMUNICATION IN AMPHIBIANS AND REPTILES

3. The discovery and characterization of splendipherin, the first anuran sex pheromone............................................................... 21
   M. A. Apponyi and J. H. Bowie

4. Chemically mediated mate recognition in the tailed frog (*Ascaphus truei*)...... 24
   M. J. Asay, P. G. Harowicz, and L. Su

5. Responses to sex- and species-specific chemical signals in allopatric and sympatric salamander species............................................. 32
   C. A. Palmer and L. D. Houck

6. The pheromonal repelling response in red-spotted newts (*Notophthalmus viridescens*)............................................................... 42
   D. Park, H. L. Eisthen, and C. R. Propper

7. The effects of cloacal secretions on brown tree snake behavior.................. 49
   M. J. Greene and R. T. Mason

# COMMUNICATION IN MAMMALS

8. Species and sub-species recognition in the North American beaver............. 56
   A. M. Peterson, L. Sun, and F. Rosell

9. Self-grooming in meadow voles...................................................... 64
   M. H. Ferkin

10. Protein content of male diet does not influence proceptive or receptive behavior in female meadow voles, *Microtus pennsylvanicus*............ 70
    A. A. Pierce, M. H. Ferkin, and N. P. Patel

11. The signaling of competitive ability by male house mice........................... 77
    N. Malone, S. D. Armstrong, R. E. Humphries, R. J. Beynon, and J. L. Hurst

12. A possible function for female enurination in the mara, *Dolichotis patagonum*................................................................. 89
    D. S. Ottway, S. J. Pankhurst, and J. S. Waterhouse

13. The evolution of perfume-blending and wing sacs in emballonurid bats........ 93
    C. C. Voigt

14. Behavioral responsiveness of captive giant pandas (*Ailuropoda melanoleuca*) to substrate odors from conspecifics of the opposite sex.................. 101
    D. Liu, G. Zhang, R. Wei, H. Zhang, J. Fang, and R. Sun

15. Chemical signals in giant panda urine (*Ailuropoda melanoleuca*)............... 110
    M. Dehnhard, T. Hildebrandt, T. Knauf, A. Ochs, J. Ringleb, and F. Göritz

16. Chemical communication of musth in captive Asian elephants, *Elephas maximus*.................................................................... 118
    N. L. Scott and L. E. L. Rasmussen

17. Chemical analysis of preovulatory female African elephant urine: A search for putative pheromones...................................................... 128
    T. E. Goodwin, L. E. L. Rasmussen, B. A. Schulte, P. A. Brown, B. L. Davis, W. M. Dill, N. C. Dowdy, A. R. Hicks, R. G. Morshedi, D. Mwanza, and H. Loizi

18. Assessing chemical communication in elephants.................................... 140
    B. A. Schulte, K. Bagley, M. Correll, A. Gray, S. M. Heineman, H. Loizi, M. Malament, N. L. Scott, B. E. Slade, L. Stanley, T. E. Goodwin, and L. E. L. Rasmussen

19. The gland and the sac – the preorbital apparatus of muntjacs..................... 152
    S. J. Rehorek, W. J. Hillenius, J. Kennaugh, and N. Chapman

20. The chemistry of scent marking in two lemurs: *Lemur catta* and *Propithecus verreauxi coquereli*........................................................... 159
    R. A. Hayes, T. Morelli, and P. C. Wright

## PRIMING IN MAMMALS

21. Soiled bedding from group-housed females exerts strong influence on male reproductive condition........................................................ 168
    S. Koyama and S. Kamimura

## PROTEINS, MUPs, MHC

22. The role of the major histocompatibility complex in scent communication..... 173
    M. D. Thom, R. J. Beynon, and J. L. Hurst

23. Characterisation of proteins in scent marks: Proteomics meets semiochemistry................................................................ 183
    D. H. L. Robertson, S. Cheetham, S. Armstrong, J. L. Hurst, and R. J. Beynon

24. The "scents" of ownership...................................................... 199
    J. L. Hurst, M. D. Thom, C. M. Nevison, R. E. Humphries, and R. J. Beynon

25. The role of scent in inter-male aggression in house mice & laboratory mice... 209
    J. C. Lacey and J. L. Hurst

## VOMERONASAL SYSTEMS AND MAIN OLFACTORY SYSTEMS

26. Chemical signals and vomeronasal system function in axolotls (*Ambystoma mexicanum*)..................................................................... 216
    H. L. Eisthen and D. Park

27. From the eye to the nose: Ancient orbital to vomeronasal communication in tetrapods?........................................................................................ 228
    W. J. Hillenius and S. J. Rehorek

28. Prey chemical signal transduction in the vomeronasal system of garter snakes   242
    M. Halpern, A. R. Cinelli, and D. Wang

29. Mode of delivery of prey-derived chemoattractants to the olfactory and vomeronasal epithelia results in differential firing of mitral cells in the main and accessory olfactory bulbs of garter snakes.................   256
    C. Li, J. Kubie, and M. Halpern

30. Communication by mosaic signals: Individual recognition and underlying neural mechanisms.............................................................................   269
    R. E. Johnston

31. Sexual dimorphism in the accessory olfactory bulb and vomeronasal organ of the gray short-tailed opossum, *Monodelphis domestica*..................   283
    J. H. Mansfield, W. Quan, C. Jia, and M. Halpern

32. The neurobiology of odor-based sexual preference: The case of the Golden hamster.........................................................................................   291
    A. Petrulis

## HUMANS

33. Retention of olfactory memories by newborn infants.............................   300
    R. H. Porter and J. J. Rieser

34. Human sweaty smell does not affect women's menstrual cycle..................   308
    L. Sun, W. A. Williams, and C. Avalos

## PART II: INTERSPECIFIC RESPONSES

## PREDATOR AND ALARM CUES IN FISH AND AMPHIBIANS

35. Local predation risk assessment based on low concentration chemical alarm cues in prey fishes: Evidence for threat-sensitivity.......................   313
    G. E. Brown

36. Learned recognition of heterospecific alarm cues by prey fishes: A case study of minnows and stickleback................................................................. 321
M. S. Pollock, D. P. Chivers, R. C. Kusch, R. J. Tremaine, R. G. Friesen, X. Zhao, and G. E. Brown

37. The response of prey fishes to chemical alarm cues: What recent field experiments reveal about the old testing paradigm........................ 328
R. J. Tremaine, M. S. Pollock, R. G. Friesen, R. C. Kusch, and D. P. Chivers

38. Response of juvenile goldfish (*Carassius auratus*) to chemical alarm cues: Relationship between response intensity, response duration, and the level of predation risk........................................................ 334
X. Zhao and D. P. Chivers

39. The effects of predation of phenotypic and life history variation in an aquatic vertebrate................................................................. 342
R. C. Kusch, R. S. Mirza, M. S. Pollock, R. J. Tremaine, and D. P. Chivers

40. Nocturnal shift in the antipredator response to predator-diet cues in laboratory and field trials............................................................ 349
A. M. Sullivan, D. M. Madison, and J. C. Maerz

41. Long-term persistence of a salamander anti-predator cue......................... 357
M. P. Machura and D. M. Madison

42. Decline in avoidance of predator chemical cues: Habituation or biorhythm shift?................................................................................. 365
D. M. Madison, J. C. Maerz, and A. M. Sullivan

43. Chemically mediated life-history shifts in embryonic amphibians............... 373
R. S. Mirza and J. M. Kiesecker

44. Latent alarm signals: Are they present in vertebrates?................................. 381
O. B. Stabell

# PREYING AND FORAGING IN REPTILES, BIRDS, AND MAMMALS

45. Blood is not a cue for poststrike trailing in rattlesnakes.......................... 389
T. L. Smith and K. V. Kardong

46. Rattlesnakes can use airborne cues during post-strike prey relocation........... 397
    M. R. Parker and K. V. Kardong

47. The sense of smell in procellariiforms: An overview and new direction........ 403
    G. B. Cunningham and G. A. Nevitt

48. Cottontails and gopherweed: Anti-feeding compounds from a spurge........... 409
    D. Müller-Schwarze and J. Giner

    Subject Index................................................................................ 417

# THIRTY YEARS ON THE ODOR TRAIL: FROM THE FIRST TO THE TENTH INTERNATIONAL SYMPOSIUM ON CHEMICAL SIGNALS IN VERTEBRATES

Dietland Müller-Schwarze[*]


This milestone $10^{th}$ volume in the series "Chemical Signals in Vertebrates" (CSV) provides a vantage point to review the past progress of our field and assess current and future trends.

## 1. EARLY DEVELOPMENTS

The first symposium in 1976 brought together vertebrate "smell researchers" for the first time. It convened to spur contacts and collaboration between researchers working separately at disparate growing points, but having in common an interest in vertebrates. At the time, pheromone research and chemical ecology of insects was well established, with many outstanding books and meetings advancing the cause. Identification of the first insect pheromones had stimulated interest in identifying chemosignals in other animals as well, greatly aided by ever more sensitive analytical techniques, among them advent of Gas Liquid Chromatography in the 1950s.

Many separate strands of pioneer work during the preceding decades were ripe to coalesce into a field with common and complementary interests:

- The first behavior studies involving the olfactory sense arose from practical needs such as the olfactory ability of tracking dogs in the 1930s to Neuhaus' papers on odor thresholds in dogs in the 1950s.
- Before the advent of serious functional studies of chemical signals, anatomy of scent-producing organs and chemistry of smelly animal secretions had well

---

[*] Dietland Müller-Schwarze, Department of Environmental and Forest Biology; College of Environmental Science and Forestry, State University of New York, Syracuse, New York 13210. dmullers@esf.edu

advanced: Skin Gland Anatomy was first comprehensively collected in a volume by Schaffer in 1940; Perfume Chemistry elucidated smelly secretions of animals such as civet cat, muskrat, or beaver by Lederer, Ruzicka and others.

- Interest in behavior led to studies of chemical communication in context: Field studies of fish alarm odor in free-ranging minnows (von Frisch, 1941); General Vertebrate Ethology, with focus on communication and courtship (summarized by Tinbergen in the 1950s); stream odor discrimination by fish (Hasler and Wisby, 1951; Scholz et al., 1976). Even birds such as turkey vultures proved to be olfactorily guided when foraging (Stager, 1964).

- During the same years, physiological laboratory studies discovered chemosensory priming of reproductive behavior in mice (van der Lee and Boot; Bruce, late 1950s). Today, we are still not sure to what extent these laboratory phenomena occur in a natural setting.

## 2. RECENT TRENDS

How has the "market share" of vertebrate research fared over the years? A look at the Journal of Chemical Ecology is revealing: The number of papers dealing with vertebrates has increased from 5-10 per year in the first five years of the Journal to about 30 currently. But so has the total number of papers: from 50 per year initially to nearly 200 now. As a percentage, vertebrate papers slightly increased their share from 10-15% "then" to 13-19% now (Figure 1).

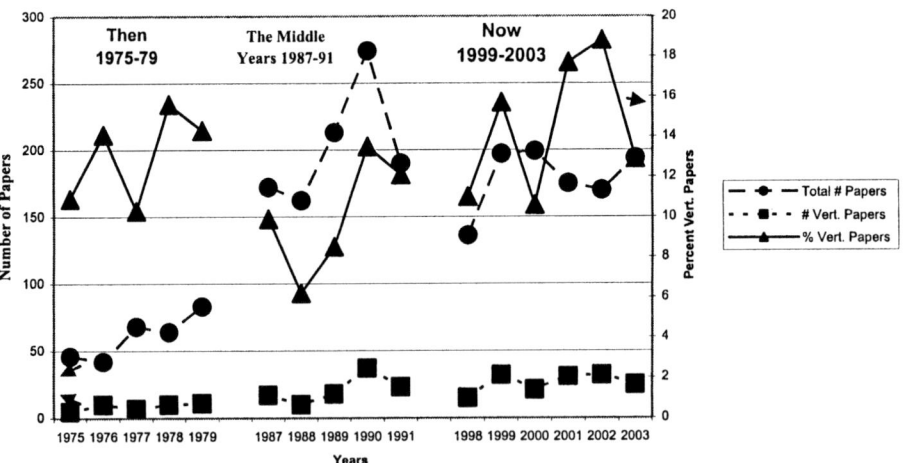

**Figure 1.** Numbers and Percent of Papers on Vertebrates in the Journal of Chemical Ecology from its Inception in 1975 to 2003.

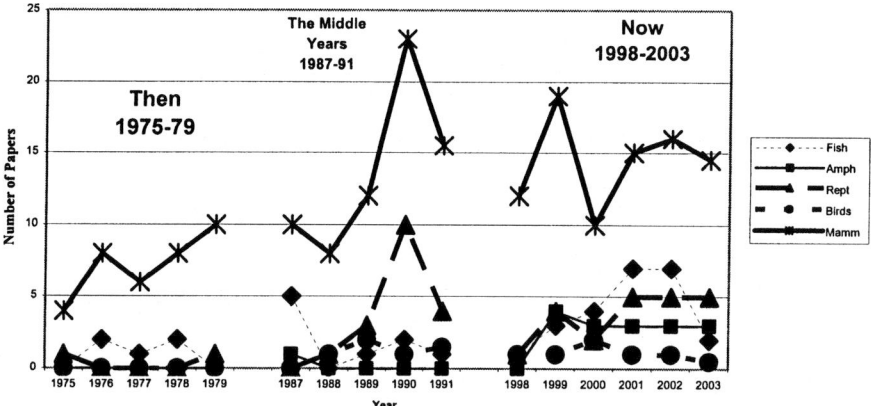

Figure 2. Numbers of Papers Dealing with Fish, Amphibians, Reptiles, Birds and Mammals in the Journal of Chemical Ecology from its Inception in 1975 to 2003.

Most vertebrate papers in the "Journal" deal with mammals. There are about as many articles on mammals as for the other four vertebrate classes combined (Figure 2). Within this distribution, the trend appears to be towards greater taxonomic diversity of studied species.

To gauge the progress of vertebrate research, let's now turn to the published CSV volumes. The first CSV symposium, staged at Saratoga Springs, New York, in 1976, stated its goals in the preface of the 1977 proceedings. It was to cover the following aspects of chemical signaling: 1) its evolution in higher vertebrates, including man, 2) social contexts in which the chemical signals are operative, and 3) its ecological determinants.

The then stated "Newly Emerging Questions" focused on intraspecific behavior:

1) Advisability of using the term "pheromone" in light of differences between insects and vertebrates
2) Effect of learning on chemical communication
3) Role of diet in origin of pheromones
4) Effect of multisensory context on chemocommunication
5) Elucidation of multicomponent pheromones.

Soon thereafter, a workshop, held at Nordwijkerhout in the Netherlands in 1978 (Ritter, 1979), agreed on the following tasks to be tackled by researchers working with vertebrates, primarily mammals:

1) Study more taxa, and wild animals, to expand on the laboratory findings so far

2) Develop better behavior assessment of the role of pheromones, consider the full repertoire of behavior, to go beyond studies with limited stimulus-response bioassays
3) Consider other sensory modalities (do not study chemical signals in isolation)
4) Use relevant cues in reception studies, rather than off-the shelf chemicals
5) Field experiments are needed
6) Efforts at practical applications should include domestic animals; pest control; primer pheromones; and animals as odor detectors
7) Interdisciplinary cooperation should be nurtured
8) We need people "truly trained in chemical ecology"
9) Attention should be paid to exchange programs and training in techniques.

At the occasion of the tenth volume in this series, we can take stock how far we have come in light of these goals stated about 25 years ago? Within vertebrate research itself, some trends are clear. Numbers of papers on applied aspects have oscillated and hovered around 5% (Figure 3), but are likely to increase, as more basic behavior is understood. Strictly sensory studies, such those on the vomeronasal organ, published in our series, peaked in the 1980s, with over 12% in volume 2 (Figure 3), in part probably due to migration to other, specialized journals and proceedings. Likewise, papers on humans in *this* series peaked in the early nineties (Figure 3). While the first symposium and volume dealt exclusively with intraspecific behaviors, over 30% of all papers now investigate interspecific relationships, such as predator alarm, predation, herbivory, or interspecific competition (Figure 4). There appears to be more collaboration now: The average number of authors per paper has doubled from 1.2 in 1975 to 2.4 in 2003.

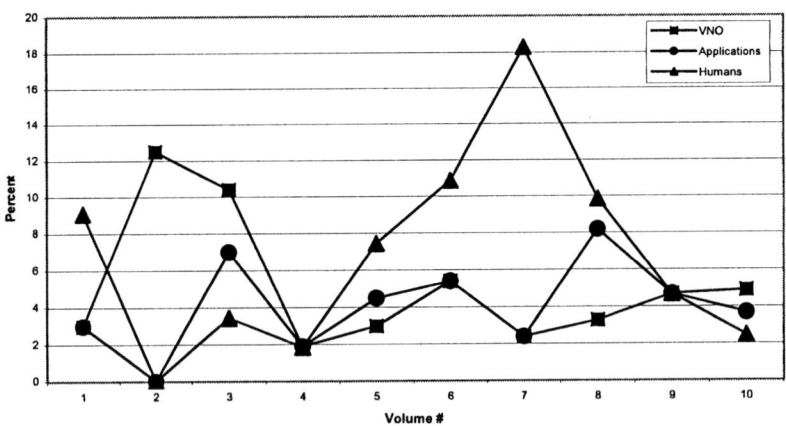

Figure 3. Numbers of Papers on Humans, the Vomeronasal Organ, and on Practical Applications in CSV Volumes 1 – 10.

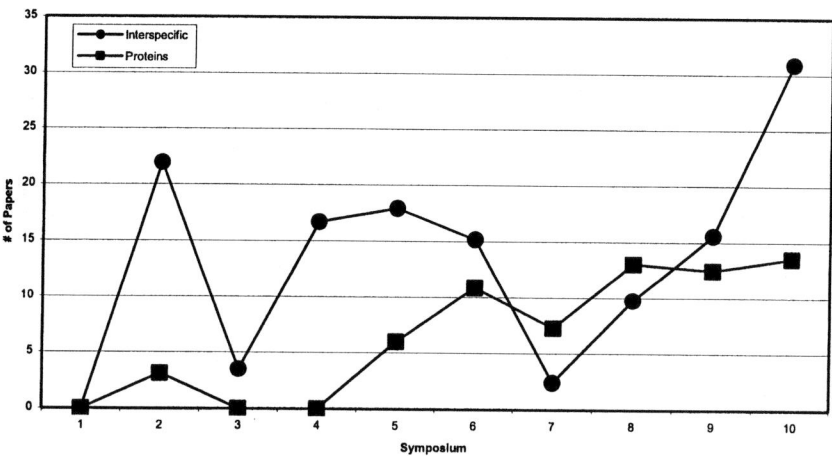

**Figure 4.** Percent papers in CSV volumes dealing with interspecific interactions or proteins.

The period between the first and tenth symposium saw the discovery of the first reptile pheromones (garter snakes; Mason et al., 1989); first fish pheromones (goldfish: Sorensen, 1992), and the first amphibian pheromones (salamanders: Kikuyama et al., 1995; frogs: Wabnitz et al., 1999).

Some developments could not have been anticipated. One is the increasingly important role of proteins as odor carriers, from 0 percent of papers to about 14% over the 10 CSV symposia (Figure 4). Hurst and Beynon and their teams pioneered this currently very active field (e.g. Beynon et al., 1999).

From here on, future needs and possible trends will include:

- We will study more complex (tritrophic and beyond) ecological relationships, encompassing ever larger pieces of an ecosystem.

- Large, non-volatile molecules will most likely be recognized to play important roles in communication in a number of organisms and contexts.

- "Post-Genomic Chemical Ecology" (Berenbaum, 2002) will elucidate pathways of genetic control on both the side of signal production and the chemoreception side.

- Applications in conservation, pest control, management and breeding of domestic animals, wildlife management, animals as chemical detectors, such as, for example, rats for mine sniffing, will become more numerous and more reliable.

- New technologies, such as electronic noses, computerized behavior recording, and remote and miniaturized recording devices will expand our horizon to hitherto inaccessible realms.

## 3. REFERENCES

Berenbaum, M. R., 2002, Postgenomic Chemical Ecology: From genetic code to ecological interactions, *J. Chem. Ecol.* **28**:873-896.

Beynon, R. J., Robertson, D. H. L., Hubbard, S. J., Gaskell, S. J., and Hurst, J. L., 1999, The role of protein binding in chemical communication: Major urinary proteins in the house mouse, in: *Advances in Chemical Signals in Vertebrates*, R. E. Johnson, D. Müller-Schwarze, and P. W. Sorensen, eds., Klewer Academic / Plenum Press, New York, pp. 137-147.

Bruce, H. M., 1959, An exteroceptive block to pregnancy in the mouse, *Nature* **184**:105.

Frisch, K. Von, 1941, Über einen Schreckstoff der Fischhaut und seine biologische Bedeutung, *Z. Vergl. Physiol.* **29**:46-145.

Hasler, A. D., and Wisby, W. J., 1951, Discrimination of stream odors by fishes and relation to parent stream behavior, *Am. Nat.* **85**:223-238.

Lederer, E., 1950, Odeurs et parfums des animaux. *Fort. Chem. Org. Nat.* **6**:87-153.

Kikuyama, S., Toyota, F., Ohmiya, Y., Matsuda, K., Tanaka, S., and Hayashi, H., 1995, Sodefrin: A female-attracting peptide pheromone in newt cloacal glands, *Science* **267**:1643-1645.

Van Der Lee, S., and Boot, L. M., 1955, Spontaneous pseudopregnancy in mice, *Acta Physiol. Pharm. N.* **4**:442-443.

Mason, R. T., Fales, H. M., Jones, T. H., Pannell, L. K., Chinn, J. W., and Crews, D., 1989, Sex pheromones in snakes, *Science* **245**:290-293.

Neuhaus, W., 1955, Die Unterscheidung von Duftquantitäten bei Mensch und Hund nach Versuchen mit Buttersäure, *Z. Vergl. Physiol.* **37**:234-252.

Neuhaus, W., 1956, Die Riechschwelle von Duftgemischen beim Hund und ihr Verhältnis zu den Schwellen unvermischter Duftstoffe. *Z. Vergl. Physiol.* **38**:238-258.

Parkes, A. S., and Bruce, H. M., 1961, Olfactory stimuli in mammalian reproduction, *Science* **134**:1049-1054.

Prelog, V., Ruzicka, L., and Wieland, P., 1945, Steroide and Sexualhormone. *Helv. Chim. Acta* **27**:66-68.

Ritter, F. J., ed., 1979, *Chemical Ecology: Odour Communication in Animals. Scientific aspects, practical uses and economic prospects*, Elsevier/North Holland, Amsterdam.

Schaffer, J., 1940, *Die Hautdrüsenorgane der Säugetiere*, Urban and Schwarzenberg, Berlin and Vienna.

Scholz, A. T., Horrall, R. M., Cooper, J. C., and Hasler, A. D., 1976, Imprinting to chemical cues: The basis for home stream selection in salmon, *Science* **192**:1247-1249.

Sorensen, P. W., 1992, Hormonally derived sex pheromones in goldfish: a model for understanding the evolution of sex pheromones in fish, *Biol. Bull.* **183**:1173-177.

Stager, K. E., 1964, The role of olfaction in food location by the turkey vulture (*Cathartes aura*), *Los Angeles County Museum Contributions to Science*, No. 81.

Tinbergen, N., 1951, *The Study of Instinct*, Oxford University Press, London.

Wabnitz, P. A., Bowie, J. H., Tyler, M. J., Wallace, J. C., and Smith, B. P., 1999, Aquatic sex pheromone from a male tree frog, *Nature* **401**:444-445.

# PHEROMONES: CONVERGENCE AND CONTRASTS IN INSECTS AND VERTEBRATES[1]

Tristram D. Wyatt[2]

## 1. INTRODUCTION

Chemical signals or pheromones are the most important signals for most of the animal kingdom. The organization of the olfactory system and brain, independently evolved across many taxa, makes it almost inevitable that chemical communication will evolve as animals are selected to respond to their chemical environment (Wyatt, 2003). As might be expected, pheromones play key roles in the lives of insects and vertebrates. However, the literature tends to separate these taxonomic groups rather than emphasising the similarities in the ways they use pheromones. For example, small molecules can be important in vertebrate signaling as well as in insects, though in terrestrial vertebrates these may be associated with proteins. In this chapter I would like to discuss the molecules used as pheromones, the paradox of signature odours in social insects and social mammals (where differences are the message), and finally the possible signal role of pheromones in complex social groups in mammals and social insects where only one female reproduces.

## 2. THE MOLECULES USED AS PHEROMONES

Elephants and moths are unlikely mates, so scientists and the general public were surprised when it was discovered that one of the world's largest living land animals, the Asian elephant (*Elephas maximus*), shares its female sex pheromone with some 140 species of moth (Rasmussen et al., 1996). The compound is a small, volatile molecule (Z)-7-dodecen-1-yl acetate. The shared use of a compound as a signal illustrates a relatively common phenomenon of independent evolution of particular molecules as

---

[1] This chapter is based upon Wyatt, TD (2003) *Pheromones and Animal Behaviour: Communication by Smell and Taste*. Cambridge University Press, Cambridge and reproduced with permission.
[2] Department of Zoology, University of Oxford, South Parks Road, Oxford OX1 3PS, United Kingdom, email: tristram.wyatt@zoo.ox.ac.uk, web: www.online.ox.ac.uk/pheromones

signals by species that are not closely related (Kelly, 1996). Such coincidences are a consequence of the common origin of life: basic enzyme pathways are common to all multicellular organisms, and most classes of molecule are found throughout the animal kingdom. However, despite sharing an attraction to (Z)-7-dodecen-1-yl acetate, male moths and elephants are unlikely to be confused. Apart from the mating difficulties should they try, male moths are unlikely to be attracted by the pheromones in female elephant urine because moth pheromones are multicomponent (Section 2.2). The (Z)-7-dodecen-1-yl acetate would be only one of perhaps five or six other similar compounds making up a precise blend for each moth species. Male elephants are unlikely to be attracted to a female moth because she releases such small quantities (picograms per hour) that they would not be noticed by a male elephant (but can be tracked by the specialised sensory system of a male moth).

The importance of small molecule pheromones in vertebrate communication should not have been a surprise after the work over many years on the role of molecules such as brevicomin in mice by Novotny and others (Novotny et al., 1999; Leinders-Zufall et al., 2000). Similarly, the crested auklet (*Aethia cristatella*), a monogamous seabird, has recently been shown to produce a distinctive tangerine-like scent in the breeding season, of the volatile molecules *cis*-4-decenal and octanal (Hagelin et al., 2003; Jones et al., 2004). The possible importance of smell for bird courtship was long ago highlighted by Darwin (1871), reporting observations of the musk duck *Biziura lobata* in Australia. These examples are an important illustration that, like insects, mammals and other vertebrates can use small molecules, singly or in simple mixtures, as pheromones for sexual signalling. It is harder to identify mammalian pheromones than those of insects but this does not necessarily mean that their pheromones are more complicated.

However, unlike small molecule pheromones in terrestrial invertebrates such as moths, many terrestrial mammals may increase the activity of their small molecule pheromones by interaction with carrier proteins. For example, in mice the small volatile molecules are presented as ligands of Mouse Urinary Proteins (MUPs). The MUPs provide a slow release of the volatile signal and the highly variable MUPs may also provide individuality to the signal (Hurst et al., 2001; Beynon and Hurst, this volume; Hurst and Beynon, this volume). As far as I know, there are no examples among the invertebrates of combining small molecules with proteins.

Whereas volatility is a key signal design feature of pheromones in air, solubility of molecules is perhaps the functional equivalent in water – and many soluble peptides are pheromones in their own right. There appear to be two main types of molecules used as pheromones in aquatic species. First, there are soluble molecules similar in size to those used as pheromones on land, such as the steroid-based pheromones used as fish sex pheromones and barnacle egg hatching pheromone. Second, large, polar molecules can be used, which despite their size can be highly soluble. For example, anthopleurine, the alarm pheromone of a sea anemone is a large cation. Many other aquatic animals use polypeptides as chemical signals. The first peptide pheromone to be identified in a vertebrate was the decapeptide, sodefrin, in the Asian red-bellied newt (*Cynops pyrrhogaster*) (Kikuyama et al., 1995) and the first peptide pheromone in anurans, splendipherin, the male aquatic sex pheromone of the tree frog *Litoria splendida* (Wabnitz et al., 1999; Apponyi and Bowie, this volume). Aquatic invertebrates also use peptides as sex pheromones, for example in the sea-slug mollusc *Aplysia* (Painter et al., 1999). Barnacle larvae settling out of the plankton ensure they settle in good sites by having a very specific response to certain peptides of their adult conspecifics.

## 3. EVOLUTION OF CHEMICAL CUES INTO SIGNALS

Chemical senses are the oldest, shared by all organisms including bacteria, so animals are pre-adapted to detect chemical signals in the environment (Wilson, 1970). Chemical information is used to locate potential food sources and to detect predators as well as to receive the chemical signals in social interactions. Signals are derived from movements, body parts or molecules already in use and are subsequently changed in the course of evolution to enhance their signal function. Thus pheromones evolve from compounds originally having other uses or significance, for example from hormones, host plant odours, chemicals released on injury, or waste products. There is selection for functional signal features such as longevity and specificity. There is also evolution in the sensory systems and response of the receiver. The original functions of the chemicals may or may not be eventually lost. The ubiquity and extraordinary diversity of pheromones are the evolutionary consequence of the powerful and flexible way the olfactory system is organised; taste does not have this flexibility. Most animal olfactory systems have a large range of relatively non-specific olfactory receptors which means that almost any chemical in the rich chemical world of animals will stimulate some olfactory sensory neurons and can potentially evolve into a pheromone. If detection of a particular chemical cue leads to greater reproductive success or survival, there can be selection for receptors more sensitive to it or expressed in greater numbers. In some cases animals may evolve a finely tuned system, including specialised sensory organs and brain circuits, such as those of male moths used to detect and respond to female pheromones.

## 4. SPECIES SPECIFICITY

There are two main ways of gaining specificity in pheromone signals. One is by the evolution of a large unique molecule. Peptide pheromones, using the 20 coded amino acids available in eukaryotic systems, offer an extraordinary variety of unique sequences; with a five amino-acid polypeptide there are $20^5$ ($\sim$ 3.2 million) (Browne et al., 1998). For example, two related species of the newt *Cynops* have species-specific decapeptide pheromones which differ by just two amino acids (Yamamoto et al., 2000). Among insects, a very few species use a unique complex molecule as a single component pheromone; for example, periplanone-B is the sex pheromone of the American cockroach (*Periplaneta americana*) (Roelofs, 1995).

More commonly, specificity is gained largely by using a unique blend of relatively simple compounds as a multicomponent pheromone. For example, female sex pheromones in moths usually consist of five to six fatty acids or their derivatives. Vertebrates may also have multi-component pheromones. For example the mouse pheromone which elicits aggression in other males, consists of two compounds dehydro-*exo*-brevicomin and 2-*sec*-butyl-4,5-dihydrothiazole (Novotny et al., 1999), each of which is inactive alone. Similarly, in the goldfish, while each of two female prostaglandin pheromones, F2α (PGF2α) and 15-keto-PGF2α, have similar effects on male behaviour when presented singly, both are needed together to stimulate a gonadotropin surge in males (Stacey and Sorensen, 1999). It is possible that other pheromone components add species specificity in these fish.

## 5. DISTANCE FOR SIGNALS

While we tend to think of pheromones as being detected by 'sniffing' air or water after travelling some distance from the signaller, many chemical cues are detected by contact chemoreception, as in the case of an ant tapping its antennae on a fellow ant to detect the complex mixtures of chemicals on its cuticle that differ between colonies and allow distinction of nestmates from strangers. In both vertebrates and invertebrates, pheromones may be transferred directly from signaller to receiver. For example, male Queen butterflies (*Danaus gilippus*) deposit crystals of the pheromone danaidone from their hair pencils directly onto the antennae of the female (Eisner and Meinwald, 1995). The male of the terrestrial salamander (*Plethodon jordani*) directly transfers his high molecular weight glycopeptide pheromone from his chin gland to the nostrils of the female (Rollmann et al., 1999). The male of the related salamander, *Desmognathus ochrophaeus*, takes this a stage further by directly 'injecting' his pheromone into her capillary blood supply, using elongated teeth to pierce the female skin, thus bypassing her chemosensory system (Houck and Reagan, 1990). Perhaps at the extreme of this continuum are the molecules passed, together with sperm, to the female during mating in many species: for example, the fruit fly *Drosophila melanogaster* and garter snakes (see references in Wyatt, 2003).

## 6. HONEST SIGNALS

Pheromones can be used as honest signals (Zahavi, 1975) which provide reliable information because they accurately reflect the signaller's ability or resources (Guilford, 1995). For example, female tiger moths (*Utetheisa ornatrix*) choose a male with the most pheromone. His pheromone is derived from the same plant poisons, used to protect the eggs, which he will pass to the female at mating. His pheromone load is correlated with the gift he will give (Eisner and Meinwald, 1995). In garter snake females, the levels of skin pheromones reflect evidence of the previous season's fertility. Male garter snakes court larger snakes, which have more pheromone (LeMaster and Mason, 2002). In mammals, production of pheromone is directly related to hormone levels and so scent marks will tend to be honest (Ferkin et al., 1994). Animals such as mammals and lizards that scent mark their territories leave signals that are inherently reliable – only if the owner does own the territory will his marks exclusively cover it (Gosling and Roberts, 2001). Where pheromones effectively have the role of badges of status as, for example, in cockroaches (Moore et al., 1997), queenless ants (Peeters, 1997), or mice (Hurst and Rich, 1999), the major cost may be that of maintaining the advertised status.

## 7. SOCIAL RECOGNITION – SIGNATURE ODOURS

One of the most important uses of odour signals in both mammals and social insects is as signature odours, chemical cues used for social recognition. Signature odours do not fit the original pheromone criterion of a defined chemical mixture eliciting particular behaviour or other response (Karlson and Lüscher, 1959). The cues used for social recognition of kin, clans, colony members and the like are complex, greatly varied mixtures of many compounds. The differences between the odour mixtures *are* the message. The resulting chemical signatures of both mammals and social insects are complex and variable mixtures, giving a forest of peaks on a gas chromatograph trace, in contrast to the small number of defined peaks for the sex pheromones of moths and other insects. These complex mixtures reflect the overlaying of many different messages.

For example, the saddle-back tamarin (*Saguinus fuscicollis*) a South American primate, produces chemical messages which identify species, subspecies, individual and gender, and may also contain information on social status (Epple et al., 1993). Social insects carry a chemical message on their cuticle that includes information about their species, colony, caste, age and gender. In both mammals and social insects the cues giving reproductive status, in particular ovarian status, may be the key to the role of pheromones in reproduction in social animals. While signals of caste, gender, life stage or species may not vary much within the species and could thus be said to be anonymous (Hölldobler and Carlin, 1987), the variability of colony and kin recognition chemical signatures is what gives them their specificity. In the case of the honeybees, at the entrance to the nest there are guard bees with a specific and characteristic posture. They will challenge any bee trying to enter the hive and if only it has the right particular signature will it be let in; strangers are attacked and killed. And of course we are familiar with similar recognition phenomena using odour cues across the vertebrates, for example among beavers (Sun and Müller-Schwarze, 1999) and hamsters (Johnston, this volume) (incidentally, people are also good at recognising their own family by smell – see Wyatt, 2003 for review).

## 8. PHEROMONES AND REPRODUCTION IN SOCIAL GROUPS: CONTROL OR SIGNALLING?

In social insect and social mammal species such as honey bees and naked mole rats, only one female reproduces. In social insects, queen pheromone signals may be honest cooperative signals, not control. Might these ideas apply to social mammals?

Pheromones are important for the many species of mammal that live in social groups with animals sharing and defending the same territory and for species of social insect that live in colonies. In some societies of mammal and social insect, cooperation is taken a stage further, with a reproductive division of labour: some individuals (helpers or workers) do not themselves reproduce but instead help to rear the offspring of other group members, usually the helpers' sisters or mothers. In the social insects (ants, termites and some bees and wasps), this is termed eusociality. The reason that such altruistic reproductive behaviour can persist is kin selection, which allows the helpers to gain their inclusive fitness indirectly by rearing copies of their genes in their brothers and sisters (Bourke, 1997). Hamilton's rule for kin selection predicts that altruistic behaviour will be more likely to be selected for if the individuals are closely related (and thus more likely to share the helper's gene for helping) and if the decrease in the actor's personal fitness is relatively small compared with the increase in the recipient's fitness (Hamilton, 1964; Keller and Chapuisat, 1999).

One way of describing the sharing of reproduction in social groups is by the term reproductive skew, which describes how much the spread of reproduction differs from an equal share for each member of that sex in the group (Keller and Reeve, 1994). Reproductive skew for males or females in a group ranges between 'zero', with equal shares (where all group members of a sex reproduce, for example female spotted hyenas), and 'one', in highly skewed animal societies in which effectively only one or a few members reproduce (for example, females in termites, ants, honeybees and the naked mole rat). Species with small colonies can be highly reproductively skewed, for example bumblebees, paper wasps, and the common marmoset.

Using reproductive skew as a measure, one can envisage social animals of all kinds placed on a eusociality continuum from no skew to high skew (Lacey and Sherman (1997), but see counterviews of Crespi and Yanega (1995) and Crespi and Choe (1997)). At the high skew end, eusocial species show cooperative care of the brood, overlap of adult generations (with offspring helping parents) together with reproductive division of labour, with some individuals specialised for reproduction (called queens or kings in social insects), and other more or less sterile individuals showing reproductive altruism (Wilson, 1971). In addition to the well-known highly eusocial insects among the Hymenoptera (wasps, bees, and ants) and Isoptera (termites), and among mammals (naked mole rats), there is a growing list of other eusocial animals, with species of eusocial spiders, aphids, gall thrips and coral reef shrimps all now recorded.

Eusocial societies have cooperative broodcare but in even the most cooperative societies, genetic conflicts of interest are inevitable (Emlen, 1997). In particular, group members will compete over who gets to reproduce. In most mammalian societies and in those social insect species in which almost all individuals could potentially reproduce, fierce fighting determines who reproduces. Perhaps surprisingly at first sight, in some of the most skewed societies, with the greatest morphological differences between the queen and workers, and in some mammals, pheromones produced by the dominant female 'settle the dispute', by appearing to stop subordinate females from reproducing. In social insects this phenomenon was traditionally viewed as pheromone control by the queen. An alternative view is gaining ground: that the queen's pheromones are cooperative signals, not control by a form of chemical aggression. The proposals were outlined first for social insects so I describe these first, but the same or similar points probably apply to cooperatively breeding (eusocial) mammals (Section 4.2).

## 9. SOCIAL INSECT QUEEN PHEROMONES

The coordination and integration of colony activities, in particular recruitment for foraging and defence, has been an essential contribution to the success of social insects: the road to sociality was paved with pheromones (Blum, 1974). Pheromones play a central role in these activities and in other functions such as recognition (of caste, sex, kin, colony, and species), caste determination, trophallaxis (mouth-to-mouth transfer of food), nest entrance marking and colony reproduction (Winston, 1992). Termites show convergent evolution of chemical signaling with the eusocial Hymenoptera.

Eusocial insect colonies are characteristically divided into two castes, reproductives and workers: a few individuals, queens or kings, are reproductive and workers reproduce little or not at all. The kin conflict between workers and the queen within social insect colonies can be over the level and timing of resources put into rearing reproductives, their sex ratio, and egg laying by workers (Keller and Reeve, 1999). However, despite the conflict, dominance with open aggression by the queen is virtually absent in more advanced insect societies, which have effectively sterile, morphologically distinct, worker castes (Wilson, 1971, p. 302). As colonies increase in size, it is hard to see how physical domination could work for more than a few tens of animals let alone the 500,000 individuals in a weaver-ant colony, controlled by a single queen (Wilson, 1971, p. 432; Hölldobler and Wilson, 1977). Instead, in advanced insect societies, pheromones take the place of fights. The phenomenon of queen pheromone influence within social insect colonies is clear in advanced ants, wasps, and bees. Pheromones play a similar role in termites. The queen's primer pheromones affect the colony production of reproductives by influencing the behaviour of workers; this is important both for maximising her reproductive fitness and that of the colony (Winston, 1992). Her pheromones also appear to cause the workers not to develop their ovaries.

News of the health of the queen is continually spread throughout a colony of social Hymenoptera, mediated by the queen's pheromones passed from one colony member to another (Winston and Slessor, 1992). The queen is surrounded by a retinue of eight to 10 workers, which constantly change as new workers approach and lick or touch her with their antennae. After picking up the queen mandibular pheromone (QMP), these workers groom themselves and then act as messengers by running through the rest of the colony for about 30 minutes, making frequent reciprocal antennal contacts with other workers, and passing on the QMP by contact, as if playing chemical tag (Seeley, 1979). These queen pheromone effects can be dramatically demonstrated by removing the queen and seeing the rapid changes in worker behaviour and physiology; these can start in as little 30 minutes in honeybees (*Apis mellifera*). Without the queen pheromone, workers start to rear new queens. Keller & Nonacs (1993), following Seeley (1985) and others, argue that the pheromone effects are not the consequence of pheromone control by the queen but instead that workers are using the queen pheromone as an honest signal that the queen is there and that the workers' response to the pheromone increases their inclusive fitness as much as that of the queen. The queen could perhaps control the colony by deception, fooling workers to act in her interest rather than theirs, but Keller and Nonacs (1993) argue that dishonest signalling in the colony is unlikely to be evolutionarily stable and conclude that queen-produced pheromones are honest messages of queen activity or presence. There are possible scenarios for the evolution of control by pheromones (Bourke and Franks, 1995, p. 239 ) but, nonetheless, if it was not in their interest to respond, workers or subordinate queens would evolve to ignore queen pheromones. The genetic variation in worker sensitivity to queen pheromones, on which selection could act, has been demonstrated in honeybees (Slessor et al., 1998). Similarly, there will be selection for queen behaviour that avoids costly queen–worker conflict that reduces colony productivity (Keller and Nonacs, 1993). Evolutionary solutions in the eusocial nest may be most stable where benefits are shared between workers and queen, pulling in the same direction (Seeley, 1995, p. 11). If queen pheromone is a signal to say that 'I am laying eggs' then one should expect the time course of pheromone production to match egg laying and to correlate with fecundity, which it does in honeybees and fire ants (Winston and Slessor, 1992; Vargo, 1998; Vargo, 1999). A honeybee queen's queen mandibular pheromone blend changes with age (Winston and Slessor, 1992). The full blend, including the aromatics, is only produced after mating and when she begins to lay eggs.

The reason that physical fights do not occur is not because the pheromone controls the workers but because their interests often match those of the queen. The strong morphological specialisation of the queen as an egg and pheromone factory and of the workers for their many colony-sustaining roles means that an individual worker gains more by helping to rear the queen's eggs than by laying its own. With specialised morphological castes, the queen pheromone may be a relatively low-cost cooperative signal (Keller and Reeve, 1999). In a cooperative signal, with benefits to both sides, evolutionarily stable signals do not have to be differentially costly to signalers with high values of the signalled attribute (which is the central assumption of the Zahavi handicap model for signals) (Keller and Reeve, 1999). A further pheromone-mediated effect is the way that any eggs that *are* produced by workers are destroyed by other workers, termed 'worker policing'. Worker honeybees have ovaries and although they cannot mate, they can lay unfertilised eggs which become males. Workers destroy the eggs laid by other workers because if the queen is multiply mated, workers are on average more related to the sons of the queen than to the sons of other workers (Ratnieks, 1993). Worker policing is made possible because the queen's eggs can be recognised by a pheromone mark from her Dufour's gland (Ratnieks, 1995). In nests with a queen, almost all the eggs produced by workers are destroyed (Ratnieks and Visscher, 1989). Workers also attack workers with well-developed ovaries (Visscher and Dukas, 1995). A genetic basis for worker policing has been found in honeybees (Montague and Oldroyd, 1998). Worker policing is evolutionarily stable because it benefits the queen *and* the average worker (Vander Meer and Morel, 1995; Bourke, 1997). Even in once-mated single queen colonies, in which workers would be more related to their own sons than to the queen's sons (Bourke, 1997), workers might not reproduce if it reduces the efficiency of the colony.

## 10. PRIMER PHEROMONES AND REPRODUCTION IN SOCIAL MAMMALS

It is in social mammals, those living in groups on shared territories and especially those breeding cooperatively, that mammal primer pheromone interactions have reached their greatest complexity and subtlety. Pheromone stimuli in social mammals can induce hormonal changes, affect the success of pregnancy, alter the course of puberty, modulate female cyclicity and ovulation, and modulate reproductive behaviour and aggression. These physiological effects include the Bruce and Whitten effects in mice.

Cooperative breeding, with alloparental care in which members of the social group assist in rearing young that are not their own, is common in some mammalian taxa, in particular rodents and canids. For example, cooperative or communal nesting and care of young have been reported for 35 species and from nine of 30 rodent families (Solomon and Getz, 1997). Cooperative breeding covers a wide range of behaviour depending on reproductive skew in the species, from plural breeders with all females reproducing through to singular breeders, social groups in which only one female breeds together with 'helpers-at-the-nest' (Solomon and French, 1997).

Most of the pioneering work on mammal primer pheromones was on social rodents such as house mice which are plural cooperative breeders (all females breed, although not all males). Female house mice suckle each other's young and cooperatively defend the nest. Characteristic of these societies is an interplay of dominance (in particular between males), sex, and population density.

In plural breeders, the effects of females on each other are mutual, but in singular cooperative breeding species, such as beavers, prairie voles or the common marmoset, the dominant female suppresses reproduction by the subordinate females. The parallels between social organization in these species and social insects are explored in Section 4.5. It is worth noting that some of these social effects, such as influences on puberty timing, are also seen in solitary rodent species under some conditions.

## 11. REPRODUCTION IN SINGULAR COOPERATIVELY BREEDING MAMMALS WITH HIGH REPRODUCTIVE SKEW

Reproductive suppression is common in singular cooperatively breeding mammals in which typically only one dominant female breeds. As in many social insects, the subordinate females are often her daughters, and in mammals, as in social insects, signals affecting the reproduction of subordinates range, in different species, from physical dominance to pheromones. Most mammal social groups do not use pheromones for this. For example, in the most eusocial mammals, naked mole rats (*Heterocephalus glaber*), with colonies of up to 300 non-breeding workers, the suppression of worker fertility by the queen is not pheromonal (Faulkes and Abbott, 1993). Instead, the queen, which is larger than other colony members, exerts her reproductive suppression on the non-breeding workers by physical dominance, 'shoving' and pushing subordinates down the tunnels (Bennett et al., 1999). In singular breeding canids the mechanism has only been identified in the grey wolf (*Canis lupus*): subordinates could reproduce but do not because their mating attempts are interrupted by their parents (Asa, 1997).

It is in some of the singular cooperatively breeding rodents and the New World primates that there are strong pheromone parallels with advanced social insects (Solomon and Getz, 1997; Carter and Roberts, 1997; Abbott et al., 1998). We know most about the reproductive biology of prairie voles (*Microtus ochrogaster*) and the common marmoset (*Callithrix jacchus*). In both species, many of the effects are mediated by odours for recognition, signal or primer pheromones.

In prairie voles, monogamous pairs and their offspring form the core of a communal breeding group. Continued breeding by the original pair and concurrent inhibition of reproduction of other members of the group is promoted by reproductive suppression of offspring, incest avoidance, social preferences for the familiar sexual partner and active defence of territory and mate (Carter and Roberts, 1997). Almost two-thirds of prairie voles young remain in their parents' nest (philopatry). These non-breeding subordinates engage in all parental behaviour except nursing (Solomon and Getz, 1997).

The young of the common marmoset also stay within their natal group into adulthood and do not breed. All group members, of both sexes, contribute to infant care, and may groom, tend (babysit) and transport young, but in addition may help with post-weaning feeding of infants (Tardif, 1997; French, 1997). The evolution of cooperative breeding may be a two-step process (Lacey and Sherman, 1997). The first step is the presence of ecological conditions that encourage natal philopatry: staying on the parental territory rather than trying to breed on one's own. This could be because of high costs, or low success, of independent breeding or dispersal, and would lead to groups containing two or more generations of related adults. The second step is the evolution of alloparental care, depending on the benefits to kin and ultimately on inclusive fitness. Long-term studies of the costs and benefits of helping in mammals and birds give widespread confirmation that helpers frequently do gain large indirect genetic benefits by helping to rear collateral kin (Emlen, 1997).

For prairie voles, indirect benefits from alloparenting could include better survival of sibling pups, faster pup development and reduced workload for parents, thus allowing the parents to produce more litters (Solomon, 1991; Wang and Novak, 1994). For common marmosets, the initial benefits of alloparenting might originally have been increased survival of young, but once set on the path of helping, it has become a requirement as the energy costs of breeding are so high that a lone pair is effectively incapable of reproducing successfully (French, 1997). Cooperation between animals may also be needed for the successful founding of new marmoset groups (Abbott et al., 1998). The importance of ecological factors for the fine balance of benefits and costs to helpers is shown by the patchy distribution of singular and plural cooperative breeding across related genera. Even in the same genus there may be species that are singular breeders and others that are plural breeders, for example prairie voles, and common voles respectively. Populations of the same species in different places, for example prairie voles (Roberts et al., 1998), may show more or less alloparental care according to local ecological conditions.

## 12. INHIBITION OR SUPPRESSION OF SUBORDINATE REPRODUCTION

The size of social groups is not the deciding factor for the transition to pheromonal control in mammals, as species that use pheromones in reproductive suppression tend to have small family groups. More species of cooperatively breeding mammals may turn out to use pheromones than is currently realised. Two effects keep subordinate female prairie voles pre-pubescent (Carter and Roberts, 1997). First, they delay puberty as long as they are exposed to only familiar males (father or male sibs) recognised by odour. Second, the stimulatory effect of urine from an unfamiliar male is overruled by inhibitory pheromones in the urine of their mother and sisters. Subordinate females thus remain functionally pre-pubescent and provide support to the communal family. The suppression of subordinate males is likely to be behavioural as they still produce sperm but do not mate. Suppression of ovulation in subordinate common marmoset females is by a combination of olfactory, visual and behavioural cues but once reproductively suppressed, this can be extended by odour alone: if a subordinate female is taken from the group, she will start her ovarian cycle but disinhibition is delayed by about 20 more days if she is exposed to the scent marks of the dominant female. A feature of reproductive suppression in mammals is the variety of mechanisms controlling singular cooperative breeding, even in closely related species. For example, unlike the case of the prairie vole, pheromone cues are not sufficient to suppress oestrus in the pine vole (*Microtus pinetorum*) (Brant et al., 1998). Similarly, in the golden lion tamarin (*Leontopithecus rosalia*), a member of the same family as the marmosets, subordinate females ovulate and are physiologically capable of mating but do not do so (French, 1997).

## 13. PARALLELS BETWEEN SOCIAL MAMMALS AND SOCIAL INSECTS

The response of subordinates in marmosets or prairie voles may be an adaptive response to signals from the principal female, analogous to the worker responses to signalling by social insect queen pheromones (Keller and Nonacs, 1993). Subordinates in marmosets, prairie voles and social insects may have evolved specific, adaptive responses to signs of subordinate status that lead them to respond with alloparental and other behaviour that increases their inclusive fitness by helping the society or family group (Abbott et al., 1998). Like workers in social insect colonies, subordinate female marmosets show many behavioural, neuroendocrinological and physiological differences from dominant females (Abbott et al., 1998). The differences include both the alloparental tasks undertaken by subordinate marmosets and also their physiological responses to pheromones and other cues from the dominant female. Abbott et al. (1998) suggest that the behaviour and physiology of subordinates seem to be a stable alternative to dominant status, not a state of generalised stress imposed by the dominant female and endured by the subordinates to their physiological detriment (there is no elevation in the circulating hormones, cortisol or prolactin, associated with stress).

A further parallel comes from developmental pathways. The spontaneous alloparenting behaviour and high likelihood of remaining in the parental nest (philopatry) of subordinate prairie voles are influenced by their prenatal hormonal environment in the uterus (Roberts et al., 1996). I wonder how different this is from developmental influences on social insect larvae as they are directed to worker or queen roles?

The roles of pheromones in influencing who reproduces in social groups of both insects and mammals are clearly complex. The interplay between pheromones and hormones, and the way that closely related species achieve similar ends by either a pheromone or behavioural dominance route, should make us reconsider rigid categories. What pheromones and behavioural dominance share in their mechanism of action is, ultimately, an effect on hormone release from the hypothalamus in mammals, and from the corpora allata in insects (Wyatt, 2003). Could one argue that pheromones and behavioural dominance are equivalent at the ultimate physiological level?

## 14. CONCLUSION

The individuals in animal societies interact via a complex web of semiochemical signals. Eusocial species of social insects and social mammals are characterized by reproductive division of labour. In some species, group members fight to establish which animals will reproduce. Other species use pheromones that act as signals rather than as coercion. The mechanisms used in social insects and in mammals have many similarities.

## 15. REFERENCES

Abbott, D. H., Saltzman, W., Schultz-Darken, N. J., and Tannenbaum, P.L., 1998, Adaptations to subordinate status in female marmoset monkeys *Comp. Biochem. Phys. C* **119**:261-274.
Asa, C. S., 1997, Hormonal and experiential factors in the expression of social and parental behavior in canids, in: *Cooperative Breeding in Mammals*, N. G. Solomon, and J. A. French, eds., Cambridge University Press, Cambridge, pp. 129-149.
Bennett, N. C., Faulkes, C. G., and Jarvis, J.U.M., 1999, Socially-induced infertility, incest avoidance and the monopoly of reproduction in the cooperatively breeding African mole-rat, Family Bathyergidae, in: *Advances in the Study of Behavior*, P. J. B. Slater, ed., 28 edn., Academic Press, New York, pp. 75-114.
Blum, M. S., 1974, Pheromonal bases of social manifestations in insects, in: *Pheromones*, M. C. Birch, ed., North-Holland, Amsterdam, pp. 190-199.
Bourke, A. F. G., 1997, Sociality and kin selection in insects, in: *Behavioural Ecology*, J. R. Krebs, and N. B. Davies, eds., Blackwell Science, Oxford, pp. 203-227.
Bourke, A. F. G., and Franks, N. R., 1995, *Social Evolution in Ants*, Princeton University Press, Princeton.
Brant, C. L., Schwab, T. M., Vandenbergh, J. G., Schaefer, R. L. and Solomon, N. G., 1998, Behavioural suppression of female pine voles after replacement of the breeding male, *Anim. Behav.* **55**:615-627.
Browne, K. A., Tamburri, M. N., and ZimmerFaust, R. K., 1998, Modelling quantitative structure-activity relationships between animal behaviour and environmental signal molecules, *J. Exp. Biol.* **201**:245-258.
Carter, C. S., and Roberts, R. L., 1997, The psychobiological basis of cooperative breeding in rodents, in: *Cooperative Breeding in Mammals*, N. G. Solomon and J. A. French, eds., Cambridge University Press, Cambridge, pp. 231-266.

Crespi, B. J., and Choe, J. C., 1997, Explanation and evolution of social systems, in: *The Evolution of Social Behavior in Insects and Arachnids*, J. C. Choe, and B. J. Crespi, eds, Cambridge University Press, Cambridge, pp. 499-524.
Crespi, B. J. and Yanega, D., 1995, The definition of eusociality, *Behav. Ecol.* **6**:109-115.
Darwin, C., 1871, *The Descent of Man and Selection in Relation to Sex*, John Murray, London.
Eisner, T., and Meinwald, J., 1995, Defense-mechanisms of arthropods and the chemistry of sexual selection, *Proc. Natl. Acad. Sci. USA* **92**:50-55.
Emlen, S. T., 1997, Predicting family dynamics in social vertebrates, in: *Behavioural Ecology*, J. R. Krebs, and N. B. Davies, eds., Blackwell Science, Oxford, pp. 228-253.
Epple, G., Belcher, A. M., Kuderling, I., Zeller, U., Scolnick, L., Greenfield, K. L., and Smith, A. B., 1993, Making sense out of scents - species-differences in scent glands, scent marking behavior and scent mark composition in the Callitrichidae, in: *Marmosets and Tamarins. Systematics, Behaviour, and Ecology*, A. B. Rylands, ed., Oxford Science Publications, Oxford University Press, Oxford, pp. 123-151.
Faulkes, C. G., and Abbott, D. H., 1993, Evidence that primer pheromones do not cause social suppression of reproduction in male and female naked mole-rats (*Heterocephalus glaber*), *J. Reprod. Fertil.* **99**:225-230.
Ferkin, M. H., Sorokin, E. S., Renfroe, M. W., and Johnston, R. E., 1994, Attractiveness of male odors to females varies directly with plasma testosterone concentration in meadow voles, *Physiol. Behav.* **55**:347-353.
French, J. A., 1997, Proximate regulation of singular breeding in callitrichid primates, in: *Cooperative Breeding in Mammals*, N. G. Solomon, and J. A. French, eds., Cambridge University Press, Cambridge, pp. 34-75.
Gosling, L. M., and Roberts, S. C., 2001, Scent-marking by male mammals: cheat-proof signals to competitors and mates, *Adv. Stud. Behav.* **30**:169-217.
Guilford, T., 1995, Animal signals - all honesty and light. *Trends Ecol. Evol.* **10**:100-101.
Hagelin, J. C., Jones, I. L., and Rasmussen, L. E. L., 2003, A Tangerine-Scented Social Odour in a Monogamous Seabird, *Proc. R. Soc. Lond. B Biol. Sci.* **270**:1323-1329.
Hamilton, W.D., 1964, The genetical evolution of social behaviour. I and II. *J Theor Biol* **7**:1-32.
Hölldobler, B., and Carlin, N. F., 1987, Anonymity and specificity in the chemical communication signals of social insects, *J. Comp. Physiol. [A]* **161**:567-581.
Hölldobler, B., and Wilson, E. O., 1977, Weaver ants, *Sci. Am.* **237**:146-154.
Houck, L. D., and Reagan, N. L., 1990, Male courtship pheromones increase female receptivity in a plethodontid salamander, *Anim. Behav.* **39**:729-734.
Hurst, J. L., Payne, C. E., Nevison, C. M., Marie, A. D., Humphries, R. E., Robertson, D. H. L., Cavaggioni, A., and Beynon, R. J., 2001, Individual recognition in mice mediated by major urinary proteins. *Nature* **414**:631-634.
Hurst, J. L., and Rich, T. J., 1999, Scent marks as competitive signals of mate quality, in: *Advances in Chemical Signals in Vertebrates*, R. E. Johnston, R.E., ed., Kluwer Academic/Plenum Press, New York, pp. 209-226.
Jones, I. L., Hagelin, J. C., Major, H. L., and Rasmussen, L. E. L., 2004, An experimental field study of the function of crested auklet feather odor, *Condor* **106**:71-78.
Karlson, P., and Lüscher, M., 1959, 'Pheromones': a new term for a class of biologically active substances, *Nature* **183**:155-156.
Keller, L., and Chapuisat, M., 1999, Cooperation among selfish individuals in insect societies, *Bioscience* **49**:899-909.
Keller, L., and Nonacs, P., 1993, The role of queen pheromones in social insects - queen control or queen signal, *Anim. Behav.* **45**:787-794.
Keller, L., and Reeve, H. K., 1994, Partitioning of reproduction in animal societies, *Trends Ecol. Evol.* **9**:98-102.
Keller, L., and Reeve, H. K., 1999, Dynamics of conflicts within insect societies, in: *Levels of Selection in Evolution*, L. Keller, ed., Princeton University Press, Princeton, New Jersey.
Kelly, D. R., 1996, When is a butterfly like an elephant? *Chem. Biol.* **3**:595-602.
Kikuyama, S., Toyoda, F., Ohmiya, Y., Matsuda, K., Tanaka, S., and Hayashi, H., 1995, Sodefrin: A female-attracting peptide pheromone in newt cloacal glands, *Science* **267**:1643-1645.
Lacey, E. A., and Sherman, P. W., 1997, Cooperative breeding in naked mole-rats: implications for vertebrate and invertebrate sociality, in: *Cooperative Breeding in Mammals*, N. G. Solomon, and J. A. French, eds., Cambridge University Press, Cambridge, pp. 267-301.
Leinders-Zufall, T., Lane, A. P., Puche, A. C., Ma, W. D., Novotny, M. V., Shipley, M. T., and Zufall, F., 2000, Ultrasensitive pheromone detection by mammalian vomeronasal neurons, *Nature* **405**:792-796.
LeMaster, M. P., and Mason, R. T., 2002, Variation in a female sexual attractiveness pheromone controls male mate choice in garter snakes, *J. Chem. Ecol.* **28**:1269-1285.
Montague, C. E., and Oldroyd, B. P., 1998, The evolution of worker sterility in honey bees: An investigation

into a behavioral mutant causing failure of worker policing, *Evolution* **52**:1408-1415.
Moore, P. J., Reagan-Wallin, N. L., Haynes, K. F., and Moore, A.J., 1997, Odour conveys status on cockroaches, *Nature* **389**:25.
Novotny, M. V., Ma, W., Zidek, L., and Daev, E., 1999, Recent biochemical insights into puberty acceleration, estrus induction and puberty delay in the house mouse, in: *Advances in Chemical Signals in Vertebrates*, R. E. Johnston, D. Müller-Schwarze, and P. W. Sorensen, eds., Kluwer Academic/Plenum Press, New York, pp. 99-116.
Painter, S. D., Clough, B., Akalal, D. B. G., and Nagle, G. T., 1999, Attractin, a water-borne peptide pheromone in Aplysia, *Invertebr. Reprod. Dev.* **36**:191-194.
Peeters, C., 1997, Morphologically 'primative' ants: comparative review of social characters, and the importance of queen-worker dimorphism, in: *The Evolution of Social Behavior in Insects and Arachnids*, J. C. Choe, and B. J. Crespi, eds., Cambridge University Press, Cambridge, pp. 372-391.
Rasmussen, L. E. L., Lee, T. D., Roelofs, W. L., Zhang, A. J., and Daves, G. D., 1996, Insect pheromone in elephants, *Nature* **379**:684.
Ratnieks, F. L. W., 1993, Egg-laying, egg-removal, and ovary development by workers in queenright honey-bee colonies, *Behav. Ecol. Sociobiol.* **32**:191-198.
Ratnieks, F. L. W., 1995, Evidence for a queen-produced egg-marking pheromone and its use in worker policing in the honey-bee, *J. Apic. Res.* **34**:31-37.
Ratnieks, F. L. W., and Visscher, P. K., 1989, Worker policing in the honeybee, *Nature* **342**:796-797.
Roberts, R. L., Williams, J. R., Wang, A. K., and Carter, C. S., 1998, Cooperative breeding and monogamy in prairie voles: Influence of the sire and geographical variation, *Anim. Behav.* **55**:1131-1140.
Roberts, R. L., Zullo, A., Gustafson, E. A., and Carter, C. S., 1996, Perinatal steroid treatments alter alloparental and affiliative behavior in prairie voles, *Horm. Behav.* **30**:576-582.
Roelofs, W. L., 1995, The chemistry of sex attraction, in: *Chemical Ecology: the Chemistry of Biotic Interaction*, T. Eisner, ed., National Academy of Sciences, Washington, D.C., pp. 103-117.
Rollmann, S. M., Houck, L. D., and Feldhoff, R. C., 1999, Proteinaceous pheromone affecting female receptivity in a terrestrial salamander, *Science* **285**:1907-1909.
Seeley, T. D., 1979, Queen substance dispersal by messenger workers in honey bee colonies, *Behav. Ecol. Sociobiol.* **5**:391-415.
Seeley, T. D., 1985, *Honeybee Ecology: a Study of Adaptation in Social Life*, Princeton University Press, Princeton, New Jersey.
Seeley, T. D., 1995, *The Wisdom of the Hive. The Social Physiology of Honey Bee Colonies*, Harvard University Press, Cambridge, Massachusetts.
Slessor, K. N., Foster, L. J., and Winston, M. L., 1998, Royal flavours: honey bee queen pheromones, in: *Pheromone Communication in Social Insects: Ants, Wasps, Bees, and Termites*, R. K. Vander Meer, M. D. Breed, K. E. Espelie, and M. L. Winston, eds., Westview Press, Boulder, Colorado, pp. 331-344.
Solomon, N. G., 1991, Current indirect fitness benefits associated with philopatry in juvenile prairie voles, *Behav. Ecol. Sociobiol.* **29**:277-282.
Solomon, N. G., and French, J.A., 1997, *Cooperative Breeding in Mammals*, Cambridge University Press, Cambridge.
Solomon, N. G., and Getz, L. L., 1997, Examination of alternative hypotheses for cooperative breeding in rodents, in: *Cooperative Breeding in Mammals*, N. G. Solomon, and J. A. French, eds., Cambridge University Press, Cambridge, pp. 199-230.
Stacey, N., and Sorensen, P. W., 1999, Pheromones, fish, in: *Encyclopedia of Reproduction*, E. Knobil, ed., 3 edn., Academic Press, New York, pp. 748-755.
Sun, L. X., and Müller-Schwarze, D., 1999, Chemical signals in the beaver: one species, two secretions, many functions? in: *Advances in Chemical Signals in Vertebrates*, R. E. Johnston, D. Müller-Schwarze, and P. W. Sorensen, eds., Kluwer Academic/Plenum Press, New York, pp. 281-288.
Tardif, S. D., 1997, The bioenergetics of parental behavior and the evolution of alloparental care in marmosets and tamarins, in: *Cooperative Breeding in Mammals*, N. G. Solomon, and J. A. French, eds., Cambridge University Press, Cambridge, pp. 11-33.
Vander Meer, R. K. and Morel, L., 1995, Ant queens deposit pheromones and antimicrobial agents on eggs, *Naturwissenschaften* **82**:93-95.
Vargo, E. L., 1998, Primer pheromones in ants, in: *Pheromone Communication in Social Insects: Ants, Wasps, Bees, and Termites*, R. K. Vander Meer, M. D. Breed, K. E. Espelie, and M. L. Winston, eds., Westview Press, Boulder, Colorado, pp. 293-313.
Vargo, E. L., 1999, Reproductive development and ontogeny of queen pheromone production in the fire ant *Solenopsis invicta*, *Physiol. Entomol.* **24**:1-7.
Visscher, P. K., and Dukas, R., 1995, Honey-bees recognize development of nestmates' ovaries, *Anim. Behav.* **49**:542-544.

Wabnitz, P. A., Bowie, J. H., Tyler, M. J., Wallace, J. C., and Smith, B. P., 1999, Aquatic sex pheromone from a male tree frog, *Nature* **401**:444-445.

Wang, Z. X., and Novak, M. A., 1994, Parental care and litter development in primiparous and multiparous prairie voles (*Microtus ochrogaster*), *J. Mammal.* **75**:18-23.

Wilson, E. O., 1970, Chemical communication within animal species, in: *Chemical Ecology*, E. Sondheimer, and J. B. Simeone, eds., Academic Press, New York, pp. 133-155.

Wilson, E. O., 1971, *The Insect Societies*, Belknap Press, Harvard, Massachusetts.

Winston, M. L., 1992, Semiochemicals and insect sociality, in: *Insect Chemical Ecology. An Evolutionary Approach*, B. D. Roitberg, ed., Chapman and Hall, New York, pp. 315-333.

Winston, M. L., and Slessor, K. N., 1992, The essence of royalty - honey-bee queen pheromone, *Am. Sci.* **80**:374-385.

Wyatt, T. D., 2003, *Pheromones and Animal Behaviour: Communication by Smell and Taste*, Cambridge University Press, Cambridge.

Yamamoto, K., Kawai, Y., Hayashi, T., Ohe, Y., Hayashi, H., Toyoda, F., Kawahara, G., Iwata, T., and Kikuyama, S., 2000, Silefrin, a sodefrin-like pheromone in the abdominal gland of the sword-tailed newt, Cynops ensicauda. *FEBS Lett.* **472**:267-270.

Zahavi, A., 1975, Mate selection: A selection for a handicap. *J. Theor. Biol.* **53**:205-214.

# THE DISCOVERY AND CHARACTERISATION OF SPLENDIPHERIN, THE FIRST ANURAN SEX PHEROMONE

Margit A. Apponyi and John H. Bowie*

## 1. INTRODUCTION

Over the course of the last decade, our research group has been studying frog skin secretions with a view to isolation of biologically active compounds. The secretions from over 25 species of Australian frogs have been studied, giving rise to over 200 biologically active compounds. To this end, the secretions from *Litoria splendida* have been studied. *L. splendida,* also known as the magnificent tree frog, was first identified in 1977 (Tyler et al., 1977). This species is immediately characterised by the presence of enlarged parotid glands at the rear of the head and the pale sulphur coloured spots with dark edging, scattered over the entire dorsal surface. *L. splendida* averages 10 cm in length, and has large discs on the end of its slightly webbed fingers, extensively webbed toes and orange or yellow flanks.

*Litoria splendida* is found in a reasonably small geographical area, in the north-western parts of the Northern Territory, and in the north-eastern parts of Western Australia (Tyler and Davies, 1986; Tyler et al., 1994). Specimens collected for these studies were obtained from the Kimberley region of Western Australia.

During these studies, monthly skin secretions were collected from male and female specimens over a period of three years (Wabnitz et al., 1999). The secretions were collected using the surface electrical stimulation method, in which an electrode of low charge is rubbed gently on the dorsal surface of the animal to produce the alarm response. The secretions are then washed off with deionised water, concentrated, and analysed by high performance liquid chromatography (HPLC). This process can be repeated monthly without causing harm to the animal. No animals were sacrificed for this study.

The chromatograms of these secretions indicated a small component present in the secretions from males only. Comparison of the chromatograms from the three-year period show that this 25 residue peptide (GLVSS IGKAL GGLLA DVVKS KGQPA-

---

* Department of Chemistry, University of Adelaide, South Australia, 5005 Australia

OH) is produced in the highest levels during the mating season, from January to March each year. During this period the peptide constitutes up to 1% of the total secretion material, dropping to as low as 0.1% from June through to November. It was therefore investigated for a possible role in the breeding cycle of this species.

## 2. BEHAVIOURAL TESTING

Behavioural tests were conducted in a glass tank containing a 2 cm depth of water totalling 1600 mL. Females of the species exposed to the pheromone at a concentration of approximately 10 pM were attracted to the source with remarkably rapid response times. An increase in alertness and change of posture was noticeable twenty seconds after introduction of the pheromone to the tank (Wabnitz et al., 1999; Wabnitz et al., 2000).

When higher concentrations of peptide were used, the female frogs became agitated, as they were unable to locate the source of the pheromone, and climbed the walls of the tank. Lower concentrations of the peptide went unnoticed.

These tests were repeated with eight different females at different times and in different directions, the pheromone being introduced to random ends of the tank, with a 100% success rate, both with the natural compound and a synthetic version.

Behavioural testing with females from a related species, *Litoria caerulea*, and with males from both species gave negative results. The pheromone was not recognised by any test subjects other than *L. splendida* females. The peptide is therefore a species specific male sex attractant pheromone.

Amphibian pheromones have been previously isolated from newt and salamander species, however, this peptide, which we have named splendipherin, is the first pheromone isolated from any anuran species. The delivery method of those previously isolated amphibian species are very clear. Sodefrin, the ten-residue peptide pheromone of *Cynops pyrrhogaster* and silefrin, the ten residue peptide pheromone of *Cynops ensicauda* are both sent through the water by the male newts by a vigorous shaking movement of the tail (Kikuyama et al., 1995; Yamamoto et al., 2000). The 20 kDa proteinaceous male courtship pheromone of *Plethodon jordani* is applied to the female's skin by direct contact (Rollman et al., 1999).

The pheromone identified from *L. splendida* does not have such an obvious delivery method. Diffusion across the distance of approximately one metre would be expected to take hours (Tanford, 1961). Instead, the female frogs found the source of the pheromone in a matter of minutes (Wabnitz et al., 2000). This infers that the molecule is likely to be moving across the surface of the water with a surfactant-like action.

Indeed, studies on the surface behaviour of the amphibian peptides maculatin and citropin show that these peptides preferentially sit at an air-water interface over bulk solution and tend towards α-helical conformation at the interface (E.E. Ambroggio, personal communication).

## 3. STRUCTURE DETERMINATION

We have determined the structure of splendipherin using nuclear magnetic resonance (NMR) spectroscopy. NMR is an extremely powerful tool in structure determination, particularly because the solvent system can be chosen to mimic a specific environment. Because we expect splendipherin to be moving across the surface of the water, the solvent used for the structure determination of splendipherin was a 1:1 mixture of trifluoroethanol and water. This solvent is a good membrane and interface mimic.

The structure of splendipherin has been determined using restrained molecular dynamics and simulated annealing calculations. The resultant structure is currently being employed in molecular modelling calculations using the CHARMM package. The structure and mechanism of movement of splendipherin will be published together when this work is complete.

## 4. ACKNOWLEDGEMENTS

We would like to thank the Australian Research Council for providing funding for this work.

## 5. REFERENCES

Kikuyama, S., Toyoda, F., Ohmiya, Y., Matsuda, K., Tanaka, S., and Hayashi, H.,1995, Sodefrin: a female-attracting peptide pheromone in newt cloacal glands, Science **267**:1643-1645.

Rollman, S. M., Houck, L. D., and Feldhoff, R. C., 1999, Proteinaceous pheromone affecting female receptivity in a terrestrial salamander, Science **285**:1907-1909.

Tanford, C., 1961, Physical Chemistry of Macromolecules, John Wiley and Sons, Inc.

Tyler, M. J., and Davies, M., 1986, Frogs of the Northern Territory, G.L. Duffield.

Tyler, M. J., Davies, M., and Martin, A. A., 1977, A New Species of Large, Green Tree Frog From Northern Western Australia, *Trans. R. Soc. S. Aust.* **101**(5):133-138.

Tyler M. J., Smith, L. A., and Johnstone, R. E., 1994, Frogs of Western Australia, Lamb Print.

Wabnitz, P. A., Bowie, J. H., Tyler, M. J., Wallace, J. C., and Smith, B. P., 1999, Animal behaviour: Aquatic sex pheromone from a male tree frog, *Nature* **401**:444-445.

Wabnitz, P. A., Bowie, J. H., Tyler, M. J., Wallace, J. C., and Smith, B. P., 2000, Differences in the skin peptides of the male and female Australian tree frog Litoria splendida, *Euro. J. Biochem.* **267**:269-275.

Yamamoto K., Kawai, Y., Hayashi, T., Ohe, Y., Hayashi, H., Toyoda, F., Kawahara, G., Iwata, T., and Kikuyama, S., 2000, Silefrin, a sodefrin-like pheromone in the abdominal gland of the sword-tailed newt, Cynops ensicauda. *FEBS letters* **472**:267-270.

# CHEMICALLY MEDIATED MATE RECOGNITION IN THE TAILED FROG (*ASCAPHUS TRUEI*)

Matthew J. Asay, Polly G. Harowicz, and Lixing Su[*]

## 1. INTRODUCTION

Amphibian communication methods are a popular field of study (Duellman and Trueb, 1994; Houck, 1998). Acoustic signals are used by the majority of adult anuran species to attract mates (Duellman and Trueb, 1994), and acoustic communication has been overwhelmingly documented in anurans. Few anurans have been reported to use visual signals for mate recognition (Summers et al., 1999). Chemical communication related to mate recognition in adult anurans has only been documented recently (Forester and Thompson, 1998; Wabnitz et al., 1999; Pearl et al., 2000), despite the evidence that chemical signals are widespread in urodeles (Houck, 1998). These alternative forms of communication can be important, especially under conditions where acoustic communication is not favored, such as in a noisy environment or in a habitat where auditorily oriented predators are abundant. The behavioral functions of non-acoustic communication in anurans are poorly understood and its ecological significance is difficult to assess due to the small number of studies available. Hence, studies of non-acoustic communication systems in anurans are of particular interest and will shed light on how different communication systems evolve under different environmental conditions.

Tailed frogs (*Ascaphus truei*) are usually found near permanent, fast flowing streams in forested areas (Metter, 1967). Because their normal habitat is usually noisy, selection may not favor the use of acoustic communication. Several morphological features of tailed frogs make it unlikely that vocal signals are used to attract mates. Tailed frogs have small lungs and lack vocal cords, a tympanic membrane, and columella, so their ability to produce and detect sound is probably poor (Noble and Putnam, 1931; Schmidt, 1970). Although other "earless" frogs do vocalize (Hetherington and Lindquist, 1999), there are no reports of any sounds produced by tailed frogs.

---

[*] Address correspondence to: Lixing Sun, Department of Biological Sciences, Central Washington University, Ellensburg, Washington, 98926.

As a silent, nocturnal animal, the tailed frog does not have many options available for mate attraction and recognition beyond chemical signals. Many studies have demonstrated the ability of anurans to recognize predators (Feminella and Hawkins, 1994; Flowers and Graves, 1997) and conspecifics (Graves et al., 1993). Olfaction has also been shown to play a role in spatial orientation, homing, and food location in anurans (Grubb, 1976). Studies of kin recognition in the larvae of several anuran species have shown that information about kin is most likely transmitted as chemical signals (Waldman, 1985; Blaustein et al., 1993). Kin recognition using chemical signals may continue after metamorphosis, as in *Rana cascadae* and *Rana sylvatica* (Blaustein et al., 1984; Cornell et al., 1989). Feminella and Hawkins (1994) showed that tailed frog tadpoles use chemical cues to detect predators, but no other information about chemical communication is available, especially in adults.

Tailed frogs are assumed to search for mates by crawling along the stream bottom (Jameson, 1955). If so, visual communication is unlikely, except at very short distances. Tailed frogs may simply grab onto the nearest moving object of appropriate size, but this could be costly if a predator is encountered or clasped by mistake. If a male or an unreceptive female is clasped, the ensuing wrestling match is a waste of energy (Wells, 1977) and may subject the pair to the risk of predation. A cue to guide tailed frogs toward a suitable mate would reduce the costs involved and increase the likelihood of successfully mating. Under this scenario, the ability to recognize a chemical cue would be advantageous.

The purpose of this study was to investigate the mate recognition ability of tailed frogs. Specifically, we tested the hypotheses that tailed frogs do not use visual signals to attract mates and that tailed frogs can detect waterborne chemical cues from conspecifics. We used two experiments to achieve this goal. The first experiment tested the visual mate recognition ability of males. The second experiment examined the ability of both sexes to detect waterborne chemical signals from conspecifics.

## 2. METHODS

### 2.1 Animal Collection and Care

Study animals were collected from Cold Creek, Kittitas County, Washington (T22N, R11E, S29, NW, elevation ~1000 m) on June 26, July 27, and September 22, 1998 by searching the stream banks at night. A total of 59 males and 36 females were captured. Frogs were housed separately in plastic boxes (33 cm x 20 cm x 10 cm) with gravel, water and a plastic cup for shelter and kept in a cold room (10-12 °C) on a 14h:10h L:D light cycle. They were fed 2-3 crickets (*Acheta domestica*) per week and given fresh water every other week.

### 2.2 Reproductive Status

Frogs used for the tests were reproductively mature to ensure the presence of any signals linked to sexual readiness. Males were considered mature if they developed black nuptial tubercles on the palm, forearm, shoulders (ventral surface), and chin (Daugherty and Sheldon, 1982a). Females were considered mature if eggs were visible through their abdominal wall or they were at least 40 mm in total length at the time of testing (Bull and Carter, 1996). Tests were performed between September 9, 1998 and November 6, 1998.

## 2.3 Visual Signals

We first tested the ability of males to distinguish between male and female conspecifics and between gravid or non-gravid females using visual cues only. We designed a two-way choice test apparatus for visual preference. The test apparatus was a 38 L aquarium that was divided into three chambers by glass partitions 10 cm from each end. These two stimulus chambers isolated the test frogs in the central test chamber from any chemical signals from the stimulus frogs in the stimulus chambers but could still receive visual signals. Stimulus frogs were size-matched for total length to the nearest millimeter to prevent any choices based on size. Tailed frogs are sexually dimorphic in size, so we had few male frogs large enough to pair with females. Therefore, stimulus frogs were used in five trials and were placed in opposite sides of the apparatus between trials. Damp paper towels were placed on each side of the test chamber. Test frogs were placed in the middle of the chamber under a circular glass dish (diameter 10 cm) and allowed to acclimate for five minutes. The dish was then removed and the frogs were videotaped for 10 minutes. Twenty different frogs were tested in each experiment and each frog was used only once. The damp paper towels were replaced, and the test chamber was wiped with wet paper towels between trials.

A central dividing line was drawn in the test chamber. We define preference based on the location of the head of a test frog on either side of this line. Data were collected from the videotapes and analyzed for time spent in the section of the test chamber adjacent to either stimulus frog. The videotapes were also analyzed for the first stimulus area chosen and the frequency that test frogs were observed in each stimulus area during the entire 10 minute session. The duration data were analyzed using the Wilcoxon matched pairs signed ranks test. The frequency data were analyzed with a sign test. The first choice data were analyzed with a binomial test. In one of these experiments, one frog did not make a choice for over 3 minutes, so that trial was not used in the first choice analysis.

## 2.4 Chemical Signals

In this experiment, we built an apparatus for a three-way choice test. The three treatments were male-conditioned water, female-conditioned water and control water. Each water treatment was conditioned by placing two reproductively mature frogs in 19 L of water for 24 hours. Control water was prepared at the same time and placed under the same conditions for 24 hours.

The experimental apparatus consisted of a test chamber constructed from a 19 L bucket attached to three other 19 L buckets. Water flowed into the chamber through three evenly-spaced holes in the sides of the bucket and out through a hole in the center of the test chamber. Treated water entered the chamber by gravity flow from the other 19 L buckets connected to the chamber with Tygon® tubing. In-line flow meters and clamps were used to keep water flowing in each tube at a rate of approximately 2 L per hour. Food coloring was placed in each treatment of signal water to make the flow more obvious to the observer. The color was rotated between each set of trials to avoid any preferences based on the food coloring. Each test subject was exposed to the three different water samples in each trial.

The frogs were videotaped for 15 minutes after a 5 minute acclimation period. The videotapes were analyzed for frequency and duration of behaviors using an ethogram. The behaviors were jump, climb, walk, and sit. Jumping is an instantaneous event, so time was not recorded for this behavior. Duration data were analyzed using the Friedman two-way analysis of variance by ranks. The frequency data were analyzed using a $\chi^2$ test. All data were analyzed based on response frequency or time for treatment type (male, female, control) and for behavioral pattern (jump, climb walk, sit). Duration data with significant results (P<0.05) were further analyzed using *post hoc* comparisons to identify which treatments were different (Daniel, 1990).

## 3. RESULTS

### 3.1 Visual Signals

Twenty trials were performed in each experiment (Experiment 1: male vs. female; Experiment 2: gravid vs. non-gravid). There was no significant difference in the mean time spent in either stimulus area for the test males (Wilcoxon matched pairs signed ranks test: Experiment 1: T=1.122, P=0.262; Experiment 2: T=0.934, P=0.350). Also, no significant differences were found for the frequency observed in either side of the test chamber for the entire 10 minute period or for first choice by test males.

### 3.2 Chemical Signals

In terms of response frequency, males were observed significantly more often in the female stimulus area than the other two stimulus areas (Figure 1, $\chi^2_2$=8.03, n=53, P=0.018). Also, males jumped while in the female stimulus area significantly more frequently than the other two stimulus areas (Figure 1, $\chi^2_2$=6.47, n=53, P=0.039). All other behaviors by males were without significant differences.

For female test subjects, jump was observed more frequently in the male stimulus area than the other two stimulus areas (Figure 2, $\chi^2_2$=17.45, n=29, P=0.0002). No significant differences were detected in the frequency of other behavioral responses.

For duration of response, males sat in the female stimulus area significantly more than the other two stimulus areas (Figure 3, W=7.283, n=53, P=0.026). Multiple comparison tests showed that the difference existed between female and male (P<0.05), but not between the other treatments. No significant differences were found in other behavioral patterns for male test subjects. There were no significant differences in the duration of any of the behaviors performed by females.

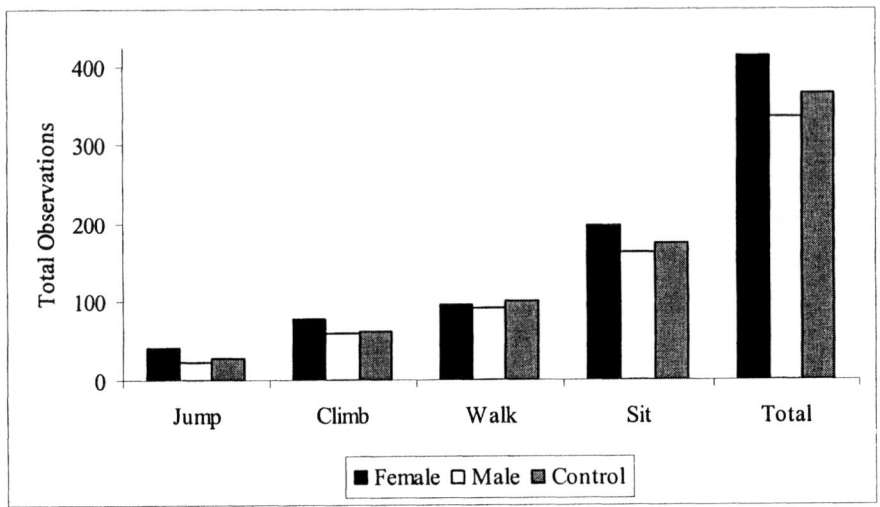

**Figure 1.** Total number of times all males were observed performing individual behaviors in each treatment area (n=53).

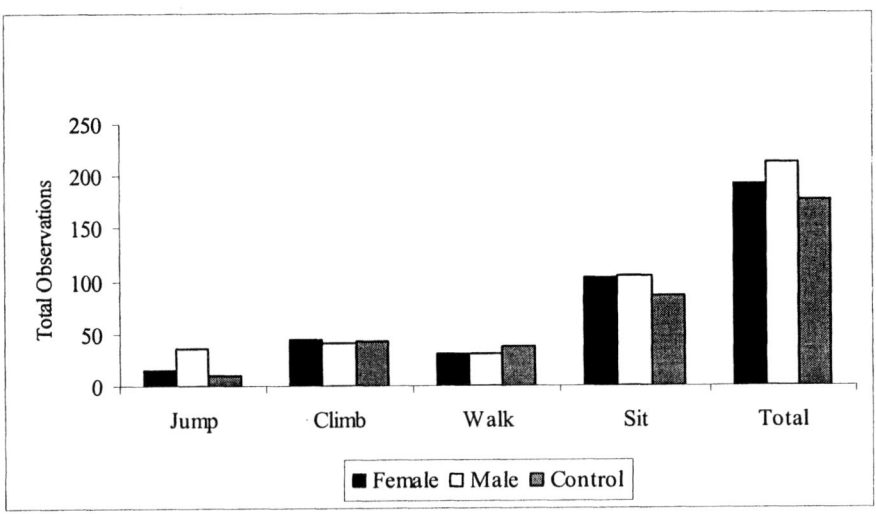

**Figure 2.** Total number of times all females were observed performing individual behaviors in each treatment area (n=29).

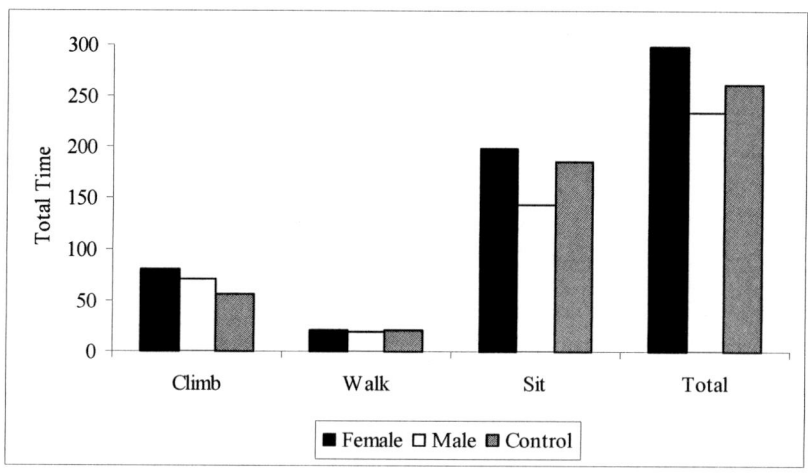

Figure 3. Total time all males were observed performing each behavior in individual treatment areas (n=53).

## 4. DISCUSSION

Recognition using chemical signals has been extensively demonstrated in a variety of animals, such as mammals, reptiles, amphibians, fish, and insects. In amphibians, it has been documented in caecilians (Warbeck et al., 1996) and salamanders (Houck, 1998). Attempts to study chemical recognition in anurans are relatively rare. To our knowledge, only three studies have investigated the possibility of mate recognition via chemical signals (Forester and Thompson, 1998; Wabnitz et al., 1999; Pearl et al., 2000). Thus, our study is among the first to methodologically investigate chemical signals as a mechanism of mate attraction or recognition in anurans.

No evidence from our study suggests that tailed frogs use visual cues to find mates. This is understandable because tailed frogs are nocturnal and are rarely found outside of a stream in the daytime (Metter, 1967), so visual signals would not be useful. Metter (1964) observed that males will attempt amplexus with the nearest available frog when placed in a mixed sex group of conspecifics. Many male anurans, even some without advertisement calls, use release calls when grasped by a male (Duellman and Trueb, 1994; Marco et al., 1998), but this is not the case with tailed frogs because they cannot produce sound. It seems that male tailed frogs can determine the suitability of potential mates after achieving amplexus. Wernz (1969) reported several males clasping non-gravid females and then releasing them after 20 minutes, presumably because they detected the reproductive status of the females. Duellman and Trueb (1994) suggest that continuation of amplexus by tailed frogs occurs based on the greater girth and firmness of gravid females. This method of mate recognition using tactile information may work, but only after amplexus is achieved. It would save a significant amount of time and energy, as well as reduce predation risk, if there is a pre-amplectic mechanism of recognizing suitable mates. Without using visual and acoustic signals for mate recognition, chemical communication seems to be the most likely alternative for tailed frogs.

Jameson (1955) states that tailed frogs find mates by swimming or crawling on the bottom of streams. Although this appears to be an assumption, no study has so far provided evidence to contradict these statements. If this assumption is true, then it would be advantageous to have the ability to detect a signal in the water from potential mates. Two results in our study support the presence of a waterborne mate attraction signal. One is that males and females jumped more frequently toward signals from the opposite sex than toward the other stimuli. The other is that males were recorded significantly more frequently in the female stimulus area than the other two stimulus areas. Because our design eliminated other possibilities, the signal for mate attraction or recognition is most likely a waterborne chemical signal.

Tailed frogs have a small home range (Daugherty and Sheldon, 1982b) and females sometimes aggregate (Brown, 1975), so large quantities of feces may be deposited in the same area. Feces may provide a cue for amphibians to find a mate. For instance, plethodontid salamanders recognize chemical cues in fecal pellets (Jaeger et al., 1986; Horne and Jaeger, 1988). Tailed frogs may use a similar system. Chemical signals are excellent for communicating information over short distances and the water/air interface is especially favorable for the transmission of these cues. Substances reach much higher concentrations at this interface than in air or water alone (Doving et al., 1993). Because tailed frogs are mainly aquatic, chemical signals may be particularly favored for communication.

In conclusion, we have found that tailed frogs probably do not rely on visual cues alone for mate recognition but are able to use waterborne chemical signals to find mates. While our major goal in this study was to determine whether tailed frogs are able to use chemical signals for mate recognition, the behavioral and ecological functions of mate recognition using chemical signals need to be further determined in their natural settings.

## 5. ACKNOWLEDGEMENTS

Comments by Dietland Müller-Schwarze and two anonymous reviewers greatly improved this paper. Financial support was provided by The American Museum of Natural History Theodore Roosevelt Memorial Fund and Central Washington University Department of Biological Sciences. Richard Lukose, James Stegan and Daniel Ray assisted with animal capture.

## 6. REFERENCES

Blaustein, A. R., O'Hara, R. K., and Olson, D. H., 1984, Kin preference behavior is present after metamorphosis in *Rana cascadae* frogs, *Anim. Behav.* 32:445-450.

Blaustein, A. R., Yoshikawa, T., Asoh, K., and Walls S. C., 1993, Ontogenetic shifts in tadpole kin recognition: loss of signal and perception, *Anim. Behav.* 46:525-538.

Brown, H. A., 1975, Temperature and development of the tailed frog, *Ascaphus truei*, *Comp. Biochem. Physiol. A* 50:397-405.

Bull, E. L., and Carter, B. E., 1996, Tailed frogs: Distribution, ecology, and association with timber harvest in northeastern Oregon, Res. Pap. PNW-RP-497 Portland, OR: U.S. Department of Agriculture, Forest Service, Pacific Northwest Research Station.

Cornell, T. J., Berven, K. A., and Gamboa, G. J., 1989, Kin recognition by tadpoles and froglets of the wood frog *Rana sylvatica*, *Oecol* **78**:312-316.
Daniel, W. W., 1990, *Applied Nonparametric Statistics*, PWS-Kent Publishing Company, Boston.
Daugherty, C. H., and Sheldon, A. L., 1982a, Age-determination, growth, and life history of a Montana population of the tailed frog (*Ascaphus truei*), *Herpetologica* **38**:461-468.
Daugherty, C. H., and Sheldon, A. L., 1982b, Age-specific movement patterns of the frog *Ascaphus truei*, *Herpetologica* **38**:468-474.
Doving, K. B., Trotier, D., Rosin, J., and Holley, A., 1993, Functional architecture of the vomeronasal organ of the frog (Genus *Rana*), *Acta Zool.* **74**:173-180.
Duellman, W. E., and Trueb, L., 1994, *Biology of Amphibians*, The Johns Hopkins University Press, Baltimore.
Feminella, J. W., and Hawkins, C. P., 1994, Tailed frog tadpoles differentially alter their feeding behavior in response to non-visual cues from four predators, *J. N. Am. Benthol. Soc.* **13**:310-320.
Flowers, M. A., and Graves, B. M., 1997, Juvenile toads avoid chemical cues from snake predators, *Anim. Behav.* **53**:641-646.
Forester, D. C., and Thompson, K. J., 1998, Gauntlet behavior as a male sexual tactic in the American toad (Amphibia: Bufonidae), *Behaviour* **135**:99-119.
Graves, B. M., Summers, C. H., and Olmstead, K. L., 1993, Sensory mediation of aggregation among postmetamorphic *Bufo cognatus*, *J Herpetol* **27**:315-319.
Grubb, J. C., 1976, Maze orientation by Mexican toads, *Bufo valliceps* (Amphibia, Anura, Bufonidae), using olfactory and configurational cues, *J. Herpetol.* **10**:97-104.
Hetherington, T. E., and Lindquist, E. D., 1999, Lung-based hearing in an "earless" anuran amphibian, *J. Comp. Physiol. A* **184**:395-401.
Horne, E. A., and Jaeger, R. G. 1988, Territorial pheromones of female red-backed salamanders, *Ethology* **78**:143-152.
Houck, L. D., 1998, Integrative studies of amphibians: From molecules to mating, *Am. Zool.* **38**:108-117.
Jaeger, R. G., Goy, J. M., Tarver, M., and Marquez, C., 1986, Salamander territoriality: pheromonal markers as advertisement by males, *Anim. Behav.* **34**:860-864.
Jameson, D. L., 1955, Evolutionary trends in the courtship and mating behavior of Salientia, *Syst. Zool.* **4**:105-119.
Marco, A., Kiesecker, J. M., Chivers, D. P., and Blaustein, A. R., 1998, Sex recognition and mate choice by male western toads, *Bufo boreas*, *Anim. Behav.* **55**:1631-1635.
Metter, D. E., 1964, On breeding and sperm retention in *Ascaphus*, *Copeia* **1964**:710-711.
Metter, D. E., 1967, Variation in the ribbed frog *Ascaphus truei* Stejneger, *Copeia* **1967**:634-649.
Noble, G. K., and Putnam, P. G., 1931, Observations on the life history of *Ascaphus truei* Stejneger, *Copeia* **1931**:97-101.
Pearl, C. A., Cervantes, M., Chan, M., Ho, U., Shoji, R., and Thomas, E. O., 2000, Evidence for a mate-attracting chemosignal in the dwarf African clawed frog *Hymenochirus*, *Horm. Behav.* **38**:67-74.
Schmidt, R. S., 1970, Auditory receptors of two mating call-less anurans, *Copeia* **1970**:169-170.
Summers, K., Symula, R., Clough, M., and Cronin, T., 1999, Visual mate choice in poison frogs, *Proc. R. Soc. Lond. B Biol. Sci.* **266**:2141-2145.
Wabnitz, P. A., Bowie, J. H., Tyler, M. J., Wallace, J. C., and Smith, B. P., 1999, Aquatic sex pheromone from a male tree frog, *Nature* **401**:444-445.
Waldman, B., 1985, Olfactory basis of kin recognition in toad tadpoles, *J. Comp. Physiol. A* **156**:565-577.
Warbeck, A., Breiter, I., and Parzefall, J., 1996, Evidence for chemical communication in the aquatic caecilian *Typhlonectes natans* (Typhlonectidae, Gymnophiona), *Memoires de Biospeologie* **23**:37-41.
Wells, K. D., 1977, The social behavior of anuran amphibians, *Anim. Behav.* **25**:666-693.
Wernz, J. G., 1969, Spring mating of *Ascaphus*, *J. Herpetol.* **3**:167-169.

# RESPONSES TO SEX- AND SPECIES-SPECIFIC CHEMICAL SIGNALS IN ALLOPATRIC AND SYMPATRIC SALAMANDER SPECIES

Catherine A. Palmer and Lynne D. Houck[*]

## 1. INTRODUCTION

Sexually reproducing organisms depend on a close association between one sex's ability to produce and transmit a signal in a given environment and the ability of the opposite sex to decipher and respond to that signal. While continuously confronted with an assortment of visual, acoustic and chemical stimuli from their surroundings, many species readily distinguish signals produced by prospective mating partners. This tightly linked signaler/receiver system enables these animals to quickly locate or attract mates, assess mate quality, and avoid the reproductive costs that may be associated with hybridization. In many organisms, reproductively isolated populations exhibit differences in their mating signals, and these signals act as barriers to gene exchange (Ryan, 1990; Coyne and Oyama, 1995; McLennan and Ryan, 1997, 1999). It is well accepted that reproductive isolation can evolve gradually in geographically separated populations as the populations diverge in response to adaptations to different environments or to other selective pressures. More controversial, however, is the idea that natural selection may rapidly increase divergence in mate recognition by way of selection against hybridization following secondary contact. The reinforcement of mate recognition by this process will result in the pattern of reproductive character displacement, whereby the signals that reduce mating between populations diverge more dramatically between sympatric populations than they do between allopatric populations (Dobzhansky, 1937; Dawley, 1987; McKinnon and Liley, 1986; Reagan, 1992).

Sexual isolation has arisen among numerous species of geographically isolated populations of salamanders (review by Arnold et al., 1993). Behavioral experiments indicate that the degree of reproductive isolation is stronger between sympatric species of large *Plethodon* than it is between allopatric populations of these species (Reagan, 1992). Sexual isolation may be due to species-specific differences in the chemosensory systems

---

[*]Department of Zoology, Oregon State University, Corvallis, OR 97331

of these terrestrial vertebrates (Arnold, 1976; Dawley, 1986; Verrell, 1989). Plethodontid salamanders rely on chemical cues for territorial advertisement, sex identification, mate assessment and, in some cases, species recognition (review by Mathis et al., 1995; Marco et al., 1998). Surprisingly, very little research has been carried out to determine if substrate-borne chemical signals are used to maintain reproductive barriers between these salamander species (but see Verrell, 1989; Verrell, 2003). The goal of the present study is two-fold. First, we aim to determine whether closely related large Eastern *Plethodon* species are capable of accurately assessing sex- and species-specific substrate-borne chemical signals. Secondly, we ask if variation in the chemical signal alone is enough to explain the existing patterns of isolation, wherein sexual isolation is stronger between sympatric species than it is between allopatric species.

## 2. CHEMICAL COMMUNICATION IN TERRESTRIAL SALAMANDERS

Salamanders from the genus *Plethodon* are a morphologically conservative group of nocturnal, terrestrial animals that are incapable of vocalizing. Visual and auditory signals therefore play minor roles in their ability to find, attract or sexually persuade mates. During the breeding season, male plethodontid salamanders detect and precisely track pheromone trails left by passing females by tapping their nasolabial grooves on the substrate (Gergits and Jaeger, 1990). Nasolabial grooves are furrows on the snout, extending from the edge of each naris down to the upper lip. When an animal taps its snout to the substrate (or the skin of another salamander), moisture containing chemosensory information is passed up the grooves and into the nares through capillary action (Brown, 1968; Dawley and Bass, 1989). The chemicals are then directed laterally to chemoreceptors in the highly specialized vomeronasal epithelium (Dawley and Bass, 1989). The vomeronasal receptor neurons project to the accessory olfactory bulb (Schmidt, 1988), and presumably the information is then transmitted via the amygdala to a region in the hypothalamus known to be involved in reproductive behavior (cf Halpern, 1987). Thus, these salamanders can easily acquire information via substrate-borne chemical signaling and these signals may play a fundamental role both in promoting conspecific interactions and in maintaining reproductive barriers between species.

This study involves three closely related species of large, Eastern salamanders within the genus *Plethodon*: *P. shermani*, *P. montanus* and *P. teyahalee*. Two of these species, *P. montanus* and *P. shermani*, are geographically separated from one another and sexual isolation has evolved in allopatry (Stalker index = 0.342, where 0 = random mating and 1 = complete isolation; Reagan, 1992). Where *P. teyahalee* and *P. montanus* are found in sympatry, morphological, molecular and behavioral evidence indicate that hybridization does not occur (Stalker index = 1.0, Reagan, 1992; Highton and Peabody, 2000). In contrast, *P. teyahalee* and *P. shermani* hybridize extensively in many areas of contact and, in one region, the parental species have been replaced entirely by hybrids (Stalker index = 0.017 - 0.586; Reagan, 1992; Highton and Peabody, 2000). There are no obvious differences in the courtship behaviors of these species (Arnold, 1976; Reagan, 1992), nor is there a correlation between genetic divergence and levels of reproductive isolation (Reagan, 1992).

Dawley (1984, 1986, 1987) demonstrated that some species of large Eastern *Plethodon* are capable of detecting and responding to sex- and species-specific air-borne chemical stimuli. When presented with substrate-borne odors, however, these *Plethodon* species were unable to distinguish sex- or species-specific odors (Dawley, 1984). This lack of detection is surprising for three reasons: (1) our understanding of the structure and function of nasolabial grooves, as well as the associated vomeronasal organ, illustrates a mechanism by which substrate-borne chemical cues can be readily acquired and processed, (2) *Plethodon* salamanders inhabit moist terrestrial environments where transport of these signals is easily facilitated and where non-volatile molecules can persist for longer periods than volatile ones, and (3) small Eastern *Plethodon* species have been shown to detect and respond to substrate-borne territorial signals (review by Mathis et al., 1995). For these reasons, it is generally well accepted that terrestrial salamanders rely on substrate-borne chemical cues when selecting reproductive partners, despite a limited amount of experimental evidence. The animals used in Dawley's (1984) experiments, however, were not in reproductive condition, and sensitivity to odors is known to increase during the mating season (Toyoda and Kikuyama, 2000). Furthermore, amphibian sex pheromones show considerable seasonal variation and are usually only produced during the breeding season (Rollmann et al., 2000; Wabnitz et al., 1999). If the chemical signals used as sex and/or species identifiers are produced and broadcast during the mating season, we should be able to detect individual preferences at that time.

## 3. METHODS AND RESULTS

### 3.1. Collection and Maintenance of Salamanders

We collected 20 adult males and 20 adult females for each of three species, *P. teyahalee*, *P. montanus* and *P. shermani*, between 2-16 August 2001. The *P. shermani* were captured from Macon County, North Carolina (lat 35°10'48", long 83°33'38"), and *P. teyahalee* and *P. montanus* were collected from a single locality in Madison County, North Carolina (lat 35°48'50", long 82°56'58"). We obtained females with enlarged ova and males with visible mental glands to ensure that the animals used in the experiment were in reproductive condition. The animals were shipped to Oregon State University where they were maintained at 13-15°C on a natural (Corvallis, OR) photoperiod. The salamanders were housed individually in plastic shoeboxes (31 X 16.5 X 9 cm) with damp paper towel substrates and fed two waxworms (*Galleria mellonella*) each week.

### 3.2. Test Protocol

The odor preference experiments were conducted in the laboratory from 27 August to 18 October 2001, a time period corresponding with the breeding season for each of these species. Odor sources were obtained by lining the bottom of a small plastic box (17 X 12 X 6 cm) with a piece of single-ply filter paper moistened with dechlorinated water. A scent-marking animal was placed on top of the filter paper and allowed to move about the box for a period of 24 hr. A blank scent was produced by moistening a piece of filter paper with dechlorinated water and storing it in a covered enclosure for 24 hr.

Fifteen salamanders of each sex and of each species were tested in a set of eight odor preference trials (Figure 1; 120 tests per species). The order in which an animal performed a particular trial was assigned randomly. For each trial, we used a clear, rectangular box (31 X 16.5 X 9 cm) as the experimental chamber. Each side of the box contained a piece of marked filter paper (or a blank) with a space of 3.5 cm between the two substrates to prevent diffusion of chemical cues between sides. The side of the box in which the odors were placed was randomized. At 1800 hr, the experimental animal was introduced to the center of the experimental box and the animal's behavior was observed under dim red illumination. Data were collected every two minutes for a period of 2 hrs (60 observations / animal). We recorded: (1) the side of the box where the animal's head was positioned, (2) whether the animal was tapping the substrate (i.e. nasolabial grooves in repeated contact with the substrate, Arnold 1976), and (3) if the animal was located on the wall of the experimental chamber (indicating lack of contact with chemical signals). Nose tapping behavior was recorded to assess the general activity level of the animals and is not considered an independent measure from which to infer odor preferences. We simultaneously tested 55 animals (on average) each trial night, and provided salamanders with at least one day of rest between trial nights.

### 3.3. Data Analysis

Individuals were tested for side preferences (right versus left) by comparing the number of times an individual animal was found on either side of the experimental chamber during each of its eight trials. Animals displaying a preference for one side of the box over the other, regardless of odor choices, were excluded from the analyses. In addition, if an animal spent more than half of the testing period on the wall of the experimental chamber it was excluded from the analysis. For each scent preference trial, we tallied the number of observations out of 60 that the animal spent on each side of the experimental arena. Data were subjected to a Wilcoxon matched-pairs signed ranks test to test the null hypothesis that a salamander would spend an equal amount of time (30 observations) on both sides of the experimental chamber if it had no odor preference.

### 3.4. Female Odor Discrimination

For each of the three *Plethodon* species, a single female exhibited a statistically significant preference for one side of the box over the other throughout her eight trials. These three females were excluded from the analysis. In addition, individuals from 22 *P. teyahalee* tests, 25 *P. shermani* tests and 19 *P. montanus* tests spent more than half of the time on the wall of the experimental chamber and so were not included in the statistical analyses. Females did not exhibit a preference for one substrate odor over the other in 22 of 24 trials (Table 1). However, when given a choice between a blank and the odor of a *P. shermani* male, *P. montanus* females spent the majority of their time on the substrate with no odor (Table 1, $P = 0.01$). Similarly, *P. shermani* females spent most of their time on the substrate with no odor when the other option was a substrate marked by a *P. teyahalee* male (Table 1, $P = 0.03$).

### 3.5. Male Odor Discrimination

Two *P. shermani* males and one *P. montanus* male demonstrated a preference for one side of the box over the other throughout the eight trials and were excluded from the study. In individual tests, 16 *P. teyahalee* tests, 11 *P. shermani* tests and 13 *P. montanus* tests had males that remained on the wall of the experimental chamber for over half of the experimental duration and were not included in the analyses. Males did not exhibit a preference during blank versus conspecific male odor trials, blank versus either heterospecific female odor trials, or trials in which two heterospecific female odors were present (Table 1). In most trials, males did not display a preference for conspecific female versus heterospecific female odors. The one exception is the *P. montanus* trial in which males exhibited a preference for conspecific female odors over *P. teyahalee* female odors (Table 1, P = 0.03). Males of all three species preferred the odor of conspecific females to that of a blank (Table 1, *P. teyahalee*: P = 0.03; *P. shermani*: P = 0.03; and *P. montanus*: P = 0.001). Males of all species also displayed a preference for conspecific female odors over conspecific male odors (*P. teyahalee*: P = 0.04; *P. shermani*: P = 0.01; and *P. montanus*: P = 0.001). In all trials, males were significantly more active (more nose taps) than females (P = 0.001).

Table 1. Preferences of three species of *Plethodon* in trials with two substrate-borne odor choices. In each trial, the female or male was given a choice of two substrates marked by another salamander. Response to substrate was measured every two minutes for two hours (for a total of 60 observations). T is the value of the Wilcoxon matched-pairs signed-ranks test, p is the two-sided probability with significance of * α <= 0.05 and ** α <= 0.01. C = conspecific, H = heterospecific. For *P. teyahalee*, H1 = *P. shermani*, H2 = *P. montanus*; for *P. shermani*, H1 = *P. montanus*, H2 = *P. teyahalee*; and for *P. montanus*, H1 = *P. shermani*, H2 = *P. teyahalee*

| | | FEMALES: | | | | | | MALES: | | |
|---|---|---|---|---|---|---|---|---|---|---|
| Species | N | Mean response to substrate | | T | p | Species | N | Mean response to substrate | T | p |
| **Trial 1** | | Blank vs. C♂ | | | | **Trial 1** | | Blank vs. C♂ | | |
| P. teyahalee | 14 | 34.2 | 25.8 | 39.5 | 0.41 | P. teyahalee | 13 | 38.5 | 21.5 | 25.0 | 0.27 |
| P. shermani | 8 | 17.1 | 42.9 | 8.0 | 0.19 | P. shermani | 11 | 34.9 | 25.1 | 21.5 | 0.31 |
| P. montanus | 10 | 22.3 | 37.7 | 14.5 | 0.16 | P. montanus | 11 | 28.3 | 31.7 | 29.0 | 0.72 |
| **Trial 2** | | Blank vs. C♀ | | | | **Trial 2** | | Blank vs. C♀ | | |
| P. teyahalee | 11 | 28.7 | 31.3 | 30.0 | 0.79 | P. teyahalee | 12 | 16.7 | 43.3 | 9.5 | 0.03* |
| P. shermani | 9 | 29.6 | 30.4 | 22.0 | 0.95 | P. shermani | 11 | 18.8 | 41.2 | 1.0 | 0.03* |
| P. montanus | 11 | 36.5 | 23.5 | 24.0 | 0.42 | P. montanus | 14 | 10.8 | 49.2 | 7.0 | 0.00** |
| **Trial 3** | | C♂ vs. C♀ | | | | **Trial 3** | | C♂ vs. C♀ | | |
| P. teyahalee | 11 | 31.8 | 28.2 | 30.0 | 0.79 | P. teyahalee | 11 | 11.0 | 49.0 | 10.0 | 0.04* |
| P. shermani | 12 | 28.5 | 31.5 | 28.5 | 0.69 | P. shermani | 13 | 16.8 | 43.2 | 5.0 | 0.01* |
| P. montanus | 14 | 27.2 | 32.8 | 52.0 | 0.98 | P. montanus | 12 | 17.4 | 42.6 | 2.0 | 0.00** |
| **Trial 4** | | Blank vs. H1♂ | | | | **Trial 4** | | Blank vs. H1♀ | | |
| P. teyahalee | 12 | 32.2 | 27.7 | 31.0 | 0.53 | P. teyahalee | 12 | 18.2 | 41.8 | 23.0 | 0.21 |
| P. shermani | 13 | 35.2 | 24.8 | 36.0 | 0.51 | P. shermani | 12 | 30.7 | 29.3 | 38.0 | 0.94 |
| P. montanus | 9 | 52.0 | 8.0 | 0.0 | 0.01* | P. montanus | 12 | 26.8 | 33.2 | 20.0 | 0.45 |
| **Trial 5** | | Blank vs. H2♂ | | | | **Trial 5** | | Blank vs. H2♀ | | |
| P. teyahalee | 12 | 28.9 | 31.1 | 38.5 | 0.97 | P. teyahalee | 13 | 33.3 | 26.7 | 33.0 | 0.64 |
| P. shermani | 10 | 47.3 | 12.7 | 6.0 | 0.03* | P. shermani | 13 | 29.4 | 30.6 | 43.0 | 0.86 |
| P. montanus | 10 | 29.4 | 30.6 | 26.5 | 0.92 | P. montanus | 10 | 41.4 | 18.6 | 14.5 | 0.19 |
| **Trial 6** | | C♂ vs. H1♂ | | | | **Trial 6** | | C♀ vs. H1♀ | | |
| P. teyahalee | 8 | 26.6 | 33.4 | 16.5 | 0.83 | P. teyahalee | 14 | 36.0 | 24.0 | 37.5 | 0.94 |
| P. shermani | 13 | 33.5 | 26.5 | 35.5 | 0.49 | P. shermani | 11 | 34.3 | 25.7 | 16.5 | 0.26 |
| P. montanus | 12 | 30.6 | 29.4 | 32.0 | 0.58 | P. montanus | 14 | 27.1 | 32.9 | 44.5 | 0.62 |
| **Trial 7** | | C♂ vs. H2♂ | | | | **Trial 7** | | C♀ vs. H2♀ | | |
| P. teyahalee | 11 | 28.5 | 31.5 | 30.0 | 0.79 | P. teyahalee | 14 | 41.8 | 18.2 | 30.0 | 0.16 |
| P. shermani | 13 | 27.1 | 32.9 | 36.0 | 0.51 | P. shermani | 12 | 34.1 | 25.9 | 25.0 | 0.27 |
| P. montanus | 14 | 25.9 | 34.1 | 39.0 | 0.40 | P. montanus | 13 | 42.2 | 17.8 | 15.0 | 0.03* |
| **Trial 8** | | H1♂ vs. H2♂ | | | | **Trial 8** | | H1♀ vs. H2♀ | | |
| P. teyahalee | 11 | 22.6 | 37.4 | 24.0 | 0.42 | P. teyahalee | 15 | 25.5 | 34.5 | 46.0 | 0.43 |
| P. shermani | 9 | 20.9 | 39.1 | 11.0 | 0.17 | P. shermani | 10 | 30.7 | 29.3 | 27.5 | 0.99 |
| P. montanus | 13 | 27.9 | 32.1 | 39.5 | 0.68 | P. montanus | 13 | 31.5 | 28.5 | 39.5 | 0.68 |

## 4. DISCUSSION

This study indicates that mating interactions for large Eastern *Plethodon* species depend largely on the male's ability to perceive, distinguish and respond to substrate-borne chemical signals produced by the female. Although females of these species produce a chemical signal that attracts males, females did not use substrate-borne odors to locate mating partners during our laboratory experiments. These females may, however, assess male chemical signals during courtship interactions as the male makes physical contact and applies courtship pheromones directly onto the female's nares.

Males of all three species displayed a strong preference for substrates previously occupied by a conspecific female when the alternative was a conspecific male odor or a substrate with no salamander odor. These males showed no interest in heterospecific female odors under the same testing conditions, suggesting that these chemical signals have diverged in allopatry. When males were given a choice between a substrate previously occupied by a conspecific female odor and a substrate containing a heterospecific female odor, however, the attraction for the conspecific was no longer apparent. Thus, the strength of the conspecific signal is dampened when the male is simultaneously introduced to a heterospecific female's scent. In one case, however, male *P. montanus* displayed a strong preference for the female conspecific when the alternate odor was that of a heterospecific *P. teyahalee* female. In this case, *P. montanus* and *P. teyahalee* exist in sympatry without hybridizing, indicating that the divergence of chemical cues is greater between sympatric species.

### 4.1. Discussion of Female Behavior

Female *P. shermani, P. montanus* and *P. teyahalee* did not respond preferentially to sex-specific or species-specific substrate-borne chemical cues in our laboratory experiments. The lack of a distinct preference and reduced activity that we observed could accurately reflect the natural behavior of these species during the mating season. In most signal/receiver systems involved in mate attraction, one sex produces a signal while the other sex detects and pursues that signal. While there have been observations of male *P. shermani* rapidly following the trails of passing females (Gergits and Jaeger, 1990; Reagan, 1992), similar behaviors have not been reported for females. In our study, male salamanders tapped the substrate much more frequently than did females, indicating that the males more actively sample the environment, presumably in search of a mate. This difference in male and female search behavior may correlate with a sexual dimorphism of the vomeronasal organ (VNO). In VNO studies of a related plethodontid salamander, *P. cinereus,* the male has a significantly larger VNO than the female during the breeding season (Dawley, 1992). Furthermore, observations from staged courtship encounters in the laboratory indicate that males pursue and initiate courtship, whereas the females are initially passive (CP, personal observations). In our tests, *P. shermani* and *P. montanus* females spent approximately twice as much time on substrates marked by conspecific males than on substrates with no salamander odor (a trend that was not statistically significant). Similarly, in four out of six trials, females visited the substrate with no salamander odor more frequently than the substrate bearing a heterospecific male odor, but in only two of these four trials was this difference statistically different.

A similar odor-choice experiment was conducted by Dawley (1986, 1987), who examined female responses to air-borne chemical signals in large Eastern *Plethodon* species. Dawley (1986) concluded that females always showed indifference to heterospecific male odors, but were attracted, repulsed or unresponsive to conspecific males. Results from two of our trials also suggest that females are repulsed by substrates bearing heterospecific male odors. The *P. shermani* females avoided *P. teyahalee,* and *P. montanus* females avoided *P. shermani* male odors, both spending significantly more time on the substrate with no salamander odor. When these females were exposed to substrates marked by the same heterospecific males in two other trials, however, the avoidance response was no longer apparent (Table 1). For this reason, evidence for a genuine avoidance response seems lacking. In fact, based on the data from all eight trials, we conclude that females of the *Plethodon* species used in this study do not use substrate-borne chemical cues to appraise and/or locate potential mating partners.

Our conclusion that females do not use substrate-borne odors to actively locate mates does not imply that these salamanders are incapable of detecting and assessing male chemical signals. Pheromones delivered during initial contact between mates, as well as during courtship interactions, may play a significant role in female mate choice. Chemical signals can be conveyed to the female directly from the surface of the male's body as the pair comes into physical contact (Arnold, 1976). Furthermore, pheromones are delivered directly to the female's nares when the male 'slaps' his mental gland onto her snout during courtship (Organ, 1958; Arnold, 1976). Experimental studies show that the application of this proteinaceous courtship pheromone results in a significant increase in female receptivity, indicating that pheromone delivery improves courtship success (Houck et al., 1998; Rollmann et al., 1999). Courtship pheromone delivery does not guarantee insemination, however, and the female may leave the courting male following pheromone delivery (Reagan, 1992). Thus, females may use chemical signals during courtship to identify and assess appropriate mating partners and these courtship pheromones may play a role in sexual isolation.

### 4.2. Discussion of Male Behavior

Male *P. shermani, P. montanus* and *P. teyahalee* are fully capable of discriminating between sex-specific odors and show a strong preference for female chemical cues. When the males were given a choice between a conspecific female odor and a substrate containing no odor, all three species showed a significant preference for the female odor. In addition, males preferred substrates marked by conspecific females to those marked by conspecific males. In a third type of trial, males presented with a conspecific male odor and a substrate with no salamander odor did not discriminate between substrates. These results provide compelling evidence that males are displaying a sex-specific response and are not merely attracted to the scent of any other salamander and/or avoiding substrates that have been marked by conspecific males.

Large eastern *Plethodon* species emerge from their underground burrows at nightfall to forage and to mate. Because these animals occupy and defend their own burrows, they do not have immediate access to a mating partner on a given night, nor do they rely on acoustic or visual signals to attract or discern an appropriate mate. Instead, the results of our laboratory tests suggest that a strong sex-specific chemical signal is produced by the female and this signal can easily be detected by a conspecific male. When the female deposits this pheromone on a moist, terrestrial substrate in a natural setting, the male should be able to locate the female relatively quickly. In Dawley's (1984) experiments, males did not respond to female substrate-borne odors outside of the breeding season.

We infer that immature and non-gravid females produce a signal that is distinct from the odor advertised by gravid females or that this sex-specific signal is absent altogether in non-breeding females.

Male preferences for the odor of a conspecific female are straightforward, but results of tests using heterospecific female odors yield mixed results. When *P. shermani*, *P. teyahalee* and *P. montanus* males were allowed to choose between a substrate marked by a heterospecific female and a second substrate with no odor, the males did not display a preference. Behavioral observations in the laboratory reveal that male terrestrial salamanders frequently fail to initiate courtships with heterotypic females (review by Arnold et al., 1993). Our results indicate that the odors of the heterospecific females are either not recognized by the male or that these chemical signals fail to carry reproductive significance and are simply ignored. Although *P. teyahalee* males spent more than twice as much time on substrates bearing *P. shermani* female odors than on the substrate with no salamander odor, the results were not statistically significant. However, *P. teyahalee* males may indeed be attracted to *P. shermani* female odors given that these two species hybridize in many areas where they come into contact with each other (Highton and Peabody, 2000).

Both *P. shermani* and *P. teyahalee* males failed to respond to a conspecific female odor when a heterospecific female odor was presented simultaneously. This result is similar to that observed for female swordtail fish, wherein the response to a conspecific odor was stronger when the alternative choice was water rather than a heterospecific odor (Crapon de Caprona and Ryan, 1990). This pattern of discrimination suggests that although large Eastern *Plethodon* males prefer the odor produced by their own females, males are capable of perceiving heterospecific female signals. The fact that males are capable of perceiving differences between these odors supports the notion that mate-recognition systems can evolve in allopatry (Crapon de Caprona and Ryan, 1990).

The *P. montanus* males, on the other hand, maintained a strong preference for conspecific female odors when the second odor was that of a *P. teyahalee* female. Similarly, P. *teyahalee* males spent more than twice as much time on conspecific female odors than on substrates marked by *P. montanus* females, but this result was not statistically significant. The asymmetry in male response to conspecific and heterospecific female odors (*P. teyahalee* showed no strong preference, *P. montanus* preferred conspecific female odors) is unexplained, as these two species are sympatric and reproductive isolation is complete (Reagan, 1992). Male *P. montanus* spent more than twice the amount of time on the substrate without a salamander scent than on the substrate containing *P. teyahalee* female odors (results were not statistically significant). Thus, there is strong evidence that female signals attract conspecific mates, and there is some suggestion that these signals may also function to repel sympatric, heterospecific males. In any event, the chemical cue produced by the female salamander provides species-specific information that may play a large role in maintaining this reproductive barrier. The mate-recognition system of large Eastern *Plethodon* species has evolved in allopatry, but the signaling system is more specialized in areas where the species co-exist.

## 5. SUMMARY AND PROPOSAL FOR FUTURE WORK

In this terrestrial salamander system, chemical signals are broadcast in the environment by reproductively active females and are detected by the males as they tap their nares to the substrate. These signals function in mate recognition and have evolved in allopatry. There is evidence that these chemical cues have diverged further in areas where closely related salamander species are sympatric, presumably to prevent hybridization. Whether the pattern of signal divergence disclosed by this study reflects true character displacement as defined by Dobzhansky (1937) has yet to be established. To date, research on hybrid viability in this system has not been conducted, presumably because of the difficulties that are associated with mass-rearing of terrestrial salamander eggs in the laboratory. A focus on post-mating isolation is essential, for pre-mating isolation may evolve in response to post-mating consequences.

Furthermore, the chemical signals involved in sexual isolation and their site of production have yet to be identified. Proteins are good candidate molecules as sex-attractants in terrestrial salamanders and warrant investigation. To date, all of the pheromones that have been characterized for amphibian mating systems have been proteins (Rollmann et al., 1999; Wabnitz et al., 1999; Toyoda and Kikuyama, 2000). In addition, a protein signal may have greater stability in terrestrial environments than other molecules. For example, rodents release proteins in their urine and these proteins encase a bound volatile molecule. In this arrangement, the volatiles are transformed into stable signals as they are time-released into the environment during the relatively slow process of protein degradation (Hurst et al., 1998). If the signals used in salamander mate-recognition are indeed proteins, the sequences for the genes encoding the proteins can be determined. Furthermore, the evolution of these sequences can be analyzed and models of the selective pressures acting on the gene (i.e., neutral, diversifying or stabilizing selection) can be tested (Yang, 2000). From this analysis, we can gain a better understanding of what processes caused the signal to change. Ultimately, this system has the potential to help us understand the selective processes involved in the evolution of reproductive isolation.

## 6. ACKNOWLEDGEMENTS

We extend our appreciation to S.J. Arnold, D. Mayers, H. Murdoch, A. Kurbis, E. Adams, M. Rudenko, and L. Mead for helping us collect salamanders in the field and/or for helping us collect data in the laboratory. Financial support was provided by a Grant-in-Aid from Highlands Biological Station, a NSF Pre-doctoral Fellowship to C.A. Palmer, and NSF IBN-0110666 to L.D. Houck.

## 7. REFERENCES

Arnold, S. J., 1976, Sexual behavior, sexual interference, and sexual defense in the salamanders *Ambystoma maculatum, Ambystoma tigrinum* and *Plethodon jordani, Zeit. Tierpsychol.* **42**:247-300.

Arnold, S. J., Reagan N. L., and Verrell, P. A., 1993, Reproductive isolation and speciation in plethodontid salamanders, *Herpetologica* **49(2)**:216-228.

Brown, C. W., 1968, Additional observations on the function of the nasolabial grooves of plethodontid salamanders, *Copeia* **1968**:728-731.

Coyne, J. A., and Oyama, R., 1995, Localization of pheromonal sexual dimorphism in *Drosophila melanogaster* and its effect on sexual isolation, *Proc. Natl. Acad. Sci. USA*. **92**:9505-9509.

Crapon de Caprona, M. D., and Ryan, M. J., 1990, Conspecific mate recognition in swordtails, *Xiphophorus nigrensis* and *X. pygmaeus* (Poeciliidae): olfactory and visual cues, *Anim. Behav.* **39**:290-296.

Dawley, E. M., 1984, Recognition of individual, sex and species odours by salamander of the *Plethodon glutinosus*-*P. jordani* complex, *Anim. Behav.* 32:353-361.
Dawley, E. M., 1986, Behavioral isolating mechanisms in sympatric terrestrial salamanders, *Herpetologica* 42(2):156-164.
Dawley, E. M., 1987, Species discrimination between hybridizing and non-hybridizing terrestrial salamanders, *Copeia* 1987:924-931.
Dawley, E. M., 1992, Sexual dimorphism in a chemosensory system: the role of the vomeronasal organ in salamander reproductive behavior, *Copeia* 1992(1):113-120.
Dawley, E. M., and Bass, A. H., 1989, Chemical access to the vomeronasal organs of a plethodontid salamander, *J. Morphol.* 200:163-174.
Dobzhansky, T., 1937, *Genetics and the Origin of Species*, Columbia University Press.
Gergits W. F., and Jaeger, R. G., 1990, Field observations of the behavior of the red-backed salamander (*Plethodon cinereus*): Courtship and agonistic interactions, *J. Herpetol.* 24:93-95.
Halpern, M., 1987, The organization and function of the vomeronasal system, *Ann. Rev. Neurosci.* 10:325-362.
Highton, R., and Peabody, R. B., 2000, Geographic protein variation and speciation in salamanders of the *Plethodon jordani* and *Plethodon glutinosus* complexes in the southern Appalachian Mountains with the description of four new species, in: *The Biology of Plethodontid Salamanders*, R. C. Bruce, R. G. Jaeger, and L.D. Houck, eds., Kluwer Academic/Plenum Publishers, New York, pp. 31-93.
Houck, L.D., Bell, A. M., Reagan-Wallin, N. L., and Feldhoff, R. C., 1998, Effects of experimental delivery of male courtship pheromones on the timing of courtship in a terrestrial salamander, *Plethodon jordani* (Caudata: Plethodontidae), *Copeia* 1998(1):214-219.
Hurst, J. L., Robertson, D. H. L., Tolladay, U., and Beynon, R.J., 1998, Proteins in urine scent marks of male house mice extend the longevity of olfactory signals, *Anim. Behav.* 55:1289-1297.
Marco, A., Chivers, D. P., Kiesecker, J. M., and Blaustein, A. R., 1998, Mate choice by chemical cues in western redbacked (*Plethodon vehiculum*) and Dunn's (*P. dunni*) salamanders, *Ethology* 104:781-788.
Mathis, A., Jaeger, R. G., Keen, W. H., Ducey, P. K., Walls, S. C., and Buchanan, B. W., 1995, Aggression and territoriality by salamanders and a comparison with the territorial behavior of frogs, in: *Amphibian Biology Vol. 2: Social Behaviour*, H. Heatwole and B. K. Sullivan, eds., Surry Beatty & Sons, Chipping Norton New South Wales, Australia, pp. 634-676.
McKinnon, J. S., and Liley, N. R., 1986, Asymmetric species specificity in responses to female sexual pheromone by males of two species of *Trichogaster* (Pisces: Belontiidae), *Can. J. Zool.* 65:1129-1134.
McLennan D. A., and Ryan, M. J., 1997, Responses to conspecific and heterospecific olfactory cues in the swordtail *Xiphophorus cortezi*, *Anim. Behav.* 54:1077-1088.
McLennan D. A. and Ryan, M. J., 1999, Interspecific recognition and discrimination based upon olfactory cues in northern swordtails, *Evolution* 53(3):880-888.
Organ, J. A., 1958, Courtship and spermatophore of *Plethodon jordani metcalfi*, *Copeia* 1958:251-259.
Reagan, N. L., 1992, Evolution of sexual isolation in salamanders of the genus *Plethodon*. Doctoral Thesis, University of Chicago.
Rollmann, S. M., Houck, L. D., and Feldhoff, R. C., 1999, Proteinaceous pheromone affecting female receptivity in a terrestrial salamander, *Science* 285:1907-1909.
Ryan, M. J., 1990, Signals, Species, and Sexual Selection, *Amer Sci.* 78:46-52.
Schmidt, A., Naujoks-Manteuffel, C., and Roth, G., 1988, Olfactory and vomeronasal projections and the pathway of the nervus terminalis in ten species of salamanders, *Cell Tissue Res.* 251:45-50.
Toyoda, F., and Kikuyama, S., 2000, Hormonal influence on the olfactory response to a female-attracting pheromone, sodefrin, in the newt, *Cynops pyrrhogaster*, *Comp. Biochem. Physiol. B* 126(2): 239-245.
Verrell, P. A., 1989, An experimental study of the behavioral basis of sexual isolation between two sympatric plethodontid salamanders, *Desmognathus imitator* and *D. ochrophaeus*, *Ethology* 80:274-282.
Verrell, P. A., 2003, Population and species divergence of chemical cues that influence male recognition of females in desmognathine salamanders, *Ethology* 109:577-586.
Wabnitz, P. A., Bowie, J. H., Tyler, M. J., Wallace, J. C., and Smith, B. P., 1999, Animal behaviour: Aquatic sex pheromone of a male tree frog, *Nature* 401:444-445.
Yang, Z., 2000, Phylogenetic analysis by Maximum Likelihood (PAML), version 3.0, University College London, London, England.

# THE PHEROMONAL REPELLING RESPONSE IN RED-SPOTTED NEWTS (*NOTOPHTHALMUS VIRIDESCENS*)

Daesik Park[1,2,3], Heather L. Eisthen[2], and Catherine R. Propper[3]

## 1. INTRODUCTION

A potential courting site can be defined as an area in which at least one sexually receptive female is present, with or without males. Although many studies have examined sexual competition within courting sites, little is known about the process by which animals choose a specific courting site. In studying the process of choosing a specific courting site, the concept of potential reproductive rate (PRR: the maximum number of independent offspring that parents produce per unit time; Clutton-Brock and Vincent, 1991) can be useful, as males would be expected to choose a courting site in which they can maximize their PRR.

When a male chooses a courting site, it should integrate information from different sensory systems to evaluate its expected PRR following the choice. Chemical cues may play a particularly important role such choices for aquatic animals, because visual signals are often limited in aquatic breeding habitats by plants and by turbid water (Dodson et al., 1994). For chemical cues to be useful in the evaluation of expected PRR, these cues should contain information about female fecundity, male-male competition, and courting progress within a courting site.

The use of chemical cues in the choice of courting sites has been suggested in insect species that display a pheromonal repelling response or pheromonal inhibitory response (Hirai et al., 1978; Bijpost et al., 1985). In wind tunnels, male tortricid moths display decreased sexual activity and an increase in lateral movement when courting males and females are placed upwind (Hirai et al., 1978). These behaviors are interpreted as adaptive responses that increase the chance of the subject male finding another courting site at which his PRR may be higher.

A clear, explicit definition of a pheromonal repelling response has not yet been offered. In this paper, we define a pheromonal repelling response as the abandonment of a courting site in favor of others based on chemical cues from conspecifics. We were the

---

[1] Department of Science Education, Kangwon National University, Chuncheon, Kangwon 200-701, Korea.
[2] Department of Zoology, Michigan State University, East Lansing, MI 48824.
[3] Department of Biological Sciences, Northern Arizona University, Flagstaff, AZ 86001.

first to describe a pheromonal repelling response in a vertebrate (Park and Propper, 2001), and to date this remains the only report of such a phenomenon. We are currently working to characterize the pheromonal repelling response in newts in greater detail.

Red-spotted newts, *Notophthalmus viridescens*, present an ideal model system for investigating the dynamics of a pheromonal repelling response in vertebrates. Red-spotted newts are subject to high levels of male-male competition during mating (Gill, 1978; Able, 1999), due to their highly male-biased operational sex ratio (OSR: the ratio of sexually active males to fertilizable females ready to mate; Emlen and Oring, 1977). Both males and females use chemical cues from two pheromone-releasing glands, the cloacal and genial glands, to discriminate conspecific sex (Pool and Dent, 1977; Dawley, 1984; Verrell, 1985) and male glandular secretions increase receptivity in females (Rogoff, 1927). These results suggest that male newts may use chemical cues from conspecifics to choose among courting sites at a distance.

Using a series of Y-maze olfactory choice tests, we found that male red-spotted newts show a pheromonal repelling response and that the occurrence of the repelling response depends on the number of males present, on male-female interactions, and on female fecundity. The purpose of the work presented here is to describe and elaborate on the main findings that we have reported in previous studies (Experiment I–IV, Park and Propper, 2001; Experiment V-VI, Park et al., submitted).

## 2. MATERIALS AND METHODS

### 2.1 Subjects and Maintenance

Eastern red-spotted newts (*N. v. viridescens*) in breeding condition were purchased from a licensed supplier (Charles Sullivan Co., Nashville, TN). Upon arrival at the laboratory, males and females were kept in separate aquaria (50 cm long x 26 cm wide x 30 cm high) containing approximately 25 L aged tap water at a density of no more than 20 individuals per tank. Half of the water was changed weekly. Water temperature was maintained at 10°C and the animals were maintained on a schedule of 16 h light: 8 h dark, with lights on at 0600 hr, to mimic the spring breeding season. Refuges were provided in the form of floating paper towels and pieces of broken pottery that were placed on the bottom of the aquaria. Animals were fed live *Tubifex* worms *ad libitum*.

All males used in the study were in full breeding condition, as evidenced by swollen cloacae, fully developed tailfins, and callosities on the hind legs (Petranka, 1998). Stimulus and subject animals were randomly selected from a pool of more than 30 females and 50 males.

### 2.2 Y-maze Olfactory Choice Tests

To investigate responses of male newts to chemical cues, we performed choice tests using a Y-maze (Park and Propper, 2001). In such mazes, the test animal begins the trial in a start box (8 cm long × 4.5 cm wide × 5 cm high) at the bottom of the stem of the "Y". Stimuli, in this case other newts, are placed behind an opaque, perforated plastic mesh barrier at the ends of the two arms of the maze. The Y-maze was constructed of Plexiglas, and each arm of the maze measured 22 cm long × 4.5 cm wide × 5 cm high.

The two arms of the maze were continuously infused with aged tap water at a flow rate of 30 ml/min from a reservoir containing 500 ml aged tap water. At this flow rate, water flow from the two arms remained separate to the drain at the end of the Y-maze; nevertheless, test males could sample the water from both arms while behind the perforated starting gate. After each trial, mazes were washed using aged tap water. All experiments were conducted between 1000 and 1500 hr.

After a 3-min adaptation period behind the starting gate, the gate was slowly raised, allowing the animal to enter the maze. Each animal was allowed 5 min to leave the start box. If the animal performed this initial response, it was allowed another 10 min to complete its choice by traveling more than half length of a given side arm. Animals that failed to leave the start box within 5 min or to complete a choice within 10 min were removed from the maze, and data for those animals were not included in the analysis. The arm of the Y-maze chosen by each subject was recorded and analyzed using one-tailed binomial tests. Each subject animal was given each choice test only once. Stimulus animals were randomly assigned to one side of the maze for each trial by tossing a coin.

## 3. RESULTS

### 3.1. Experiment I

We first verified that males have no preference when exposed to chemical cues from two same-sized females. An average, sexually-mature adult female newt has a mass of approximately 3 g. We used matched pairs of females that differed by less than 0.05 g in mass, placing one into each side arm of Y-maze. When subject males were allowed to choose between two same-size females, no preference was observed (Table 1, I; one-tailed binomial test, $P = 0.37$).

### 3.2. Experiment II

To determine whether a pheromonal repelling response occurs, we first placed one of two same-sized females into each arm of the Y-maze, and then added three males to one side to mimic a state of high male-male competition. In this experiment, the stimulus males and female were allowed to interact for 5 min in the side arm before each test begins. We found that subjects strongly preferred the arm containing a single female compared with that containing a female plus three males (Table 1, II; one-tailed binomial test, $P = 0.017$).

### 3.3. Experiment III

To determine whether chemical cues from three males alone are sufficient to repel conspecific males, we placed three males in one arm of the Y-maze and left the other arm empty. Subject males were then allowed to choose between the two arms. Subjects significantly preferred the arm containing three males to the arm containing nothing but plain water (Table 1, III; one-tailed binomial test, $P = 0.003$).

**Table 1.** Results of the Y-maze olfactory choice tests.

| Experiment (no. of subjects) | Sources of chemical cues used | No. of times chosen | One-tailed binomial test ($P$) |
|---|---|---|---|
| I (37) | One female | 17 | 0.37 |
| | One female | 20 | |
| II (22) | One female + three males | 5 | 0.017 |
| | One female | 17 | |
| III (26) | Three males | 21 | 0.003 |
| | Plain tap water | 5 | |
| IV (24) | One female/ three males * | 10 | 0.27 |
| | One female | 14 | |
| V (25) | One large female | 21 | <0.001 |
| | One small female | 4 | |
| VI (29) | One large female + four males | 11 | 0.13 |
| | One small female | 18 | |

* No visual, tactile, or olfactory interactions were allowed between the female and the males.

### 3.4. Experiment IV

In the next experiment, we determined whether behavioral interactions between the stimulus males and females are necessary to produce a repelling response. The set-up for this experiment was similar to that described for Experiment II, above, except that we inserted a solid barrier between the stimulus female and the stimulus three males that were placed in one arm of the maze. In this experiment, subject males displayed no preference between the two arms of the maze; that is, no repelling response occurred (Table 1, IV; one-tailed binomial test, $P = 0.27$).

### 3.5. Experiment V

To determine whether female fecundity affects male choice, we determined whether or not subject males display a preference when presented with large and small females. We placed one large ($3.75 \pm 0.57$ g, n = 5) and one small ($2.67 \pm 0.18$ g, n = 5) female into each arm of Y-maze and allowed subject males to choose between them. Males showed a strong preference for the side containing the large female (Table 1, V; one-tailed binomial test, $P < 0.001$).

### 3.6. Experiment VI

To determine whether female fecundity affects the repelling response, we placed one large and one small female (used in experiment V, above) into each arm of the Y-maze. We then added four males to the arm containing the large female and allowed the stimulus animals to interact for 5 min before the subject male was placed in the start box.

In this test, subject males did not show a preference for either side (Table 1, VI; one-tailed binomial test, $P = 0.13$).

## 4. DISCUSSION

To test whether red-spotted newts display a pheromonal repelling response and to further characterize the response, we conducted a series of Y-maze choice tests. Male red-spotted newts displayed a pheromonal repelling response by avoiding an arm containing one female plus three males that are interacting in favor of an arm containing one female alone. Our experiments demonstrate that the occurrence of a pheromonal repelling response depends on the number of stimulus males, on interactions between the female and male stimulus animals, and on female fecundity. These results support the existence of a pheromonal repelling response in vertebrates, and suggest that male red-spotted newts may use chemical cues from conspecifics to evaluate their expected PRRs before choosing a courting site.

Three different variables that affect the occurrence of a pheromonal repelling response emerged from our experiments. First, the number of males present within a courting site affects the probability that the subject will display a pheromonal repelling response. The expected PRR for a male at a courting site containing a single female will be determined by the female's PRR and by the male's ability to compete successfully with rivals. For example, if one male completely out-competes the others, his PRR should be the same as that of the female. Conversely, a high level of male-male competition should result in the reduction of each male's PRR. In a previous study, we found that the minimal sex ratio that induces a significant pheromonal repelling response when compared with chemical cues from one female alone is 3 males: 1 female (Park and Propper, 2001). These results suggest that subject males prefer a courting site in which few or no other males are present, to maximize their PRR by avoiding male-male competition.

Second, behavioral interactions between stimulus males and females are necessary for the production of a pheromonal repelling response. That is, although the number of males present at the courting site is an important factor in determining whether or not a pheromonal repelling response occurs, the response will only occur if the potential rivals are interacting with the female present at the courting site. The results of experiment IV therefore suggest that chemical cues from males and females that are interacting may contain information about the progress of courtship as well as the number of males within a courting site. This outcome is logical given that males that are not courting a female would probably not change the expected PRR of approaching males. Data from field studies support this inference, as latecomers have been shown to achieve lower general mating success than early-arriving animals (Candolin and Voigt, 2003).

Finally, female fecundity appears to affect the occurrence of the pheromonal repelling response: the result of experiment VI demonstrates that subject males do not display a preference when given a choice between one large female plus four males versus one small female. Because the expected PRR of an approaching male depends in part on the PRR of the female within the courting site, the high PRR expected for a male mating with a large female may be balanced by the low PRR expected when male-male competition is strong.

Taken together, the results of our experiments suggest that males simultaneously analyze information about the probability and strength of male-male competition, the expected fecundity of different females, and the progress of courtship within a courting site to make a decision that will maximize the males' expected PRR. The use of Y-mazes in our experiments demonstrates that chemical cues released by both male and female newts contain sufficient information about these different factors for males to be able to choose an appropriate courting site.

What are possible benefits of the pheromonal repelling response to the individuals involved? For approaching males, the benefits seem obvious: they can save their time and energy by avoiding a visit to a courting site that is not likely to result in successful mating, and can instead find another courting site where their expected PRR is higher. Males that have already chosen a courting site and may be interacting with a female can maximize their expected PRR by preventing an increase in male-male competition. For large females who are being courted, the benefit may be that they are more likely to mate successfully by avoiding sexual interference among rival males, which is the most common cause of mating failure in the field (Massey, 1988). Finally, as a by-product, small females that are not as attractive as large females may benefit from an increased probability of attracting males; males repelled from other courting sites may be attracted to a courting site in which small females are present without males.

In conclusion, the choice of a specific courting site should be a complex process, because individuals should consider many factors that affect their expected PRR. The details of the process of choosing a potential courting site are poorly understood. The existence of a pheromonal repelling response in red-spotted newts will allow us to investigate the relative importance of the factors involved. In future experiments, we plan to further explore the factors that contribute to the pheromonal repelling response, as well as the nature of the chemical cues involved. Finally, this paradoxical phenomenon, in which a mixture of two different odorants is less attractive than either alone, provides a unique opportunity to examine the processing of pheromones and odorant mixtures in the olfactory system.

## 5. ACKNOWLEDGMENTS

The experiments described here were supported by grants from the International Rotary Foundation Fellowship and Sigma-Xi to DP, the National Science Foundation (IBN 9982934) to HLE, and the Council for Tobacco Research (4661R1) to CRP.

## 6. REFERENCES

Able, D. J., 1999, Scramble competition selects for greater tailfin size in male red-spotted newts (Amphibia: Salamandridae), *Behav. Ecol. Sociobiol.* **46**:423-428.
Bijpost, S. C. A., Thomas, G., and Kruijt, J. P., 1985, Olfactory interactions between sexually active males in *Adoxophyes orana* (F. v. R.) (Lepidoptera: Tortricidae), *Behaviour* **95**:121-137.
Candolin, U., and Voigt, H. R., 2003, Size-dependent selection on arrival times in sticklebacks: why small males arrive first, *Evolution* **57**:862-871.
Clutton-Brock, T. H., and Vincent, A. C. J., 1991, Sexual selection and the potential reproductive rates of males and females, *Nature* **351**:58-60.

Dawley, E. M., 1984, Identification of sex through odors by male red-spotted newts, *Notophthalmus viridescens*, *Herpetologica* **40**:101-105.
Dodson, S. I., Crowl, T. A., Peckarsky, B. L., Kats, L. B., Covich, A. P., and Culp, J. M., 1994, Non-visual communication in freshwater benthos: an overview, *J. North Am. Benthol. Soc.* **13**:268-282.
Emlen, S. T., and Oring, L. W., 1977, Ecology, sexual selection and the evolution of mating systems, *Science* **197**:215-223.
Gill, D. E., 1978, Meta-population ecology of red-spotted newt, *Notophthalmus viridescens* (Rafinesque), *Ecol. Monogr.* **48**:145-166.
Hirai K., Shorey, H. H., and Gaston, L. K., 1978, Competition among courting male moths: male-to-male inhibitory pheromone, *Science* **202**:644-645.
Massey, A., 1988, Sexual interactions in red-spotted newt populations, *Anim. Behav.* **36**:205-210.
Park, D., and Propper, C.R., 2001, Repellent function of male pheromones in the red-spotted newt, *J. Exp. Zool.* **289**:404-408.
Park, D., McGuire, J. M., and Eisthen, H. L., Differential responses of large and small male red-spotted newts, *Notophthalmus viridescens*, to conspecific chemical cues, *Copeia, submitted*.
Petranka, J. W., 1998, *Salamanders of the United States and Canada*, Smithsonian Institution Press, Washington D.C.
Pool, T. B., and Dent, J. N., 1977, Neuronal regulation of product discharge from the hedonic glands of the red-spotted newt, *Notophthalmus viridescens*, *J. Exp. Zool.* **201**:203-220.
Rogoff, J. L., 1927, The hedonic glands of *Triturus viridescens*; a structural and functional study, *Anat. Rec.* **34**:132-133.
Verrell, P. A., 1985, Male mate choice for large, fecund females in the red-spotted newt, *Notophthalmus viridescens*: how is size assessed? *Herpetologica* **41**:382-386.

# THE EFFECTS OF CLOACAL SECRETIONS ON BROWN TREE SNAKE BEHAVIOR

Michael J. Greene[4] and Robert T. Mason[5]

All snakes possess paired cloacal glands that release secretions through ducts along the cloacal orifice (Graves and Duvall, 1988; Mason, 1992; Price and LaPoint, 1981). As these secretions represent an obvious source of glandular, volatile, and potentially meaningful chemical information in snakes their role in the regulation of snake physiology and behavior has been examined for many years, however few studies have shown measurable effects of cloacal secretions on snake behavior (Gelbach et al., 1968; Graves and Duvall, 1988; Noble, 1937; Mason, 1992; Mason et al., 1998; Watkins et al., 1969). Garter snake (Thamnophis) cloacal secretions are composed mainly of nonvolatile lipids and proteins as well as volatile components including trimethylamine, acetic acid and propanoic acid (Weldon and Leto,. 1995; Wood et al., 1995). Blindsnake (Leptotyphlops dulcis) cloacal secretions are composed primarily of glycoproteins and free fatty acids including linoleic, palmitic and oleic acids (Blum et al., 1971).

As snakes often release cloacal secretions when disturbed and the secretions are volatile and odorous, it has been hypothesized that they act either as predator deterrents or as alarm pheromones (Graves and Duvall, 1988; Mason, 1992; Price and LaPointe, 1981). Graves and Duvall (1988) showed that exposure to cloacal secretions caused the rattlesnake, Crotalis viridis, to be more responsive to provocation. It rattled more and increased the increased heart rate, a physiological response correlated to defensiveness in this species. The blind snake, L. dulcis, is a predator of ant and termite brood that uses cloacal secretions as a defense against ant attacks and also as an ant repellent (Gelbach et al., 1968; Watkins et al., 1969). There is evidence that blindsnake cloacal secretions are attractive to conspecifics but repellent to some ophiophagous snakes (Watkins et al., 1969). An early hypothesis for the function of cloacal secretions was that they acted as the female sex pheromone in snakes (Noble, 1937; Price and LaPointe, 1981), although more recent work has shown that these pheromones are present in snake skin lipids (Greene and Mason, 1998; Mason, 1992; Mason et al., 1998).

---

[4] Department of Biology, University of Colorado at Denver, Denver, CO 80217 U.S.A.;
   michael.greene@cudenver.edu
[5] Department of Zoology, Oregon State University, Corvallis, OR 97331 U.S.A.

The aim of this paper is to discuss the effects of cloacal secretions on brown tree snake (Boiga irregularis) behavior. The brown tree snake, Boiga irregularis, is a rear-fanged colubrid native to Australia, Papua New Guinea, and the Solomon Islands (Cogger, 1992). The snake is an invasive pest species of Pacific islands where it has caused economic damage as well as a dramatic decrease in the abundance and diversity of the native fauna it preys upon (Rodda et al., 1992, 1997). The snake has most notably been a pest on Guam where it has caused economic problems such as power outages and ecological damage including the extirpation or extinction of 9 of Guam's 12 native forest bird species (Rodda et al., 1997). The snakes can reach lengths of over 3 meters and reach masses of up to approximately 1.5 kilograms (Cogger, 1992).

Female brown tree snakes release cloacal secretions to inhibit male courtship (Greene and Mason, 2003). Female brown tree snakes are particularly active in the process of courtship compared to reports for other snakes (Greene and Mason, 2000). Females display overt courtship behaviors that appear to solicit courtship from males and may represent mechanisms to asses to quality of potential mates (Greene and Mason, 2000). Also, females lift their tails, gape their cloacae, and deposit secretions during courtship to which males respond by ceasing courtship activity (Greene and Mason, 2000). These secretions are released from females' paired cloacal glands and are combined with liquid from the urogenital tract. Laboratory-based bioassays showed that female cloacal secretions inhibit male courtship, causing both a decrease in the amount of time spent courting and in the courtship intensity (Greene and Mason, 2003). The pheromone is only present in female cloacal secretions, not in male secretions. Also, the response of males to this courtship inhibition pheromone is context specific; males only responded to female cloacal secretions during courtship, not during bouts of male-male ritualized combat. The secretions had no effect on female courtship behavior. It appears that this pheromone is a mechanism for females to reject unsuitable males or to signal that they are not in a proper condition to mate (Greene and Mason, 2003).

Only freshly collected cloacal secretions were used in the courtship inhibition study (Greene and Mason, 2003), however in the process of running that study we serendipitously made the observation that brown tree snakes responded to secretions that had been aged in a sealed vial for over 30 minutes with behaviors normally seen when snakes are irritated or in a defensive situation. Brown tree snakes on Guam are known for their willingness to strike and bite when provoked, although they initially use crypsis and a decrease in locomotion as defensive behaviors (Rodda et al., 1999). Other defensive behaviors include lateral compression of the body, head hiding, forming and S-shaped coil of the anterior portion of its body that is used in lunging strikes, and flaring the quadrate bones to form a more triangular shaped head (Fritts, 1988; Johnson, 1975; Rodda et al., 1999). Prolonged provocation can lead to fleeing by the snakes, accompanied by lashing of their bodies and tails. As with other snakes, brown tree snakes commonly release cloacal secretions, often while lashing their tails, in response to disturbance.

We tested the effects of cloacal secretions that had been aged in a sealed vial on brown tree snakes. Our bioassay measured changes in locomotion, avoidance behavior and the expression of defensive behaviors in response to touch stimuli. A blank control, freshly collected cloacal secretions and aged cloacal secretions were the stimuli used in the study. An applied aim of this study was to investigate chemical cues that might be of use in controlling the brown tree snake in its introduced habitat (Mason and Greene, 2001).

We used randomly chosen snakes from our captive colony at Oregon State University; the snakes were all mature adults in breeding condition during the study (Greene and Mason, 1998, 2000, 2001, 2003). The animals were originally collected from the introduced population on Guam and were housed in our laboratory for approximately six years prior to the study under an established laboratory protocol (Greene et al., 1997). Temperatures cycled from 25°C to 30°C and relative humidity ranged between 75% and 80% in the snake room, simulating conditions on Guam. Lighting (approximately 14L:10D) was provided by overhead fluorescent lights and ambient sunlight entering the room through windows (Greene and Mason, 1997; 2000; 2003).

The experiment was conducted during scotophase when the snakes were most active. Each snake (N = 6; 3 males and 3 females) was tested with three stimuli in a random order on different nights: 1) a volatile control composed of a 2% solution of Aqua Velva cologne (JB Williams Company, White Plains, New York, U.S.), 2) freshly collected cloacal secretions, and 3) aged cloacal secretions. The dilute cologne control was chosen as it volatilizes at a rate much like cloacal secretions and the snakes do not respond to it adversely. On each night, cloacal secretions were collected by applying light pressure both anteriorly and posteriorly to a snake's cloaca and collecting the fine spray emanating from the paired cloacal glands along with the bolus of liquid from the urogenital opening into a clean, although not sterilized, 15 ml glass screw top tube. The cloacal secretions from three to five snakes not being used in the study as subjects were pooled for use in the trials. Freshly collected cloacal secretions were used in the experiment approximately 10 minutes after collection. Aged cloacal secretions were allowed to sit in the sealed vial for approximately 60 minutes at room temperature before being used in the experiment. All of the snakes used in the study were tested with all three stimuli on different nights and in a random order.

Our behavioral bioassay was designed to take advantage of the high degree of thigmotaxis displayed by the snakes and their tendency to actively explore novel environments. Snakes were tested in a clear cube-shaped Plexiglas arena, measuring 1.25m on each side. The floor of the arena was evenly divided into four quadrants with tape. Before a snake was added to the arena, a randomly chosen quadrant was treated

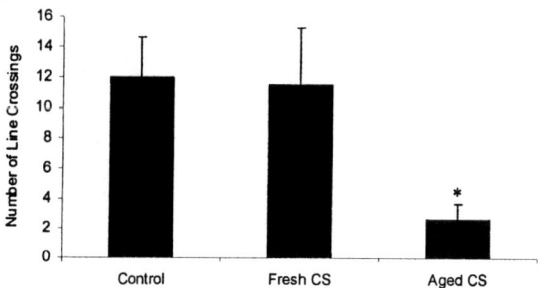

**Figure 1.** The number of line crossings exhibited by focal snakes during the observation period in response to the dilute cologne control, freshly collected cloacal secretions (Fresh CS) and aged cloacal secretions (Aged CS).

by rubbing approximately 2 ml of the stimulus evenly over the entire quadrant. Immediately after the stimulus was applied, a snake was placed in the center of the arena and observed for 30 minutes. In each case, the snakes quickly moved towards the walls of the arena and slowly explored the arena edges (Greene and Mason, 2003). After a trial ended, the snake was removed to its home cage and the arena was washed with soap and water, dried with paper towels and allowed to air out for 30 more minutes before used in the next trial.

During the observation period an observer, who was not aware of the treatment added to the arena, measured the number of line crossing as an indicator of activity and the amount of time the snake spent in each quadrant as an indicator of avoidance of the treated quadrant. At the end of the observation period a second person lightly touched each snake at its mid-body with the rubber handle of a snake hook two times separated by 5 seconds while the observer recorded the behaviors displayed by the snakes. The number of line crossings and the amount of time the snakes spent in the treated quadrant were compared using a Friedman ANOVA. The number of snakes responding with defensive behaviors was compared using a Cochran Q test. Aged cloacal secretions caused a significant decrease in the number of line crossings by the experimental snakes in comparison to freshly collected cloacal secretions and the dilute cologne control (Friedman ANOVA: $Fr = 9.0$; $df = 2$; $p < 0.011$; Figure 1), indicating that the snakes do not actively avoid the aged cloacal secretion stimulus. The snakes crossed a mean of 2.667 ($\pm$ 1.022 SE) lines when treated with aged cloacal secretions versus 11.500 ($\pm$ 3.819) for freshly collected cloacal secretions and 12.000 ($\pm$ 2.646) for the control treatment. There were no significant differences in the mean times that the snakes spent in the treated quadrant of the arena during the trials between the treatments ($Fr = 1.2$; $df = 2$; $p < 0.549$; Figure 2), indicating that the snakes did not avoid treated quadrants. However, we noted that while exploring arenas treated with aged cloacal secretions snakes typically elevated their heads a few centimeters above the treated quadrant, an uncommon behavior during brown tree snake locomotion and exploration.

There were significant differences in the number of trials in which defensive behaviors were displayed after being touched by the snake hook handle between treatments (Cochran's Q-test: $Q = 7.6$; $df = 2$; $p < 0.022$). Snakes displayed defensive

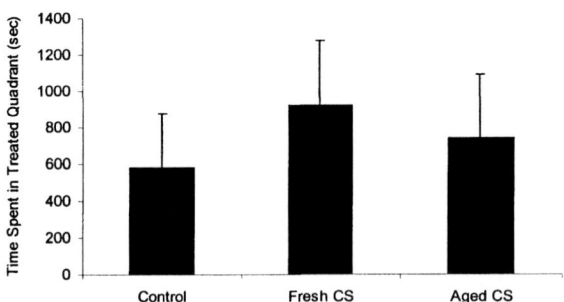

Figure 2. The mean amount of time focal snakes spent in the treated quadrant of the testing arena when treated with dilute cologne control, freshly collected cloacal secretions (Fresh CS) and aged cloacal secretions (Aged CS).

behaviors in 2 of 6 trials to the control treatment; in one case a single defensive strike at the snake hook handle and in the other case a weak bout of tail lashing (Table 1). There were no defensive behaviors displayed to the freshly collected cloacal secretion treatment. Snakes displayed defensive behaviors in 5 of 6 trials when exposed to the aged cloacal secretion treatment (Table 1). In all cases where defensive behaviors were displayed in the aged cloacal secretion treatment, the head shape of the snakes was triangular as observed during defensive displays. In four of these cases, huffing behavior was displayed in which the snakes audibly exhaled air from their lungs in a rhythmic manner. Huffing bouts lasted up to 20 seconds; we have never observed this behavior so obviously in another context. During these huffing displays, the snakes' bodies were laterally compressed. In four of the cases, the snakes displayed body lashing, in which the snakes erratically flailed their heads and tails, and in three cases snakes displayed fleeing, in which the snakes quickly moved away from the snake hook while lashing their bodies. No strikes at the snake hook were observed with the aged cloacal secretion treatment.

Aged cloacal secretions elicited behaviors in brown tree snakes normally associated with defense, including a decrease in locomotion, lateral compression of the body and fleeing. To the human nose, aged cloacal secretions smell rancid in comparison to the already malodorous freshly collected cloacal secretions, although this clearly does not indicate biological activity. However, differences in the behavioral responses of snakes to aged cloacal secretions versus fresh cloacal secretions and the control suggested that changes in the chemical composition of the cloacal secretions had occurred. The compounds that elicited a defensive response in the brown tree snakes studied were produced from chemical changes in the cloacal secretions over time. Oxidation or bacterial degradation of compounds in fresh cloacal secretions may have created the compound or compounds that elicit the defensive response in the snakes. The vials used to collect cloacal secretions were clean, but not sterilized, providing a source of microbes along with bacteria present in the cloacal secretions itself. Further work is needed to characterize the rate of production for the active compounds and to determine if these chemical changes can occur on the natural substrate, or if the reaction is limited to a sealed vial that prevents complete evaporation of the sample.

The experimental snakes did not display avoidance of the treated quadrants in the study. However, this is not unexpected as the experimental design did not test this directly as each of the three stimuli evaporated within the first few minutes of each trial.

**Table 1.** Behaviors elicited from snakes by touch after a 30 minute exposure to the control, freshly collected cloacal secretions, and aged cloacal secretions.

| Snake | Behaviors Displayed To: | | |
|---|---|---|---|
| | Control | Fresh Cloacal Secretions | Aged Cloacal Secretions |
| 1 | — | — | — |
| 2 | Strike | — | Body-lashing; Fleeing |
| 3 | — | — | Huffing; Body-lashing; Fleeing |
| 4 | Tail-lashing | — | Huffing; Body-lashing |
| 5 | — | — | Huffing; Body-lashing |
| 6 | — | — | Huffing |

The snakes also tended to remain motionless in one spot for long periods of time in response to aged cloacal secretions. Responding to aged cloacal secretions with a reduction in locomotion may be somewhat mutually exclusive to avoidance of the cue as brown tree snakes do not generally flee when disturbed but instead attempt to use crypsis. However, the snakes maintained an elevated head posture only over areas of the arena treated with aged cloacal secretions, indicating an aversion to the aged cloacal secretion stimulus. This particular behavior is not normally observed during brown tree snake locomotion and exploration (MJG, personal observation).

In this study, we have documented an effect of aged cloacal secretions on brown tree snake behavior. Except for huffing behavior, the behaviors elicited from the snakes after the aged cloacal secretions treatment were similar to those observed under two other contexts: 1) after a male loses a bout of ritualized combat behavior (Greene and Mason, 1998) and 2) after a prolonged bout of defensive behavior by a captive animal, including repeated striking. After long bouts of defensive behavior the snakes respond to touch with body lashing and by quickly jerking the part of their body touched away from the stimulus (MJG, personal observation). Although the snakes did respond to the control treatment with two displays of defensive behavior, the behaviors seen under the aged cloacal secretions treatment were of a much greater duration and intensity. Although have never observed huffing behavior in the brown tree snake, we believe that it reflects a high degree of agitation. Much like the study by Graves and Duvall (1988) it appears that exposure to aged cloacal secretions made the focal snakes more likely to respond to an aversive stimulus with defensive behaviors. Our results are similar to an anecdotal report of a king snake (*Lampropeltis getulus*) responding to cloacal secretions left for over 15 minutes on a countertop with behaviors described by the authors as "extremely disturbed" (Brisbin, 1968).

In conclusion, cloacal secretions have the following behavioral effects on brown tree snakes: 1) they act as a pheromone that inhibits male courtship behavior, and 2) when aged, they act as a chemical signal and/or chemical irritant that that elicits defensive behaviors from the snakes. Our results represent a significant part of only a small handful of studies that have empirically demonstrated behavioral effects of cloacal secretions, despite the attention they have historically received in the literature (Mason, 1992). The data also demonstrate the effects of volatile chemical signals on snake behavior; the effects of non-volatile signals on snake behavior have been more commonly studied (Mason, 1992).

## 1. ACKNOWLEDGEMENTS

The authors would like to thank Shantel Stark for assisting with the experiments and Wes Mizuno for designing and building the experimental arena. This work was supported by a grant-in-aid award from the Sigma Xi Scientific Research Society to MJG and by grants from the U.S. Fish & Wildlife Service (USDI 14-16-0009-1577), the National Science Foundation (INT-9114567), as well as a National Science Foundation National Young Investigator Award (IBN-9357245) to RTM. This research was conducted under the authority of U.S. Fish & Wildlife Permits No. PRT-769753 and Oregon State University Institutional Animal Care and Use Committee Protocol No. LAR-932.

## 2. REFERENCES

Blum, M. S., Byrd, J. B., Travis, J. R., Watkins, J. F. II, and Gehlbach, F. R, 1971, Chemistry of the cloacal sac secretion of the blind snake *Leptotyphlops dulcis*, *Comp. Biochem. Phys. B* **38**:103-107.
Brisbon, I. L., 1968, Evidence for the use of post-anal musk as an alarm device in the king snake *Lampropeltis getulus*, *Herpetologica* **24**:169-170.
Cogger, H. G, 1992, *Reptiles and Amphibians of Australia*, Comstock/Cornell University Press, Cornell, New York.
Fritts, T. H., 1988, The brown tree snake, *Boiga irregularis*, a threat to Pacific Islands, *US Fish and Wildlife Biological Report 88(31)*.
Gehlbach, F. R., Watkins, J. F., and Reno H. W., 1968, Blind snake defensive behaviour elicited by ant attacks, *BioScience* **18**:784-785.
Graves, B., and Duvall, D., 1988, Evidence of an alarm pheromone from the cloacal sacs of prairie rattlesnakes, *Southwest. Nat.* **3**:339-45.
Greene, M. J., and Mason, R. T., 1998, Chemical mediation of reproduction in the brown tree snake, *Boiga irregularis*, *Ecoscience* **5**:405-409.
Greene, M. J., and Mason, R. T., 2000, The courtship, mating and combat behaviour of the brown tree snake, *Boiga irregularis*, *Herpetologica* **56**:166-175.
Greene, M. J., and Mason, R. T., 2003, Pheromonal inhibition of male courtship behaviour in the brown tree snake, *Boiga irregularis* - A mechanism for the rejection of potential mates. *Anim. Behav.* **65**:905-910.
Greene, M. J., Nichols, D. K., Hoyt, R. J., Jr., and Mason, R. T. 1997. The brown tree snake (*Boiga irregularis*) as a laboratory animal. *Lab Animal.* **26**:28-31.
Greene, M. J., Stark, S. L., and Mason, R. T., 2001, Pheromone trailing behavior of the brown tree snake, *Boiga irregularis*, *J. Chem. Ecol.* **27**:2193-2201.
Johnson, C. R., 1975, Defensive display behavior in some Australian and Papuan-New Guinean pygopodid lizards, boid, colubrid, and elapid snakes, *Zool. J. Linn. Soc-Lond.* **56**:265-282.
Mason, R. T., 1992, Reptilian pheromones, in: *Biology of the Reptilia, Vol. 18.*, C. Gans and D. Crews eds., University of Chicago Press, Chicago, pp. 114-228.
Mason, R. T., and Greene, M. J., 2001, Invading pest species and the threat to biodiversity: Pheromonal control of Guam brown tree snakes, *Boiga irregularis*, in: *Chemical Signals in Vertebrates, IX*, A. Marchlewska-Koj, J. Lepri and D. Müller-Schwarze, eds., Kluwer Academic/Plenum Publishers, New York, pp. 361-368.
Mason, R.T., Chivers, D., Mathis, A., and Blaustein, A.R., 1998, Bioassays with reptiles and amphibians, in: *Methods in Chemical Ecology*, J. Millar and K. Haynes, eds., Chapman & Hall, New York, pp. 271-325.
Noble, G. K., 1937, The sense organs involved in the courtship of *Storeria*, *Thamnophis* and other snakes, *B. Am. Mus. Nat. Hist.* **73**:673-725.
Price, A. H., and LaPointe, J. L., 1981, Structure-functional aspects of the scent glands in *Lampropeltis getulus spendida*, *Copeia* **1981**:138-146.
Rodda, G. H., Fritts, T. H., and Chiszar, D., 1997, Disappearance of Guam's Wildlife, *BioScience* **47**:565-74.
Rodda, G. H., Fritts, T. H., and Conry, P. J., 1992, Origin and population growth of the brown tree snake, *Boiga irregularis*, on Guam, *Pac. Sci.* **46**:46-57.
Rodda, G. H., Fritts, T. H, McCoid, M. J., and Campbell III, E. W., 1999, An overview of the biology of the brown tree snake (*Boiga irregularis*), a costly introduced pest on Pacific islands, in: *Problem Snake Management: the Habu and Brown tree snake*, G. H. Rodda, U. Sawai, D. Chiszar, and H. Tanaka, eds, Cornell University Press, Ithaca, New York, pp. 44-80.
Watkins, J. F. II, Gehlback, F. R., and Kroll J. C., 1969, Attractant-repellent secretions of blind snakes (*Leptotyphlops dulcis*) and their army ant prey (*Neivamyrmex nigrescens*), *Ecology* **50**:1098-1101.
Weldon, P. J., and Leto, T. L., 1995, A comparative analysis of proteins in the scent gland secretions of snakes. *J. Herp.* **29**:474-476.
Wood, W. F., Parker J. M., and Weldon, P. J., 1995, Volatile components in scent gland secretions of garter snakes (*Thamnophis* spp.), *J. Chem. Ecol.* **21**:213-219.

# SPECIES AND SUB-SPECIES RECOGNITION IN THE NORTH AMERICAN BEAVER

Anne Marie Peterson, Lixing Sun, and Frank Rosell[*]

## 1. INTRODUCTION

Many mammals have complex chemical signals for communicating such information as species, family membership, individuality and physiological state (Müller-Schwarze, 1974). Studies on how closely related species or subspecies recognize each other using chemical signals are of particular significance because, when closely related species are reproducing in the same area, species-specific pheromones can be essential for the formation and maintenance of a precopulatory isolating mechanism among them (Moore, 1965; Doty, 1972; Kotenkova and Naidenko, 1999). Study of recognition between subspecies can provide us vital information about how this mechanism is formed during the speciation process.

Studies of species or subspecies recognition using chemical signals or cues have shown a complicated picture. The subterranean mole rat (*Spalax ehrenbergi*) can discriminate between conspecific and heterospecific individuals (Todrank and Heth, 1996). Tufted capuchins (*Cebus apella*) can recognize three species of New World monkeys but did not show discrimination between two species of Old World macaques (Ueno, 1994). The mule deer (*O. h.. hemionus*) discriminate between its own subspecies and the black-tailed deer (*O. h. columbianus*) through the tarsal gland secretion (Müller-Schwarze, 1974). Male bank voles (*Clethrionomys glareolus*) prefer the scent of females of their own species (Rauschert, 1963), but at the subspecies level, the results vary (Godfrey, 1958; Rauschert, 1963). *Peromyscus maniculatus* males prefer females of their own species as opposed to the congeneric *P. polionotus*. Female *P. m.*, however, equally prefer both species. *P. polionotus.*, on the other hand, did not show any discrimination between the two species (Moore, 1965). Discrimination can depend on the innate capability, contingent physiological conditions of the subject (e.g., Doty, 1972), or on the

---

[*] Anne Marie Peterson and Lixing Sun, Department of Biological Science, Central Washington University, Ellensburg, WA 98926-7537. Frank Rosell, Department of Environmental Sciences and Health Studies, Telemark University College, N-3800 Bøi Telmark, Norway.

experimental procedure designed to detect the discrimination (e.g., Blaustein et al., 1987). To confirm whether species and subspecies recognition happens, the methodological issue should be resolved first. Unfortunately, few studies are available to show that the choice of behavioral patterns is crucial for demonstrating whether there is a discrimination or preference.

There are two allopatric species of beavers in the world, *C. canadensis* and *C. fiber*. From 1935 to 1937, *C. canadensis* was introduced into several European countries (Halley and Rosell, 2002). In Finland, dispersal of *C. canadensis* to Scandinavia is a serious threat to the endemic *C. fiber* populations (Rosell and Sun, 1999). The ability to discriminate subspecies and species affects dispersal pattern, mate choice, and other social interactions of beavers. Therefore a study of this type can help assess the potential ecological and genetic consequences of re-introductions.

Among the 24 subspecies of *C. canadensis*, reintroductions in the past have resulted in different subspecies intermixed or distributed near each other (Hall, 1981). Coexistence of several subspecies artificially brought together may have profound ecological, genetic and evolutionary consequences for the beaver. One fundamental question is whether different subspecies interbreed. This question can be at least partially answered by investigating whether beavers discriminate between subspecies.

In this study, we mimicked beavers' natural scent mound building behavior to examine whether beavers can discriminate between species or subspecies through an olfactory playback experiment. If this discrimination does occur, we predict that beavers should respond more strongly toward individuals of sympatric conspecifics than those of allopatric conspecifics or heterospecifics.

## 2. METHODS

Beavers live in family units that are usually composed of a mated pair, yearlings, and kits. They occupy and defend territories (Schulte, 1993). Beavers rely heavily on chemical signals for social interaction and recognition. They use anal gland secretion (AGS) and castoreum to communicate many types of information, including family membership, kinship, sex, individuality, and territoriality (Svendsen, 1980; Sun and Müller-Schwarze, 1997; Schulte, 1998). To do these, they build scent mounds on the bank, within 2 meters from the water, in areas of high activity around their territory, and then apply castoreum and/or AGS to the top (Svendsen, 1978, 1980; Rosell and Nolet, 1997; Rosell and Sundsdal, 2001).

The two subspecies of *C. canadensis* used in this study are far apart, and there is no record indicating that either has been introduced to the other. To collect secretion samples from *C. canadensis*, we trapped both subspecies of the beaver using Hancock live traps baited with aspen (*Populus tremuloides*) in Allegany State Park, New York, and Ellensburg, Washington, between 1995 and 1998. Beavers were sexed based on AGS color (Schulte et al., 1995) by the presence or absence of the os penis (Osborn, 1955). They were aged based on their size and weight (Schulte, 1993). Samples from *C. f.* were collected from beavers killed during the hunting season of 1997 in Bø municipality, Norway. All samples were immediately stored at -20°C until use. Past research has shown that the chemicals found in AGS and castoreum remain intact using the above procedure for collection and storage (Sun and Müller-Schwarze, 1997).

To make a species or subspecies scent, rather than individual scent, we first mixed AGS or castoreum from several adults of the same species or subspecies. Specifically, we blended AGS from 14 males and eight females to make the solutions for this study. Sixteen males and seven females provided castoreum. Each individual contributed approximately an equal amount (in volume) to each mixture of the combined secretion (either AGC or castoreum). Next, we took 0.3 ml of either blended AGS or blended castoreum and dissolved it into 6 ml of methylene chloride (= 1:20 volume ratio). Then, 0.25 ml of this solution was applied to the cork for each treatment on a given evening in the field. Sun and Müller-Schwarze (1997) found that these concentrations are far above the response threshold and can elicit observable territorial response in the beaver.

We mimicked beavers' natural scent mounding behavior for the field playback bioassay with a randomized block ANOVA design with three levels of treatment, blocking on secretion type, AGS or castoreum. The three treatment levels were secretions from 1) *C. c. leucodontus*, 2) *C. c. acadicus*, and 3) *C. fiber*. The dependent (measured) variable was response frequency over a 6-night trial session. The subjects of the experiment were eight *C. c. leucodontus* families at Ellensburg, Washington. Each beaver family was considered to represent one subject. This is because overnight response does now allow us to discriminate responses from different family members. For the same subspecies, the donors and recipients were at least 20 km away so as to avoid possible previous contacts between them.

During the playback, we used the procedure of Sun and Müller-Schwarze (1997) and built three experimental scent mounds (ESMs) for each secretion type each night before beavers emerge from the lodge. With latex gloves, we built ESMs (30 cm apart, 20 cm high, 20 cm wide and 30 cm from the shoreline) to mimic beavers' natural scent mounds. A cork (top diameter: 8 cm) was inserted into each ESM. We then applied 0.25 ml of one of the three treatments in random order to the cork of each ESM. A total of six scent mounds were constructed at each site each night, a group of three scented with AGS and another group of three scented with castoreum. The two sets of ESMs were separated by at least 10 m. Because beavers in our study area rarely emerge before dark, it was difficult to observe them directly. Instead, we recorded the beavers' overnight responses to the ESMs. During a trial (6 consecutive nights), we recorded the state of the ESMs on the next day after response, eliminated all residue from the previous ESMs, rebuilt new sets of ESMs and applied fresh samples to the cork every day. We used the same response patterns (e.g. sniffing, pawing, etc.) as described by Sun and Müller-Schwarze (1997).

Scent mound construction in beavers shows a seasonal pattern. Because it is most intense from April to June, and gradually tapers off (Svendsen 1980), our study started in June to avoid this seasonal effect in beaver mound construction and response. A total of 108 nights of data were collected from June to November in 1998. Five of the families were tested with two to three (6-night) trials and three of the families were used in one trial. Because of this inconsistency, we included only data from the first trial for each family. Because there were many nights that beavers did not respond, we only include data with a minimal response frequency of 40% as measured by the category "ESM Responded" for data analysis to avoid these blanks. Families that were used in more than one trial were given at least three weeks off between trials to avoid possible habituation from repeated use (Sun and Müller-Schwarze, 1997).

Frequency of response was calculated by adding up the number of times a particular response category occurred in each night of the 6-night trial and dividing by six. Because the data were proportional values, they were arcsine transformed to meet the normality and equal variance prerequisites in our ANOVA analysis (Zar, 1996) using the Minitab software (McKenzie and Goldman, 1999). The pre-decided level of significance was 0.05 for all statistical tests. For representation in the figures, all response frequency data were transformed from arcsine back to percentages.

## 3. RESULTS

Beavers did not respond significantly differently to different types of secretion (AGS and castoreum) in different months for four of the five sites. For the exception site, there was a significant interaction between secretion type and month ($F_{1,12}=15.98$, $P<0.001$). A simple effects analysis indicated that the beavers at this site responded significantly more to AGS than castoreum in August (73 ± 1% compared to 0 ± 1%, $F_{1,4}=248.80$, $P=0.0001$), but responded significantly more to castoreum in June and July (June: 61.25 ± 1.20 %, July: 62.0 ± 1.20 %, $F_{2,6}=45.34$, $P<0.001$). Also, there was no significant difference in response to different taxa and over time for all response categories for the sites used in more than one trial. Because there was little evidence to indicate that seasonality played a significant role in beavers' response, we combined all results in the following analyses.

The overall response follows the expected trend of greatest frequency for the weaker patterns and lower frequency for the stronger territorial responses. Beavers showed no significant difference in response frequency to secretions from their own subspecies, a different subspecies, or *C. fiber* in any response category, based on either AGS (Figure 1) or castoreum (Figure 2). Response to *C. fiber* was consistently higher than to either of the subspecies, but the difference was not significant. However, in six of the ten response patterns, beavers responded significantly stronger to castoreum than to AGS ($F_{1,24}=5.47$, $P=0.028$ for ESM Removed; $F_{1,24}=7.80$, $P=0.010$ for ESM Flattened; $F_{1,24}=11.36$, $P=0.003$ for ESM Obliterated; $F_{1,24}=4.80$, $P=0.038$ for Cork Touched; $F_{1,24}=5.77$, $P=0.024$ for Cork Removed). For Cork Dug Out, the secretion effect was significant ($F_{1,24}=4.59$, $P=0.043$), but there was an interaction between treatment and secretion ($F_{1,24}=3.78$, $P=0.037$).

## 4. DISCUSSION

Our results rejected our prediction that beavers would respond more strongly to conspecifics versus heterospecifics and to the same subspecies versus different subspecies. Thus, we do not have evidence that beavers are able to recognize species and subspecies by AGS or castoreum. The two beaver species had been separate from the Oligocene until the introduction of the North American species in Europe in 1935 (Rosell and Sun, 1999). It appears that they have been allopatric for so long that there has been no selection force for a differential response toward their own species. Even between-species mating was observed in captivity, though no hybrid offspring were born (Lavrov and Orlov, 1973). In the wild, however, there has been no report that the two species are interbreeding or any hybrid has been produced in Finland where the two species came

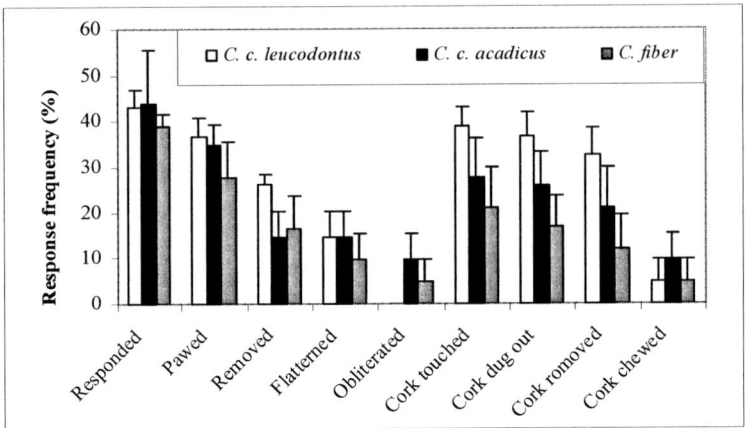

**Figure 1.** Overnight response of *C. c. leucodontus* to anal gland secretions (ACG) of *C. c. leucodontus*, *C. c. acadicus*, and *C. fiber*. Bars are standard errors.

into contact (Nolet and Rosell, 1998). Failure to show discrimination between the two subspecies found in our study may be another piece of evidence that there is little selection force favoring discrimination of individuals of allopatric populations. Beaver do not show behavioral isolation whether speciated completely (*C. canadensis* versus *C.*

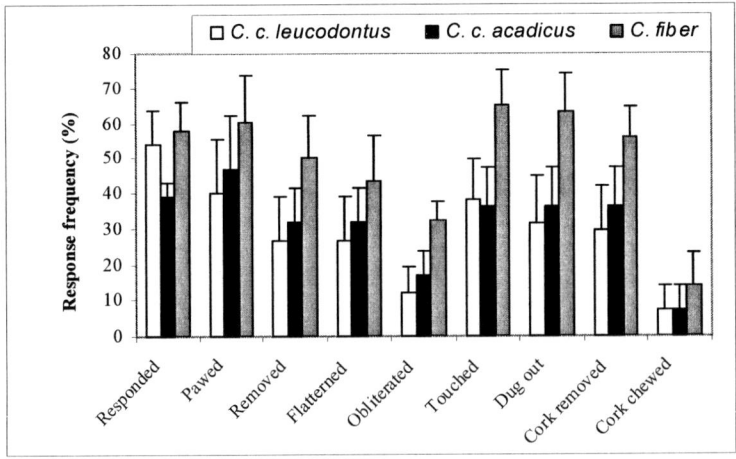

**Figure 2.** Overnight response of *C. c. lecodontus* to castoreum of *C. c. leucodontus*, *C. c. acadicus*, and *C. fiber*. Bars are standard errors.

*fiber*) or incompletely (*C. c. leucodontus* versus *C. c. acadicus*). These results further support the allopatric hypothesis (Mayr, 1970) for the origin of behavioral isolating mechanisms, which states that premating isolating mechanisms arise as by-products of genetic divergence in geographically isolated populations.

In our study, the seasonality of beavers' scent mounding behavior did not significantly affect the response of the subject. This is most likely due to the fact that we started the experiment in June and successfully avoided the peak of the scent mounding activity. Hence, between-trial habituation did not occur, although within-trial habituation may be likely, especially for castoreum (Sun and Müller-Schwarze, 1998a). Thus, it is legitimate to use data from the first night of each trial in the analysis for the main effects of taxon and secretion type.

There are two possibilities that could result in the failure of showing differential response to the two species and two subspecies. One is that *C. c. leucodontus* cannot recognize different species and subspecies of beavers. The other is that they recognize species and/or subspecies but they do not show explicit discrimination in the behavioral categories that were used in our study. Castoreum is derived from food (Müller-Schwarze 1992). When diet changes, it will necessarily result in a difference in the chemical constitution of the castoreum. This would provide information as fine as those from neighbors versus those from non-neighbors for beavers to discriminate (Schulte, 1998; Rosell and Bjørkøyli, 2002). Therefore, there is no reason to believe that beavers cannot recognize the difference between sympatric conspecific individuals and allopatric conspecific or heterospecific individuals where the differences in castoreum compounds are much larger. For AGS, the similarity in the chemical composition is positively related to the genetic relatedness (Sun and Müller-Schwarze, 1998b). Beavers are able to detect slight differences for kin recognition among individuals of the same population, in addition to sex (Sun and Müller-Schwarze, 1999). When the chemical composition of AGS is more different between subspecies, let alone between species (Rosell, 2002), it is highly unlikely that beavers are unable to detect the difference between species, or subspecies while they are able to recognize information as detailed as an individual's scent. Thus, a more convincing argument is that *C. c. leucodontus* are able to recognize different species and subspecies, but this study failed to detect the discrimination. The overnight response used in our study may not be sensitive enough to show beavers' differential responses. Lack of observable discrimination does not mean lack of recognition. Thus, our study demonstrates that not all behavioral patterns can be used in choice tests to show discrimination.

## 5. ACKNOWLEDGEMENTS

We thank J. Daiker, Frode Bergan, T. Bjørkøyli, B. Hovde, J. I. Sanda, Ø. Steifetten and A. Sukhu, who helped us in the study. The study was supported by grants from Central Washington University, The Norwegian Directorate for Nature Management, the Conservation Commissions in Telemark, Aust-Agder, Vest-Agder, Oslo & Akershus, Østfold, Vestfold, Oppland, Buskerud, Hedmark and Sør-Trøndelag Counties and the Department of Environmental and Health Studies, Telemark University College, Norway.

## 6. REFERENCES

Blaustein, A. R., Bekoff, M., and Daniels, T. J., 1987, Kin recognition in vertebrates, in: *Kin Recognition in Animals*, D. J. C. Fletcher and C. D. Michener, eds, John Wiley and Sons, New York, pp. 287-257.
Doty, R. L., 1972, Odor preferences of female Peromyscus maniculatus bairdi for male use odors of P. M. bairdi and P. leucopus noveboracensis as a function of estrous state, *J. Comp. Physiol. Psychol.* **81**:191-197.
Godfrey, J., 1958, The origin of sexual isolation between bank voles, *Proc. R. Physiol. Soc.* (Edinburgh) **27**:47-55.
Hall, E. R., 1981, The Mammals of North America, Vol. II, 2$^{nd}$ edition, John Wiley and Sons, New York.
Halley, D., and Rosell, F., 2002, The beaver's reconquest of Eurasia: status, population development and management of a conservation success, *Mammal Review* **32**:153-178.
Kotenkova, E. V., and Naidenko, S. V., 1999, Discrimination of con- and heterospecific odors in different taxa of the *Mus musculus* species group, in: *Chemical Signals in Vertebrates*, R. E. Johnston, D. Müller-Schwarze, and P. W. Sorensen, eds., Kluwer Academic / Plenum Publishers, New York, pp. 299-308.
Lavrov, L. S., and Orlov, V. N., 1973, Karyotypes and taxonomy of modern beavers (*Castor, Castoridae, Mammalia*), *Zoologicheskii Zhurnal* **52**:734-742.
McKenzie, J. D. Jr., and Goldman, R., 1999, *The Student Edition of: Minitab for Windows 95 and Windows NT* (manual and software). Addison-Wesley, New Jersey.
Mayr. E., 1970, *Populations, Species and Evolution*. Belknap Press of Harvard University, Cambridge, Massachusetts.
Moore, R. E., 1965, Olfactory discrimination as an isolation mechanism between *Peromyscus maniculatus* and *Peromyscus polionotus*, *Am. Midl. Nat.* **73**:85-100.
Müller-Schwarze, D., 1974, Olfactory recognition of species, groups, individuals and physiological states among mammals, in: *Pheromones*, M. C. Birch, ed., North-Holland Publishing Company, Amsterdam, pp. 316-326.
Müller-Schwarze, D., 1992, Castoreum of beaver (*Castor canadensis*): function, chemistry and biological activity of its components, in: *Chemical Signals in Vertebrates VI*, R. L. Doty and D. Müller-Schwarze, eds., Plenum Press, New York, pp. 457-464.
Nolet, B. A., and Rosell, F., 1998, Comeback of the beaver *Castor fiber*: an overview of old and new conservation problems, *Biol. Conserv.* **83**:165-173.
Osborn, D. J., 1955, Techniques of sexing beaver, *Castor canadensis, J. Mammal.* **36**:141-142.
Rauscher, K., 1963, Sexuelle Affinität zwischen Arten und Unterarten von Rötelmäusen (*Clethrionomys*), *Biol. Zentralbl.* **82**:653-664.
Rosell, F. 2002, The function of scent marking in beaver (*Castor fiber*) territorial defence, PhD thesis, Norwegian University of Science and Technology, Trondheim, Norway.
Rosell, F., and Bjørkøyli, T., 2002, A test of the dear enemy phenomenon in the Eurasian beaver (*Castor fiber*), *Anim. Behav.* **6**:1073-1078.
Rosell, F., and Nolet, B. A., 1997, Factors affecting scent-marking behavior in Eurasian beaver (*Castor fiber*), *J. Chem. Ecol.* **23**:679-690.
Rosell, F., and Sun, L., 1999, Use of anal gland secretion to distinguish the two beaver species *Castor canadensis* and *C. fiber*, *Wild. Biol.* **5**:119-123.
Rosell, F., and Sundsdal, L.J., 2001, Odorant source used in Eurasian beaver territory marking, *J. Chem. Ecol.* **27**:2471-2491.
Schulte, B. A., 1993, *Chemical Communication and Ecology of the North American Beaver (Castor canadensis)*, Ph.D. thesis, State University of New York, Syracuse, New York.
Schulte, B. A., 1998, Scent marking and responses to male castor fluid by beavers, *J. Mammal.* **79**:191-203.
Schulte, B. A., Müller-Schwarze, D., and Sun, L., 1995, Using anal gland secretion to determine sex in beaver, *J. Wildl. Manage.* **59**:614-618.
Sun, L., and Müller-Schwarze, D., 1997, Sibling recognition in the beaver: a field test for phenotype matching. *Anim Behav.* **54**:493-502.
Sun, L., and Müller-Schwarze, D., 1998a, Beaver response to recurrent alien scents: scent fence or scent match? *Anim Behav.* **55**:1529-1536.
Sun, L., and Müller-Schwarze, D., 1998b, Anal gland secretion codes for relatedness in the beaver, *Castor canadensis, Ethology* **104**:917-927.
Sun, L., and Müller-Schwarze, D., 1999, Chemical signals in the beaver: One species, two secretions, many functions? in: *Chemical Signals in Vertebrates*, R. E. Johnston, D. Müller-Schwarze, and P. W. Sorensen, eds., Kluwer Academic / Plenum Publishers, New York, pp. 281-287.
Svendsen, G. E., 1978, Castor and anal glands of the beaver (*Castor canadensis*), *J. Mammal.* **59**:618-620.

Svendsen, G. E., 1980, Patterns of scent-mounding in a population of beaver (*Castor canadensis*), *J. Chem. Ecol.* **6**:133-147.
Todrank, J., and Heth, G., 1996, Individual odours in two chromosomal species of blind, subterranean mole rate (*Spalax ehrenbergi*): conspecific and cross-species discrimination, *Ethology* **102**:806-811.
Ueno, Y., 1994, Olfactory discrimination of urine odors from five species by tufted capuchin (*Cebus apella*), Primates **35**:311-323.
Zar, J. H., 1996, *Biostatistical Analysis*. 3rd edition, Prentice Hall, New Jersey.

# SELF-GROOMING IN MEADOW VOLES

Michael H. Ferkin[*]

## 1. INTRODUCTION

Self-grooming is ubiquitous among mammals (Spruijt et al., 1992). However, no single study has determined the role of self-grooming when an individual encounters a conspecific or its scent. A review of the literature provides three different hypotheses to explain the occurrence of self-grooming. First, self-grooming in response to the odor of a conspecific is a general response to any olfactory stimuli (Roth and Katz, 1979; Meyerson and Hoglund, 1981; Wolff et al., 2002). Second, it is a redirected behavior that indicates the groomer's fear or anxiety (Fentress, 1968; Delius, 1979; van Erp et al., 1994). Third, self-grooming in response to the odor of a conspecific is a form of olfactory communication that broadcasts scents, indicating the groomer's heightened interest for a particular conspecific (Thiessen, 1977; Ferkin et al., 1996).

The goal of this study is to determine which of the three hypotheses best explains the role of self-grooming in meadow voles, *Microtus pennsylvanicus*. The above hypotheses generate specific predictions. If it is a general response to odors of conspecifics and heterospecifics, voles should self-groom at similar rates in response to the odors of male and female conspecifics and to those of heterospecifics independent of gender. If it is a redirected behavior, indicating stress or conflict between flight and fight, voles should self-groom more when they encounter odors of same-sex conspecifics and those of heterospecifics of either sex as compared to odors of opposite-sex conspecifics. Lastly, if self-grooming is a form of olfactory communication related to reproduction, voles should self-groom more in response to odors of opposite-sex conspecifics as compared to odors of same-sex conspecifics and heterospecifics. I distinguish among these three hypotheses by comparing the amount of time meadow voles self-groom in response to odors of different conspecifics and those of heterospecifics.

---

[*] M. H. Ferkin, Department of Biology, University of Memphis, Memphis, TN 38152

## 2. METHODS

### 2.1. Animals

In this study, I used meadow voles and prairie voles, *Microtus ochrogaster*. Adult male meadow voles were selected as subjects and scent donors and prairie voles were selected only as donors. Meadow and prairie voles were housed from birth under long-photoperiod conditions (14:10h L:D, lights on at 0700h CST). This photoperiod is characteristic of a daylength prevalent during their breeding season.

At 21 days of age voles were weaned and housed with littermates in clear plastic cages (26 x 32 x 31 cm) containing wood chip bedding and a 12-gram piece of cotton nesting material. At 35 days of age voles were separated from littermates and singly housed in clear plastic cages (13 x 16 x 13 cm) until the start of the tests. Meadow voles were 90-150 and prairie voles were 80-130 days of age. I used 114 meadow voles in this study. Of these meadow voles, 84 males were used as subjects (groomers), another 14 males and 14 females were used as odor donors. I also used 14 male and 14 female prairie voles as scent donors.

Cages containing voles were cleaned once a week. Voles were provided food (Laboratory Rodent Diet # 5015, PMI Inc., St. Louis, MO, U.S.A.) and water *ad libitum*. The cotton nesting material was replaced every two weeks. Cotton nesting material had been in the home cage of donor voles for 14 days before it served as odor stimuli. Stimulus odors were from a) male meadow voles, b) female meadow voles, c) female prairie voles, d) male prairie voles, e) vanilla extract, and f) clean cotton bedding. In the positive control condition, 250 µl of vanilla extract was placed on an 8-gram clean piece of cotton nesting material 1 hour before it was used as a scent stimulus. In the negative control condition, 250 µl of water was placed on an 8-gram piece of clean cotton 1 hour before it was used as a scent stimulus.

### 2.2. Testing Procedure

Meadow voles to be tested were placed into a covered cage of the same dimension as their home cages that contained clean wood chip bedding and 8 grams of cotton bedding from an odor donor. I recorded the amount of time voles spent self-grooming for the next 5 minutes. Typically voles begin self-grooming within 30 seconds of exposure to a scent stimulus (Ferkin et al., 1996). We considered as self-grooming a cephalocaudal progression that begins with rhythmic movements of the paws around the mouth and face, over the ears, descending to the ventrum, flank, anogenital area, and tail (Ferkin et al., 1996). Self-grooming was recorded when subjects rubbed, licked or scratched any of these body areas during its exposure to the scented cotton bedding. For each test, the experimenter was blind to the identity of the scent donor. After each test, I cleaned the test cage with 70% ethyl alcohol, dried it, and replaced the soiled wood chip bedding with fresh chips.

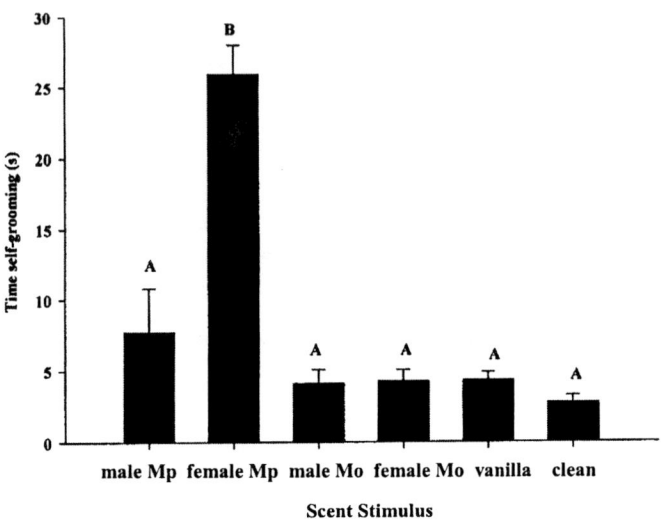

**Figure 1.** Time that male meadow voles spent self-grooming when they were exposed to different odors. On the X-axis, the abbreviations Mp and Mo refer to meadow voles and prairie voles, respectively, vanilla refers vanilla extract and clean refers to clean bedding. Histograms capped with different letters are significantly different from one another ($P < 0.05$).

## 3. RESULTS

The data were not normally distributed. Thus, we used a Kruskal-Wallis H-test on ranks to determine if significant differences existed in the amount of time male voles self-groomed when they were exposed to the different odor stimuli ($H = 45.63$, $df = 5$, $P < 0.001$). Male meadow voles spent more time self-grooming when they were exposed to odors of female meadow voles as compared to the odors of male meadow voles, male and female prairie voles, vanilla, and clean cotton (Dunn's test, $p < 0.05$ for all comparisons, Figure 1). Male meadow voles, however, spent similar amounts of time self-grooming in response to odors of male meadow voles, male and female prairie voles, vanilla, and clean cotton (Dunn's test, $P > 0.05$ for all comparisons; Figure 1).

## 4. DISCUSSION

The goal of this paper was to determine which of three hypotheses, redirected behavior, general arousal, or olfactory communication, best explained the response of meadow voles to the odors of conspecifics and heterospecifics. Briefly, voles self-groomed more in response to odors of opposite-sex conspecifics as compared to those of same-sex conspecifics odors, to those of heterospecifics, vanilla extract, and clean cotton. The data were not consistent with the predictions of the redirected behavior hypothesis (Fentress 1968; Delius 1979; van Erp et al. 1994). If self-grooming was a redirected behavior, voles would self-groom at similar rates when they encountered the odors of heterospecifics, same- or opposite-sex conspecifics, or a novel odor like vanilla extract.

The present results were not consistent with the general arousal hypothesis (Roth and Katz, 1979; Meyerson and Hoglund, 1981; Wolff et al., 2002). Wolff et al. (2002) concluded that self-grooming behavior was not associated with reproduction in prairie voles because they self-groomed at similar rates in response to male and female odors. This conclusion must be tempered by the methods used in their study. First, they used pairs of tethered males and a single free-moving female. Thus, it is not clear whether males self-groomed in response to the odors of the other tethered male or to those of the female. Second, tethering an animal may be stressful and the amount of time that the male prairie voles self-groomed may be in response to being tethered and not able to move freely (Thor et al., 1988; van Erp et al., 1994). Interestingly, male prairie voles self-groom more in response to bedding scented by male conspecifics than to that of opposite-sex conspecifics (Ferkin et al., 2001).

Meadow voles self-groomed more in response to odors of opposite-sex conspecifics as compared to odors of same-sex conspecifics and heterospecifics. Thus, the results of the study supports the hypothesis that self-grooming may be a specialized form of olfactory communication that voles use to communicate with target conspecifics (Thiessen, 1977; Ferkin et al., 1996). In that the amount of time males self-groomed was biased toward odors of female conspecifics rather than those of male, it may be involved in reproductive behavior. There are several lines of evidence that support this idea. First, several terrestrial male mammals self-groomed more in response to odors of female conspecifics than to those of male conspecifics (Steiner, 1973; Brockie, 1976; Wiepkema, 1979; Ferkin et al., 2001). Second, male meadow voles self-groom more in response to odors of reproductively active females as compared to those of reproductively quiescent females (Ferkin et al., 1996). Third, male meadow and prairie voles self-groomed more in response to odors of unfamiliar and unrelated females than to those of female siblings (Paz-y-Mino C. et al., 2002). Fourth, intact male rats and meadow voles self-groom more than gonadectomized males do when they are exposed to odors of female conspecifics (Moore 1986; Leonard & Ferkin unpubl data, respectively). Finally, by self-grooming a male may reduce intersexual aggression or increase the attractiveness of his odor to nearby female conspecifics (Thiessen, 1977; Harriman and Thiessen, 1985). Female meadow and prairie voles spend more time investigating the odors of male conspecifics that recently self-groomed as compared to those of males that did not recently self-groom (Ferkin et al., 1996, 2001).

Individuals that self-groom when they encounter the odors of particular opposite-sex conspecifics may be more likely than those that do not to indicate their identity and presence to nearby opposite-sex conspecifics. By self-grooming when they encounter the odors of opposite-sex conspecifics, individuals communicate their interest to potential partners (Ferkin et al., 1996). For example, female prairie voles self-groomed more in response to their mate's odors than to those of an unfamiliar male (Witt et al., 1990). It is also possible that individuals do not self-groom more to odors of opposite-sex conspecifics but that they self-groom less to the odors of same-sex conspecifics and heterospecifics. By reducing the amount of time that they self-groom in response to such odors, individuals may be attempting to reduce the likelihood that they get the attention of these competitors or predators. An alternative explanation is that self-grooming by male voles may affect whether social interactions between them and that nearby male conspecifics that receive these odors will behavior either in an agonistic or amicable manner towards them (Shanas and Terkel, 1997; Bursten et al., 2000). Thus, odors conveyed by self-grooming may have different meanings to same- and opposite-sex conspecifics that receive them. Alternatively, the groomer, depending upon the intended recipient, may alter the message or chemical signal. At present, we distinguish between these two hypotheses. The present data, however, are consistent with the idea that individuals adjust the amount of time that they spend self-grooming when they encounter the odors of particular conspecifics and that the adjustment is context specific. For example, individuals may self-groom more when they encounter the odors of a potential mate, but self-groom less when they encounter the odors of a potential competitor or heterospecific.

It is important to realize that self-grooming also has other functions, such as care of the body surface, removal of ectoparasites, thermoregulation, self-stimulation, and spreading antibacterial agents from the saliva onto the skin (Spruijt et al., 1992). Nevertheless, the present data are consistent with the view that self-grooming has secondary functions outside of those listed above (Tinbergen, 1952; Spruijt et al., 1992). Self-grooming appears to be involved in the transfer of chemical information from the groomer to its audience (Thiessen, 1977; Ferkin et al., 2001). In that males spend more time self-grooming in response to odors of reproductively active female conspecifics as compared to those of male conspecifics, it may increase the likelihood that groomers are able to signal their presence in an area to nearby female conspecifics (Steiner, 1973; Brockie, 1976; Thiessen, 1977; Ferkin et al., 1996, 2001; this study). These observations also suggest that odors produced by male groomers may be a sexual signal to females.

To understand the potential communicative role of self-grooming in social contexts we need to develop approaches that focus on why an individual engages in this behavior. One approach may be to ask questions that allow us to understand the message in the odors projected by groomers to their audience. This requires making inferences about the groomer's motivational state and its association, if any, with its audience. Another approach may be to examine the response of the audience to a groomer. Such information may help us to determine the meaning of the conveyed odor to the audience. We also need to examine the proximate and ultimate reasons why voles self-groom at different rates when they encounter the odors of different conspecifics. For example, this may include studying the physiological changes that this behavior has on the groomer and its audience. Alternatively, we could examine the effects of self-grooming on the survival and reproductive success of the groomer and its audience. Finally, we need to determine the benefits and costs of self-grooming in response to odors of conspecifics. Self-grooming and responding to the produced odors may represent a tradeoff between the costs of increased detection by competitors or predators and the benefits in facilitating reproduction.

## 5. ACKNOWLEDGEMENTS

I thank A. Pierce, J. delBarco-Trillo, and S. Schoech for commenting on earlier drafts of this manuscript. This work was supported by National Science Foundation grant IBN 9421529 and National Institutes of Health AG-16594-01.

# 6. REFERENCES

Brockie, R., 1976, Self-anointing by wild hedgehogs, *Erinaceus europaeus. Anim. Behav.* 24:68-71.
Bursten, S. N., Berridge, K. C. & Owings, D. H., 2000, Do California ground squirrels *(Spermophilus beecheyi)* use ritualized syntactic cephalocaudal grooming as an agonistic signal? *J. Comp. Psych.* 114:281-290.
Delius, J. D., 1979, Irrelevant behaviour, information processing and arousal homeostasis, *Psychol. Forsch.* 33:165-188.
Fentress, J. C., 1968, Interrupted ongoing behaviour in two species of vole *(Microtus* and *Clethrionomys brittanicus) II.* Extended analysis of motivational variables underlying fleeing and grooming behaviour, *Anim. Behav.* 16:154-167.
Ferkin, M. H., Leonard, S. T., Heath, L. A. & Paz-y-Mito C. G., 2001, Self-grooming as a tactic used by prairie voles *Microtus ochrogaster* to enhance sexual communication, *Ethology* 107:939-949.
Ferkin, M. H. Sorokin, E. S., & Johnston, R. E., 1996, Self-grooming as a sexually dimorphic communicative behaviour in meadow voles, *Microtus pennsylvanicus, Anim. Behav.* 51:801-810.
Harriman, A. E. & Thiessen, D. D., 1985, Harderian letdown in male Mongolian gerbils *(Meriones unguiculatus)* contributes to proceptive behavior, *Horm. Behav.* 19:213-219.
Meyerson, B. J., and Hoglund, A. U., 1981, Exploratory behaviour and socio-sexual behaviour in the male laboratory rat: a methodological approach for the investigation of drug action, *Acta Pharmacol. Toxicol.* 48:168-180.
Moore, C. L., 1986, A hormonal basis for sex differences in the self-grooming of rats, *Horm. Behav.* 20:155-165.
Paz-y-Mino, C. G., Leonard, S. T., Ferkin, M. H., and Trimble, J. F., 2002, Self-grooming and sibling recognition in meadow voles *(Microtus pennsylvanicus)* and prairie voles *(M. ochrogaster), Anim. Behav.* 63:331-338.
Roth, K. A., and Katz, R. J., 1979, Stress, behavioural arousal and open-field activity - a reexamination of motionality in the rat, *Neurosci. Biobehav. Rev.* 3:247-263.
Spruijt, B. M., Van Hooff, J. A. R. A. M., and Gispen, W. H., 1992, Ethology and neurobiology of grooming behavior, *Physiol. Rev.* 72:825-852.
Steiner, A. L. 1973. Self- and allo-grooming behavior in some ground squirrels (Sciuridae), a descriptive study, *Can. J. Zool.* 51:151-161.
Thiessen, D. D., 1977, Thermogenetics and the evolution of pheromonal communication, *Prog. Psychobiol. Physiol. Psychol.* 7:91-191.
Tinbergen, N., 1952, "Derived" activities: their causation, biological significance, origin, and emancipation during evolution, *Q. Rev. Biol.* 27:1-32.
Thor, D. H., Harrison, R. J., and Schneider, S. R., 1988, Sex differences in investigatory and grooming behaviors of laboratory rats *(Rattus norvegicus)* following exposure to novelty, *J. Comp. Psychol.* 102:188-192.
van Erp, A. M. M., Kruk, M. R., Meelis, W., and Willekens-Bramer, D. C., 1994, Effect of environmental stressors on time course, variability, and form of self-grooming in the rat: Handling, social contact, defeat, novelty, restraint and fur moistening, *Brain Behav. Research* 64:47-55.
Wiepkema, P. R., 1979, The social significance of self-grooming in rats, *Netherlands J. Zoo.* 29:622-623.
Witt, D. M., Carter, C. S., Chayer, R., and Adams, K., 1990, Patterns of behavior during postpartum estrous in prairie voles, *Anim. Behav.* 39:528-534.
Wolff, J. O., Watson, M.H., and Thomas, S.A., 2002, Is self-grooming by male prairie voles a predictor of mate choice? *Ethology* 108:169-179.

# PROTEIN CONTENT OF MALE DIET DOES NOT INFLUENCE PROCEPTIVE OR RECEPTIVE BEHAVIOR IN FEMALE MEADOW VOLES, *MICROTUS PENNSYLVANICUS*

Andrew A. Pierce, Michael H. Ferkin, and Nerav P. Patel[*]

## 1. ABSTRACT

Sexual behavior is comprised of the three components: attractivity, proceptivity, and receptivity. Previous work has shown that a meadow vole's odor attractivity to the opposite sex is positively correlated with the amount of protein in the diet that it consumed. We determined whether protein content of a male vole's diet (high, medium, or low) affected proceptivity and receptivity of a female vole. The protein content of the diet of male voles did not affect female proceptivity or receptivity. These findings suggest that the protein content of a male's diet may affect a female's initial interest in a male but not whether she will mate with him.

## 2. INTRODUCTION

In mammals, the particular diet of an individual affects the chemical signals that it produces and the responses of conspecifics to those signals (Beauchamp, 1976; Porter et al., 1977; Porter and Doane, 1977; Skeen and Thiessen ,1977; Sastry et al., 1980; Ferkin et al., 1997). For example, changes in diet influence olfactory communication between Australian cashmere goats (*Capra hircus*) and between meadow voles (*Microtus pennsylvanicus*). Cashmere goats investigated the odors of opposite sex conspecifics fed high-protein diets more than those of opposite-sex conspecifics fed low-protein diets (Walkden-Brown et al., 1994). Similarly, female meadow voles were most attracted to odors of males fed a diet of 25% protein content, least attracted to odors of males fed a 9% protein diet, and moderately attracted to those of males fed a 15% protein diet (Ferkin

---

[*] Department of Biology, University of Memphis, Memphis, TN 38152

et al., 1997). This finding suggests that differences in protein content of the diet are sufficient to produce graded odor signals among male voles. Ferkin et al. (1997) hypothesized that protein content of the diet may be a mediating factor in mate attraction.

Mate attraction, or attractivity, is only a single component of sexuality. Sexuality also includes proceptivity and receptivity (Beach, 1976). Proceptivity is the individual's interest in sex. Proceptivity is a precopulatory behavior that usually is the individual's response to an attractive individual. Typically, proceptive behaviors include orientation towards an attractive individual, approaching it, or self-grooming in response to its odors. The latter is a proceptive behavior for both male and female meadow voles (Gray and Dewsbury, 1975; Ferkin et al., 1996). Receptivity is the act of coitus, and represents the consumatory act of sexual behavior, involving mounting, intromission, and ejaculation (Beach, 1976). In general, receptivity will not occur if individuals do not display proceptive behaviors to one another, and proceptive behavior does not occur if an individual is not attractive to members of the opposite-sex.

Much work has focused on answering questions about attractivity and proceptivity in meadow voles (Ferkin et al., 1991, 1992; Ferkin and Gorman, 1992; Ferkin and Kile, 1996; Leonard and Ferkin, 1999), while relatively less is known about their receptivity (Gray and Dewsbury, 1975) and mate choice. Meadow voles are promiscuous, solitary animals that deposit scent marks in their home ranges, and compete for mates (Madison and McShea, 1987, Boonstra et al., 1993). Voles live in habitats that contain ephemeral and often patchy resources (Batzli, 1985). Their diet, which is mainly grasses and forbs, is relatively low in protein content. Studies by Bergeron and Jodin (1987, 1989) suggest that voles seek out protein-rich vegetation, eschewing vegetation with high concentrations of phenols. High quality foodstuffs, such as those rich in protein, may be difficult to obtain on a regular basis (Bergeron and Jodin, 1989). This is compounded by the fact that they tend not to cache quantities of food (Batzli, 1985). Securing a protein-rich diet may indicate an individual's quality, making them attractive to the opposite sex (Ferkin et al., 1997) and potentially good mates (Kodrick-Brown and Brown, 1984). Females may benefit both directly and indirectly by choosing mates that consume protein-rich diets (Kodrick-Brown and Brown, 1984; Kodrick-Brown, 1989). The goal of the present study was to test the hypothesis that protein content of a male's diet affects the proceptive and receptive behaviors displayed by opposite-sex conspecifics. Specifically, we determined whether the percentage of dietary protein in a male's diet would affect (1) the amount of time that females self-groomed when exposed to odors from males who had been fed diets that varied in protein content, (2) whether or not an entire copulation (including mounting, intromission, and ejaculation) occurred (Gray and Dewsbury, 1975), and (3) their primary intromission latency (PIL). There were three predictions for this hypothesis. First, females will self-groom more in response to odors of males fed higher protein diets than males fed lower protein diets, and that females would be more likely to copulate with and have shorter PILs with males fed higher protein diets than those fed lower protein diets. Second, males fed a lower protein diet would be more likely than those fed a higher protein diet to intromit and have shorter PILs, and that females will self-groom more in response to odors of the former rather than the latter males. Third, protein content of the male diet does not affect the amount of female self-grooming, females' willingness to copulate, or the PIL.

## 3. METHODS

### 3.1 Animals

We used 28 male and 28 female voles that were 4-10 months old at the start of the experiment, from a breeding colony maintained at the University of Memphis. The colony was descended from wild stock captured in northern Kentucky, USA. The voles were born under long photoperiod (LP) conditions (14:10 h, L:D, lights on at 07:00 CST). Voles were weaned at 21 days of age, housed with litter mates until they were 42 days of age, and thereafter housed singly in clear polyethylene cages (30.5 x 35.5 x 22.8 cm, LxWxH). When voles were between 60 and 100 days old, they were paired with an opposite-sex conspecific and allowed to interact freely. Individuals from pairs that produced offspring were used in these experiments. All voles used in this experiment were unfamiliar with the vole with which they were paired. The cages contained hardwood shavings and cotton nest material; wood shavings were changed weekly; cotton-bedding material was changed every 14 days. Food (PMI foods 5015, St. Louis, MO) and tap water were supplied ad libitum until the start of the experiment.

### 3.2 Diets

At the start of the experiment, we assigned each vole to a randomly selected protein group, high protein (HP), mid-protein (MP), or low protein (LP). The three experimental diets were custom milled by Test Diet (PMI foods, INC., St. Louis, MO, USA). The foods were iso-caloric but varied in protein content. The protein content of the diet was 8.5%, 16.5% or 23.5% (Ferkin et al., 1997). The LP, MP, and HP diets contain similar percentages of protein as found in hay, alfalfa, and legumes, respectively (Lindroth et al., 1984; Batzli, 1985). Male voles were fed a LP, MP, or HP diets for 28 days before the start of the experiments. All male voles remained on their respective diet for the duration of the study. We measured body mass of males in the three groups prior to being fed the diet and at the end of the study. Female voles remained on their natal diet (see section 3.1), which was approximately 23.5% protein.

### 3.3 Experiment 1: Self-grooming

In this experiment, female voles were exposed in their home cages to approximately 8 grams of soiled cotton-bedding taken from the home cages of males fed a LP, MP, or HP diets or fresh cotton bedding that served as a control. The soiled cotton nesting material had been in the male's home cage for at least 10 days prior to these tests. We removed the female's cotton nesting material from her home cage and replaced it with either one type of male-scented cotton nesting material or unscented cotton. After the first instance of self-grooming (Ferkin et al., 1997), we recorded for the next 5 minutes the amount of time they self-groomed during their exposure to the cotton-bedding stimuli. In this experiment, there were 5 females exposed to LP male odors, 5 females exposed to males fed MP diets, 6 females exposed to males fed HP diets, and 10 females exposed to unscented bedding. We used a Kruskal-Wallis test to test for difference among groups, and Mann-Whitney U-tests to make post-hoc comparisons, with significance levels set at $\alpha=0.05$.

## 3.4 Experiment 2: Copulatory Behavior

In this experiment, female voles were allowed to interact directly with a LP, MP or HP male. Twelve LP, 8 MP, and 9 HP male voles were paired with a like number of females. The pairing was random. We removed the female from her home cage and placed her into a larger cage (37x21x15 cm). The female remained in this cage for 5 minutes before she was joined by a male. As soon as the male was placed into the cage with the female, we videotaped the pair for 6 hours, filming them in real-time. After the 6-hour observation, the males were returned to their home cage and females were left in their breeding cage; their roles in the study were terminated. We used a VCR to playback the videotape, and recorded the occurrence of successful copulation, including at least 1 ejaculation (Gray and Dewsbury, 1975) and the primary intromission latency (PIL). Only pairs that copulated (mounting, intromission, and at least 1 ejaculation) were included in the analyses PIL analysis. We used a Fisher's exact test to evaluate differences in the occurrence of copulation between groups, and a Kruskal-Wallis test to for differences in PIL.

## 4. RESULTS

### 4.1 Experiment 1

Females self-groomed more in response to male-scented bedding than in response to the unscented cotton control. The amount of time spent self-grooming in response to bedding from males that had been fed different amounts of dietary protein varied between the control and treatment groups (Kruskal Wallis H= 13.847, d.f.=3, p=0.003), as shown in Figure 1.

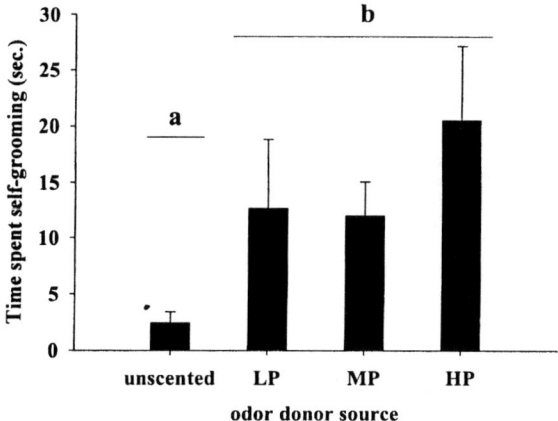

**Figure 1.** The amount of time (mean and standard error) females spent self-grooming to unscented bedding, LP male bedding, MP male bedding, or HP male bedding.

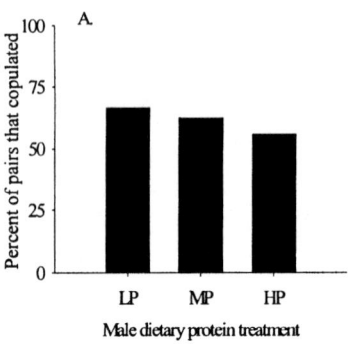 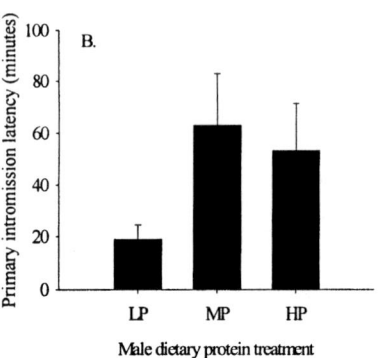

**Figure 2.** Panel A depicts the percent of pairs that copulated in each of the three treatment groups. Panel B depicts the means and standard errors for the primary intromission latency in the three treatment groups. There were no differences among groups in either the occurrence of copulation or the primary intromission latency.

There were differences when we compared the control vs. LP (Mann Whitney U=8.0, 2-sided p=0.034), control vs. MP (Mann Whitney U=3.0, 2-sided p=0.006), control vs. HP groups (Mann Whitney U=2.00, 2-sided p=0.001). Females spent similar amounts of time self-grooming in response to odors of males in the LP, HP, and MP groups. LP and MP (Mann Whitney U=11, 2-sided p=0.754), LP and HP (Mann Whitney U=12.0, 2-sided p=0.372), or MP and HP (Mann Whitney U= 17.0, 2-sided p=0.685). Females spent similar amounts of time self-grooming in response to odors of males in the LP, HP, and MP groups.

### 4.2 Experiment 2

Eighteen pairs of voles copulated: 5 of 9 on the HP diet, 5 f 8 on the MP diet, and 8 of 12 on the LP diet. There were no differences between the protein treatment groups in terms of the occurrence of copulation (Fisher's exact test $X^2$=36.00, df= 34, two-sided p= 0.375; Figure 2A). We also found no differences among the males in the three groups in their latency to intromit (Kuskal-Wallis H= 4.99, d.f.= 2  p=0.082; Figure 2B).

### 5. DISCUSSION

Protein content of the diet of male meadow voles had no effect on female proceptivity and receptivity. In experiment 1, female voles spent similar amounts of time self-grooming to bedding scented by males fed HP, MP, and LP diets. In experiment 2, protein content of the diet of males did not affect the sexual behavior of females. Specifically, the female's willingness to copulate and the PIL did not vary for males fed diets with different protein contents. The data do not support the hypothesis that the protein content of a male's diet affects the proceptive and receptive behaviors displayed by female conspecifics towards them. These results are not consistent with the predictions that females would respond preferentially to males fed any particular protein diet.

We suggest three explanations as to why the protein content of the diet of males affected their attractivity to females (Ferkin et al., 1997), but did not affect the proceptivity and receptivity of female voles. First, females may detect differences in odors, making for graded odor signals, but use this information to respond preferentially to the different odor donors. When they encounter male conspecifics, females may use visual, tactile or other sensory cues to assess potential mates. Second, because they are promiscuous (Madison and McShae, 1987; Boonstra et al., 1993), females may choose to mate with all male conspecifics that want to mate with them. Third, preferences of females for the odors of males fed high protein content diets may not be associated directly with mate choice, but more closely associated with finding a male that has located a rich food that they may be able to use (Galef, 1994). Females may be more interested in finding high quality food sources by investigating males that consumed such foodstuffs (Galef, 1994). This idea, however, is not consistent with previous findings in that same-sex conspecifics did not show a preference for the odors of individuals on HP diets, relative to those of individuals on either LP or MP diets (Ferkin et al., 1997).

Despite finding no differences in female proceptive and receptive behaviors as a result of dietary difference among males, differences in quality of signal could be important in determining the reproductive success of individuals. If the individual leaves scent-marks around its home range, an opposite-sex conspecific may spend a longer time investigating those odors raising the likelihood that the individual who deposited the odor will return to that area, increasing their chances for copulation. Next, the individual may leave an odor trail which could be followed by an interested opposite-sex conspecific; the odor trails of more attractive individuals may be followed more frequently than those of less attractive individuals. Finally, the protein content of the diet may affect male proceptivity towards females. Males fed a diet rich in protein may self-groom more and investigate female odors more than do males that were fed a low in protein. In this case, female proceptivity and receptivity would be a consequence of male behavior and not a cause (Ferkin et al., 1991, 1992). At present, we have no data to determine the degree that male sexual behavior affects female responses and female sexual behavior affects male responses.

## 6. ACKNOWLEDGMENTS

We thank J. M. Macedonia for the inspiration to carry out this experiment and for help with the experimental design. Funding for this research was provided by a Sigma Xi GIAR grant to A. A. P. and NIH grant to M. H. F.

## 7. REFERENCES

Batzli, 1985, Nutrition, in: *Biology of the New World Microtus*, R.H. Tamrin, ed., American Society of Mammalogists Special Publication, Number 8, pp. 779-811.
Beach, F. A., 1976, Sexual attractivity, proceptivity, and receptivity in female mammals, *Horm. Behav.* 7:105-138.
Beauchamp, G. K., 1976, Diet influences attractiveness of urine in guinea pigs, *Nature* **263**:587-588.

Bergeron, J. M., and Jodin, L., 1987, Defining 'high quality' food resources of herbivores: the case for meadow voles (*Microtus pennsylvanicus*), *Oecologia* 71:510-517.

Bergeron, J. M., and Jodin, L., 1989, Patterns of resource use, food quality, and health status of voles (*Microtus pennsylvanicus*) trapped from fluctuating populations, *Oecologia* 79:306-314.

Boonstra, R., Xia, X., and Pavone, L., 1993, Mating system of the meadow vole, *Microtus pennsylvanicus*, *Behav. Ecol.* 4:83-89.

Ferkin, M. H., Gorman, M. R., and Zucker, I., 1991, Ovarian hormones influence odor cues emitted by female meadow voles, *Microtus pennsylvanicus*, *Horm. Behav.* 25:572-81.

Ferkin, M. H., Gorman, M. R., and Zucker, I., 1992, Influence of gonadal hormones on odours emitted by male meadow voles (*Microtus pennsylvanicus*). *J. Reprod. Fertil.* 95:729-36.

Ferkin, M. H., Gorman, M. R., 1992, Photoperiod and gonadal hormones influence odor preferences of the male meadow vole, *Microtus pennsylvanicus*, *Physiol. Behav.* 51:1087-91.

Ferkin, M. H., and Kile, J. R., 1996, Melatonin treatment affects the attractiveness of the anogenital area scent in meadow voles (*Microtus pennsylvanicus*). *Horm. Behav.* 30:227-235.

Ferkin, M. H., Sorokin, E. S., and Johnston, R. E., 1996, Self grooming as a sexually dimorphic communicative behaviour in meadow voles, *Microtus pennsylvanicus*. *Anim. Behav.* 51:810-810.

Ferkin, M. H., Sorokin, E. S., Johnston, R. E., and Lee C. J., 1997, Attractiveness of scents varies with protein content of the diet in meadow voles, *Anim. Behav.* 53:133-141.

Galef, B. G., Jr., 1994, Olfactory communication about food among rats: a review of recent findings, in: *Behavioral of Feeding: Basic and Applied Research in Mammals*, B. G. Galef, Jr., M. Mainardi, and P. Valsecchi, eds., Harwood Academic, Chur, Switzerland, pp. 83- 102.

Gray, G. D., and Dewsbury, D., 1975, A quantitative analysis of copulatory behavior in the meadow vole, *Microtus pennsylvanicus*, *Anim. Behav.* 23:260-267.

Kodrik-Brown, A., 1989, Dietary carotinoids and male mating success in the guppy: an environmental component to mate choice. *Behav. Ecol. Sociobiol.* 25:393-401.

Kodrik-Brown, A., and Brown, J. H., 1984, Truth in advertising: the kinds of traits favored by sexual selection. *Am. Nat.* 124:309-323.

Leonard, S. T., and Ferkin, M. H., 1999, Prolactin and testosterone affect seasonal differences in male odor preferences for female odors in meadow voles. *Physiol. Behav.* 68:139-143.

Lindroth, R. L. Batzli, G. O., and Guntenspergen, G. R., 1984, Artificial diets for use in studies with Microtine rodents. *J. Mammal.* 65:139-143.

Madison, D. M., and McShae, W. J., 1987, Seasonal changes in reproductive tolerance, spacing, and social organization of meadow voles: a microtine model, *Am. Zool.* 27:899-908.

Porter, R. H., and Doane, H. M., 1977, Dietary-dependent cross-species similarities in maternal chemical cues, *Physiol. Behav.* 19:129-131.

Porter, R. H., Deni, R., and Doane, H. M., 1977, Responses of *Acomys cahirinus* pups to chemical cues produced by a foster species, *Behav. Biol.* 20:244-251.

Sastry, S. D., Buck, K. T., Janak, J., Dressler, M., and Preti, G., 1980, Volatiles emitted by humans, *Biochem. Appl. Mass Spectrom.*, Supplement, 1085–1129.

Skeen, J. T., and Thiessen, D. D., 1977, Scent of Gerbil Cuisine, *Physiol. Behav.* 19:11-14.

Walkden-Brown, S. W., Restall, B. J. Noton, B. W., Scaramuzzi, R. J., and Martin, G. B., 1994, Effects of nutrition on seasonal patterns of LH, FSH, testosterone concentration, testicular mass, sebaceous gland volume, and odor in Australian cashmere goats, *J. Reprod. Fertil.* 102:351-360.

# THE SIGNALLING OF COMPETITIVE ABILITY BY MALE HOUSE MICE

Nicholas Malone, Stuart D. Armstrong, Richard E. Humphries, Robert J. Beynon, and Jane L. Hurst[*]

## 1. INTRODUCTION

Female mammals generally prefer males of high competitive ability as mates (Cox and Le Boeuf, 1977; Wolff, 1985; Hurst, 1987; D'Amato, 1988; Anderson, 1994; Potts et al., 1994) as such males are able to secure the best resources (including those useful to breeding females) and may pass on their competitive ability to their offspring. Indeed, in deer mice (*Peromyscus maniculatus*) sons of dominant males often become dominant themselves (Dewsbury, 1990), a finding that suggests a genetic basis for the qualities needed to achieve dominant status. Thus males of high competitive ability may gain mating opportunities if they are able to advertise their competitive ability to females. Males of high competitive ability could also benefit from communicating this to rival males. If males of lower competitive ability are deterred from challenging, this will reduce the fighting required to defend a territory or to maintain dominance, thereby reducing the risk of injury and other fighting related costs (Gosling, 1982). Males may thus gain both reproductive and competitive advantages from advertising that they are of high competitive ability to conspecifics.

Animals can benefit by responding to signals of competitive ability from conspecifics, but only if the signal is a reliable indicator of the signaller's ability (Zahavi, 1979, 1987). There is strong selection pressure for receivers to respond only to reliable signals. For example, males deterred from escalating aggression against an opponent that falsely signals itself to be of a high competitive ability would be disadvantaged against such 'cheats'. Some kind of reliability check needs to be built into the signalling system, manifest through cost of signalling. Here we discuss the different ways in which signals might indicate a signaller's competitive ability reliably. We then consider the potential costs involved in signalling competitive ability through scent marks by reviewing

---

[*] Nicholas Malone, Richard E. Humphries, and Jane L. Hurst, Animal Behavior Group, Faculty of Veterinary Science, University of Liverpool, Neston, UK. Stuart D. Armstrong and Robert J. Beynon, Protein Function Group, Faculty of Veterinary Science, University of Liverpool, Neston, UK.

competitive scent mark signalling in the house mouse, in which both scent marking behaviour and the chemical basis of scent signals have been widely studied.

## 2. SIGNALLING COMPETITIVE ABILITY

Signals of competitive ability can usefully be categorised as signals of competitive potential (i.e. likelihood of success in competitive interactions) or signals of competitive success (i.e. reflecting the outcome of previous competitive encounters). Signals of competitive potential may reflect some aspect of male quality that strongly influences but does not guarantee competitive success. For example, in the toads *Bufo bufo* and *Bufo calamita* vocalisation pitch correlates with body size, a physical trait that confers a competitive advantage in this species (Davies and Halliday, 1978; Arak, 1983). Thus, these toads use vocal pitch to signal competitive potential. Countersinging in territorial songbirds may also be an example of signalling competitive potential. Male songbirds actively attempt to follow or overlap the songs of rivals with their own (Lemon, 1968; Dabelsteen et al., 1997). This matching and overlapping of songs correlates with male aggression (McGregor et al., 1992; Dabelsteen et al., 1997) and is used by males to assess competitive ability (Naguib et al., 1999).

Signals that are costly to produce are another way of signalling competitive potential (Zahavi, 1987; Johnstone, 1997). In theory, males of high quality should be able to invest more in expensive signalling materials whilst pursuing necessary day-to-day activities (foraging, grooming, territorial defence, etc). Males that can demonstrate an ability to continue their daily activities despite the energy and time handicaps imposed by signalling should be serious opponents or desirable mates. However, as outlined below, other aspects of signals may provide direct proof of competitive success, not just competitive potential.

### 2.1 Material Costs of Signal Production

The material cost of producing a signal is perhaps the most obvious type of signalling cost. The jumping behaviour or 'stotting' of gazelles in front of predators depletes their bodily resources (e.g. glycogen reserves). Gazelles stot at a low rate when in poor condition, and predators prefer to hunt these individuals rather than those that stot at high rates (Fitzgibbon and Fanshaw, 1988). Thus stotting behaviour provides an example of a signal that is both honest (poor quality animals cannot maintain a high stotting frequency) and reliable (stotting frequency is proportional to the quality of the animal). The flashy tail of the peacock may have several material costs associated with it; the production of preening oils, the investment of keratin and pigments in feather development and the effort required to haul such a heavy tail around, for example. Gazelles and peacocks provide visual examples of signals with material costs but it is possible that the production of olfactory signals may also have significant material costs.

Many mammal studies have highlighted the difference in scent marking frequency seen between the dominant members of a group and their less competitive subordinates (e.g. rabbits, Mykytowycz, 1965; sugar gliders, Schultze-Westrum, 1965; mouse lemurs, Perret, 1995; house mice, Desjardins et al., 1973; reviewed by Ralls, 1971 and Brown and Macdonald, 1985). Scent marks require constant replenishment as they degrade and diminish over time. However, little is known about the energetic costs of manufacturing scent marks and other odiferous secretions, and it is not known whether increased deposition of scent involves a significant increase in production costs. As dominant males have larger secretory glands than their subordinates (e.g. rabbit chin and anal glands, Mykytowycz, 1965; house mouse preputial gland, Lombardi and Vandenbergh, 1977; Harvey et al., 1989; Gosling et al., 2000), highly competitive animals may be able to afford an extra investment in glandular tissue, allowing them to output a greater amount of scent. However, the synthesis of signalling molecules may not be the only cost to a signaller. Males may need to defend a space in which to deposit their scent marks, to attain and maintain high social rank if the production of signals is monopolised by dominant individuals, and to replenish their scent marks to maintain a high level of fresh scent signals. Thus, in addition to the material costs of scent mark production, males have to bear the costs of competing for spatial, social and temporal opportunities to deposit scent marks. In theory, securing such opportunities to deposit scent marks may in itself signal the competitive success of a male over those competing against it. If the opportunity to signal depends on the outcome of competition between males, signalling will reflect the competitive success achieved by a male rather than the male's unproven competitive potential or likelihood of success. When assessing how scent marks may signal competitive ability, we thus need to consider the spatial, social and temporal context of scent deposition in addition to the chemical composition of scents.

## 2.2 Spatially Related Signalling Opportunities

Examples of spatially-related signalling opportunities include competition for the best lekking position (Wiley, 1991; Höglund and Alatalo, 1995) or priority of access to the most noticeable scent marking sites within a territory (e.g. rabbits, Mykytowycz, 1965; sugar gliders, Schultze-Westrum, 1965). Males that achieve such objectives can only do so if they are the best competitors. Thus the ability to signal at a particular location is reliable proof of competitive success. Rival males and potential mates can compare the competitive success of different males by comparing the relative value of their achievements. For example, males that have won access to the most sought after lekking positions or most resource-rich territories are likely be the most competitive in the area. By comparing the achievements of different males, observers can assess the degree of competition that each male has been able to endure (e.g. injury, loss of foraging time, diversion of energy).

## 2.3 Socially Related Signalling Opportunities

Socially determined signalling opportunities might occur in species where males produce a signal of dominant social status and females preferentially mate with dominant males. High social status may determine the opportunity to signal. Thus, if aggressive retaliation from dominant males prevents subordinates from giving out competitive signals, any male that manages to signal is likely to be dominant. Furthermore, if only dominant males are able to signal in a competitive environment, these signals offer proof that a male has overcome competition to become dominant. Breast badge formation in male Harris sparrows provides a visual example of a socially related signalling opportunity (Rohwer and Rohwer, 1978). Males of this species use breast plumage colour to signal social dominance over other males. Dark breast coloration is only maintained by continual victory over aggressive challengers. The more socially dominant the male, the darker his breast 'badge' becomes. Thus the badge provides proof to females and rivals that the bearer is a successful competitor. House mice also compete for high social status and may use status related scent marking patterns to communicate their competitive success. Desjardins et al. (1973) demonstrated that house mice in a competitive environment only retain the prolific scent marking behaviour seen in isolated males if they are socially dominant. Subordinate males produce scent marks that are far fewer in number and could arguably be considered urination by necessity rather than a signal of competitiveness, or even a signal of non-competitive status (Malone, 2003).

## 2.4 Temporally Related Signalling Opportunities

There is also a temporal aspect to the cost of signalling competitive ability. Maintaining a signal of competitive potential can signal competitive ability if sustaining a signal over time carries additional costs. If females and rivals were able to detect signals over a long enough time period, low quality males would be identified eventually by their inability to match the continuity and strength of signals transmitted by high quality males (e.g., the roaring behaviour of red deer in rut, Clutton-Brock and Albon, 1979). Potential costs of prolonged signal maintenance include the depletion of bodily resources via signal production but also less direct costs such as an increased probability of detection by predators. Thus the ability to maintain a signal that outlasts those of rival males might signal physical fitness and competitive potential. In the case of scent marks, the ability to continually replenish the scent deposited around a territory might indicate a male is able to meet the material production costs of the scent and has the physical fitness to cover the whole territory before previous deposits fade away. The latter is an example of scent marks as signals of competitive potential but replenishment of scent can also provide proof of competitive success.

If rivals are successfully chased from a territory the resident reduces the attraction of any scent marks deposited by those rivals with the addition of his own scent close to ('countermarking', e.g. house mice) or across ('overmarking', e.g. golden hamsters and meadow voles) those of the intruder (Johnston et al., 1994, 1997; Hurst and Rich, 1999; Hurst et al., 2001). The ability to ensure one's own scent marks are the most recent in the territory, by rapidly countermarking or overmarking all non-self marks, will signal competitive success as only males able to evict rivals could maintain such territorial signals.

## 3. ASSESSING THE COSTS OF SCENT MARK SIGNALLING

In order to understand the costs of signalling competitive success via scent marks and assess the reliability of such signals, it is necessary to identify the costs involved. One obvious cost is the amount of signal material deposited by different males. However, this is only one component of the signal. As discussed above, a scent mark signal must also be considered in the context of its placement relative to those of rivals, and in the context of the signaller's social rank. Scent mark signals also need to be considered not just in terms of quantity but also for persistence over time. Each of these aspects of a scent mark can carry a cost to the signaller, and all of these costs must be included when comparing the signals of different males. This holistic approach to understanding scent mark signals requires consideration of both scent marking behaviour and the molecular basis of chemical signals, and how each influences the other.

Although it is well established that many mammals use scent marks in the context of competitive signalling, and that females appear to use male scent signals when assessing potential mates, the information provided by scent signals concerning male competitive potential or competitive success is poorly understood. In particular, we have little knowledge of the precise costs involved in scent signalling in any species. The house mouse provides our most thoroughly studied model of competitive scent mark signalling. Below, we examine the likely costs involved in urine mark signalling by male mice and discuss how these costs may allow scent marks to be used as a reliable signal of male competitive ability.

### 3.1 Material Costs of Signal Production in House Mice

The urine of male house mice contains numerous chemicals that play a role in olfactory communication (Jones and Nowell, 1973; Jones and Nowell, 1974; Huck, 1982; Novotny et al., 1990, 1999). Although many components of urine are excretory by-products of metabolism that may contribute to olfactory signals (e.g. MHC, Singer et al., 1997), many volatile and involatile components are deliberately synthesised and released in urine to act as chemical signals. If the synthesis of these chemicals involves significant energetic expense to males, they might act as the olfactory equivalent of a peacock's tail. In theory, the elaborate plumage and patterning of a peacock's tail is a handicap (Zahavi, 1975; Manning & Hartley, 1991). For the peacock, the cost of its visual signal might include the material costs of producing the large feathers, their pigments and the preening oil necessary to maintain them. In the case of the house mouse, the cost of its scent signals might include the material cost of the signalling compounds. As in many other species, dominant male house mice deposit a much higher frequency of scent marks compared to subordinates (Desjardins et al., 1973). In theory, the more scent a male produces, the higher the cost of synthesising the signalling chemicals. If synthesis costs are significant, the more difficult it will be for a male to divert sufficient energy away from more critical life processes. Hence males of high quality should be able to produce the strongest chemical signals.

The urine of house mice contains four volatile compounds known to mediate aggression between males. Two of these, 2-*sec*-butyl-4,5-dihydrothiazole (thiazole) and 3,4-dehydro-*exo*-brevicomin (brevicomin), are particularly prevalent in adult male urine but not in female urine (Bacchini et al., 1992; Robertson et al., 1993; Novotny et al., 1999). Two additional chemicals unique to adult male mice are the highly labile sesquiterpenes, E,E-α-farnesene (α-farnesene) and E-β-farnesene (β-farnesene), both of which discourage prolonged investigation by subordinate males (Novotny et al., 1990; Jemiolo et al., 1992). Both α-farnesene and β-farnesene are thought to be major constituents of the 'aversive substance' postulated by Jones and Nowell (1973), and may have a role in signalling male dominance (Harvey et al., 1989, Novotny et al., 1990). Little is known about the metabolic costs of producing thiazole, brevicomin, α-farnesene and β-farnesene. The farnesenes are produced in the preputial gland and males of high competitive ability can have preputial glands up to twice the weight of less competitive males (Lombardi and Vandenbergh, 1977; Novotny et al., 1990; Gosling et al., 2000). If the maintenance of a large gland requires a greater share of physiological resources than a small gland, large preputial glands could be the privilege of high quality males, who are also likely to be the best competitors. However, despite the difference in preputial gland size, there is little evidence so far to suggest that males of higher competitive ability excrete considerably higher levels of thiazole, brevicomin or farnesenes in their urine compared to males of lower competitive ability (Harvey et al., 1989, discussed in Malone, 2003).

Mouse urine also contains high concentrations of protein involved in chemical signalling, particularly that of males in which concentrations are typically around 30 mg/ml but can be as high as 60 mg/ml (Finlayson and Baumann, 1958; Payne, 2001; Beynon et al., 2001; Malone, 2003). Major urinary proteins (MUPs) account for 99% of this urinary protein. MUPs bind male signalling pheromones, having a high affinity for thiazole and brevicomin (Humphries et al., 1999; Bacchini et al., 1992; Robertson et al., 1993). MUPs are also highly polymorphic, providing the individual ownership signal in male scent marks (Hurst et al., 2001; see also Hurst et al., this volume). If MUPs are energetically expensive to synthesise, low quality males might be unable to invest in such a large amount of expensive proteins as high quality males. Thus the amount of protein in male scent marks might be used as an indicator of the owner's competitive potential.

In a recent study of wild-derived male house mice (Drickamer, 2001), the density of scent marks deposited by a male reflected the male's competitive rank relative to that of 3 rivals. A study by Malone (2003) also showed that relative competitive success among a large number of male house mice (n = 36) could be estimated accurately by comparing individual scent mark frequencies and surface area covered. Among males that encountered each other on a regular basis (but were singly housed to avoid the formation of dominant and subordinate relationships), there was a strong positive correlation between scent marking and individual competitive behaviour (Malone, 2003). Males that invested in a high rate and coverage of scent marking attacked opponents faster and more frequently compared to males that invested less in scent marking. Scent marking behaviour also correlated strongly with relative success in competitive encounters: the higher a male's competitive success in eleven encounters with different males, the greater his scent marking in terms of both number of scent marks and area covered. Indeed, the most competitive males were found to produce a scent marking pattern that covered eight times the area of that covered by the least competitive males.

Thus other mice might use the amount of scent that a male deposits to assess his competitive potential. Such signals could be reliable if the material cost of scent production prohibits extensive signalling by low quality males while males of high quality are better able to sustain the costs of scent production. However, as yet, there has been no investigation of the material costs of scent production and whether this is sufficient to prevent low quality males from depositing as much scent as those of high competitive ability. It is also necessary to establish whether other mice use the quantity of scent deposited by males in competitive assessment. In support of the hypothesis that scent production might be energetically expensive, Gosling et al. (2000) found evidence that suggested a trade-off between investment in scent marking and growth among TO laboratory mice, although this was apparent only in dominant-subordinate pairings in which the dominant was initially the smaller of the two males. The authors speculated that MUP production might be the cost that impaired growth in dominant males that invested heavily in scent marking, although this was not tested. However, Malone (2003) measured daily MUP production and loss among house mice bred from wild derived stock and found no evidence of a trade-off between MUP output and weight gain among males that were exposed regularly to competitive interactions with other males. Although individual investment in scent marking was manipulated experimentally, those that invested more in scent marking did not show a lower growth rate than matched individuals of the same weaning weight that invested less in scent marking. Furthermore, males that synthesized and excreted a large quantity of MUP did not suffer a reduction in the rate of body growth. The lack of a negative effect of increasing MUP output on growth suggested that the cost of MUP synthesis is not significant among wild house mice under normal conditions where food is generally abundant. However, as discussed earlier, the material cost of the chemicals within a scent mark may not be the only costs of scent mark signalling. Male house mice are likely to incur other types of costs from competing for opportunities to deposit scent marks.

### 3.2 Opportunities for the Spatial Deposition of Scent Marks

The scent marks of a territory owner will predominate around its territory if the owner is able to evict competitors from the defended area (Gosling, 1982). Any scent marks that intruding male house mice manage to deposit are rapidly surrounded by countermarks from the resident male (Hurst, 1990, 1993). In theory, this spatial aspect of territorial scent marking behaviour can provide a reliable signal of competitive success. Only owners that successfully defend their territories are able to restrict the deposition of scent marks from rival males within the defended area and can countermark with impunity any scent marks deposited by competitors while the owner was elsewhere. Thus, covering a territory with scent marks will signal competitive success because only a successful territory owner could achieve this spatial deposition of scent marking, regardless of any material costs of producing the scent.

Among house mice, both the number of scent marks deposited and the total area covered with scent correlate strongly with a male's proven competitive success against rivals (Drickamer, 2001; Malone, 2003). This correlation would be consistent with a high material cost of scent production indicating competitive potential, as discussed in the previous section. However, it is also consistent with the hypothesis that males signal competitive success through their ability to mark out a defended area. In addition to the total amount of scent marks deposited, Malone (2003) found that the shape of scent marks is another strong correlate of competitive behaviour in house mice. Males with very streaky marks (measured as length to breadth ratio) attacked their opponents faster and more frequently compared to males with less streaky scent marks. Scent mark streakiness also correlated with the number of encounters won or lost by each male; males with the greatest competitive success had the streakiest scent marks. The significance of scent mark streakiness in communication has not yet been investigated. One possibility is that the relative extension of scent mark length observed among winners might increase the probability of overmarking or depositing scent marks very close to those of losers, helping to demonstrate that the winner is the current owner of a disputed territory. This might be particularly important immediately after an encounter when competing scent marks are fresh, as females will be unable to use the difference in age of scent marks to discriminate between winners and losers (Rich and Hurst, 1999).

Competitive experience appears to be very important in determining the scent mark deposition patterns of individual male mice. When scent marking was measured after the first competitive encounter experienced by adult males, winners deposited a greater frequency and coverage of marks than losers, although there was no difference in scent mark streakiness between winners and losers. However, prior to the first experience of direct competition with another adult male, there were no differences in scent marking between subsequent winners and losers (Malone, 2003). Once males had experienced several encounters, scent marks became streakier immediately after a male had won an encounter compared to scent marks from the same male immediately after he had lost an encounter. Scent mark streakiness thus appears to reflect the short-term competitive success of a male.

These findings, together with those of Drickamer (2001), are consistent with the idea that mice might assess the relative competitive success of males from the spatial distribution of their scent marks. Males of low competitive ability avoid the scent marks of other males (Gosling et al., 1996) and are less likely to attack males whose scent marking patterns indicate that they are territory holders (Gosling and McKay, 1990; Hurst et al., 1994) However, further research is needed to investigate whether the spatial coverage or density of scent marks deposited by a territory owner, or the streakiness of scent marks, influence the owner's attractiveness to females or challenges from other males.

### 3.3 Scent Signal Age

Countermarking introduces a temporal aspect to scent marking as a signal of competitive success. Unlike other types of visual and acoustic signals, scent marks gradually age through the loss of volatile or unstable components, thus providing information on the time since each signal was deposited in the environment. Scent marks from a territory owner that are fresher than those from intruders in the territory provide evidence to rivals and females that the owner successfully evicted intruders from the territory or otherwise stopped them from depositing competing scent marks (Rich and Hurst, 1999). In house mice, the relative age of scent marks from males competing for territory ownership appears to indicate the competitive success of each scent owner. Competitors increase challenges against the territory owner if the freshest scent marks they encounter in a male's territory are not from the territory owner (Hurst, 1993). Females are more strongly attracted to territory owners whose scents have not been countermarked by intruder males (Rich and Hurst, 1998) or to owners whose scent marks are fresher than those of an intruder (Hurst and Rich, 1999; Rich and Hurst, 1999).

A high rate of scent marking and scent replenishment will maximise the freshness of scent marks around a territory, while repeated countermarking near the gradually ageing scent marks of competitors maximises the difference in scent freshness between the winner and the loser's scent (Hurst et al., 2001). The very high frequency of scent marks deposited by male mice of high competitive ability, together with their immediate increase in scent marking in the vicinity of scent marks from competitors (Hurst, 1990; Humphries et al., 1999) may thus be to ensure that their own scent is always the freshest throughout the territory. Investment in MUPs is also likely to play a role in maintaining signals once deposited, as MUPs bind male signalling pheromones and significantly slow their release from scent marks (Hurst et al., 1998). Competitive success may thus be signalled by the ability to maintain fresh scent throughout a defended territory in the face of competition from rivals and the degradation of scent marks through time.

## 3.4 Socially Related Signalling in House Mice

When male house mice are housed together they fight to establish social rank, with one becoming dominant whilst the others become subordinates. The majority of scent marks within a territory are deposited by the dominant; subordinates void urine in a few large pools that are spatially restricted in contrast to the numerous small marks deposited throughout the home area by dominant males (Desjardins et al., 1973). Females find the urine of dominant males more attractive than an equivalent amount of urine from subordinates, suggesting that the latter produce a signal of lower quality (Malone, 2003). It is generally thought that the aggressive behaviour of the dominant male inhibits the scent marking behaviour of subordinates, though the actual mechanism is not understood. Male house mice must compete for a social opportunity to deposit scent marks that signal high quality to mates; in order to deposit the only attractive scent marks within a territory, a male must inhibit such signalling among other males in the local area. Subordinates appear to benefit from lowering the quality of their scent and restricting scent marking activity through increased tolerance from the local dominant male territory owner (Hurst et al., 2001). Thus the patterns and quality of scent produced by males may not depend on the direct costs of material production and deposition but have a cost in terms of stimulating aggression from males willing to compete for the opportunity to signal in a particular area.

## 4. CONCLUSIONS

At present, there is very little evidence that the chemical synthesis of scents places sufficient energetic limitations on animals to suggest that the material production of scent marks could indicate owner quality. We instead suggest that scent marks are used to signal competitive ability through the spatial and temporal pattern of scent marks deposited by individual males. Furthermore, since these patterns depend on the outcome of competitive interactions between males, scent marks indicate the proven and sustained competitive success of a male, not just his competitive potential. The reliability of these signals comes from the effort required to obtain the opportunity to deposit scent marks in the face of competition from other males. In house mice, the rate of scent marking, the area covered by scent and the streakiness of scent marks all correlate with male competitive ability and are likely to play different roles in the dynamics of competitive signalling. A high rate of territorial scent marking ensures the freshness of the owner's scent marks, particularly in the vicinity of scent marks from competitors. Thus the most competitive males are able to signal not only dominance over a territory but also up to the minute proof of current ownership and dominance. Widespread coverage of the territory with scent also means that females and rivals are more likely to encounter freshly deposited scent marks from highly competitive males. This dynamic aspect of scent mark deposition patterns may be an important aspect of signalling relative competitive ability in other species that compete for scent marking opportunities. However, further work is required to understand how the behavioural dynamics of scent deposition interact with investment in the material components of scents. Further empirical evidence is also needed to establish the mechanisms underlying the use of scent marks by conspecifics to assess male competitive ability in mice and in other species.

## 5. ACKNOWLEDGEMENTS

We thank the members of the Animal Behaviour Group and Protein Function Group for their helpful comments and criticism of the concepts discussed in this paper. This work was supported by BBSRC grants to JLH and RJB while NM was supported by a postgraduate studentship from the University of Liverpool.

## 6. REFERENCES

Andersson, M., 1994, *Sexual Selection*, Princeton, NJ, Princeton University Press.
Arak, A., 1983, Sexual selection by male-male competition in natterjack toad choruses, *Nature* 306:261-262.
Bacchini, A., Gaetani, E., and Cavaggioni, A., 1992, Pheromone binding proteins of the mouse, *Mus musculus*, *Experientia* 48:419-21.
Beynon, R. J., Hurst, J. L., Gaskell, S. J., Hubbard, S. J., Humphries, R. E., Malone, N., Marie, A. D., Martinsen, L., Nevison, C. M., Payne, C. E., Robertson, D. H. L., and Veggerby, C., 2001, Mice, MUPs and myths: structure-function relationships of the major urinary proteins, in: *Chemical Signals in Vertebrates*, A. Marchlewenska-Koj, J.J. Lepri, and D. Muller-Schwarze, eds., Plenum Press, New York, pp. 149-156.
Brown, R. E., and Macdonald, D. W., 1985, *Social Odours in Mammals*, Clarendon Press, Oxford.
Clutton-Brock, T. H., and Albon, S. D., 1979, The roaring of red deer and the evolution of honest advertisement, *Behaviour* 69:145-170.
Cox, C. R., and Le Boeuf, B. J., 1977, Female incitation of male competition: a mechanism in sexual selection, *Am. Nat.* 111:317-335.
D'Amato, F. R., 1988, Effects of male social status on reproductive success and on behavior in mice (*Mus musculus*), *J. Comp. Psychol.* 102:146-151.
Dabelsteen, T., McGregor, P. K., Holland, J., Tobias, J. A., and Pedersen, S. B., 1997, The signal function of overlapping singing in male robins, *Anim. Behav.* 53:249-256.
Davies, N. B., and Halliday, T. R., 1978, Deep croaks and fighting assessment in toads *Bufo bufo*, *Nature* 274:683-685.
Desjardins, C., Maruniak, J. A., and Bronson, F. H., 1973, Social rank in house mice: differentiation revealed by ultraviolet visualization of urinary marking patterns, *Science* 182:939-941.
Dewsbury, D. A., 1990, Fathers and sons: genetic factors and social dominance in deer mice, *Peromyscus maniculatus*, *Anim. Behav.* 39:284-289.
Drickamer, L. C., 2001, Urine marking and social dominance in male house mice, *Behav. Process.* 53:113-120.
Finlayson, J. S., and Baumann, C. A., 1958, Mouse proteinuria, *Am. J. Physiol.* 192:69-72.
Fitzgibbon, C. D., and Fanshaw, J. H., 1988, Stotting in Thompson's gazelles: an honest signal of condition, *Behav. Ecol. Sociobiol.* 23:69-74.
Gosling, L. M., 1982, A reassessment of the function of scent marking in territories, *J. Comp. Ethol.* 60:89-118.
Gosling, L. M., and McKay, H. V., 1990, Competitor assessment by scent matching - an experimental test, *Behav. Ecol. Sociobiol.* 26:415-420.
Gosling, L. M., Atkinson, N. W., Collins, S. A., Roberts, R. J., and Walters, R. L., 1996, The response of subordinate male mice to scent marks varies in relation to their own competitive ability, *Anim. Behav.* 52:1185-1191.
Gosling, L. M., Roberts, S. C., Thornton, E. A., and Andrew, M. J., 2000, Life history costs of olfactory status signalling in mice, *Behav. Ecol. Sociobiol.* 48:328-332.
Harvey, S., Jemiolo, B., and Novotny, M., 1989, Pattern of volatile compounds in dominant and subordinate male mouse urine, *J. Chem. Ecol.* 15:2061-2071.
Höglund, J., and Alatalo, R. V., 1995, *Leks*, Princeton University Press, Princeton.
Huck, U. W., 1982, Pregnancy block in laboratory mice as a function of male social status, *J. Reprod. Fertil.* 66:181-184.
Humphries, R. E., Robertson, D. H. L., Beynon, R. J., and Hurst, J. L., 1999, Unravelling the chemical basis of competitive scent marking in house mice, *Anim. Behav.* 58:1177-1190.
Hurst, J. L., 1987, Behavioural variation in wild house mice *Mus domesticus* Rutty: a quantitative assessment of female social organisation, *Anim. Behav.* 35:1846-1857.
Hurst, J. L., 1990, Urine marking in wild house mice *Mus domesticus* Rutty I: Communication between males, *Anim. Behav.* 40:209-222.

Hurst, J. L., 1993, The priming effect of urine substrate marks on interactions between male house mice, *Mus musculus domesticus* Schwarz and Schwarz, *Anim. Behav.* **45**:55-81.
Hurst, J. L., Hayden, L., Kingston, M., Luck, R., and Sorensen, K., 1994, Responses of the aboriginal house mouse *Mus spretus* Lataste to tunnels bearing the odours of conspecifics, *Anim. Behav.* **48**:1219-1229.
Hurst, J. L., Robertson, D. H. L., Tolladay, U., and Beynon, R. J., 1998, Proteins in urine scent marks of male house mice extend the longevity of olfactory signals, *Anim. Behav.* **55**:1289-1297.
Hurst, J. L., and Rich, T. J., 1999, Scent marks as competitive signals of mate quality, in: *Chemical Signals in Vertebrates*, R. E. Johnston, D. Muller-Schwarze, and P. Sorensen, eds., Plenum Press, New York, pp. 209-226.
Hurst, J. L., Beynon, R.J., Humphries, R.E., Malone, N., Nevison, C.M., Payne, C.E., Robertson, D.H.L., and Veggerby, C., 2001, Information in scent signals of competitive social status: the interface between behaviour and chemistry, in: *Chemical Signals in Vertebrates*, A. Marchlewenska-Koj, J.J. Lepri, and D. Muller-Schwarze, eds., Plenum Press, New York, pp. 43-52.
Jemiolo, B., Xie, T. M., and Novotny, M., 1992, Urine marking in male mice: responses to natural and synthetic chemosignals, *Physiol. Behav.* **52**:521-526.
Johnston, R. E., Chiang, G., and Tung, C., 1994, The information in scent over-marks of golden hamsters, *Anim. Behav.* **48**:323-330.
Johnston, R. E., Sorokin, E. S., and Ferkin, M. H., 1997, Scent counter-marking: female meadow voles discriminate individual's marks and prefer top-scent males, *Anim. Behav.* **54**:679-690.
Johnstone, R. A., 1997, The evolution of animal signals, in: *Behavioural Ecology: An Evolutionary Approach*, J. R. Krebs and N. B. Davies, eds., Blackwell Science, Oxford, pp. 155-178.
Jones, R. B., and Nowell, N. W., 1973, Aversive and aggression-promoting properties of urine from dominant and subordinate male mice, *Anim. Learn. Behav.* **1**:207-210.
Jones, R. B., and Nowell, N. W., 1974, A comparison of the aversive and female attractant properties of urine from dominant and subordinate male mice, *Med Weter* **2**:141-144.
Lemon, R. E., 1968, The relation between organization and function of song in cardinals, *Behaviour* **32**:158-178.
Lombardi, J. R., and Vandenbergh, J. G., 1977, Pheromonally induced sexual maturation in females: regulation by the social environment of the male, *Science* **196**:545-6.
Malone, N., 2003, *The signalling of competitive ability by male house mice*, Ph.D. thesis, University of Liverpool.
Manning, J. T., and Hartley, M. A., 1991, Symmetry and ornamentation are correlated in the peacock's train, *Anim. Behav.* **42**:1020-1021.
McGregor, P. K., Dabelsteen, T., Shepherd, M., and Pedersen, S. B., 1992, The signal value of matched singing in great tits: evidence from interactive playback experiments, *Anim. Behav.* **43**:987-998.
Mykytowycz, R., 1965, Further observations on the territorial function and histology of the submandibular cutaneous (chin) glands in the rabbit, *Oryctolagus cuniculus* (L.), *Anim. Behav.* **13**:400.
Naguib, M., Fichtel, C., and Todt, D., 1999, Nightingales respond more strongly to vocal leaders of simulated dyadic interactions, *P. Roy. Soc. Lond. B. Bio.* **266**:537-542.
Novotny, M., Harvey, S., and Jemiolo, B., 1990, Chemistry of male dominance in the house mouse, *Mus domesticus*, *Experientia* **46**:109-13.
Novotny, M. V., Ma, W. D., Wiesler, D., and Zidek, L., 1999, Positive identification of the puberty-accelerating pheromone of the house mouse: the volatile ligands associating with the major urinary protein, *P. Roy. Soc. Lond. B. Bio.* **266**:2017-2022.
Payne, C. E., 2001, *Urinary proteins and their ligands in wild house mice: modulation, heterogeneity and response*, Ph.D. Thesis, University of Liverpool.
Perret, M., 1995, Chemocommunication in the reproductive function of mouse lemurs, in: *Creatures of the Dark: The Nocturnal Prosimians*, L. Alterman, G. A. Doyle, and M. K. Izard, eds., Plenum Press, New York.
Potts, W. K., Manning, C. J., and Wakeland, E. K., 1994, The role of infectious disease, inbreeding and mating preferences in maintaining MHC genetic diversity - an experimental test, *Philos. T. Roy. Soc. B* **346**:369-378.
Ralls, K., 1971, Mammalian scent marking, *Science* **171**:443-449.
Rich, T. J., and Hurst, J. L., 1998, Scent marks as reliable signals of the competitive ability of mates, *Anim. Behav.* **56**:727-735.
Rich, T. J., and Hurst, J. L., 1999, The competing countermarks hypothesis: reliable assessment of competitive ability by potential mates, *Anim. Behav.* **58**:1027-1037.
Robertson, D. H. L., Beynon, R. J., and Evershed, R. P., 1993, Extraction, characterization, and binding analysis of two pheromonally active ligands associated with major urinary proteins of the house mouse (*Mus musculus*), *J. Chem. Ecol.* **19**:1405-1416.

Rohwer, S., and Rohwer, F. C., 1978, Status signalling in Harris sparrows: experimental deceptions achieved, *Anim. Behav.* **26**:1012-1022.

Schultze-Westrum, J., 1965, Innerartliche Verständigung durch Düfte beim Gleitbeutler *Petaurus breviceps papuanus* Thomas (Marsupialia, Phalangeridae), *Z. Versuchstierkd,* **50**:151-220.

Singer, A. G., Beauchamp, G.K., and Yamazaki, K., 1997, Volatile signals of the major histocompatibility complex in male mouse urine, *P. Natl. Acad. Sci. U S A* **94**:2210-2214.

Wiley, R.H., 1991, Lekking in birds and mammals: behavioral and evolutionary issues, *Adv. Stud. Behav.* **20**:201-291.

Wolff, P. R., 1985, Mating behaviour and female choice: their relation to social structure in wild caught house mice (*Mus domesticus*) housed in semi-natural environment, *Anim. Behav.* **207**:43-51.

Zahavi, A., 1975, Mate selection - a selection for a handicap, *J. Theor. Biol.* **53**:205-14.

Zahavi, A., 1979, Ritualisation and the evolution of movement signals, *Behaviour* **72**:77-81.

Zahavi, A., 1987, The theory of signal selection and some of its implications, in: *International Symposium of Biological Evolution* V. P. Delfino, ed., Adriatica Editrice, Bari, pp. 305-327.

# A POSSIBLE FUNCTION FOR FEMALE ENURINATION IN THE MARA, *DOLICHOTIS PATAGONUM*

Deborah S. Ottway, Sheila J. Pankhurst and John S. Waterhouse[*]

## 1. INTRODUCTION

The mara (*Dolichotis patagonum*, Zimmerman, 1758) is a large (6 – 10kg) caviomorph rodent from the dry, scrub deserts of central and southern Argentina. In this socially monogamous species, scent marking appears to play an important role in the maintenance of the pair bond (Dubost and Genest, 1974, Taber and Macdonald, 1984). Sexually mature adult maras practise two forms of scent marking. Firstly, both sexes perform anal marking, whereby secretions from paired ano-genital scent glands are distributed by the action of pressing, dragging and rocking the anal area against the ground (Pankhurst, 1998). Secondly, both sexes also use enurination or urine spraying. Males enurinate by rearing up on their hind legs and spraying urine forwards from a bipedal stance. Females, and some juvenile males, enurinate by spraying urine backwards from a quadripedal stance. The target of enurination is usually a conspecific. Taber and Macdonald (1984) suggest that female enurination could function to provide the male with a sample from which female reproductive status could be assessed, thereby advertising receptivity. This hypothesis generates the predictions that female enurination frequency will increase significantly during oestrus (0-8 hours post-partum, Dubost and Genest, 1974), and that the target will be the female's mate. As part of a wider, ongoing study, we aimed to test this hypothesis by determining the targets of female enurination and assessing the frequency of enurination in relation to female reproductive status.

---

[*] Environmental Sciences Research Centre, Anglia Polytechnic University, East Road, Cambridge CB1 1PT, United Kingdom.

## 2. METHODS

The subjects of our study were 41 socially monogamous pairs of adult maras belonging to a free-ranging population at Whipsnade Wild Animal Park, U.K. To facilitate individual identification, we fitted every animal with a coloured ear tag printed with a unique number. During capture of the animals for tagging, each individual was sexed. Males were tagged in the right ear, females in the left ear. Focal animal follows were then used to determine the identity of established pairs as well as the reproductive status of each female within a pair. Enurination behaviour is relatively infrequent and of short duration, so all instances of enurination were recorded by one observer during focal follows, each lasting at least three hours. For each incidence of enurination, we recorded the date, time, female reproductive status and target of the urine spray. To record reproductive status, each female was assigned to one of five categories:

1. Non-reproductive (neither pregnant nor lactating).
2. Pregnant (obviously pregnant, with swollen belly).
3. Heavily pregnant (days immediately prior to parturition, belly extremely swollen).
4. 0-8 hours post-partum (in oestrus).
5. Lactating (seen suckling pups, teats distended/prominent).

As maras experience a post-partum oestrus, females may be simultaneously pregnant and lactating. For the purpose of this study, we scored females as pregnant only when they were known not to be lactating. It is possible, however, that a number of the females scored as lactating were also pregnant.

## 3. RESULTS

Females enurinated over their mates, strange males, and over both their own and strange pups (Table 1). We also observed females enurinating over other females, over free-ranging peafowl, and over small children who had approached the animals from the rear while the maras were feeding.

### 3.1 Female Enurination in the Mara

Enurination frequency did not differ significantly with female reproductive status (Kruskal-Wallis test, ns, $\chi^2 = 4.36$, df = 4, p = 0.36 (Figure 1).

**Table 1.** Targets of female enurination, as a proportion of all observations (total number of episodes = 144; n = 41 females, all reproductive categories).

| Target of female enurination | Male mate | Strange male(s) | Own pup(s) | Strange pup(s) | Other |
|---|---|---|---|---|---|
| Proportion of all episodes observed | 0.33 | 0.34 | 0.10 | 0.19 | 0.04 |

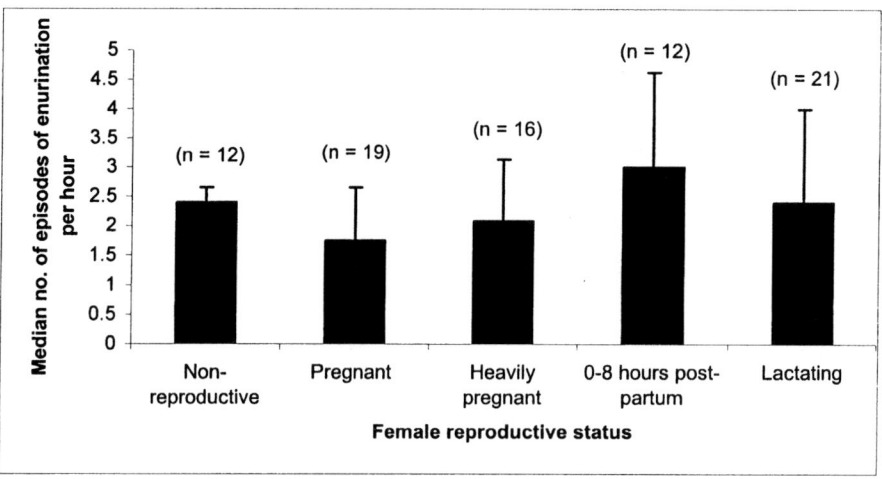

Figure 1. Median frequency of enurination per hour by female maras, in relation to female reproductive status. Bars show median values and vertical lines show inter-quartile ranges. Some females were sampled more than once, at different stages of their reproductive cycles. The sum of the sample sizes for each category (n=80) is therefore greater than the actual sample size (n=41).

## 4. DISCUSSION

The hypothesis that female maras use enurination to advertise receptivity was not supported by this study, as female enurination frequency did not differ significantly throughout the oestrus cycle. In addition, females sprayed any mara (and some other species, including humans) that approached too closely, although male maras, either the female's mate or a stranger, were the females' targets in 67 percent of all observed enurination incidences. Male maras often try to gain close proximity to females. An established pair will remain spatially and temporally close throughout the female's oestrus cycle, and males will threaten and chase off other males that try to approach their female (Pankhurst, 1998). This does not prevent strange males from attempting to gain access to a paired female. However if the newcomer succeeds, he can sometimes separate a female from her mate for several hours (Pankhurst, 1998). It is not surprising, therefore, that male maras are the target of female enurination in the majority of cases. The slight but insignificant increase in enurination frequency during oestrus may be similarly explained; it is during oestrus that female maras are likely to receive the highest number of possibly unwelcome approaches from males.

It could be argued that, since the majority of enurination targets are male maras, females do spray to advertise sexual receptivity. Maras are socially monogamous, but there is some genetic evidence to show that they do not always mate within their pair-bond (Pankhurst, 1998). The data showing that females target strange males as often as their established mates could be interpreted as an indication that females are receptive to extra-pair copulations. Determining the context of female enurination – the events that precede it and the response from the targets – should increase clarification. We are currently working to obtain the necessary data and early results indicate that female enurination occurs in a variety of circumstances, often when maras are feeding in large groups at spatially concentrated supplementary food sources. Furthermore, enurination appears to result in decreased proximity between the target and the spraying female, rather than the reverse.

We suggest that enurination by female maras is used primarily as a deterrent behaviour, serving to repulse unwelcome advances, usually by males or pups. This is largely in agreement with Dubost & Genest (1974) who observed that enurination by female maras caused the recipient to draw back, and have suggested that female enurination actually serves to indicate a non-receptive state.

## 5. ACKNOWLEDGEMENTS

Grateful thanks to the Zoological Society of London and Whipsnade Wild Animal Park for their help and cooperation with this study.

## 6. REFERENCES

Dubost, G. and Genest, H., 1974, Le comportement social d'une colonie de maras *Dolichotis patagonum* Z. dans le Parc de Branféré, *Z. Tierpsychol.* 35:225-302.

Pankhurst, S. J., 1998, The social organisation of the mara at Whipsnade Wild Animal Park, Ph.D. thesis, Cambridge University.

Taber, A., and Macdonald, D. W., 1984, Scent dispensing papillae and associated behaviour of the mara, *J. Zool. Lond.* **203**:298-301.

Zimmerman, E., 1758, Geographische Geschichte des Menschen und der vierfüssigen Thiere. Bd. 2, Leipzig.

# THE EVOLUTION OF PERFUME-BLENDING AND WING SACS IN EMBALLONURID BATS

Christian C. Voigt[*]

## 1. INTRODUCTION

Although bats comprise the second largest mammalian group and despite the fact that bats have diverse organs for the display of chemical signals (Bloss, 2003), studies on olfaction in bats are scarce. Some members of the family Emballonuridae are remarkable among bats in having highly complex, sexual dimorphic organs for the storage and display of scents. Emballonurid bats include the Old World sheath-tailed and tomb bats and the New World sac-winged bats (Voigt in press). Among the New World members, only males of the genus *Saccopteryx*, *Balantiopteryx*, *Peropteryx*, and *Cormura* have a sac in each of their front wing membrane (= antebrachium) and these sacs usually contain odoriferous liquids (Starck, 1958; Quay, 1970; Bradbury and Emmons, 1974; Voigt and von Helversen, 1998; Scully et al., 1999). Position and size of the wing sacs vary between the males of different species; females have only rudiments of this organ and do not use it in a behavioural context.

## 2. WING SAC MORPHOLOGY IN EMBALLONURID BATS

Here, I describe the wing sac morphology of five species of Emballonuridae, namely *Saccopteryx bilineata*, *Saccopteryx leptura*, *Balantiopteryx plicata*, *Peropteryx macrotis*, and *Cormura brevirostris*, and discuss possible phylogenetic relationships and origins.

Male *Saccopteryx bilineata* weigh about 7.5 g and have relatively large wing sacs of 8 to 10 mm length that protrude the antebrachium ventrally (Figure 1). The opening is formed as two lips: a lateral lip adjacent to the forearm and a medial lip at the thoracal side. In adult males, the sac interior lacks pigments and the membrane is thicker than the adjacent antebrachial membrane. In males that are less than a year old, the interior of the

---

[*] Christian C. Voigt, Leibniz-Institute for Zoo and Wildlife Research, Alfred-Kowalke Str. 17, 10315 Berlin, Germany (voigt@izw-berlin.de)

sac is usually brown and the epithelium is as thick as the wing membrane. The interior of the sac is structured by a large fold of the medial lips and 8 to 14 smaller folds at both sides of the sac. Two muscle ligaments attach to the sac opening: ligament 1 extends from the ventral lip to the thorax and functions as an opener of the sac and ligament 2 extends from the distal tip of the sac to the wrist and functions as a closure of the sac. The scent of male *S. bilineata* has been described as sweetish, with a touch of bitter almond. Despite the presence of scents, neither Starck (1958) nor Scully et al. (1999) could detect any glandular tissue in the sac interior. Starck postulated that the scents of the wing sacs are either produced by the cornified epithelial layers or that they serve as holding or storage organs for some external scents.

Wing sacs of male *Saccopteryx leptura* are almost identical to those of *S. bilineata*. Male *S. leptura* weigh about 5 to 6 g and have relatively smaller wing sacs than males of *S. bilineata* (Figure 2A) (length of sac opening: 5 to 7 mm). Within the genus *Saccopteryx*, the overall structure of the wing sacs seems to be similar. As in the previous species, the wing sac lays parallel to the forearm, the inner epithelium lacks pigments, one major fold and several minor folds divide it into smaller cavities, and two ligaments are present.

Male *Balantiopteryx plicata* (Figure 2B) differ in the wing sac morphology from the genus *Saccopteryx* in several aspects: the wing sacs are not attached to the forearm, the opening and also the cavity is relatively small, ligaments are not present, and the epithelium of the wing sac is almost as thin as the wing membrane. In addition, the interior of the wing sac is not structured by folds. The opening of the wing sacs is about 3 to 5 mm long. Wing sacs of male *B. plicata* caught in July in Northern Costa Rica did not contain any odoriferous liquids (pers. observation). Males weighed 4 to 6 g.

The wing sacs of the following two emballonurid species differ from those of the previous species in several basic features: the anterior tips of the wing sacs touch the front edges of the antebrachium and they are flat without a cavity. The 7 to 11 g heavy males of *Cormura brevirostris* have large wing sacs that extend almost over the whole antebrachium from the anterior edge to almost the upper arm (approximately 10 mm length, Figure 2C). The inner epithelium is white and the scent has been described as pleasant and sweetish. The wing sacs of the 4 to 7 g heavy males of *Peropteryx macrotis* are similar to those of *C. brevirostris* in respect to the distal tip of the wing sacs touching the edge of the antebrachium (Figure 2D). However, the wing sacs of *Peropteryx* are much smaller than those of *Cormura* and reach only halfway through the antebrachium. The inner epithelium of the wing sac is lighter than the wing membrane in *Peropteryx* but not as distinct white as in *Cormura*.

The phylogeny of the family Emballonuridae was studied using chromosomal banding (Hood and Baker, 1986), protein electrophoresis (Robbins and Sarich, 1988), and hyoid morphology (Griffiths and Smith, 1991). These phylogenetic analyses are currently based on a few selected taxa and therefore too coarse to permit a reconstruction of the evolution of wing sacs.

Based on wing sac morphology it is likely that *Saccopteryx* and *Balantiopteryx* shared a common ancestor with wing sacs forming a cavity as the synapomorphic trait. In contrast to these two taxa, *Cormura* and *Peropteryx* have flat wing sacs. The numerous folds found in the wing sacs of *Saccopteryx* are probably an autapomorphic trait for this genus. According to Griffiths and Smith (1991) *Saccopteryx* and *Cormura* are more closely related to each other than to *Balantiopteryx* and *Peropteryx*. This conclusion is not supported by wing sac morphology.

PERFUME-BLENDING IN EMBALLONURID BATS 95

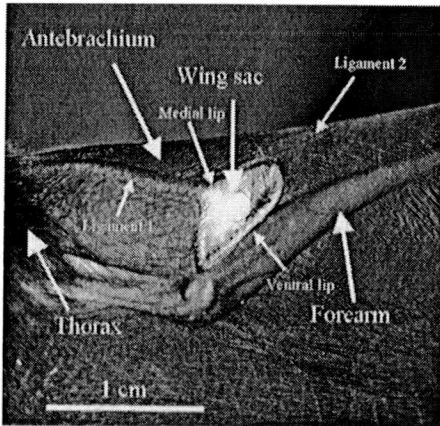

**Figure 1.** Dorsal view of a wing sac from a male *Saccopteryx bilineata*. The thorax of the bat is at the left side of the picture and the wing is fully extended.

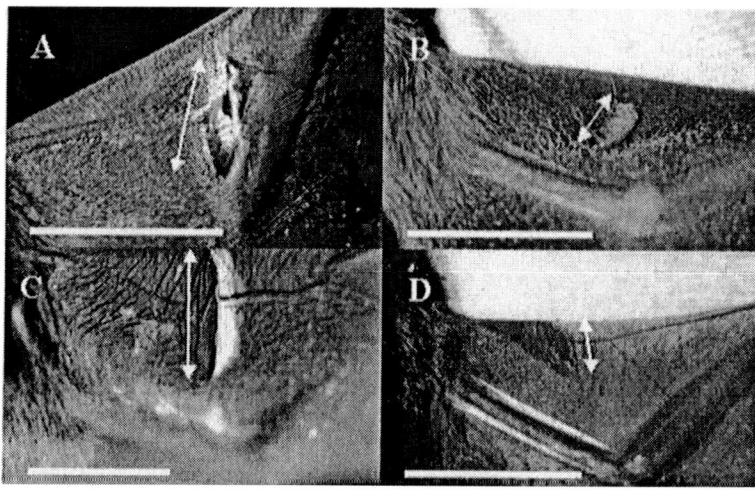

**Figure 2.** Dorsal view of wing sacs of four emballonurid species: *Saccopteryx leptura* (A), *Balantiopteryx plicata* (B), *Cormura brevirostris* (C), and *Peropteryx macrotis* (D) (the thick white line indicates 1 cm and the thin line with the arrow heads the length and position of the sac opening). In all pictures, the position of the bats' thorax, wings, forearms, and sacs is similar to that of Figure 1. In contrast to *Balantiopteryx* and *Saccopteryx*, wing sacs of *Peropteryx* and *Cormura* touch the front edge of the antebrachium. *S. leptura* and *C. brevirostrirs* were caught at the biological stations "La Selva" (Costa Rica, province Heredia), *B. plicata* at "Palo Verde" (Costa Rica, province Guanacaste), and *Peropteryx macrotis* at "Tiputini Biodiversity Station" (Ecuador, province Orellano).

## 3. THE BLENDING OF PERFUME IN THE GREATER SAC-WINGED BAT

Although several emballonurid species possess wing sacs with an odoriferous content, only *Saccopteryx bilineata* has been studied in relation to the origin of the scent. Earlier histological studies revealed that wing sacs of *Saccopteryx* do not contain any glandular tissue (Starck, 1958; Scully et al., 1999). Instead of releasing scent secretions from epithelial cells directly into the sac, male *S. bilineata* transfer fluids from other body regions into them. Thus, wing sacs could also be called holding sacs or perfume containers (Starck, 1958; Scully et al., 1999; Voigt et al., in press).

Each day, male sac-winged bats perform a complex perfume-blending behaviour that lasts on average 31 minutes (max. 1 h). The behavioural sequence is divided into two distinct phases. Firstly, males take up urine orally and lick the sac interior extensively and afterwards males transfer secretions of the genital and gular region into the wing sacs.

During the mating season, phase 1 starts on average at 1552 hours and lasts 6.8 min. During this time, males bend towards the genital region on average 6 times for the oral uptake of urine. After each bending movements males lick the interior of the wing sacs extensively. At the end of phase 1, males rest for a few minutes.

Phase 2 lasts on average 21.5 min and ends at 1623 hours (Voigt, 2002). During phase 2, males bend towards the genital region and press their chin against the penis (Figure 3A). Males persist in this position for a fraction of a second and then bend upwards again to their normal roosting posture. With this movement, males take up a white droplet from the genital region. This droplet is smeared into one of the wing sacs with a sideward movement of the head (Figure 3B). Afterwards, males rest for a few seconds and then transfer a second droplet from the gular gland into the wing sac (Voigt, 2002). Ejaculate was not found in the wing sacs. Therefore, the droplet from the genital region probably originates from a preputial gland or some accessory gland. The behavioural sequence is repeated several times during phase 2 and during subsequent filling movements males switch between the right and left sac. On average, males transfer 11 droplets from the genital region into the wing sacs (Voigt, 2002).

Perfume-blending behaviour has not been observed in any other emballonurid species as yet. However, as (1) histological examinations indicate the absence of glandular tissue in the wing sac also in other species and as (2) odoriferous liquids are present at least in some of these species, it is very likely that other emballonurid bats perform a behaviour similar to that of *Saccopteryx bilineata*.

## 4. THE USE OF WING SACS FOR SCENT DISPLAYS IN THE GREATER SAC-WINGED BAT

The greater sac winged bat (*Saccopteryx bilineata*) has the most complex wing sacs among emballonurid bats (Starck, 1958; Scully et al., 1999) and is the species where we know most about its mating system and olfactory communication (Voigt et al., in press). *Saccopteryx bilineata* has a broad geographical distribution from northern Argentina to southern Mexico. Daytime roosts are located in well-lit portions of tree cavities, in buttress cavities of large rain forest trees and in abandoned buildings. *S. bilineata* roosts on vertical surfaces, supporting its body with the thumbs of the folded forearms. Individuals of a harem maintain minimum distances of 5 to 10 cm to each other. Violations against this spacing rule are punished by attacks with the folded forearm.

**Figure 3.** Male *Saccopteryx bilineata* transferring a droplet of secretion from the genital region into the right wing sac during phase 2. First, the male presses his chin against the penis (A) and then smears the droplet with a sideward movement of the head into the slightly opened right wing sac (B).

Colonies may count up to 60 individuals and are subdivided into smaller units, each including a single adult male and several females (Bradbury and Emmons, 1974; Bradbury and Vehrencamp, 1976, 1977). For reasons of simplicity, these social subunits are called harems in the remainder of this paper and the male defending the unit the harem holder. In the neighbourhood of harems, non-territorial males queue for the access to territories (Voigt and Streich, 2003).

Recent work of our group has focused on the mating system, especially on questions of reproductive tactics and mate choice (summarized in Voigt et al., in press). We have mainly worked with colonies that roost close to the biological station "La Selva" (Costa Rica, province Heredia) inside abandoned buildings. Using genetic methods, it was shown that harem males are not able to monopolize females within their territory: on average 70% of the offspring are not sired by the harem holder. This high percentage of extra-harem paternities indicates a high potential for free mate selection in the mating system of *S. bilineata*. Despite this high percentage of extra-harem paternities it pays for harem holders to defend a territory because they sire on average more offspring than non-territorial males (Heckel and von Helversen, 2002). In addition, reproductive success of harem males increases with increasing harem size (Heckel and von Helversen, 2003). Our behavioural observations indicate that harem holders use multiple signals for courtship and territory defence.

Whereas physical contests between males are rare, males predominantly use territorial songs and olfactory signals to mark territory boundaries (Voigt et al., in press). Gular gland secretions are used for marking of territory boundaries, especially in the morning or late afternoon. Males transfer the secretions of the sebaceous gular gland to the substrate at harem boundaries by rubbing their chin against the substrate. This is done by quick sideward movements of the head. In addition, wing sacs are also used for transferring scents across territory boundaries. Bradbury and Emmons (1974) called this behaviour salting because it resembles the shaking movement that is performed when people salt a large piece of food. Salting *Saccopteryx* males shake the folded forearm into the direction of the receiver with the wing sacs opened.

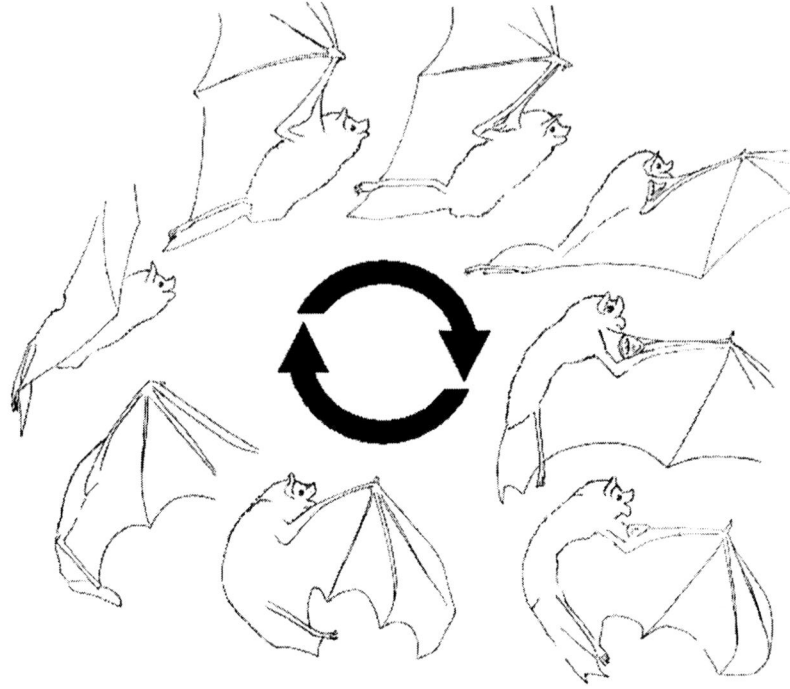

Figure 4. Wing stroke cycle of a scent-fanning male (each picture separated by 40 ms). During scent-fanning, both wings are pushed forward (instead of downward) and the tail membrane is pushed upwards towards the female (not shown in this picture sequence). Both movements probably direct the air towards the roosting female. During scent-fanning, males are at an almost constant distance to the roosting female (ca. 15 cm). Whereas normal wing beat cycles last only ca. 200 ms, cycles with scent-fanning last ca. 320 ms.

Harem males use courtship songs and scent displays to attract females to their harem (Voigt et al., in press). The wing sacs of male *Saccopteryx* play a key role during aerial courtship displays. Mostly at dawn and dusk harem males hover in front of females when the latter return to their roosting site (Figure 4). These hovering flights can last up to 14 seconds. The normal hovering wing beat cycle is interrupted at approximately every 1.4 s to push the scent of the wing sacs towards the females (Voigt and von Helversen, 1998). During fanning movements, males push the stretched wings towards the recipient female instead of downwards. Thus, hovering males lose buoyancy during each fanning movement. To continue aerial scent-fanning, a male has to regain its original position in front of the female during subsequent wing beat cycles.

## 5. WING SAC MORPHOLOGY AND MATING SYSTEMS

Both the specific morphology of the wing sacs in the antebrachium (Starck, 1958; Scully et al., 1999) and the fragrances differ between species (pers. observation). Most likely the holding sacs are homologous organs with common characters, such as the location in the antebrachium, the cornified epithelium and the lack of secreting cells (Starck, 1958; Scully et al., 1999).

The specific morphology and location of olfactory organs may on the other hand present an adaptation to the particular needs of a species' social system. Males of *Balantiopteryx* and *Saccopteryx* for example have holding sacs with a cavity located either in the centre of the antebrachium or parallel to the forearm. Both taxa are polygynous (Bradbury and Emmons, 1974) and in both taxa males perform aerial displays; hovering flights in *Saccopteryx bilineata* and mating swarms in *Balantiopteryx plicata* during which several males and females fly around landmarks as a group (Bradbury and Vehrencamp, 1976). Possibly, the location of the antebrachial sac in *Saccopteryx* and *Balantiopteryx* facilitates the emission of fragrances during flight.

By contrast, *Peropteryx* and *Cormura* possess holding sacs that are flat and reach the front edge of the antebrachium. Aerial displays have not been observed in these taxa. Male *Peropteryx kappleri* (Bradbury and Vehrencamp, 1976) and *Cormura brevirostris* (Voigt, unpublished observation) guard single females at least occasionally by roosting on top of them. Thus far, behavioural observations indicate that emballonurid species with flat holding sacs roost in smaller groups with a presumably lower potential for polygyny.

## 6. THE EVOLUTION OF PERFUME-BLENDING AND WING SACS IN EMBALLONURID BATS

The wing sacs of emballonurid bats are unique within the order Chiroptera. In bats, scents are most often used for scent-marking of either territories or other individuals. Male bats for example frequently mark their potential mating partners by rubbing secretions of chest or gular glands onto them (*Molossus molossus* – Häussler, 1987). Such a marking behaviour is impossible at least in *Saccopteryx bilineata*, as individuals defend a minimum distance to their neighbour. Males are also smaller than females and therefore harem males cannot enforce the deposition of scent-marks on females. The only alternative for *Saccopteryx* males is to display scents from a distance.

Gular gland secretions, which are part of the sac perfume, are used by male *Saccopteryx* for the scent-marking of their territory. As these secretions are also involved during perfume-blending and hence during courtship display, it is likely that some compounds of the sac perfume indicate harem ownership. Possibly an ancestral form of *Saccopteryx* deposited gular gland secretions on its unstructured antebrachium. The antebrachium is easy to reach with the gular gland and it probably facilitated the emission of scents during normal flight or aerial displays. Sexual selection may have further modified the behaviour of ancestral males and thus also the specific morphology of the antebrachium. Several other advantages may have accelerated the evolution of wing sac morphologies: (1) wing sacs facilitate the transfer of scents during aerial displays while minimizing the weight associated with an actual gland (Scully et al., 1999), (2) closed wing sacs minimize the evaporation of volatiles, and (3) via ligaments wing sacs enable a neuromuscular-controlled release of scent whose production may be under hormonal control.

Considering the fact that male *S. bilineata* spend up to an hour each day of their adult lives cleaning, moistening and refilling of their wing sacs, it appears that the wing sacs are an important sex-specific organ for communication in *S. bilineata* and possibly also in other Emballonuridae.

## 7. ACKNOWLEDGMENTS

This study was financed by grants from the German National Science Foundation (DFG; VO 890/6-1) and from the German Academic Exchange Program (DAAD).

## 8. REFERENCES

Bloss, J. M., 1999, Olfaction and the use of chemical signals in bats, *Acta Chiropt.*, **1**:31.
Bradbury, J. W., and Emmons, L., 1974, Social organization of some trinidad bats, I Emballonuridae, *Z. Tierpsych.* **36**:137.
Bradbury, J. W., and Vehrencamp, S. L., 1976, Social organization and foraging in emballonurid bats I. Field studies, *Behav. Ecol. Sociobiol.* **1**:337.
Bradbury, J. W., and Vehrencamp, S. L., 1977, Social organization and foraging in emballonurid bats II. Mating systems, *Behav. Ecol. Sociobiol.* **2**:1.
Griffiths, T. A., and Smith, A. L., 1991, Systematics of emballonurid bats (Chiroptera: Emballonuridae and Rhinopompatidae), based on hyoid morphology, *Bull. American Mus. Nat. Hist.* **206**:62.
Häussler, U., 1987, Male social behaviour in *Molossus molossus* (Molossidae), in: *European Bat Research*, V. Haňák, I. Horáček, and J. Gaisler, eds., Charles University Press, Prague, pp. 125-130.
Heckel, G., and von Helversen, O., 2002, Male tactics and reproductive success in the harem polgygynous bat *Saccopteryx bilineata*, *Behav. Ecol.* **13**:750.
Heckel, G., and Helversen, O. von, 2003, Genetic mating system and the significance of harem association in the bat *Saccopteryx bilineata*, *Mol. Ecol.* **12**:219.
Heckel, G., Voigt, C. C., Mayer, F., and von Helversen O., 1999, Extra-harem paternity in the white-lined bat *Saccopteryx bilineata* (Emballonuridae), *Behaviour* **136**:1173.
Hood, C. S., and Baker, R. J., 1986, G- and C-banding chromosomal studies of bats of the family Emballonuridae, *J. Mammal.* **67**:705.
Quay, W. B., 1970, Integument and derivatives, in: *Biology of Bats*, W. A. Wimsatt, ed., Academic Press, New York, pp. 1-56.
Robbins, L. W., and Sarich, V. M., 1988, Evolutionary relationships in the family Emballonuridae (Chiroptera), *J. Mammal.* **69**:1.
Scully, W. M, Fenton, M. B., and Saleuddin, A. S. M., 1999, A histological examination of holding sacs and scent glandular organs of some bats (Emballonuridae, Hipposideridae, Phyllostomidae, Vespertilionidae and Molossidae*), Can. J. Zool.* **78**:613.
Starck, D., 1958, Beitrag zur Kenntnis der Armtaschen und anderer Hautdrüsenorgane von *Saccopteryx bilineata* Temminck 1838 (Chiroptera, Emballonuridae), *Gegenbaur Morphol. Jahrb.* **99**:3.
Voigt, C. C., 2002, Individual variation of perfume-blending in male sac-winged bats, *Anim. Behav.* **63**:907.
Voigt, C. C., in press, Sac-winged bats, sheath-tailed bats and ghost bats (Emballonuridae). *Grizmek's Life Encyclopedia: Mammals*, Gale Group.
Voigt, C. C., and Streich, J. W. 2003. Queuing for harem ownership in colonies of the sac-winged bat. *Anim. Behav.* **65**:149.
Voigt, C. C. and Helversen, O. von, 1999, Storage and display of odor by male *Saccopteryx bilineata* (Chiroptera; Emballonuridae), *Behav. Ecol. Sociobiol.*, **47**:29.
Voigt, C. C., Heckel, G., Helversen, O. von, in press, Conflicts and strategies in the mating system of the sac-winged bat, in: *Functional and Evolutionary Ecology of Bats*, G. McCracken, Zubaid, and T. H. Kunz, eds., Oxford University Press, Oxford.

# BEHAVIORAL RESPONSIVENESS OF CAPTIVE GIANT PANDAS (*AILUROPODA MELANOLEUCA*) TO SUBSTRATE ODORS FROM CONSPECIFICS OF THE OPPOSITE SEX

Dingzhen Liu, Guiquan Zhang, Rongping Wei, Hemin Zhang, Jiming Fang, Ruyong Sun[*]

## 1. INTRODUCTION

Chemical signals are supposed to play major roles for the mammalian social and reproductive behavior. Conover and Gittleman (1989) have classified scent-markings' potential functions into six categories: 1) identity--information regarding individual, group, or sexual identification (Rasa, 1973; Gorman, 1976); 2) status--information on dominance status (MacDonald, 1979; Erlinge et al., 1982); 3) reproductive status--related to courtship and breeding (Gorman, 1980; Gorman and Trowbridge, 1989); 4) spatial information--serves a territorial function (Peters and Mech, 1975; Kruuk, 1978); 5) temporal information--reveals when individuals are at particular locations (Rasa, 1973); and 6) foraging--serves a bookkeeping function by informing an individual of whether it has previously looked for food in a particular area (Henry, 1977; MacDonald, 1979). Indeed, there is a wealth of evidence that the anal gland secretion contains information concerning individual identity, such as gender, age and family (for beavers, Sun and Müller-Schwarze, 1998a,b; for giant pandas, Hagey and MacDonald, 2003; Yuan et al., 2003; for wolves, Raymer et al., 1984; for lions, Andersen and Vulpius, 1999). Secretions of the anogenital gland in the giant panda (*Ailuropoda melanoleuca*) contain information about male pandas' sexual ability and sexual performance (Liu et al., 2003).

The giant panda is a rare animal that inhabits and feeds on dense bamboo forests among isolated mountains in Sichuan, Shananxi and Gansu provinces in China (Hu et al., 1985). Captive giant pandas had a low rate of 30% breeding success before 1991

---

[*] Ministry of Education Key Laboratory for Biodiversity Sciences and Ecological Engineering & Institute of Ecology, Beijing Normal University, Beijing 100875. Q. Zhang, R. Wei and H. Zhang, China Conservation and Research Center for the Giant Panda, Wolong 623006.

(Wolong Nature Reserve and China Conservation and Research Center for the Giant Panda, 1993). One reason is the unsuccessful technique for hand raising the abandoned cub by unsuccessful mother or mother of the twin; another reason is the low rate of successful mating and pregnancy. More than two-thirds of males are poor in sexual performance and about one-third of females do not show typical estrous behavior in captivity (Zhang et al., 1994). Giant pandas are typically solitary and only get together during mating season. Olfaction is believed to play an important role in giant pandas' communication, especially during the mating season. Both male and female pandas recognize individual odors in anogenital gland secretions, and the gender and reproductive conditions have significant effects on discrimination tests (Swaisgood et al., 1999, 2000). Yet little is known about their behavioral responsiveness to substrate odors from conspecifics of the opposite sex, the effects of sexual experiences of females on their behavioral responsiveness, and mating success at the time of being first paired with males. To answer these questions, a behavioral experiment on 11 captive sub-adult and adult giant pandas was conducted during the mating and non-mating season.

## 2. METHODS

### 2.1 Subjects and Study Site

We conducted this study at the China Conservation and Research Center at Wolong, Sichuan province, China during the mating season (March – June) and non-mating season (August-November) of 1995, using two males and nine females at least 4.5 years of age. The subjects were adult and sub-adult giant pandas. Table 1 shows the numbers of subjects in our study. They were housed individually in cages. A cage contained a night pen (5.8 x 2.3 m) and an outdoor yard (5.8 x 13 m) with grass, climbing apparatuses, and a small pond as water source. Each outdoor enclosure adjoined two others via a cement wall on which there was a small wire mesh fence door (1 x 1m). Therefore, the subjects could see, smell, hear, and even have some limited physical contact through the mesh fence with neighboring animals of the opposite sex. Based on their reproductive records, all females were divided into sexually experienced and inexperienced groups. Those females who had been paired with males or gave birth in the past were grouped into the experienced group. Otherwise they would be grouped into the sexually inexperienced group. There were two sexual experienced females and seven sexual inexperienced females in the mating season of 1995. Besides, females and males housed as neighbors were thought to be familiar. Otherwise they were thought to be strange to each other.

### 2.2 Experimental Design, Behavioral Observation, Recording, and Analysis

The subjects were exposed to the opposite sex's substrate odor for 30 minutes. The exposure experiment was repeated twice a week. Methods of focal sampling and continuously recording were used throughout the study. Pandas' behaviors were observed and recorded by mini-tape cassette recorder. Pandas' behavior in their own cages was used as the control. The data were then transferred into computer by OBSERVER 3.0 (Noldus Company).

Table 1. Subjects used in the study.

| Observation stages | Sub-adults | | Adults | |
|---|---|---|---|---|
| | Female | Male | Female | Male |
| Mating season | 3 | | 6 | 2 |
| Non-mating season | 3 | | | 2 |

The durations of the following behavioral categories were recorded: scent-marking - rubbing the anogenital area around or up and down on the surface of an object or on the wall; ingesting, handling and eating steamed bread, apples and grass and drinking water and milk; investigating - investigating an object with a distance of 0.1m or longer between the end of nose and the object; sniffing - investigating an object with the distance of less than 0.1 m between the end of nose and the object and with a response of Flehmen; grooming - scratching and licking of the pelage; resting – inactive behaviors including sleeping and non-sleeping; urinating/defecating – urinating in a squat, leg-cock, handstand, or standing posture on the wall or ground and/or defecating at the meantime; locomotion - rapidly pacing and moving around the enclosure without placing feet in the same position each time and following the same path; and playing - rolling and somersaulting with manipulation of objects, such as food dishes, bamboo stalks, tree-branches or toys provided by the keeper.

Data distribution was examined by *One-Sample Kolmogorov-Smirnov test*, and transformed by square root to fit the normal distribution (Sokal and Rohlf, 1995). The behavioral comparisons between the control and experiment were analyzed by Wilcoxon Rank test, and comparisons between sexual experience and odor effects on pandas' behavioral responsiveness were analyzed by two-way ANOVA. Finally, the mating duration of familiar mates and strange mates were analyzed by Mann-Whitney U test. Significance level was 0.05.

## 3. RESULTS

During the non-mating season, we found both the male and female pandas exposed to the opposite sex substrate odor showed similar behavioral patterns in scent-marking, ingesting, investigating, sniffing behavior, but different patterns in locomotion, grooming, playing, resting and urinating/defecating. Females spend more time playing and resting instead of urinating or defecating in their own cages, while males spend more time in locomotion and grooming. We found no significant effects of substrate odor both on male and female pandas' behavioral responsiveness (Figure 1).

During the mating season, however, the females exposed to males substrate odor spent significantly more time locomotion, scent-marking and sniffing ($P<0.05$, $P<0.01$ and $P<0.05$, respectively), and remarkably less time resting than when they were in their own cages ($P<0.05$). Females also spent less time feeding and drinking when exposed to the male's substrate odor, yet the difference did not reach the statistical level ($P=0.069$). Besides, females also communicated with males more vocally. They showed significantly less chomping (0.07±0.07 vs. 1.28±1.14 times/30 minutes, $P<0.05$), and more bleating

(29.48±12.62 vs. 25.72±14.22 times/30 minutes) and chirping (23.06±7.69 vs. 9.06±3.89 times/30 minutes) which are indicative of courtship, being receptive and invitation to mating. However, the differences of the last two vocal behaviors did not reach statistical level (both $P>0.05$). Males' substrate odor has significant effects on females' behavior. Males, on the other hand, did not show any significant changes in behavior when they were exposed to females' substrate odor (Figure 2).

Females' responsiveness to males' substrate odor was greatly affected by the females' sexual experience (Table 2). Both the sexually experienced and inexperienced females exposed to the male substrate odor showed significantly more locomoting, whereas the sexually inexperienced females spent less time in locomoting and more time

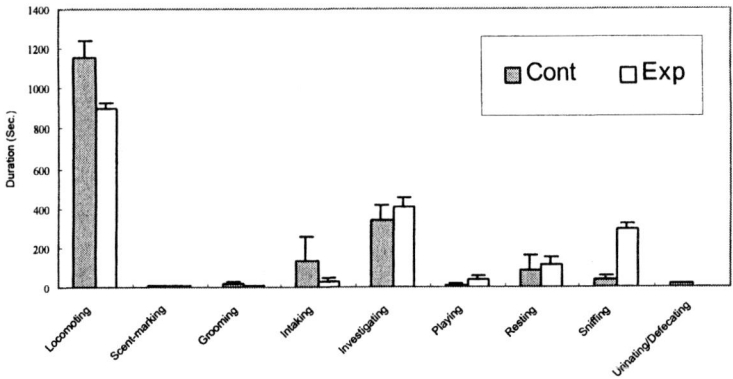

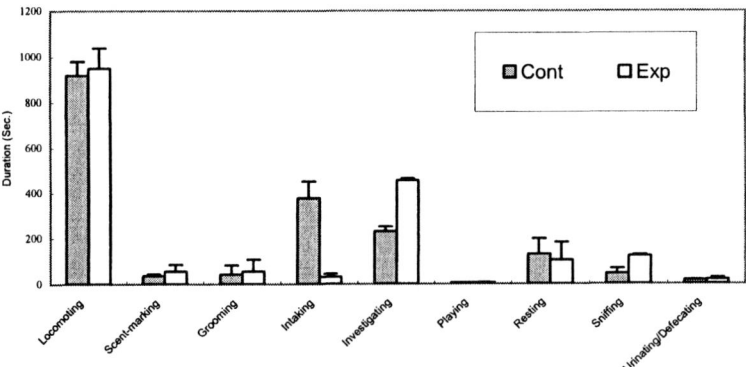

Figure 1. Behavioral responsiveness of female (upper panel) and male (lower panel) pandas to the substrate odor of the opposite sex (exp) and controls (cont) during the non-mating season.

investigating, resting and sniffing than sexually experienced females. Moreover, the sexually inexperienced females also uttered more chirping ($P=0.07$) while the experienced females displayed more chomping ($P<0.05$). We also found an interaction between sexual experience and odor in locomoting, chomping and investigating behavior. These results suggest that females' scent marking during mating season might be mainly to male pandas for advertising their state of estrus, and the happening of chomping behavior might indicate the females' sexual experiences and sexual ability.

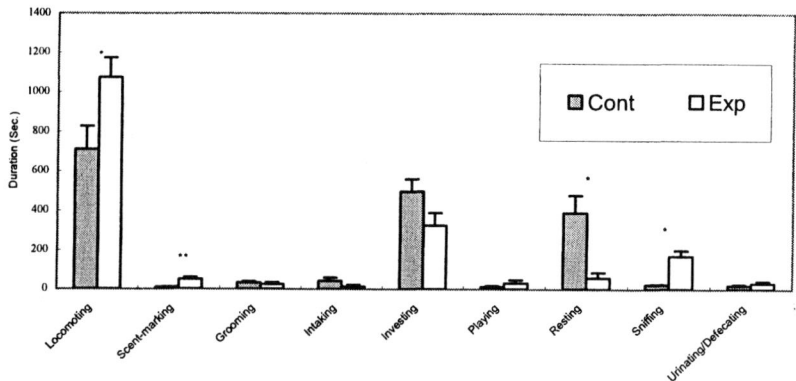

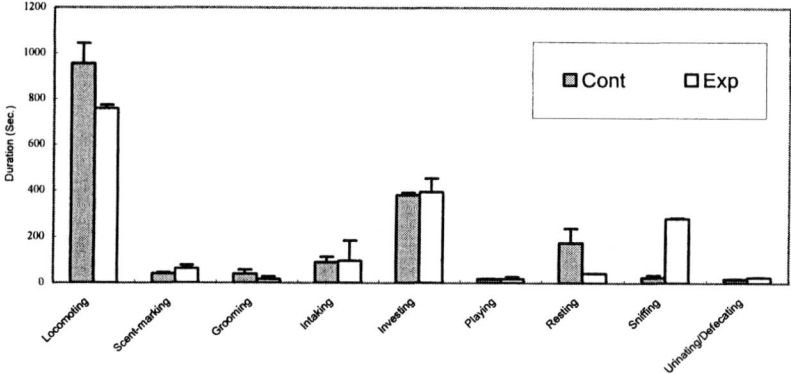

**Figure 2.** Behavioral responsiveness of female (upper panel) and male (lower panel) pandas to the substrate odor of the opposite sex during the mating season.

There were four females in the mating season that exhibited receptive state and paired with male pandas. We found that only one of the four females could naturally mate with a male at the time when she was paired with a male for the first time in 1995, and the mating lasted 998 seconds after intromission. She gave birth to twins finally in the fall. Surprisingly, we found that the average mating duration after intromission for familiar pairs is 517.00±312.00 seconds (n=2), while the average mating duration for strange pairs is 307.75±112.25 seconds (n=2) ($P>0.05$, Mann-Whitney $U$ test). The percentage of natural mating success for familiar pairs is 60%, while it is 42.85% for strange pairs. The communication between the male and female pandas by neighboring or pre-experiencing before being paired might have important impact on the success of natural mating in captive giant pandas.

Table 2. The effects of sexual experience and odor on behavioral responsiveness of female pandas during the mating season.

| Behavioral Measure | Sexually inexperienced (N=7) | | Sexually experienced (N=2) | | SE[a] | OE[b] | SE-OE |
|---|---|---|---|---|---|---|---|
| | Cont. | Exp. | Cont. | Exp. | p | p | p |
| Locomotion | 765.45 ± 145.07[c] | 953.88 ± 76.07 | 516.33 ± 12.63 | 1492.05 ± 110.81 | ns | ** | * |
| Scent-Mark | 9.47 ± 2.56 | 53.95 ± 12.46 | 8.39±1.85 | 31.64 ± 19.69 | ns | * | ns |
| Bleat[d] | 29.24 ± 18.35 | 35.47 ± 15.58 | 13.39 ± 3.39 | 8.50 ± 8.50 | ns | ns | ns |
| Chirp[d] | 10.58 ± 4.86 | 29.57 ± 8.34 | 3.71 ± 3.71 | 0.25 ± 0.25 | ns | ns | ns |
| Chomp[d] | 0.04 ± 0.02 | 0 | 5.62 ± 4.75 | 0.30 ± 0.30 | * | * | * |
| Groom | 26.28 ± 4.69 | 29.13 ± 11.82 | 51.15 ± 22.54 | 4.30 ± 4.30 | ns | ns | ns |
| Intake | 41.39 ± 18.45 | 14.39 ± 12.78 | 41.87 ± 40.44 | 1.38 ± 1.38 | ns | ns | ns |
| Investigate | 451.69 ± 66.48 | 378.41 ± 67.97 | 654.31 ± 129.28 | 135.94 ± 112.89 | ns | * | * |
| Play | 15.13 ± 5.96 | 38.81 ± 19.55 | 0.02 ± 0.02 | 1.65 ± 0.41 | ns | ns | ns |
| Rest | 388.03 ± 110.35 | 74.79 ± 33.15 | 400.67 ± 146.40 | 0.00 ± 0.00 | ns | ** | ns |
| Sniff | 24.82 ± 3.00 | 193.94 ± 34.55 | 14.06 ± 4.17 | 81.35 ± 16.90 | ns | ** | ns |
| Urinate/Defecate | 20.83 ± 6.09 | 24.08 ± 10.42 | 17.20 ± 1.80 | 45.59 ± 46.17 | ns | ns | ns |

Note: [a] Mean±SE
[b] OE: Odor Exposure Experiment
[c] SE: Sexual Experience
[d] Behavioral categories measured by frequencies (times per 30minutes)
(ns---- not significant, *--- $p<0.05$, **-- $p<0.01$)

## 4. DISCUSSION

Chemical communication plays major roles in mammalian social life and reproduction (Ralls, 1971). In many species, males can detect the reproductive status of females either by urine or scent-marks, and promote a synchrony of estrus and therefore facilitating a successful mating (American bison, *Bison bison*, see Berger et al., 1992; golden hamster, *Mesocricetus auratus* see Tang-Martinez et al., 1993; house mice, *Mus musticulous*, see Sipos et al., 1995). Giant pandas also use scent-markings to maintain social contacts and territorial demarcation in the wild (Schaller et al., 1985). The males have a complex of four distinct postures to deposit scent-marks all year round, while the females leave scent-marks frequently only during mating season (Kleiman, 1985; Schaller et al., 1985; Liu, 1996). The composition of those scent-marks deposited on the ground was identical to that of anogenital glands, and the scent-marks contain information about gender, age group and individual membership (Hagey and MacDonald, 2003; Liu et al., 2003). The females' scent-marks during the mating season also contain some vaginal secretions (Liu, unpublished data). The substrate odor from captive pandas contains a combination of anogenital gland secretions, urines/vaginal and body odors. Females displayed more locomotion and scent-marking during the mating season and thus, might convey estrous information to their potential mates. More sniffing in females might be related with the discrimination of the conspecific membership, gender and age. Female pandas can discriminate between males individually by males' scent-marks (Swaisgood et al., 1999). In our study, male pandas exposed to the females' substrate odor also performed more sniffing though the differences did not reach statistical level. Moreover, we found no significant differences in any of the behavioral categories of male and females before and after being exposed to the odor of the opposite sex during the non-mating season. These results indicate that both females' reproductive condition and substrate odor have significant effects on their chemosensory responsiveness. The increase in locomotion, scent-marking and sniffing was mainly related to sexual activities and breeding, and females' scent-marking at this time is for sexual advertisement (Hudson and Voldermager, 1992). A similar result was found in the subsequent study of Swaisgood et al. (2000) at the same place and on the same subjects.

Giant pandas also use the visual and auditory channels to communicate in the wild besides chemical communication, and their behavioral patterns differ with different seasons and reproductive status (Schaller et al., 1985). For captive pandas, females exposed to the male's substrate odor also exhibited less chomping, more chirping and bleating although the differences in the later two behavioral categories were not significant. These changes in vocal behaviors showed that females' amicable behavior to substrate odor donors may be an invitation of mating. This is supported by the observation that females show the most frequent chirping and bleating when they are in the receptive state (Liu, 1996).

Unlike other mammals such as tigers and wolves, male giant pandas seldom patrol their territories and demarcate territory by depositing scent-markings. Their social dominance is determined and maintained mainly by scent-markings and occasionally fighting (Hu et al., 1985). The height of odor deposition may be associated with body size, a major determinant of competitive ability (White et al., 2002). We found no significant changes in scent-marking in male subjects exposed to females' substrate odor both in the mating season and non-mating season. One reason for this result is that the males' scent-marking behavior might be directed to other males to show their competitive ability although the females may detect the male's social status and competitive ability by scent-marks. Another reason is the small sample size for males in our experiment. The exposure of sexually inactive male pandas to the females' substrate odor was avoided because of management limits.

The giant panda is not typically sexually dimorphic but we could find many sexual differences both in morphology, ecology and behavioral ecology (Hu et al., 1985). In captivity, male and female pandas exposed to the opposite substrate odor displayed a different behavioral responsiveness in locomotion and play behavior (Figures 1, 2). Those differences might represent sexually dimorphic behavior. Other behavioral differences in chemosenory responsiveness to the substrate odor of the opposite sex may be due to seasonal effects and sexual experience. The different behavioral responsiveness between sexually experienced and inexperienced females indicates that the sexual experience or chemical communication with males may have an effect on successful natural mating. The significant behavioral responsiveness of females with sexual experiences might also be related to the females' mate choice. Both captive males and females were found to prefer certain mates during the mating season (Zhang Hemin, personal communication). A management technique for keeping pandas individually and alternatively by sex was recommended and conducted for promoting the chemical communication at Wolong Breeding Center since 1992. Both our current results and the subsequent breeding success records at Wolong show that the more pandas are allowed to communicate chemically between males and females, the larger the possibility would be for a successful natural mating and breeding in captivity.

## 5. ACKNOWLEDGEMENT

We are grateful to staff at China Conservation and Research Center for the Giant Panda for their assistance in the experiment there. We are thankful to Dr. Sue A. Mainka for her helpful suggestions for this work. The authors also appreciated Dr. Lixin Sun's efforts in improving the quality of this manuscript. This work was supported by grants from National Natural Science Foundation of China (Grant numbers are 30070107, 30170169, 30230080).

## 6. REFERENCES

Andersen, K. F., and Vulpius, T., 1999, Urinary volatile constituents of the lion, *Panthera leo*, *Chem. Senses* 24:179-189.
Berger, J., 1992, Facilitation of reproductive synchrony by gestation adjustment in gregarious mammals: a new hypothesis, *Ecology* 73:323-329.
Conover, G. K, and Gittleman, J. L., 1989, Scent-marking in captive red pandas (*Ailurus fulgens*), *Zoo Biology* 8:193-205.
Erlinge, S., Sandell, M. & Brinck, C. 1982: Scent-marking and its territorial significance in stoats, *Mustela erminea*, *Anim. Behav.* 30:811-818.
Gorman, M. L., 1976, A mechanism for individual recognition by odour in *Herpestes auropunctatus* (Carnivora: Viverridae), *Anim. Behav.* 24:141-145.
Gorman, M. L., 1980, Sweaty mongooses and other smelly carnivores, *Symp. Zool. Soc. London* 45:87-105.
Gorman, M. L., and Trowbridge, B. J., 1989, The role of odor in the social lives of carnivores, in: *Carnivore, Behavior, Ecology and Evolution*, J. L. Gittleman, ed., Cornell University Press, Ithaca, New York, pp. 55-88.

Hagey, L., and Macdonald, E, 2003, Chemical cues identify gender and individuality in giant pandas (*Ailuropoda melanoleuca*), *J. Chem. Ecol.* **29**:1479-1488.
Henry, J. D., 1977, The use of urine marking in the scavenging behavior of the red fox (*Vulpes vulpes*), *Behaviour* **61**:82-105.
Hu, J. C., Schaller, G. B., Pan, W. S. & Zhu, J., 1985, *Wolong's Giant Panda*, Sichuan Publishing House of Science and Technology, Chengdu, pp. 225.
Hudson, R. & Vodermager, T. 1992: Spontaneous and odour-induced chin marking in domestic female rabbits, *Animal Behaviour* **43**, 329-336.
Kleiman, D. G. 1985: Social and reproductive behaviors of the giant panda (*Ailuropoda melanoleuca*), Paper pres. Proc. Int. Symp. Giant Panda, Bongo.
Kruuk, H., 1978, Spatial organization and territorial behaviour of the European badger *Meles meles*, *J. Zoo.* **184**, 1-19.
Liu, D. Z., 1996, Studies on the behavioral ecology of captive giant pandas (*Ailuropoda melanoleuca*), PhD dissertation, Beijing Normal University, Beijing.
Liu, D. Z., Tian, H., Yuan, H., Wei, R. P., Zhang, G. Q., Yun, Z. H., and Sun, R. Y., 2003, Possible cues for sexual ability and age of giant pandas by scent-markings. *J. Chem. Ecol. (in press)*.
MacDonald, D. W., 1979, Some observations and field experiments on the urine marking behavior of the red fox, *Vulpes vulpes*, *Z. Tierpsychol.* **51**:1-22.
Peters, R. P., and Mech, L. D., 1975, Scent-Marking in Wolves. *Am. Sci.* **63**:628-637.
Ralls, K., 1971: Mammalian scent marking, *Science* **71**:443-449.
Rasa, O. A. E., 1973, Marking behavior and its social significance in the African dwarf mongoose *Helogale undulata rufula*. *Z. Tierpsychol.* **32**:449-488.
Raymer, J. D. W., Novotny, M., Asa, C., Seal, U. S., and Mech, L. D., 1984, Volatile constituents of wolf (*Canis canis*) urine as related to gender and season, *Experientia* **40**:707-709.
Schaller, G. B., Hu, J., Pan, W. S., and Zhu, J., 1985, *The giant pandas of Wolong.*, University of Chicago Press, Chicago, pp. 298.
Sipos, M. L., Perry, A. B., Nyby, J. G. & Vandenbergh, J. G. 1995: An ephemeral pheromone of female house mice: degradation by oxidation. *Animal Behaviour* **50**, 113-120.
Sokal, R. R., and Rohlf, J. F., 1995, *Biometry: The Principles and Practice of Statistics in Biological Research*, W. H. Freeman Co., New York, pp. 887.
Sun, L. X., and Muller-Schwarze, D., 1998a, Anal gland secretion codes for family membership in the beaver. *Behav. Ecol. Sociobiol.* **44**:199-208.
Sun, L. X., and Muller-Schwarze, D., 1998b, Anal gland secretion codes for relatedness in the beaver, *Castor canadensis*, *Ethology* **104**.
Swaisgood, R., R., Lindburg, D. G., Zhou, X., and Owen, M. A., 2000, The effects of sex, reproductive condition and context on discrimination of conspecific odours by giant pandas, *Anim. Behav.* **60**:227-237.
Swaisgood, R., R., Lindburg, D. G., and Zhou, X., 1999, Giant pandas discriminate individual differences in consepcific scent, *Anim. Behav.* **57**:1045-1053.
Tang-Martinez, Z., 2001, The mechanisms of kin discrimination and the evolution of kin recognition in vertebrates: a critical re-evaluation, *Behav. Process.* **53**:21-40.
White, A., M., Swaisgood, R. R., and Zhang, H., 2002, The highs and lows of chemical communication in giant pandas (*Ailuropoda melanoleuca*): effect of scent deposition height on signal discrimination, *Behav. Ecol. Sociobiol.* **51**:519-529.
Wolong Nature Reserve and China Conservation and Research Center for the Giant Panda, 1993, *Hand-Raising an Infant Giant Panda--A Study at the Wolong Nature Reserve*, Sichuan Publishing House of Science & Technology, Chengdu, pp. 138.
Yuan, H., Wei, R. P., Zhang, G. Q., Sun, L. X., Sun, R. Y., and Liu, D. Z., 2003, Possible coding for recognition of age and sex in anogenital gland secretions of giant panda, *Ailuropoda melanoleuca*, *Chem. Senses (in press)*.
Zhang, H. M., Zhang, K. W., Wei, R. P., and Chen, M., 1994, Studies on the reproduction of captive giant pandas and artificial den at the Wolong Nature Reserve., in: *Minutes of the International Symposium on the Protection of the giant panda.* A. J. Zhang and G. X. H. He, eds., Sichuan Publishing House of Science & Technology, Chengdu, pp. 221-225.

# CHEMICAL SIGNALS IN GIANT PANDA URINE (*AILUROPODA MELANOLEUCA*)

M. Dehnhard[*], T. Hildebrandt, T. Knauf, A. Ochs, J. Ringleb, F. Göritz

## 9. INTRODUCTION

Behavioral and non-behavioral cues regulate many aspects of social interactions related to vertebrate reproduction. Among the non-behavioral cues, visual and chemical (pheromones; Karlson and Lüscher, 1959) stimuli release specific behavioral or endocrine reactions in the recipient. Mechanisms of their release into the environment are manifold and secretion with urine represents one principle. The giant panda, an endangered species with a low reproductive rate in captivity (Hu and Wei, 1990) and an uncertain future in the wild, is a solitary mammal with only infrequent visual and vocal contact with conspecifics (Schaller et al., 1985). However, both inter- and intrasexual overlapp of territories is substantial, affording frequent opportunities for chemical communication. The panda is a mono-estrus, seasonal breeder with a brief single receptive period of 24 – 72 hours each spring (Kleiman et al., 1979; Kleiman, 1983; Schaller et al., 1985). Because of their solitary way of life, female individuals have to attract males in time. Therefore, the female panda has an active role in initiating and managing the interactions of the pair prior to mating (Kleiman et al., 1979), and marking behaviour of females occurs predominantly during the mating season (Kleiman, 1979; Schaller et al., 1985).

The endocrine events associated with the periovulatory period have been described by measuring urinary steroid hormone metabolites. The excretions of estrogens by the female increased approximately eight days prior to observed matings, was at maximum during the proceptive period, and decreased to basal levels during the period of receptivity (Bonney et al., 1982; Hodges et al., 1984). In captivity, successful reproduction depends on the ability of animal managers to detect estrus for the accurate timing of pairing and artificial insemination. For that purpose the estrogen course as well as behavioural and morphological indices were used for reproductive management.

---

[*] Martin Dehnhard, Institute for Zoo and Wildlife Research, Alfred-Kowalke-Str. 17, 10315 Berlin, Germany (dehnhard@izw-berlin.de)

Proceptive behaviour can include initiation and maintenance of proximity and establishment of contact. Scent undoubtedly plays an important role in coordination of mating, e.g. to attract males, to signal estrus, and to act as an aphrodisiac. Both sexes possess a large anogenital gland that secretes volatile components which include a series of short-chain fatty acids (C2–C6; Swaisgood et al., 1999). Urine also appears to function as a chemical signal (Swaisgood et al., 2000)

Male pandas show intensive vocalisation and marking behaviour in response to urine from estrus females (Swaisgood et al., 1999). Scent marking of females was present six days before the estrogen peak but not afterwards (McGeehan et al., 2002). Thus we hypothesised that female panda urine might possess estrus-related signals and that urinary volatiles, which might serve as pheromones in panda communication, could probably be used to evaluate ovarian function. Pheromones may represent more informative indicators of behavioural and physiological conditions than measurements of circulating or excreted hormones, which do not have a communication function. Using a combined approach of hormonal analysis and solid-phase microextraction (SPME; Arthur and Pawliszyn, 1990) in combination with GC-MS analyses of urinary volatiles, it should be possible to detect reproductive stage-related substances. The present study aimed to identify estrous related substances and assess their reliability for monitoring ovarian function.

## 2. MATERIALS AND METHODS

### 2.1 Animal and Sample Collection

One adult female (17 years) and one male giant panda (25 years) maintained at the Berlin Zoo were involved in the study. Urine samples were collected over a periof of four years (2000 – 2003). From the female panda, urine samples were collected weekly and daily during the mating period in spring. From the male panda, urine samples were obtained at irregular intervals. All samples were immediately frozen after collection and stored at -20°C until analysed.

### 2.2 Determination of Female Reproductive Status

Conjugated steroids were hydrolysed, extracted and measured with two enzyme immuno assays (EIA) specific for total estrogens and pregnanediol, respectively. Determinations of hormone concentrations were carried out as described by Meyer et al. (1997).

### 2.3 Behavioural Recording

In 2002 detailed observations of behavior during estrus were conducted. They covered a period of 14 days from 12 to 25 April. Daily recording time was 6 hours, divided into three sets, from 8:00 – 10:00, 11:00 – 13:00 and 15:00 – 17:00. Estrus behaviour referred to five behaviours (Schaller et al., 1985): masturbating with paws or sticks, licking the vulva, rubbing the vulva on the ground floor, sitting in a water pond (to bathe the vulva) and smearing with the backside on the ground.

## 2.4 Solid-phase Microextraction (SPME) Sampling

In principle, this method uses a fiber coated with an adsorbent that can extract organic compounds from the headspace above a liquid or solid sample. Extracted compounds are desorbed upon exposure of the SPME fiber in the heated injector port of a gas chromatograph (GC). SPME was carried out with a CTC Combi Pal system autoinjector at 70°C for 60 min using a fibre with an 85 µm polyacrylate coating. SPME sampling was done in the headspace above the surface of 5 ml diluted, acetate-buffered urine (containing 2 ml urine; 2.5 ml water; 0.5 ml 2 M acetate-buffer, pH 4.8; 1.83 g NaCl, and 2 µg undecanoic acid as internal standard) in 20 ml headspace vials (Shimadzu).

GC-MS determinations were conducted with a Shimadzu GCMS-QP 5050. MS acquisition was performed in TIC. The samples were analyzed using a 50 m SE-54 capillary column (0.32 i.d. and 1 µm film thickness, CS, Langerwehe, Germany). Ultrapure helium was used as carrier gas, with a column head pressure setting of 41.2 kPa. Injector temperature was 300°C; the interface temperature was maintained at 300°C and ionizing voltage was at 1.2 kV. Splitless injection mode was used; the purge valve was turned on 15 min after injection, with a split flow of 8 ml/min during the GC run. The GC oven was kept at 45°C for 2 min, increased to 105°C at 15°C min$^{-1}$, from 105 to 165°C at 10°C min and then from 165 to 290°C at 4°C min$^{-1}$. The mass spectra were identified by computer MS library research and compared with those of the authentic standards.

## 2.5 Quantification of Decanoic Acid

A calibration line was prepared by adding various amounts of decanoic acid (Sigma: 0.63, 1.25, 2.5, 5, 10 µg/ml) and a constant amount of 2 µg/ml undecanoic acid to pooled urines. The peak areas of the calibration samples were determined based on the m/e 60 fragment and the ratio was used to evaluate the biological samples. The calibration curve showed a good linearity in the range of 0 to 10 µg/ml ($r^2 = 0.992$). Urinary concentrations were calculated from peak ratios.

## 3. RESULTS

### 3.1 Seasonal Profile of Reproductive Hormones

Figure 1 (center) shows the course of urinary steroid metabolites in 2002. During January to March basal excretion of estradiol remained essentially constant at approximately 20 ng/ml. The onset of behavioural estrus in April was associated with a distinct rise in estrogen secretion up to 430 ng/ml on 17 April. Thereafter urinary estrogen concentrations fell to baseline levels by 19 April and remained low. Artificial insemination on 16 April did not lead to pregnancy. The slight increase of the gestagen metabolite pregnanediol during May to August might reflect the period of delayed implantation, the distinct increase in September probably reflects pseudo-pregnancy, a common phenomenon in many carnivore species.

## 3.2 Behaviour

Estrus behaviour of the animal was observed during elevated estrogenic activity. Particularly bleating and genital touching was most intense on the day before maximum elevation of estrogen secretion. Associated with the decline in estrogen excretion estrus behaviour became undetectable.

## 3.3 Identification of Urinary Volatiles Related to Reproduction

The analysis of the composition of urinary volatiles was carried out throughout the entire year and covered all phases of panda reproduction. Four selected samples are shown in Figure 1. The pattern of urinary volatiles remained relatively constant during the first two months. It consists of approximately seven basic substances (with a retention

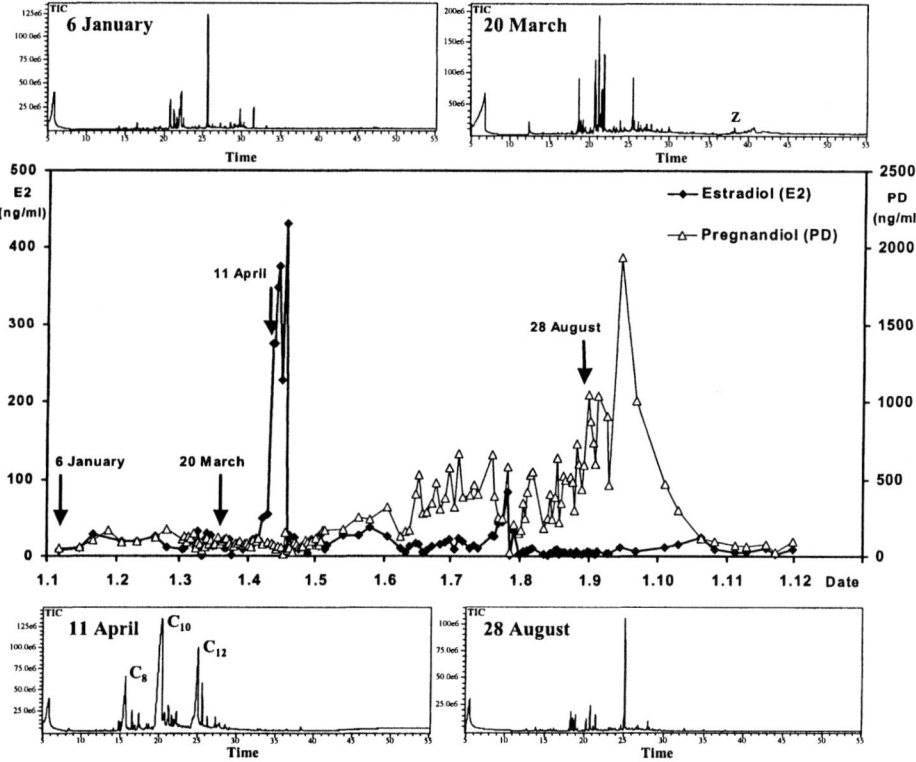

**Figure 1.** Profiles of urinary estradiol and pregnanetriol (center) througout the year 2002. The composition of urinary volatiles was compared by SPME and GC-MS. The total ion chromatogramms (TIC) of four selected samples are shown including the sample taken on the day of the estrogen increase (11 April), which is dominated by the appearance of fatty acids ($C_8$, $C_{10}$, $C_{12}$).

time of between 20 and 23 minutes) varying seemingly in a random fashion during the year. In March cis-civetone (Z) became detectable at a retention time of 38 minutes as a minor compound of urinary volatiles. It could be clearly identified based on a sample of civet as a reference which contained high amounts of cis-civetone (Figure 2B). A group of substances, however, increased dramatically in April when estrogens increased. They were identified as octanoic, decanoic and dodecanoic acid; their mass spectra fit those of synthetic fatty acids (for decanoic acid see Figure 2A). On peak excretion coinciding with the estrogen increase, amounts of 10.7, 8.9, 5.0 µg/ml urine (C8, C10, and C12, respectively) were reached, whereas its concentrations remained low throughout the rest of the year (Figure 3), except for two elevations in June and August.

In contrast to the fatty acids, the seasonal course of civetone revealed temporal variations that displayed oscillations during April to August, but afterwards remained at a lower level until the end of the year (Figure 3). There was a significant positive correlation of civet and pregnanediol (Spearman's rho, $r = 0.389$, $p < 0.001$).

The measurements of urinary volatiles were continued in 2003. Comparable to 2002 estrogenic activity remained low during the first three months. On 6 April estrogens started to increase, reaching a maximal secretion on 10 April followed by an unexpected drop and a second increase with a peak at five days later (15 April). Both estrogen increases were accompanied by the appearance of urinary fatty acids (Figure 4), which

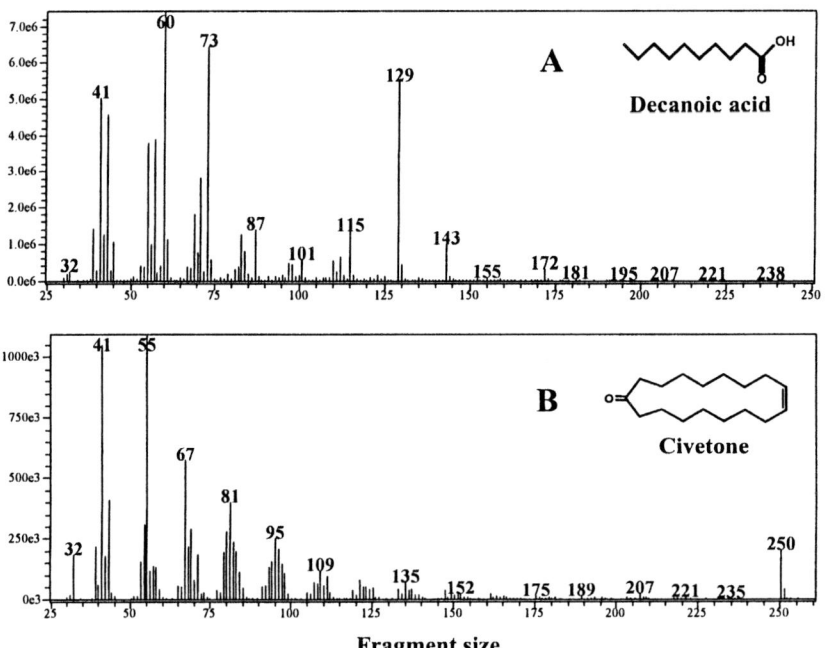

Figure 2. Mass spectra of urinary decanoic acid (A) and civetone (B).

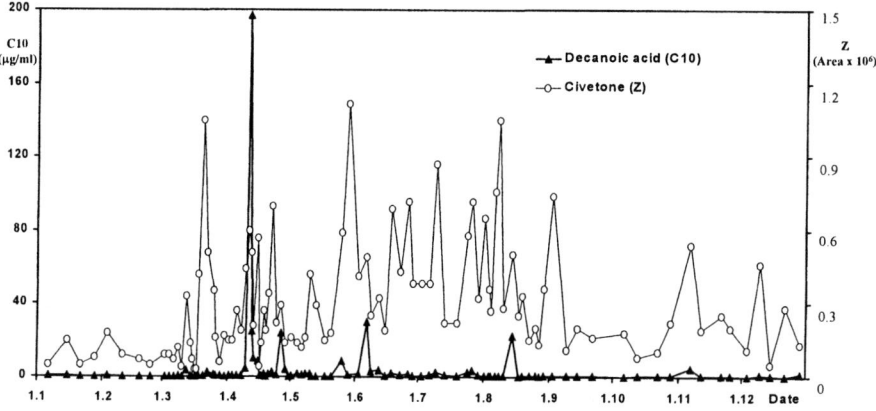

**Figure 3.** Course of urinary decanoic acid and civetone throughout the year 2002 in the female panda Yan Yan from Berlin Zoo.

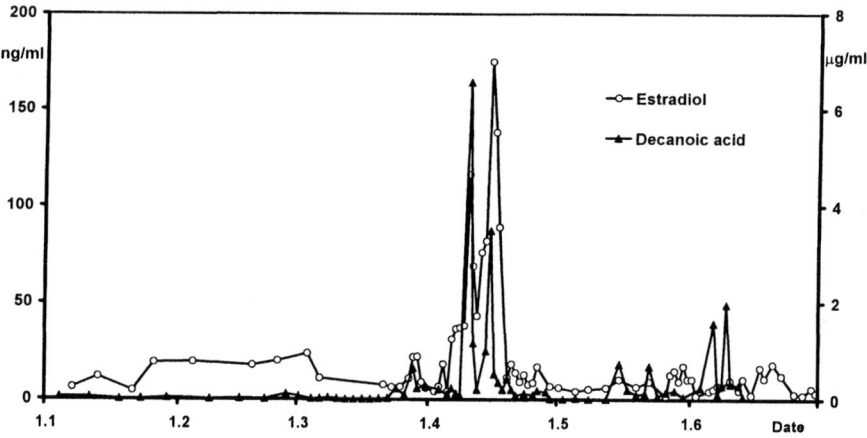

**Figure 4.** Course of urinary decanoic acid and estradiol throughout the first half of the year 2003.

did not reach the magnitude of those measured in the preceding year (6.5, 4.7, 0.5 µg/ml urine on 10 April).

## 4. DISCUSSION

The giant panda is a rare animal both in the wild and in captivity, with an uncertain prospect regarding the maintenance of this species both in its natural habitat in southeast

China and in zoos. Females are mono-estrus with a single receptive period of 1–3 days each spring (Kleiman et al., 1979; Kleiman, 1983; Schaller et al., 1985). Thus, in captivity successful reproduction depends on the ability of animal managers to detect estrus for the accurate management of pairing and artificial insemination. We hypothesised that female panda urine might possess estrus-related signals and their measurement might be used as an analytical tool to detect estrus and perhaps as a stimulant to enhance sexual behaviour of captive males.

Our systematic comparisons of the composition of urinary volatiles in the female panda throughout the year revealed a characteristic increase of urinary free fatty acids (C8, C9, C10) associated with the seasonal estrogen increase that triggers estrus behaviour. The appearance of fatty acids was demonstrated in 2002 and 2003 and, based on gas chromatographical analyses, in 2000. In addition we were able to demonstrate the presence of civetone in panda urine, a substance which was absent in urines of 20 other mammal species we have investigated so far. Fatty acids and civetone were also present in urine samples from the male panda, randomly fluctuating throughout the study period with no significant correlation to their courses in the female. In the female, urinary civetone did not reflect a course attributable to reproductive activity. However, there was a tendency towards higher civetone concentrations during the summer.

The origin of fatty acids and civetone in panda is unknown. Based on four samples of bladder urine obtained between 1999 and 2003, which were not from the days of peak excretion, decanoic acid and traces of octanoic and dodecanoic acid were also detectable in bladder urine. In addition, civeton was also demonstrated in bladder urine. Those findings might rule out the possibility of an extra-renal source from acessory glands of the reproductive tract. Octanoic and decanoic acid were also detected in a cotton wool swab from the female´s anal gland three days after urinary peak excretion.

The role of the substances as pheromones remains to be investigated. There is a distinct possibility that the triplet of fatty acids might serve as a pheromone to attract males to the female which should be emitted in time before ovulation. This is concluded from the temporal relationship of fatty acid secretion to the reproductive hormones. Peak receptivity coincided with the estrogen decrease and female's lordosis was at maximum two days from estrogen peak (McGeehan et al., 2002). Considering our data, the increase in fatty acids appeared eight days prior to the putative day of peak receptivity in 2002 and 2003 which was confirmed by an ultrasound examination in 2003. In addition, the fatty acid increase closely agreed with the start of follicular activity indicated by a first increase of FSH activity (Monfort et al., 1989) and the observation of first scent marking activity six days prior to the estrogen maximum (McGeehan et al., 2002). Unfortunately we were limited to one animal and peak fatty acid levels were variable. Compared with 2003 the higher estrogen secretion in 2002 might explain the higher amount of urinary fatty acid levels in 2002.

Until recently a signal function of civetone was unknown. Its continuous but variable presence in female panda urine and its absence in urine from other investigated species might indicate a biological function. Civetone belongs to the macrocyclic musk compounds and in the wild the powerful musky odour of panda urine had been detected from as far as 5m (Schaller, 1985). The exocrine odour glands of the musk deer (*Moschus moschiferus*) in the proximity of the male genitalia also produce an intensive musk odour (muscone) which is used for both territorial boundary marking and for attracting female partners over great distances during the rutting season (Ravi et al., 2001). The same is obvious from the African Civet (*Civetticus civetta*), a mainly solitary and territorial animal marking their territory using dung and anal secretions. There is a distinct possibility that civetone in panda urine might serve as an ingredient for territorial marking. However, until civetone and the fatty acids may be portrayed as pheromones, the investigations should be extended to other female pandas, and studies on their behavioral relevance are required.

## 5. ACKNOWLEDGMENTS

We thank J. Streich for statistical evaluations and A. Engelhardt (Aeroxon Insect Control GmbH) for the gift of a sample of civetone.

## 6. REFERENCES

Arthur, C. L., and Pawliszyn, J., 1990, Solid-phase microextraction with thermal desorption used fused silica optical fibres, *Anal. Chem.* **62**:2145-2148.
Bonney, R. C., Wood, D. J., Kleimann, D., 1982, Endocrine correlates of behavioral estrous in the female giant (*Ailuropoda melanoleuca*) and associated hormonal changes in the male, *J. Reprod. Fert.* **64**:209-215.
Hodges, J. K., Bevan, D. J., Celma, M., Hearns, J. P., Jones, D. M., Kleimann, D. G., Knight, J. A., Moore, H. D. M., 1984, Aspects of the reproductive endocrinology of the female giant panda (*Ailuropoda melanoleuca*) in captivity with special reference to the detection of ovulation and pregnancy, *J. Zool. Lond.* **203**:253-267.
Hu, J. C. & Wei, F., 1990, Development and progress of breeding and rearing giant pandas in captivity within China, in: *Research and Progress in Biology of the Giant Panda*, J. Hu, F. Wei, C. Yuan, and Y. Wu, eds., Sichuan Publishing, Sichuan, pp. 322–325.
Karlson, P., and Lüscher, M., 1959, "Pheromones": a new term for a class of biologically active substances, *Nature* **183**:55-56.
Kleiman, D. G. 1983, Ethology and reproduction of captive giant pandas (*Ailuropoda melanoleuca*). *Zeitschrift für Tierpsychologie*, **62**:1–46.
Kleiman, D. G. Kleiman, D. G., Karesh, W. B. & Chu, P. R. 1979, Behavioural changes associated with oestrus in the giant panda (*Ailuropoda melanoleuca*) with comments on female proceptive behaviour, *International Zoo Yearbook*, **19**:217–223.
McGeehan, L., Li, X., Jackintell, L., Huang, S., Wang, A., Czekala N.M., 2002, Hormonal and behavioural correlates of estrus in captive giant pandas. *Zoo Biol.* **21**:449-466.
Meyer, H. H., Rohleder, M., Streich, W. J., Göltenboth, R., Ochs, A., 1997, Sex steroid profiles and ovarian activities of the female panda Yan Yan in the Berlin Zoo, *Berl. Münch. Tierärztl. Wochenschr.* **110**:143-147.
Monfort S. L., Dahl, K. D., Czekala, N. M., Stevens, L., Bush, M., Wildt, D. E., 1989, Monitoring ovarian function and pregnancy in the giant panda (*Ailuropoda melanoleuca*) by evaluating urinary bioactive FSH and steroid metabolites. *J. Reprod. Fert.* **85**:203-212.
Schaller, G. B., Hu, J., Pan, W. & Zhu, J., 1985, *Giant Pandas of Wolong*, University of Chicago Press, Chicago.
Ravi, S., Padmanabhan, D., Mandapur, V. R., 2001, Macrocyclic musk compounds: synthetic approaches to key intermediates for exaltolide, exaltone and dilactones, *J. Indian Inst. Sci.* **81**:299-312.
Swaisgood, R. R., Lindburg, D. G., Zhou, X., 1999, Giant pandas discriminate individual differences in conspecific scent, *Anim. Behav.* **57**:1045-1053
Swaisgood R. R, Lindburg D. G., Zhou X., Owen M. A., 2000, The effects of sex, reproductive condition and context on discrimination of conspecific odours by giant pandas, *Anim Behav.* **60**:227-237.

# CHEMICAL COMMUNICATION OF MUSTH IN CAPTIVE MALE ASIAN ELEPHANTS, *Elephas maximus*

Nancy L. Scott and L.E.L. Rasmussen[*]

## 1. INTRODUCTION

### 1.1 Social Systems

Asian elephants have tightly knit social systems maintained by several modes of communication (Eisenberg et al., 1971; Sukumar, 2003). Female Asian elephants group together as family units; whether these matriarchal groups coalesce into larger herds based on genetic relationships has recently been examined (Fernando and Lande, 2000). Subadult males gradually leave the familial unit. As adults, they live either as solitary elephants with indistinct home ranges or in temporary associations with other males often at the fringes of female herds (McKay, 1973; Eisenberg, 1980), where they are attracted in by preovulatory females (Rasmussen et al., in prep). The physiological phenomenon of musth in male Asian elephants apparently increases reproductive success either by increased social dominance (Eisenberg et al., 1971; Jainudeen et al., 1972a) or expanded home ranges allowing more potential encounters with females (Joshua and Johnsingh 1995). Often males accepted by females are males in musth (Sukumar, 2003).

### 1.2 Musth

The exact reproductive significance of musth in male Asian elephants, although documented since the stories of Buddha c. 200 BC (Lahiri-Choudhury, 1992), still remains unclear. Musth, both in the wild and in captivity, first occurs in young males who are sexually mature but socially and physically not fully developed. Musth is usually an annual event with characteristic physiological changes, but males in poor health may not exhibit musth for up to 4 years (Jainudeen et al., 1972b).

---

[*] Nancy L. Scott, Portland State University, Portland, OR, 97207. L.E.L. Rasmussen, Oregon Health and Sciences University, Beaverton, OR, 97006.

Elevated serum testosterone levels, increased aggression between males, secretions from the paired facial temporal glands and dribbling of urine from the penile prepuce characterize musth in the Asian elephant (Jainudeen et al., 1972a; Cooper et al., 1990). Musth is usually not synchronous among males, nor is it always dependent on females in estrus. In wild Asian elephants, although musth is thought to be an annual event for each male, males in a given population tend to have non-overlapping musth periods so that throughout the year there are various males in musth (Eisenberg et al., 1971; Jainudeen et al., 1972b). Data on wild Asian elephant mating behavior are just becoming available with the advent of new genetic methods, radiotelemetry, and sophisticated recording devices and analytical techniques. For example, it is now known that during musth the home range of males expands, thus allowing encounters with more female groups (Joshua and Johnsingh, 1995). Among African savannah elephants (*Loxodonta africana*) mate choice by females and mate guarding by males appear to play crucial roles in reproductive behavior, as males in musth are more successful in competitions with subordinate males for gaining access to estrous females (Moss, 1983; Poole, 1989a). Male Asian elephants usually associate with female units when at least one female is in estrus, and only one male typically associates with a particular family unit or group of family units (McKay, 1973).

## 1.3 Role of Musth

We hypothesize that status in male elephants and mate guarding could be communicated by chemical signals. Asian elephants rely heavily on their chemical senses, acquiring information from chemical signals present in urine, feces, and temporal gland secretions (Rasmussen et al., 1986; Rasmussen, 1988). Previous work has shown that musth chemical signals are detected by female Asian elephants and may be important in female mate choice (Rasmussen et al., 1994; Schulte et al., 1994; Schulte and Rasmussen, 1997). Chemical signals indicative of musth could also relay the physiological status of the male elephant to male conspecifics.

Zahavi and Zahavi (1997) proposed that aggressive behavior in any species can often be avoided with "honest signals," indicating an individual's condition and rank in a hierarchy. A state such as musth that affects the physiology of an elephant could be used as an honest signal to conspecifics, especially as only well-nourished males are capable of experiencing musth (Gale, 1974). Increased serum testosterone may drive increased aggression between males. The ability of a male elephant to detect the musth state of another male could prevent aggressive interactions with a less dominant nonmusth male. Such encounters have been documented behaviorally and chemically in the wild (Rasmussen et al., 2002). Therefore, a male who is able to experience musth could be emitting reflectively honest signals to others, indicating a male of substantial condition and heightened reproductive readiness. This could temporarily increase his relative dominance among males, as seen in African savannah elephants (Poole, 1989b), and perhaps be used to assess the quality of a male by female conspecifics.

During musth, males rub the sides of their faces against trees, leaving temporal gland secretions as stationary olfactory markers that other conspecifics investigate (Eisenberg et al., 1971; Rasmussen et al., 2002). The temporal glands are larger and more active during musth than nonmusth in the male Asian elephant (Gale, 1974). As demonstrated in other mammals, this sexually dimorphic trait suggests a reproductive function (Thiessen and Rice, 1976). The enlargement of the glands in male Asian elephants prior to external secretions (Gale, 1974) coincides with emission of volatile chemical compounds just prior to the onset of musth (Rasmussen et al., 1990; Rasmussen and Perrin, 1999). Analyses of these compounds have shown that the glands become active days before the other signs of musth are apparent (e.g., urine dribbling and temporal gland secretion) (Rasmussen and Perrin, 1999).

In addition to temporal gland secretions (TGS), males in musth also dribble urine. Rasmussen (1988) showed that constituents of musth urine elicit visible chemosensory responses (e.g., the flehmen response) in elephants. Chemical signals present in musth urine may be used to indicate physiological status and relative dominance rank. In several mammalian species, males scent-mark conspecific females with urine or secretions during courtship or mating, e.g., goats, *Capra* spp., (Shank, 1972; Coblentz, 1976); Gray's waterbuck, *Kobus megaceros*, (Walther, 1979); and badgers, *Meles meles*, (Kruuk et al., 1984). This scent marking is thought to discourage other males from mating with the female long enough to ensure that the female's ova are fertilized only by the male that marked her (Jewell et al., 1986). Reindeer, *Rangifer tarandus*, also dribble urine on their hind legs as a male-repellent and female-attractant. The self-urine-marking in reindeer is thought to occur because the male himself constitutes the center of his moving territory (Espmark, 1964).

Dribbling of urine while in musth could also function as a scent marker for male Asian elephants. The inner sides of the hind legs of musth males are often stained with pungent-smelling urine. During the mating act, odiferous compounds could be transferred to females, thus signaling other males that a recent mating occurred by a musth male who may still be nearby, ready to defend his mate. The urine is also dribbled in a trail behind the musth male elephant as he walks and may be later detected by another male who encounters the urine even in the absence of the musth male. The chemosensory information obtained by another male may indicate the dominance, musth status, health, and age of the scent-marking male. Since males in musth are more aggressive and defend estrous females from other males, subordinate males may benefit from discriminating between musth and nonmusth males *via* chemical cues in the urine.

## 1.4 Chemical Senses

Elephants have a highly developed chemosensory system consisting of the main olfactory (MO) and vomeronasal organ (VNO) systems (Boas and Pauli, 1925; Eales, 1926; Rasmussen et al., 1986; Rasmussen and Hultgren, 1990). In most terrestrial mammals, MO is used for detecting airborne, volatile substances whereas the VNO system is used for detecting liquid substances such as urine and glandular secretions (Stoddart, 1980; Keverne, 1982), although increasing evidence shows crossovers between the two sensory systems. The two systems each have characteristic receptors and discrete neuronal pathways to the olfactory and accessory olfactory bulbs, respectively (Stoddart, 1980; Barlow and Mollon, 1982; Farbman, 1992). In elephants, sensory epithelia for the primary olfactory system primarily cover the well-developed ethmoturbinals (Rasmussen and Hultgren, 1990; Rasmussen et al., 1993). Airborne substances pass over the surface of these deeply protected bipolar sensory neurons after the inspired air has been well warmed. The trunk of an elephant has several important sensory roles: it transports chemical signals to the openings of the VNO ducts, thus carrying VNO-destined information; it acts as a conduit for inhalant air that will pass into the turbinate area of the main olfactory system; and the high density of several types of tactile and pressure sensory corpuscles in its tip demonstrates the tip is a major region of sensory epithelium (McKay, 1973; Rasmussen and Munger, 1996; Rasmussen et al., 2003). Elephants place this finger-like projection at the end of the trunk into a substance and then transfer the material from the trunk tip to the openings of the VNO (located on the palate), a behavior termed the "flehmen" response (Rasmussen, 1982). In the elephant the vomeronasal organ is a paired tubular structure containing both sensory and respiratory epithelium; many mucous glands are associated; long, convoluted paired ducts from the VNO open into dorsal anterior mouth cavity (Rasmussen and Hultgren, 1990; Johnson and Rasmussen, 2002). Thus, the muscular trunk acts both as a vehicular tube to allow airborne volatiles access to the olfactory turbinal bones and as the mechanical transfer agent to place liquid substances on the openings of the vomeronasal ducts during the flehmen response.

## 2. URINE BIOASSAYS WITH MALE ASIAN ELEPHANTS

### 2.1 Overview

To begin the assessment of possible chemical messages between male Asian elephants, responses of captive males to conspecific musth and nonmusth urine were measured. Some obvious disadvantages of captive studies are substantially offset by some real advantages: (1) close, accurate behavioral observations in a limited-size enclosure, (2) precise placement of test samples on substrate free of conflicting signals, (3) safe collection of test samples of elephant origin from males whose hormonal status can subsequently be determined.

### 2.2 Sample Collection

For this study 500 ml urine samples were collected both during musth and nonmusth periods from two mature male Asian elephants and frozen for subsequent tests. Concurrent serum samples retroactively established circulatory testosterone concentrations of the donor male pertinent to the collected urine aliquots (Table 1) by Dr. David Hess of the Oregon Regional Primate Center using a previously described methodology (Rasmussen et al., 1984).

### 2.3 Bioassay Procedure

Our previous studies established that a 150 ml aliquot is sufficient to elicit chemosensory responses (Scott et al., 1997). For a bioassay series, two 500 ml aliquots—one from a male in musth and the second from a male not in musth—were thawed at 20°C and then subdivided into three equal volumes. These sub-aliquots were subsequently presented to one male at a time, as paired samples, plus the water control for a total of three samples per assay. Thus, presentations were aliquots of musth urine, nonmusth urine, and tap water, a visual control. Logistically we did not have the financial resources to freeze the urine collected at –80°C. Freezing at –20°C certainly maintained the integrity of most bioactive urinary compounds, and the purpose of this study was to compare responses to musth versus nonmusth urine. The tight control

Table 1: Musth states and their parameters.

| | Serum Testosterone | Temporal Gland Secretion | Urine Dribbling |
|---|---|---|---|
| Nonmusth | < 20 ng/ml | No | No |
| Pre-musth | > 20 ng/ml | Yes/No | No |
| Musth | > 20 ng/ml | Yes | Yes |
| Post-musth | > 20 ng/ml | No | No |
| Moda Musth* | >20 ng/ml* | Yes* | No |

*The first several musth episodes of young teenage adult males (usually between age 10–18 y in captivity and age 14–19 y in the wild) are characterized by widely fluctuating testosterone levels, sporadic temporal gland secretions that often smell like honey (Rasmussen et al. 2002). Two of the test males were approaching the onset of this moda age.

exerted over the unfreezing process insured the equivalency of the samples. In the future, as specific classes of compounds are targeted for study, lower temperature freezing conditions will be utilized. Thus bioassays, which tested urine for bioactivity, were performed by presenting paired samples (musth and nonmusth) of previously frozen urine and a water control to each test animal.

The three samples were each presented in discrete areas 1m apart on a rinsed concrete slab. These urine bioassays were performed when the ambient temperature was 5–25°C to avoid freezing or rapid evaporation of samples. From the time of sample placement and subsequent release of the test animal, each bioassay was performed for the duration of one hour with a single focal animal. The subject was not present when the samples were placed. In addition, the elephant could not see the observer, thus preventing any bias in behavior that might be caused by the observer's presence. The visible physiological state of the elephant and weather conditions were recorded for each bioassay. Each elephant's behavior was videotaped as he approached the sample areas and while he was within a body length either of the urine samples or the water control sample.

Chemosensory responses indicated by trunk interaction with the sample were recorded for all occurrences (Table 2). The sequence of behaviors for each bout was recorded, as was the duration of time spent at each sample. Separate bouts of chemosensory investigation were distinguished when the trunk tip was greater than 0.5 m distant from the sample for longer than 5 s. Each behavior required the trunk to touch the sample or to be within 0.5 m of it.

Bioassays were performed in 1995, 1996, and 1998. Test males included the two donor males (age 33–36 y), an immature male familiar with the donors (age 12–13 y) and two males at different locations (age 10 and 26 y). Three of the test males, (age 26–36 y) experience musth once annually. Two young teenage males (age 10–13 y) had not yet experienced their first musth episode.

## 3. RESULTS

The relative amount of time each elephant spent investigating samples with visible olfactory behaviors was compared to overall investigation time. This resulted in a relative

**Table 2:** Monitored olfactory behaviors

| |
|---|
| **Main Olfaction:**<br>SNIFF: trunk tip directed towards sample and hovering within centimeters of it; inhalation of sample volatiles<br>BLOW: forced exhalation from trunk<br>TRUNK SHAKE: trunk movement back and forth, often accompanied by exhalation<br>**Contact/Olfaction (pre-flehmen):**<br>CHECKS: dorsal trunk tip projection placed in contact with sample<br>PLACE: entire distal surface of trunk tip placed in contact with sample<br>SCRUB: movement within the sample with the entire distal surface of trunk tip placed in contact with sample<br>SUCK: inhalation of sample into trunk with the entire distal surface of trunk tip placed in contact with sample<br>**Flehmen:** dorsal trunk tip projection is placed in sample and then the trunk swings to bring the tip into contact with the VNO ducts |

rate of olfactory investigation for each of the three samples (musth, nonmusth urine, and water). The pooled results for the three familiar male elephants show that they spent significantly more time investigating musth urine (68.6%) than nonmusth urine (31.0%) or water (0.4%) (Figure 1; Sign test, $p<0.05$). While both musth and nonmusth urine elicited observable olfactory behavioral responses, the water control was rarely investigated by the elephants.

To monitor vomeronasal organ system responses, we examined the rate of flehmen responses relative to each approach to the samples. Flehmen responses per approach to each of the urine samples by subadults (n=2) and the adults (n=3) in the study were compared. The subadults performed flehmens at a greater rate to musth compared to nonmusth urine and at an overall greater rate compared to the adult responses (Figure 2).

The subadult presented with urine form an unfamiliar donor performed several-fold more olfactory and pre-flehmen responses than did the subadult who was familiar with the urine donor (Figure 3). The range of responses from the adult presented with unfamiliar urine was greater than the 2 familiar adults, although the means for both groups were similar.

## 4. DISCUSSION

Results from the relative time spent investigating each sample demonstrate that the captive male Asian elephants in this study used chemical signals in urine to distinguish between musth and nonmusth urine. The males spent more time investigating the musth urine samples, perhaps to assess the relative dominance of the donor and status as a potential competitor for mates. Olfactory investigation of nonmusth urine shows that there are additional chemical signals other than those related to musth state present in excretions.

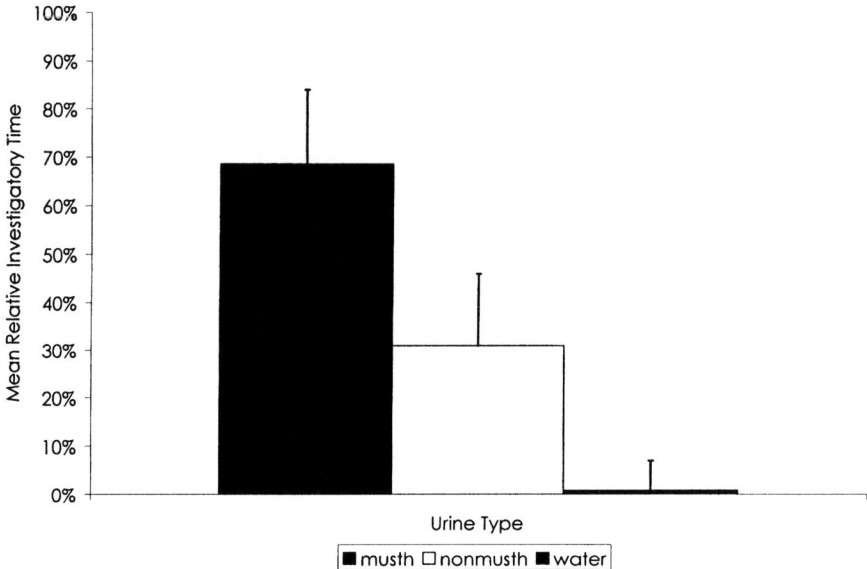

**Figure 1:** Mean relative percent of bioassay time spent investigating *familiar* urine samples; pooled results for 1 subadult male and 2 adult male Asian elephants (+/- SE)

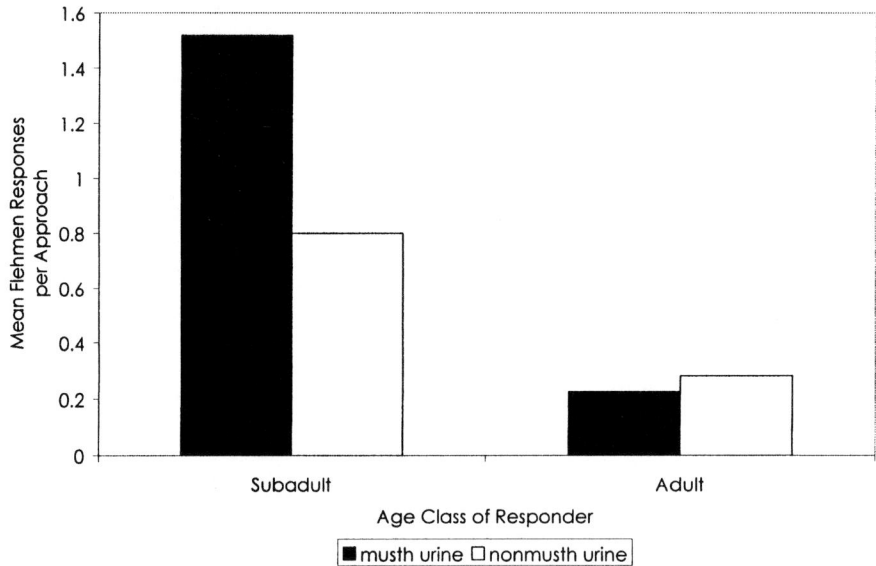

**Figure 2:** Mean flehmen responses per approach to musth and nonmusth urine by two age classes of male elephants

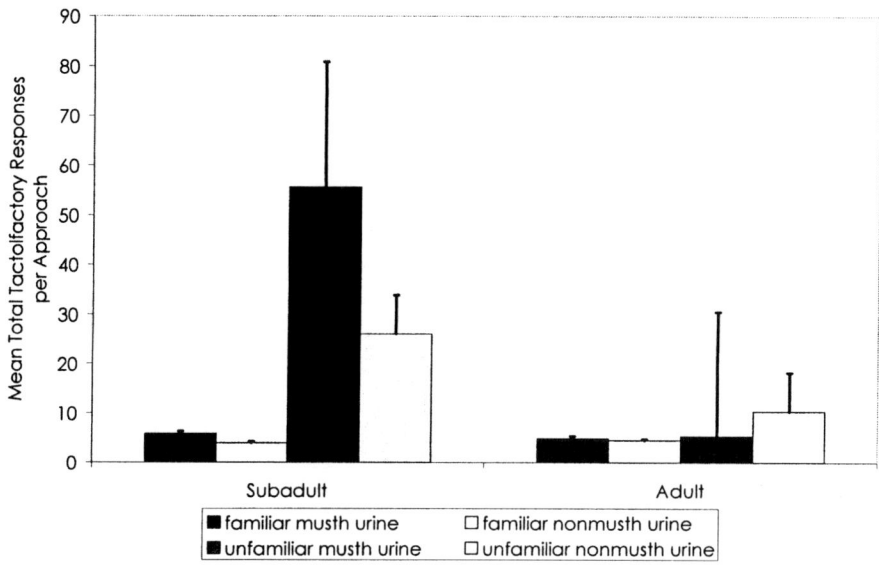

**Figure 3:** Mean sum of olfactory responses (C+P+F) per approach to musth and nonmusth urine from *familiar* and *unfamiliar* donors and a control by 2 subadult and 3 adult male elephants (+/- SE)

Subadult males may perform more flehmen responses than adults as they acquire information for their olfactory memory and learn more of their surroundings. Adult males may easily recognize the urinary chemical signals indicative of musth, while subadult males need repeated exposure to their VNO system to fully assess a urine sample for the musth state and identity of the donor.

In addition to musth state, chemical signals may include individual recognition or recognition of familiarity. Responders unfamiliar to the urine donor performed more olfactory and pre-flehmen responses than familiar responders. This was most dramatic in the subadult age class. Future research could examine how long this increased response is maintained before the responses decrease to the response frequencies observed to familiar urine. Further examination of urine bioactivity could reveal any influence a male's own musth state has on his olfactory interest in musth urine from male conspecifics. The temporary increase in relative dominance acquired by a male in musth may alter a male's olfactory investigation of musth urine compared to when he is not in musth.

## 5. ACKNOWLEDGEMENTS

This research was made possible by a generous grant from Disney's Animal Kingdom and support from Biospherics Research Corporation. Many thanks to the participating facilities: Oregon Zoo (Portland, OR), Riddle's Elephant and Wildlife Sanctuary (Greenbrier, AR), and Tulsa Zoo and Living Museum (Tulsa, OK). None of this research would have been possible without the support and assistance of all of the elephant keepers and handlers at these facilities. Thanks to Dr Bruce Schulte (Georgia Southern University) for design advice and inspiration. Eternal gratitude to Sharon Glaeser for her assistance with transcribing data from video tapes. Sincere appreciation to Dr Deborah Duffield and my graduate committee at Portland State University.

## 6. REFERENCES

Barlow, H., and Mollon, J., eds., 1982, The senses, *Cambridge Texts in the Physiological Sciences*, Cambridge University Press, Cambridge.
Boas, T., and Pauli, S., 1925, *The Elephant's Head*, Copenhagen, Denmark.
Coblentz, B. E., 1976, Functions of scent-urination in ungulates with special reference to feral goats (*Capra hircus* L.), *Am. Nat.* **110**(974):549-557.
Cooper, K. A., Harder, J. D., Clawson, D. H., Fredrick, D. L., Lodge, G. A., Peachey, H. C., Spellmire, T. J., and Winstel, D. P., 1990, Serum testosterone and musth in captive male African and Asian elephants, *Zoo Biol.* **9**:297-306.
Eales, N. B., 1926, The anatomy of the head of a foetal African elephant (*L. africana*), *T. Roy. Soc. Edin-Earth* **54**:491-546.
Eisenberg, J. F., 1980, Ecology and behavior of the Asian elephant, *Supplement to Elephant* **1**:36-55.
Eisenberg, J. F., McKay, G. M., and Jainudeen, M. R., 1971, Reproductive behavior of the Asiatic elephant (*Elephas maximus maximus* L.), *Behaviour* **38**:193-225.
Espmark, Y., 1964, Rutting behaviour in reindeer (*Rangifer tarandus* L.), *Anim. Behav.* **12**(1):159-163.
Farbman, A. I., 1992, *Cell Biology of Olfaction*, Cambridge University Press, Cambridge.
Fernando, P., and Lande, R., 2000, Molecular genetic and behavioral analysis of social organization in the Asian elephant (*Elephas maximus*), *Behav. Ecol. Sociobiol.* **48**:84-91.
Gale, U. T., 1974, *The Burmese Timber Elephant*. Trade Corporation Merchant Street, Rangoon.
Jainudeen, M. R., Katongole, C. B., and Short, R. V., 1972a, Plasma testosterone levels in relation to musth and sexual activity in the male Asiatic elephant, *Elephas maximus*, *J. Reprod. Fertil.* **29**:99-103.
Jainudeen, M. R., McKay, G. M., and Eisenberg, J. F., 1972b, Observations on musth in the domesticated Asiatic elephant (*Elephas maximus*), *Mammalia* **36**:247-261.
Jewell, P. A., Hall, S. J. G., and Rosenberg, M. M., 1986, Multiple mating and siring success during natural oestrus in the ewe, *J. Reprod. Fertil.* **77**:81-89.
Johnson, E. W., and Rasmussen, L., 2002, Morphological characteristics of the vomeronasal organ of the newborn Asian elephant (*Elephas maximus*), *Anat. Rec.* **267**(3):252-9.
Joshua, J. and Johnsingh, A., 1995, Ranging patterns of elephants in Rajaji National Park: Implications for reserve design. *A Week With Elephants*, J. Daniel and H. Datye, eds., Oxford University Press, Oxford, pp. 256-260.
Keverne, E. B., 1982, *The Senses*, Cambridge University Press, Cambridge.
Kruuk, H., Gorman, M., and Leitch, A., 1984, Scent-marking with the subcaudal gland by the European badger, *Meles meles* L., *Anim. Behav.* **32**:899-907.
Lahiri-Choudhury, D. K., 1992, Musth in Indian elephant lore. *Elephants: Majestic Creatures of the Wild*, J. Shoshani, ed., Rodale Press, Emmaus, pp. 82-84.
McKay, G. M., 1973, *Behavior and Ecology of the Asiatic Elephant in Southeastern Ceylon*, Smithsonian Institution Press, Washington, D. C.
Moss, C. J., 1983, Oestrous behaviour and female choice in the African elephant, *Behaviour* **86**:167-196.
Poole, J. H., 1989a, Announcing intent: the aggressive state of musth in African elephants, *Anim. Behav.* **37**:140-152.
Poole, J. H., 1989b, Mate guarding, reproductive success and female choice in African elephants, *Anim. Behav.* **37**:842-849.
Rasmussen, L., Krishnamurthy, V., and Sukumar, R., in prep, Behavioral and chemical confirmation of the preovulatory pheromone, (Z)-7-dodecenyl acetate, in wild Asian elephants and its relationship to musth.
Rasmussen, L., Lazar, J., and Greenwood, D., 2003, The olfactory adventures of elephantine pheromones, *Trans. J. Biochem.* **31**:137-141.
Rasmussen, L., Riddle, H. and Krishnamurthy, V., 2002, Mellifluous matures to malodorous in musth, *Nature* **415**:975-976.
Rasmussen, L. E., 1982, Asian bull elephants: flehmen-like responses to extractable components in female elephant estrous urine, *Science* **217**:159-162.

Rasmussen, L. E., Buss, I. O., Hess, D. L., and Schmidt, M. J., 1984, Testosterone and dihydrotestosterone concentrations in elephant serum and temporal gland secretions, *Biol. Reprod.* **30**:352-362.
Rasmussen, L. E., Schmidt, M. J., and Daves, G. D., 1986, Chemical communication among Asian elephants. *Chemical Signals in Vertebrates IV*, D. Duvall, D. Muller-Schwarze and R. M. Silverstein, eds., Plenum Press, New York, pp. 627-644.
Rasmussen, L. E. L., 1988, Chemosensory responses in two species of elephants to constituents of temporal gland secretion and musth urine, *J. Chem. Ecol.* **14**(8):1687-1711.
Rasmussen, L. E. L., Hess, D. L., and Haight, J. D., 1990, Chemical analysis of temporal gland secretions collected from an Asian bull elephant during a four-month musth episode, *J. Chem. Ecol.* **16**(7):2167-2181.
Rasmussen, L. E. L., and Hultgren. B., 1990, Gross and microscopic anatomy of the vomeronasal organ in the Asian elephant (*Elephas maximus*). *Chemical Signals in Vertebrates V*, D. W. MacDonald, D. Muller-Schwarze and S. E. Natynczuk, eds., Oxford University Press, Oxford, pp. 155-161.
Rasmussen, L. E. L., Johnson, E. W., and Jafek, B. W., 1993, Preliminary observations on the morphology of the vomeronasal organ of a newborn Asian elephant, *Chem. Senses* **18**(5):618.
Rasmussen, L. E. L., and Munger, B. L., 1996, The sensorineural specializations of the trunk tip (finger) of the Asian elephant, *Elephas maximus*, *Anat. Rec.* **246**:127-134.
Rasmussen, L. E. L., and Perrin, T. E., 1999, Physiological correlates of musth: lipid metabolites and chemical composition of exudates, *Physiol. Behav.* **67**(4):539-549.
Rasmussen, L. E. L., Perrin, T. E., and Gunawardena, R., 1994, Isolation of potential musth-alerting signals from temporal gland secretions of male Asian elephants (*Elephas maximus*), a new method, *Chem. Senses* **19**(5):540.
Schulte, B. A., Perrin, T. E., Scott, N., Slade, B., and Rasmussen, L. E. L., 1994, *Reproductive Condition of Female Asian Elephants and their Responsiveness to Male Secretions*, Northeastern Regional Animal Behavior Society, University of New Hampshire.
Schulte, B. A., and Rasmussen, L. E. L., 1997, *Do Female Asian Elephants (Elephas maximus) Chemosensorily Prefer Musth Males?* American Society of Mammalogists, Stillwater, Oklahoma.
Scott, N. L., Schulte, B. A., Mellen, J. D., and Rasmussen, L. E. L., 1997, *Do Male Asian Elephants (Elephas maximus) Advertise Musth in their Urine?* American Society of Mammalogists, Stillwater, Oklahoma.
Shank, C. C., 1972, Some aspects of social behaviour in a population of feral goats (*Capra hircus* L.), *Z. Tierpsychol.* **30**:488-528.
Stoddart, D. M., 1980, *The Ecology of Vertebrate Olfaction*, Chapman and Hall, New York.
Sukumar, R., 2003, *The living Elephants: Evolutionary Ecology, Behavior, and Conservation*, Oxford University Press, New York.
Thiessen, D., and Rice, M., 1976, Mammalian scent gland marking and social behavior, *Psychol. Bull.* **83**(4):505-539.
Walther, F. W., 1979, Das Verhalten der Hornträger (Bovidae), *Handb. Zool. VIII* **10**(30):1-184.
Zahavi, A., and Zahavi, A., 1997, *The Handicap Principle: A Missing Piece of Darwin's Puzzle*, Oxford University Press, Oxford.

# CHEMICAL ANALYSIS OF PREOVULATORY FEMALE AFRICAN ELEPHANT URINE: A SEARCH FOR PUTATIVE PHEROMONES

Thomas E. Goodwin, L. E. L. Rasmussen, Bruce A. Schulte, Patrick A. Brown, Ben L. Davis, Whitney M. Dill, Nichole C. Dowdy, Adam R. Hicks, Richard G. Morshedi, Daniel Mwanza, and Helen Loizi [*]

## 1. INTRODUCTION

Many similarities exist between the lifestyles and behaviors of *Loxodonta africana* and *Elephas maximus*, two of the three extant species of elephants (Sukumar, 2003; Poole, 1987, 1989a,b; Rasmussen and Krishnamurthy, 2000; Rasmussen and Schulte, 1998). While the roles of olfaction and the chemical senses in Asian elephant society have been extensively investigated (Rasmussen and Greenwood, 2003; Rasmussen et al., 1997, 2002), similar investigations in the African species are limited to recent studies of chemical signals among males (Rasmussen and Wittemyer, 2002).

Male elephants face reproductive challenges not only of locating females, as the sexes live somewhat separated, but also of detecting the most fertile period of females, i.e. as they approach ovulation during the 13-17 week estrous cycle. The Asian species has been shown to utilize a urinary pheromone (Rasmussen et al., 1996). Commencing fairly early in the follicular phase, low concentrations of an acetate pheromone attract males (Rasmussen, 2001). The available urinary concentration of this ligand elevates gradually as ovulation approaches. Males apparently can measure quantitatively the pheromone concentration and thus the female's proximity to ovulation. This is evidenced by high frequencies of flehmen responses and premating behaviors. Based on selected field observations and unpublished data on captive elephants, our group recently began a rigorously designed multiple-elephants and multiple-sites study to establish whether male

---

[*] Thomas E. Goodwin, Patrick A. Brown, Ben. L. Davis, Whitney M. Dill, Nichole C. Dowdy, Adam R. Hicks, Richard G. Morshedi, and Daniel Mwanza, Hendrix College, Conway, AR, 72032. L. E. L. Rasmussen, Oregon Health and Sciences University, Beaverton, OR, 97006. Bruce A. Schulte and Helen Loizi, Georgia Southern University, Statesboro, GA, 30460.

African elephants are chemosensorily attracted toward females; in particular whether they are attracted by urinary chemical signals at specific periovulatory periods.

The African species has a similar three-to-four month estrous cycle, characterized by an ovulatory elevation of luteinizing hormone (LH2) coinciding with the initial rise, slight decrease, then sustained elevation of serum progestins, and an anovulatory luteinizing hormone peak (LH1) about three weeks earlier (Brown, 2000). We have hypothesized that related to these two hormonal elevations, urinary chemical signals are released by female African elephants that attract males and signal impending ovulation.

Our research strategy began with the acquisition of estrous-defined urine samples from captive female African elephants. After such collections, chemical analyses were initiated to search for a putative pheromone employing four differently emphasized search strategies: (1) Identify compounds known to be pheromones in other species; (2) Identify compounds seen exclusively or largely around LH1 or LH2 elevations; (3) Identify compounds seen exclusively or largely in protease- or acid-treated samples; (4) Identify compounds that are predominant in urine samples demonstrated to be bioactive. Because our behavioral study is long-term and of a design and sample size appropriate to tease out robust chemosensory responses, we were not able to use the fourth strategy in the current report. This paper presents some of our initial chemical findings as examples typical of the data being gathered.

## 2. MATERIAL AND METHODS

### 2.1. Urine Collections

Our defined requirements for urine acquisition included collections from regularly cycling females whose serum progesterone (P4) was monitored weekly and serum LH monitored daily during the three-week period immediately preceding ovulation. Collection days were selected after reviewing the chronology of serum LH and P4 concentrations during previous estrous cycles. At least one cycle and sometimes multiple cycle histories were required to anticipate the timing of the two LH peaks. On each day of collection, concurrent serum samples were obtained to confirm subsequently the hormonal status of the female. Our schedule for each female included collection on the following days: two luteal phase days; day of LH1 and day before and after; day of LH2 and day before and after; a day midway between LH1 and LH2. On each collection day multiple mid-stream urine aliquots were collected in clean glass containers and immediately frozen on dry ice or placed in a -80 °C freezer and stored at –80 °C until analysis. Institutions participating in these collections are listed in the Acknowledgments section.

### 2.2. Chemical Analyses: Sampling Methods and Analysis Procedures

We utilized two sampling methods combined with gas chromatography-mass spectrometry (GC-MS) to identify urinary compounds. Our adsorption and absorption sampling and capture techniques allowed us to encompass a substantial molecular weight (MW) range. We spanned in a somewhat selective manner from small compounds such as acetaldehyde (MW = 44) and trimethylamine (MW = 59) up to large ones such as steroids (MW over 300). For the more volatile compounds, evacuated canister capture followed by cryogenic trapping (ECC/CT) prior to GC separation was employed. Less volatile compounds were trapped by solid phase microextraction (SPME) by two slightly different procedures, an automated one conducted at the laboratory of Dr. Goodwin and his students, and a manual one at the laboratory of Dr. Rasmussen. These procedures are described in more detail below.

*2.2.1. Sampling: ECC/CT*

One hundred mL of urine was placed in a 500-ml clean glass jar fitted with a special lid that contained two Swagelok fittings. One fitting was connected to the jar via ultra-clean Nupro SS-4H4 bellow-stem valves to a special stainless steel receiving bottle (0.85 or 6 L) evacuated to ~30 inches Hg vacuum. Similar valves connected the other fitting to a source bottle of pure air pressured to 40 psig. Prior to starting the experiment, pure air was flushed into the system. Next, the jar samples were heated to 37 °C (mimicking elephant body temperature) and allowed to equilibrate for 30 min, thus allowing the development of headspace volatility. Subsequently, at 30-min intervals for 2.5 h (and on occasion for longer inter-sample intervals and up to a total time of 24 h), the stainless steel evacuated receiving bottle was briefly opened to allow the entry of compounds developed in the headspace. At the end of the collection time, the receiving canisters were pressurized with helium to 30 psig to ensure long-term storage at room temperature and to facilitate GC/MS analyses. Further details are provided in Perrin et al. (1996) and Rasmussen and Perrin (1999).

The sample introduction system for subsequent GC/MS analyses of urine headspace volatiles contained within these pressurized stainless steel canisters involved the initial release of the volatiles from the canister and their adsorption onto an in-line Tenax trap. In turn, desorption from the Tenax, employing a six-port valve in line with a U-tube cryogenic trap (0.125 in OD x 9 in) containing 60/80 mesh glass beads, was followed by cryogenic focusing on this loop. Compounds were then released from the loop by heat, and separated by gas chromatography.

*2.2.2. Analysis: GC/MS Following ECC/CT*

The ECC/CT samples were analyzed on a Hewlett-Packard 5890A GC and a Hewlett-Packard 5970B MS. The GC used a DB-1, 0.25-mm ID x 60 m x 1.0 μm film thickness, polymethyl silicone-coated capillary column (J & W Scientific). The gas chromatograph oven was temperature programmed from –60 to 200 °C at 4°C/min. The mass spectrometer was programmed for a mass scan of 33–300, which allowed for identification of compounds from C3 through C14. Compounds were identified using an NBS 75 K Hewlett-Packard Mass Spectrometer ChemStation library search and were manually rechecked with the NIST/EPA/NIH Mass Spectral Data Base Version 4.01 and the Wiley Library Version 6-275. In addition to assigning compound identity based on mass (molecular) ion and dominant ion patterns, a number of internal standards of authentic compounds were employed for compounds of interest. Although 75% match was the minimum criterion, most compound matches were greater than 90%. ECC/CT-GC/MS was utilized primarily to identify the dominant volatile compounds and although quantitation was not conducted for this study, this method, with greater numbers of duplicate samples, will be used in the future for quantitation of lower molecular weight volatiles.

## 2.2.3. Sampling: Manual Solid Phase Microextraction (SPME)

Aliquots (500 μl) of urine were sampled by manual SPME prior to GC/MS. The vials used for SPME were steam-cleaned by hot distilled water, rinsed three times with triple-distilled water, and air-dried prior to the addition of urine. Either a reverse-direction insert top, conditioned for several days in a GC oven at 250 °C, or an aluminum foil cover through which the fiber and its holder were inserted was used. Based on the results from Asian female urine (Rasmussen, 2001), 100-μm polydimethylsiloxane (PDMS) SPME fibers were employed. No fibers were immersed; rather, they were exposed in the headspace above the liquid sample, which was gently stirred by a tiny magnetic stirring bar. For each sample, the adsorption of volatile compounds on the fiber was conducted first at native pH and at ambient temperature (25 °C), and then heated to 37 °C. For selected duplicate samples, 1 mg/ml of non-specific bacterial protease (Sigma cat. no. P-5147) was added (Poon et al., 1999; Yamazaki et al., 1999). For some of these samples, the pH was adjusted to 4.0, the pH demonstrated to result in release of ligands from urinary albumin (Lazar et al., 2002). Selected samples were also reduced further to pH 1.0.

Adsorption times were 1 h or more for each condition, at which time the fiber was retracted into the protective needle. Immediately, the SPME needle was inserted into the injector port of the GC outfitted with a special glass liner insert, and the fiber was exposed for 10 min to allow desorption at 250 °C, the temperature of the injector port. Thus, compounds adsorbed on the PDMS fiber were desorbed and focused onto the beginning of the 40 °C GC column. Because of the tendency of some analytes to adhere to glass or plastic surfaces (Prestwich, 1987), three blank samples were run prior to the analyses of samples, and fibers were reconditioned twice between analyses with a blank analysis conducted after the second conditioning. The glass liner was also cleaned and replaced weekly.

## 2.2.4. Analysis: GC/MS Following Manual SPME

GC/MS analyses for the manual SPME experiments were conducted using a Hewlett-Packard 6890A GC and a Hewlett-Packard 5973 mass selective detector. The GC column was identical to that used for ECC/CT. The GC oven was temperature programmed from a 4 min hold at 40 °C to 200 °C at 6 °C/min, then ramped at 2 °C/min to a final temperature of 235 °C.

The mass spectrometer was programmed at 0.83 scans/sec for a mass scan of 33–550, which allowed for identification of compounds from C3 through C18. Most compounds were identified using an NBS 75 K Hewlett-Packard Mass Spectrometer ChemStation library search and were manually rechecked with the NIST/EPA/NIH Mass Spectral Data Base Version 4.01 and the Wiley Library Version 6-275. Mass (molecular) ion and dominant ion patterns by mass spectrometry were of primary importance in assigning identity. The criterion of matching was at least 75% and most matches were in the 90% range. Selected internal standards also were analyzed. SPME followed by GC/MS was used for identification of higher molecular weight compounds, but there was no attempt at quantitation.

## 2.2.5. Sampling: Automated SPME

Aliquots of urine were sampled by automated SPME prior to GC/MS. Sampling was carried out with a Gerstel Multipurpose Sampler (MPS2). New 20 mL vials (Gerstel part no. GC 93640-06) were rinsed three times with reagent grade acetone, rinsed three times with deionized water, and oven-dried (110 °C) prior to the addition of urine (1 mL). Normally the urine was saturated with solid NaCl, and then the vial was crimp-sealed using magnetic crimp caps (Gerstel part no. 093640-008-00). In some instances, urine samples were treated with 1 mg/mL of non-specific bacterial protease (Sigma, cat. no. P-5147) (Poon et al., 1999; Yamazaki et al., 1999). For other samples, 1M HCl was used to adjust the pH to 3 or 4 (down from the natural pH of 7-8), the pH demonstrated to result in release of ligands from urinary albumin in Asian elephants (Lazar et al., 2002).

Based on the results from Asian female urine (Rasmussen, 2001), 100-μm polydimethylsiloxane (PDMS) SPME fibers were used (23 gauge needle for autoholder and Merlin Microseal™ septum, Supelco part no. 57341-U; 1.0 mm ID GC inlet liner, Resek part no. 20973). The SPME fiber was not immersed, but rather exposed for 30 min in the headspace above the liquid sample that was heated at 37 °C and agitated at 250 rpm. The MPS2 then retracted the SPME fiber into the protective needle and transferred it to the GC inlet where it was exposed and desorbed at 250 °C for 10 min under splitless conditions, then for an additional 10 min under split conditions to clean the fiber. In trial runs and test cases, no cross-contamination (ghosting) from run to run was detected when using this protocol. Normally, five samples were analyzed prior to cleaning the fiber and column with a blank run. The glass inlet liner was cleaned as needed.

## 2.2.6. Analyses: GC/MS Following Automated SPME

GC/MS analyses were conducted using an Agilent 6890N GC and 5973N mass selective detector. The capillary GC column was an SPB-1 (bonded; poly(dimethylsiloxane)), 60 m x 0.32 mm ID, 1 μm film thickness (Supelco cat. no. 24047). The GC oven was temperature programmed from a 20-min hold at 45 °C (20 min is the total SPME fiber desorption time) to 110 °C at 6 °C/min, held at that temperature for 10 min, then ramped at 2 °C/min to a final temperature of 210 °C where it was held for 45 min. The mass spectrometer was programmed at 2.86 scans/sec for a mass scan of 35–550. Most compounds were identified using the NIST02 mass spectral library. Additional searching was carried out with the Wiley Library Version 6-275. Mass (molecular ion) and dominant fragmentation ions were of primary importance in assigning identity. Although 75% match was the minimum criterion, most compound matches were greater than 90%.

## 3. RESULTS AND DISCUSSION

Two new techniques for the analysis, i.e. identification and quantification, of chemical compounds in animal secretions and excretions offer decisive advancements in investigations of mammals whose social actions and reproduction are well-known behaviorally. The largest terrestrial herbivores, elephants, with an enormous impact on other nearby animals and on their habitat, have been the focus of extensive behavioral studies involving both wild and captive populations. In both the Asian and African savannah species, males are known to be attracted to estrous females (Eisenberg et al., 1971; Moss, 1983). In the Asian species, the multiplicity of flehmen responses by males to preovulatory females was well-documented for eight years and correlated with the establishment through serum hormone measurements of the approximately three to four month cyclicity of the estrous cycle (Hess et al., 1983). Recently a similar female estrous cyclicity has been demonstrated for African female elephants (Brown, 2000). For the Asian species, the extensive record allowed the initiation in 1980 of an attempt to identify the active chemical signal(s) that elicited the flehmen responses by males (Rasmussen et al., 1982). The resultant investigation involving bioassay-guided fractionations was long and tedious, sustained, however, by the robust bioresponses by all male elephants tested (Rasmussen et al., 1986).

Such bioassays of urinary components separated by organic chemical extraction, flash chromatography, and high pressure liquid chromatography resulted in identification of a female-to-male preovulatory pheromone of the Asian elephant (Rasmussen et al., 1982, 1986, 1996). Many compounds in the urine from the luteal and follicular phases of captive females were identified by GC-MS and assayed. Elevated, or prominent, in preovulatory urine were numerous aldehydes, ketones, alcohols, and phenols; 4-methylphenol and 4-ethylphenol especially were a high proportion of these organic extracts (Rasmussen et al., 1997). However, many active fractions were dominated by an unknown entity with a mass of 166. It was suspected that this mass was a fragment of a larger molecule, perhaps an acetate. Final identification, in 1996, was confirmed through the use of a new technique, solid phase microextraction (SPME) (Arthur and Pawliszyn, 1990) to identify the whole molecule, (Z)-7-dodecenyl acetate (Z7-12:Ac). Only Z7-12:Ac exhibited two features: (1) a 1000-fold increase in concentration occurring primarily between two preovulatory luteinizing hormone elevations, and (2) robust bioactivity (Rasmussen et al., 1997; Rasmussen, 2001). With the new techniques available and our four-pronged research approach, we are now able to demonstrate characteristic compounds in female African elephant urine and to suggest some potentially fruitful research directions.

Tables 1 and 2 provide one example of GC-MS results from analysis of the same urine sample by ECC/CT and SPME, respectively. Clearly, ECC/CT facilitates the observation of the more volatile organic compounds. Tables 3 and 4 illustrate typical results from automated SPME/GC-MS analysis of a urine sample at physiological pH 8 versus pH 3. It is noteworthy that the compounds listed in Table 4 (pH 3) at retention times of 64.49, 65.40, 65.69, 66.24, 66.49, 67.24, and 67.89 minutes constitute a total of 62.48 area percent of the total products observed, while none of these compounds appear at all in the pH 8 sample. The compound at 64.49 minutes is a known synthetic spirocycle (**1**) that has not been observed previously as a natural product (Ehrenfreund et al., 1974; Renold et al., 1975; Schulte-Elte, et al., 1978). This is not the first time that we have observed unique natural products in African elephant secretions and excretions (Goodwin et al., 1999, 2002).

The compounds at retention times 65.40, 65.69, and 67.89 minutes (Table 4) represent three of the four possible diastereomers of the natural products commonly known as dihydroedulans (**2**). The compounds at retention times 66.24 and 67.24 minutes are the diastereomeric theaspirane (**3**) natural products. Finally, the compound at a retention time of 66.49 minutes represents one of the two diastereomers of the natural product edulan (**4**).

(1) (2) (3) (4)

(*chirality centers)

**Table 1.** Compounds identified (>74% match quality) in LH1 urine (3/11/03) of Alice[a] by ECC/CT and GC-MS (physiological pH = 8, 37 °C, NaCl added).

| Compound | Ret. Time (min.) | Compound | Ret. Time (min.) |
|---|---|---|---|
| Acetaldehyde | 12.25 | 3-Penten-2-one | 35.74 |
| Ethanol | 18.97 | Dimethyl disulfide | 36.11 |
| Acetone | 20.13 | Pyrrole | 36.38 |
| 2-Propanol | 21.38 | 3-Methylthiophene | 37.85 |
| Dimethyl sulfide | 22.25 | 3-Hexanone | 38.21 |
| 2-Methylpropanal | 24.57 | 2-Hexanone | 38.36 |
| 2-Butenal | 25.31 | Hexanal | 38.89 |
| 3-Buten-2-one | 26.25 | Isopropyl isothiocyanate | 40.45 |
| Butanal | 26.64 | 3-Methylcyclopentanone | 41.11 |
| 2-Butanone | 26.97 | 4-Heptanone | 42.86 |
| 2-Methyl-3-buten-2-ol | 28.51 | 2-Heptanone | 43.75 |
| 2-Methyl-1-propanol | 29.67 | Heptanal | 44.18 |
| 3-Methylbutanal | 30.64 | 3-Ethylcyclopentanone | 46.74 |
| 3-Methyl-2-butanone | 31.00 | 2,3-Octanedione | 47.99 |
| 2-Methylbutanal | 31.28 | Acetophenone | 51.52 |
| 3-Methyl-3-buten-2-one | 31.81 | 2-Octen-4-one | 54.39 |
| 1-Penten-3-one | 32.43 | 2-Cyclohexen-1-one, 2-methyl-5-(1-methylethenyl)- | 59.36 |
| 2-Pentanone | 32.57 | 2-Cyclohexen-1-one, 3-methyl-6-(1-methylethyl)- | 59.85 |
| Pentanal | 33.10 | 2-Buten-1-one, 1-(2,6,6-trimethyl-1,3-cyclohexadien-1-yl)-, (E)- | 64.79 |
| 4-Methyl-2-pentanone | 35.73 | 3-Buten-1-one,4-(2,6,6-trimethyl-1-cyclohexen-1-yl) | 65.93 |

[a]Female African elephant (age 33 yrs. at time of sampling), Wildlife Safari Park, Winston, OR.

**Table 2.** Compounds identified (>74% match quality) in LH1 urine (3/11/03) of Alice[a] by automated SPME and GC-MS (physiological pH = 8, 37 °C, NaCl added).

| Compound | Ret. Time (min.) |
| --- | --- |
| Isopropyl isothiocyanate | 29.02 |
| 2,3-Octanedione | 38.70 |
| Benzene, 1-methyl-4-(1-methylethyl)- | 43.31 |
| 2-Cyclohexen-1-one, 3-methyl- | 43.96 |
| Phenol, 4-methyl- | 45.86 |
| Phenol, 2-ethyl-4,5-dimethyl- | 50.10 |
| 5-Isopropyl-3,3-dimethyl-2-methylene-2,3-dihydrofuran | 52.91 |
| Phenol, 3-ethyl- | 53.49 |
| 4,7-Dimethylbenzofuran | 58.73 |
| Cyclohexanone, 5-methyl-2-(1-methylethylidene)- | 60.15 |
| 2-Cyclohexen-1-one, 3-methyl-6-(1-methylethyl)- | 61.14 |
| 2-Cyclohexen-1-one, 3-methyl-6-(1-methylethylidene)- | 67.36 |
| 2-Buten-1-one, 1-(2,6,6-trimethyl-1,3-cyclohexadien-1-yl)-, (E)- | 70.99 |
| 5,9-Undecadien-2-one, 6,10-dimethyl-, (E)- | 75.20 |

[a]Female African elephant (age 33 yrs. at time of sampling), Wildlife Safari Park, Winston, OR.

The edulans and dihydroedulans were first identified in passionfruit (Whitfield and Stanley, 1977; Prestwich et al., 1976). Subsequently, edulans were seen in human urine (Mills and Walker, 2001), while dihydroedulans have also been found in male scent organs of African butterflies (Schulz et al., 1993). The theaspiranes have been identified in green and black tea, as well as in a number of fruits and berries (Schmidt et al., 1992). Later, a theaspirane was observed by SPME/GC-MS in urine from a female Asian elephant (Rasmussen, 2001). Recently, a dihydroedulan and a theaspirane were reported from giant panda urine (Dehnhard et al., 2003).

At this point neither the origin of compounds 1-4 in elephant urine, nor their role, if any, in chemical signaling among elephants is clear. It is noteworthy that in a previous study of the isolation of volatiles from quince fruit, high vacuum distillation/extraction followed by GC-MS yielded different results if one started with homogenized fruit at its natural pH (3.7), versus homogenate at pH 7 to which an enzyme inhibitor had been added. Specifically, the pH 3.7 sample evidenced large amounts of the theaspiranes, but the pH 7 sample showed only trace amounts (Winterhalter et al., 1987).

Additionally, it has been demonstrated that under acidic conditions, certain monocyclic diols can be converted to the theaspiranes (**3**), or to the dihydroedulans (**2**), and that the latter (**2**) may be isomerized to the former (**3**) on strong acid treatment (Schulte-Elte et al., 1978; Winterhalter et al., 1987; Schmidt et al., 1995; Young et al., 2000). Thus it is possible that in African elephant urine at pH 3-4, we are observing products (**1-4**) that are formed from as yet unidentified precursors in the native pH 8 urine. If so, it is likely that the precursors are degradation products of carotenoids, as has been suggested for similar bisnorsesquiterpenes (Francke et al., 1989; Kaiser and Lamparsky, 1979).

**Table 3.** Compounds identified (>74% match quality) in LH2 urine (9/11/01) of Timba[a] by automated SPME and GC-MS (physiological pH = 8, 37 °C, NaCl added).

| Compound | Ret. Time (min.) |
| --- | --- |
| Benzene, 1,3-dimethyl- | 32.51 |
| p-Xylene | 32.79 |
| Benzaldehyde | 36.86 |
| 2,3-Octanedione | 38.82 |
| Benzyl chloride | 41.38 |
| Benzene, 1-methyl-4-(1-methylethyl)- | 43.43 |
| Phenol, 4-methyl- | 45.97 |
| Benzene, 1-methyl-4-(1-methylethenyl)- | 48.60 |
| 3-Methyl-4-isopropylphenol | 50.22 |
| 1,3,8-p-Menthatriene | 50.79 |
| Acetophenone, 4'-hydroxy- | 53.47 |
| Benzene, 1-methyl-3-(1-methylethyl)- | 54.29 |
| 3-Methyl-2,3-dihydro-benzofuran | 55.39 |
| Bicyclo[3.2.0]hept-3-en-2-one | 64.91 |
| 2-Buten-1-one, 1-(2,6,6-trimethyl-1,3-cyclohexadien-1-yl)-, (E)- | 71.10 |
| 2H-1-Benzopyran-2-one | 73.07 |
| 3-Buten-1-one, 4-[2,6,6-trimethyl-1(or 2)-cyclohexen-1-yl]- | 73.26 |
| 1,6,6-Trimethyl-7-(3-oxobut-1-enyl)-3,8-dioxatricyclo[5.1.0.0(2,4)]octan-5-one | 74.67 |

[a]Female African elephant (age 24 yrs. at time of sampling), Seneca Park Zoo, Rochester, NY.

In two pheromonal systems in the Asian elephant, proteins play transport and sequestering roles prior to pheromonal interactions with sensory receptors. Frontalin, a chemical signal of musth in older males, is linked to elephant albumin in the temporal gland secretion at appropriate pHs (Rasmussen et al., 2003). Likewise, Z-7-dodecenyl acetate (Z7-12:Ac), the preovulatory urinary pheromone, is bound to urinary albumin, maximally at alkaline pH. The latter pheromone has been more thoroughly investigated, revealing an interesting synchrony of events in the urine of the female prior to ovulation (Rasmussen, 2001). Not only does the concentration of Z7-12:Ac rise during this periovulatory period between transitory elevations of serum luteinizing hormone, but also urinary pH becomes more alkaline and protein content/creatinine levels elevate. Surprisingly, there are few low molecular mass, lipocalin-like proteins in the female urine (Lazar, 2001). Instead, a 66-kDa urinary protein is preferentially bound to the pheromone as studied by SDS and native gel electrophoresis (Lazar et al., 2002). Our preliminary studies suggest that female African elephant urine may undergo similar changes. We are modeling the demonstrated binding of Asian elephant pheromones to urinary albumin as an exploratory tool in our search for chemical signals operational from female African elephants toward conspecific males.

## 4. ACKNOWLEDGMENTS

Elephant urine samples were provided by the following organizations: Cameron Park Zoo, Indianapolis Zoo, Louisville Zoo, Nashville Zoo (R. and C. Pankow), Riddle's Elephant Sanctuary, Sedgwick County Zoo, Seneca Park Zoo, Six Flags Marine World,

**Table 4.** Compounds identified (>74% match quality) in LH2 urine (9/11/01) of Timba[a] by automated SPME and GC-MS (pH = 3, 37 °C).

| Compound | Ret. Time (min.) |
|---|---|
| 4-Heptanone | 31.82 |
| Oxepine, 2,7-dimethyl | 35.89 |
| 1-Hexen-3-yne, 2,5,5-trimethyl- | 40.56 |
| Cyclohexene, 3-methyl-6-(1-methylethylidene)- | 41.34 |
| Benzene, 1-methyl-2-(1-methylethyl)- | 43.31 |
| 1,4-Cyclohexadiene, 1-methyl-4-(1-methylethyl)- | 46.43 |
| Benzene, 1-methyl-4-(1-methylethenyl)- | 48.49 |
| Bicyclo[4.1.0]hept-2-ene, 3,7,7-trimethyl- | 49.01 |
| 1,3,8-p-Menthatriene | 50.66 |
| Benzofuran, 2,3-dihydro-2-methyl- | 52.80 |
| Benzene, 4-ethyl-1,2-dimethyl- | 54.16 |
| Benzenemethanol, .alpha.,.alpha., 4-trimethyl- | 55.62 |
| Benzene, 2-(2-butenyl)-1,3,5-trimethyl- | 58.12 |
| Naphthalene, 1,2,3,4-tetrahydro-1,1,6-trimethyl- | 59.46 |
| 1-Oxaspiro[4.5]deca-3,6-diene, 2,6,10,10-tetramethyl- | 64.49 |
| 2-Oxabicyclo[4.4.0]dec-9-ene, 1,3,7,7-tetramethyl- | 65.40 |
| 2H-1-Benzopyran, 3,4,4a,5,6,8a-hexahydro-2,5,5,8a-tetramethyl-(2.alpha.,4a.alpha.,8a.alpha.)- | 65.69 |
| 2,6,10,10-Tetramethyl-1-oxa-spiro[4.5]dec-6-ene | 66.24 |
| 2H-1-Benzopyran, 3,5,6,8a-tetrahydro-2,5,5,8a-tetramethyl-, trans | 66.49 |
| 2,6,10,10-Tetramethyl-1-oxa-spiro[4.5]dec-6-ene | 67.24 |
| 2H-1-Benzopyran, 3,4,4a,5,6,8a-hexahydro-2,5,5,8a-tetramethyl-(2.alpha.,4a.alpha.,8a.alpha.)- | 67.89 |
| 1H-Indene, 2,3-dihydro-1,1,5,6-tetramethyl- | 69.22 |
| Naphthalene, 1,2-dihydro-1,1,6-trimethyl- | 69.44 |
| Benzene, 1,1'-(1,1,2,2-tetramethyl-1,2-ethanediyl)bis[4-methyl- | 69.88 |
| Benzene, 1,4-dimethyl-2,5-bis(1-methylethyl)- | 70.03 |
| Naphthalene, 1,2-dihydro-1,4,6-trimethyl- | 71.84 |
| Benzene, 2-(1,3-butadienyl)-1,3,5-trimethyl- | 72.08 |
| Naphthalene, 2,6-dimethyl- | 72.84 |

[a]Female African elephant (age 24 yrs. at time of sampling), Seneca Park Zoo, Rochester, NY.

and Wildlife Safari Park. We thank the National Science Foundation for financial support. We are grateful to Scott and Heidi Riddle for their advice and for facilitating observations at Riddle's Elephant Sanctuary. We appreciate the gift of several chemical samples from Augustus Oils Limited. Mary Wiese provided valuable assistance in the preparation of this manuscript.

## 5. REFERENCES

Arthur, C. L., and Pawliszyn, J., 1990, Solid phase microextraction with thermal desorption using fused silica optical fibers, *Anal. Chem.* **62**:2145-2148.

Brown, J., 2000, Reproductive endocrine monitoring of elephants: an essential tool for assisting captive management, *Zoo Biol.* **19**:347-367.

Dehnhard, M., Hildebrand, T. B., Knauf, T., Ochs, A., Ringleb, J., and Goritz, F., 2003, Chemical signals in giant panda urine (*Ailuropoda melanoleuca*), Chemical Signals in Vertebrates X meeting, Oregon State University, see chapter in this book.

Ehrenfreund, J., Zink, M. P., Wolf, H. R., 1974, Oxydation von Allylalkoholen mit Blei(IV)-acetat, *Helv. Chim. Acta* **57**:1098-1116.
Eisenberg, J. F., McKay, G. M., and Jainudeen, M. K., 1971, Reproductive behavior of the Asiatic elephant (*Elephas maximus maximus* L.), *Behaviour* **38**:193-225.
Francke, W., Schulz, S., Sinnwell, V., Konig, W. A., and Roisin, Y., 1989, Epoxytetrahydroedulan, a new terpenoid from the hairpencils of *Euploea* (Lep.: Danainae) butterflies, *Liebigs Ann. Chem.* 1195-1201.
Goodwin, T. E., Rasmussen, L. E. L., Guinn, A. C., McKelvey, S. S., Gunawardena, R., Riddle, S. W., and Riddle, H. S., 1999, African elephant sesquiterpenes, *J. Nat. Prod.* **62**:1570-1572.
Goodwin, T. E., Brown, F. D., Counts, R. W., Dowdy, N. C., Fraley, P. L., Hughes, R. A., Liu, D. Z., Mashburn, C. D., Rankin, J. D., Roberson, R. S., Wooley, K. D., Rasmussen, E. L., Riddle, S. W., Riddle, H. S., and Schulz, S., 2002, African elephant sesquiterpenes. II. Identification and synthesis of new derivatives of 2,3-dihydrofarnesol, *J. Nat. Prod.* **65**:1319-1322.
Hess, D. L., Schmidt, A. M., and Schmidt, M. J., 1983, Reproductive cycle of the Asian elephant (*Elephas maximus*) in captivity, *Biol. Reprod.* **28**:767-773.
Kaiser, R., and Lamparsky, D., 1979, Volatile constituents of osmanthus absolute, in: *Essential Oils*, B. D. Mookherjee, and C. J. Mussinan, eds., American Chemical Society, Washington, DC, p. 195.
Lazar, J., 2001, Elephant sex pheromone transport and recognition, Ph.D. Thesis, University of Utah, Salt Lake City, Utah.
Lazar, J., Greenwood, D., Rasmussen, L. E. L., and Prestwich, G. D., 2002, Molecular and functional characterization of the odorant binding protein of the Asian elephant, *Elephas maximus*: Implications for the role of lipocalins in mammalian olfaction, *Biochemistry* **41**:11786-11794.
Mills, G. A., and Walker, V., 2001, Headspace solid-phase microextraction profiling of volatile compounds in urine: application to metabolic investigations, *J. Chromatogr. B*, **753**:259-268.
Moss, C. J., 1983, Oestrous behaviour and female choice in the African elephant, *Behaviour* **86**:167-196.
Perrin, T. E., Rasmussen, L. E. L., Gunawardena, R., and Rasmussen, R. A., 1996, A method for collection, long-term storage, and bioassay of labile volatile chemosignals, *J. Chem. Ecol.* **21**:207-221.
Poole, J. H., 1987, Rutting behavior in African elephants: the phenomenon of musth, *Behaviour* **102**:283-316.
Poole, J. H., 1989a, Announcing intent: the aggressive state of musth in African elephants., *Anim. Behav.* **37**:140-152.
Poole, J. H., 1989b, Mate guarding, reproductive success and female choice in African elephants, *Anim. Behav.* **37**:842-849.
Poon, K. F., Lam, P. K., and Lam, M. H., 1999, Determination of polychlorinated biphenyls in human blood serum by SPME, *Chemosphere* **39**:905-912.
Prestwich, G. D., 1987, Chemical studies of pheromone reception and catabolism, in: *Pheromone Biochemistry*, G. D. Prestwich and G. L. Bloomquist, eds., Academic Press, Orlando, pp. 473-527.
Prestwich, G. D., Whitfield, F. B., and Stanley, G., 1976, Synthesis and structures of dihydroedulan I and II trace components from the juice of *Passiflora edulis* Sims, *Tetrahedron* **32**:2945-2948.
Rasmussen, L. E. L., 2001, Source and cyclic release pattern of (Z)-7-dodecenyl acetate, the preovulatory pheromone of the female Asian elephant, *Chem. Senses* **26**:611–624.
Rasmussen, L. E. L., Greenwood, D. R., 2003, Frontalin: a chemical message of musth in Asian elephants, *Chem. Senses* **28**:433-447.
Rasmussen, L. E. L., and Krishnamurthy, V., 2000, How chemical signals integrate Asian elephant society: the known and the unknown, *Zoo Biol.* **19**:405–423.
Rasmussen, L. E. L., and Perrin, T. E., 1999, Physiological correlates of musth: lipid metabolites and chemosignal composition, *Physiol. Behav.* **67**:539–549.
Rasmussen, L. E. L., and Schulte, B. A., 1998, Chemical signals in the reproduction of Asian and African elephants, *Anim. Reprod. Sci.* **53**:19-34.
Rasmussen, L. E. L., and Wittemyer, G., 2002, Chemosignaling of musth by individual wild African elephants (*Loxodonta africana*): implications for conservation and management, *Proc. Royal Soc. London* **269**:853-860.
Rasmussen, L. E. L., Riddle, H. S., and Krishnamurthy, V., 2002, Mellifluous matures to malodorous in musth, *Nature* **415**:975-976.
Rasmussen, L. E. L., Lazar, J., and Greenwood, D. R., 2003, Olfactory adventures of elephantine pheromones, *Biochem. Soc. Trans.* **31**:137-141.
Rasmussen, L. E., Schmidt, M. J., and Daves, G. D., 1986, Chemical communication among Asian elephants, in: *Chemical Signals in Vertebrates: Evolutionary, Ecological, and Comparative Aspects*, D. Duvall, M. Silverstein, and D. Muller-Schwarze, eds., Plenum Press, New York, pp. 627–646.
Rasmussen, L. E., Schmidt, M. J., Henneous, R., Groves, D., and Daves, G. D., Jr., 1982, Asian bull elephants: flehmen-like responses to extractable components in female elephant estrous urine, *Science* **217**:159–162.

Rasmussen, L. E. L., Lee, T. D., Roelofs, W. L., Zhang, A., and Daves, G. D., Jr., 1996, Insect pheromone in elephants, *Nature* **379**:684.

Rasmussen, L. E. L., Lee, T. D., Zhang, A., Roelofs, W. L., and Daves, G. D., Jr., 1997, Purification, identification, concentration and bioactivity of (Z)-7-dodecen-1-yl acetate: sex pheromone of the female Asian elephant, *Elephas maximus, Chem. Senses* **22**:417–438.

Renold, W., Skorianetz, W., Schulte-Elte, K. H., Ohloff, G., 1975, Spiranderivate, Verfahren zu deren Herstellung, und deren Verwendung, Ger. Offen. 2504618 (*Chem. Abstr.* **84**:43818).

Schmidt, G., Full, G., Winterhalter, P., and Schreier, P., 1992, Synthesis and enantiodifferentiation of isomeric theaspiranes, *J. Agric. Food Chem.* **40**:1188-1191, and references contained therein.

Schmidt, G., Full, G., Winterhalter, P., and Schreier, P., 1995, Synthesis and enantiodifferentiation of isomeric 3,5,6,8a-tetrahydro-2,5,5,8a-tetramethyl-2H-1-benzopyrans (edulans I and II), *J. Agric. Food Chem.* **43**:185-188.

Schulte-Elte, K. H., Gautschi, F., Renold, W., Hauser, A., Fankhauser, P., Limacher, J., and Ohloff, G., 1978, Vitispiranes, important constituents of vanilla aroma, *Helv. Chim. Acta* **61**:1125-1133.

Schulz, S., Boppré, M., and Vane-Wright, R. I., 1993, Specific mixtures of secretions from male scent organs of African milkweed butterflies (Danainae), *Phil. Trans. R. Soc. Lond.* B **342**:161-181.

Sukumar, R., 2003, *The Living Elephants: Evolutionary Ecology, Behavior and Conservation*, Oxford University Press, New York.

Whitfield, F. B., and Stanley, G., 1977, The structure and stereochemistry of edulan I and II and the stereochemistry of the 2,5,5,8a-tetramethyl-3,4,4a,5,6,7,8,8a-octahydro-2H-1-benzopyrans, *Aust. J. Chem.* **30**:1073-1091.

Winterhalter, P., Lander, V., and Schreier, P., 1987, Influence of sample preparation on the composition of quince (*Cydonia oblonga*, Mill.) flavor, *J. Agric. Food Chem.* **35**:335-337.

Yamazaki, K., Beauchamp, G. K., Singer, A., Bard, J., and Boyse, E. A., 1999, Odor types: their origin and composition, *Proc. Natl. Acad. Sci. USA*, **96**:1522-1525.

Young, J-j., Jung, L-j., and Cheng, K-m., 2000, Amberlyst-15-catalyzed $S_N2'$ oxaspirocyclization of secondary allylic alcohols. Application to the total synthesis of spirocyclic ethers theaspirane and theaspirone, *Tetrahedron Lett.* **41**:3415-3418.

# ASSESSING CHEMICAL COMMUNICATION IN ELEPHANTS

Bruce A. Schulte, Kathryn Bagley, Maureen Correll, Amy Gray, Sarah M. Heineman, Helen Loizi, Michelle Malament, Nancy L. Scott, Barbara E. Slade, Lauren Stanley, Thomas E. Goodwin, and L.E.L. Rasmussen[*]

## 1. INTRODUCTION

The study of chemical communication in elephants has resulted in startling and exciting new discoveries in the past decade (Rasmussen and Schulte, 1998; Rasmussen et al., 2003). To date, the highlight of this research has been the identification of two compounds that serve as pheromones (as defined by Karlson and Lüscher, 1959). Rasmussen et al. (1996, 1997) identified (Z)-7-dodecen-1-yl acetate (Z7-12:Ac) in female Asian elephant urine collected during the pre-ovulatory period. This estrus pheromone signals approaching ovulation to conspecific males but elicits little interest from female conspecifics. The compound is not unique to elephants; Z7-12:Ac is a component of the mating pheromones for numerous lepidopteran species. More recently, Rasmussen and Greenwood (2003) isolated frontalin, a known aggregation pheromone in bark beetles (Kinzer et al., 1969), from male Asian elephant temporal gland secretion. Male and female Asian elephants exhibit a range of behaviors when exposed to frontalin. The responses depend on the age, sex, and status of the receiver. Male Asian elephants release temporal gland secretion profusely only during musth (Jainudeen et al., 1972a,b), suggesting that frontalin carries a musth-alerting message. Thus, two single compounds, likely acting in conjunction with other chemical components such as proteins (Lazar et al., 2002), mediate sexual and social interactions among Asian elephants.

The success of such investigations hinges upon readily observable behaviors that reveal chemical reception, that is, a successful bioassay protocol (Mackintosh, 1985;

---

[*] Bruce A. Schulte, Kathryn Bagley, Amy Gray, Helen Loizi, Lauren Stanley, Georgia Southern University, Statesboro, GA, 30460. Maureen Correll, College of William and Mary, Williamsburg, VA, 23186. Sarah M. Heineman, Thomas E. Goodwin, Hendrix College, Conway, AR, 72032. Michelle Malament, Miami University, Oxford, OH, 45056. Nancy L. Scott, Barbara Slade, Portland State University, Portland, OR, 97207. L.E.L. Rasmussen, Oregon Health and Sciences University, Beaverton, OR, 97006.

Wyatt, 2003). Understanding independent variables, such as age, sex, and physiological status that affect receiver responsiveness enhances the repeatability of bioassay results. Elephants use their trunks extensively in the exploration of their environment. The trunk is a modified nose and upper lip with a terminus composed of tactile hairs and free nerve endings (Rasmussen and Munger, 1996). We have established a small set of easily observable and distinguishable behaviors that indicate chemosensory interest by Asian and African elephants. Over years of conducting studies with captive and wild elephants, we have related variation in expression of these significant behaviors to critical independent variables as noted above. Through intensive observations we have identified numerous additional behaviors associated with chemosignal detection and response. The expansion of our bioassay ethogram provides us with information that may expose more independent variables of importance to understanding elephant communication. Finally, these additional dependent variables, the response behaviors, provide insight into the receiver's psychological landscape (Guilford and Dawkins, 1991).

To be effective signals, chemicals must be detectable, recognizable, and remembered. Factors affecting responses by a receiver are part of its psychological landscape and thus must be considered to decipher signal meaning (Guilford and Dawkins, 1991). We are examining the variation in response behavior by elephants to assist in our interpretation of the chemical released-chemical detected-information acquired complex. This complex could easily involve multiple sensory channels because visual, vocal, and tactile components often accompany chemical signal reception and response (Rasmussen, 1988; Rowe, 1999; Langbauer, 2000). In this paper, we primarily discuss communication in elephants via chemical signals.

## 2. MAIN BEHAVIORS OF CHEMOSENSORY BIOASSAY

An elephant's trunk leads the animal through its habitat. Whether an elephant is standing still or in transit, the trunk moves almost constantly. Over 20 years ago, Rasmussen et al. (1982) documented a flehmen-like behavior exhibited by male Asian elephants. The frequency of this behavior over years of bioassays led to the discovery of the estrous pheromone (Rasmussen et al., 1996). Explicit proof that the preovulatory pheromone moves from a urine puddle to the male Asian elephant's vomeronasal organ (VNO) is in progress but has not been yet obtained. We do know that $Z7$-$12$:Ac binds to a protein in the VNO ductal mucus (Rasmussen et al., 2003). Hence, this distinctive trunk movement by elephants appears to transfer chemical signals to the VNO, and we term this behavior flehmen rather than "flehmen-like" (Schneider, 1930; Estes, 1972). Preceding a flehmen, elephants sniff and contact the chemical source (Rasmussen et al., 1986; Rasmussen, 1988). Signal detection or transfer occurs by two primary contact responses. A check response occurs when the very end of the trunk touches the sample (Asian elephants have a single anterior "finger" while African elephants have anterior and posterior fingers). A place contact transpires when the end of the trunk flattens and forms a seal with the substrate. Olfaction and probably tactile reception accompany both behaviors. Four behaviors - sniff, check, place, and flehmen - form the major response sequence of our elephant semiochemistry bioassay (Figure 1).

The frequency of these four behaviors as well as the duration of time spent near the samples reflects level of interest (Rasmussen et al., 1982, 1996; Schulte and Rasmussen, 1999; Scott, 2002; Slade et al., 2003). The sequence of response is somewhat linear in that the order of responses is often conserved and the frequency decreases from olfactory exploration (sniff) to putative VNO input (flehmen). Sniffs may provide identification of critical sample features; sometimes, no additional information seems required. At other times, the initial information sets in motion a response sequence suggestive of a requirement for additional olfactory information, obtained via checks, places and flehmen. However, the sequence is not invariable. After the first bout of investigation, elephants often directly contact a sample with little hovering of the trunk over the sample. Although olfactory input probably is still occurring, it is not as pronounced as an overt sniff. Checks also may be reduced in frequency with "places" becoming more prominent (especially in females). In some situations, the place response is omitted and the rate of check-flehmen increased (common in males if the sample is fresh such as newly released or placed urine). The neuronal feedback loops that control the observed sequence are not well understood. Current methods to assess sensory input, such as electroolfactograms, neural pathway analysis, and brain activity are not readily applicable to elephants (Hildebrand, 1995; Johnston, 1999; Keverne, 1999; Firestein, 2001). Hence, we rely on a combination of chemical, molecular, biochemical, and behavioral analyses to identify signals, to determine the route of detection, and to decipher the meaning of the message. Because signal meaning is receiver-dependent, knowledge about differences in receivers is essential.

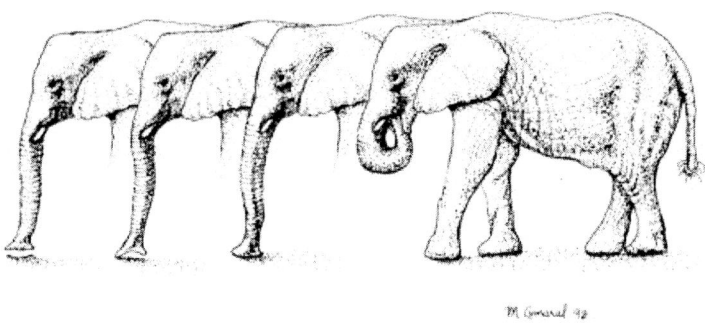

**Figure 1.** Main sequence of chemosensory trunk responses to a putative chemical signal. From left to right, sniff, check, place and flehmen (see also Schulte and Rasmussen, 1999). Drawing by Mary Amaral.

## 3. RECEIVER VARIATION

Reproductive condition, sex, age, experience, and dominance status (i.e., receiver "states") are some of the primary variables that affect receiver behavior when exposed to a chemical signal (Doty and Ferguson-Segal, 1989; Schulte and Rasmussen, 1999). Alteration in behavior with state could result for one of two reasons. First, the receiver cannot detect the signal except in certain states (e.g., some combination of the above variables). Hypothetically, a male elephant in musth with highly elevated testosterone levels might detect chemicals or concentrations that males not in musth cannot detect. Second, detection occurs but signal relevance changes. In this case, the nonmusth male may perceive that the signal is from a receptive female, but his relatively lower competitive condition may obviate high interest in the signal source. Some results suggest the latter may be applicable (Rasmussen et al., 1997). To distinguish among these hypotheses, more definitive records of behavioral responses are required because the lack of detectable behavioral response will not permit resolution between these two factors. However, the observation of particular responses excludes hypothesis one. The challenge is to determine what variables are responsible for reduced or different subsequent behaviors upon confirming detection.

Experiments using Z7-12:Ac show that Asian male elephants detect the chemical signal but exhibit very different responses depending on breeding experience or dominance (Rasmussen et al., 1997). Experienced and generally more dominant males displayed high flehmen rates and sometimes exhibited erections. However, sexually inexperienced males, generally considered subordinate, typically sniffed then backed away. Referring to the two hypotheses above, in the absence of signal detection, males would continue their forward movement. The avoidance responses by these males support hypothesis two. For captive female Asian elephants, differences in behavior to conspecific female urine reflect stage of the estrous cycle (Slade et al., 2003). Only females in the follicular phase responded more to follicular than luteal stage urine (Figure 2). Females in the follicular phase demonstrated greater interest in follicular urine by both greater rates of sniffs and checks and by the occurrence of place and flehmen responses. Females in their follicular compared to luteal phase also were more responsive to male urine, especially from a male in musth (Schulte and Rasmussen, 1999).

Our standard control in these studies is vanilla extract in water. The control generally elicits a low level of interest, mainly sniffs and checks. We interpret this as interest in an odor source but not one with a meaningful signal (i.e., with evolutionarily evolved meaning). Female Asian elephants often show about the same interest in luteal urine as our control. However, luteal urine does evoke place and flehmen response on occasion, whereas, the control does not (Slade et al., 2003). Luteal urine probably contains cues on sender identity, but chemicals either specific to or more abundant in follicular urine reveal reproductive readiness.

For female-to-female communication, the relevance of these chemicals appears restricted to females in the follicular phase. However, reproductive state is not an on-off condition. Hormone levels change cyclically and so responses are likely to be graded rather than discrete. The frequency of the main behaviors and the duration of time near the sample provide some degree of graded responsiveness. Yet, individual variation in these main response behaviors and the quantification of additional behaviors related to

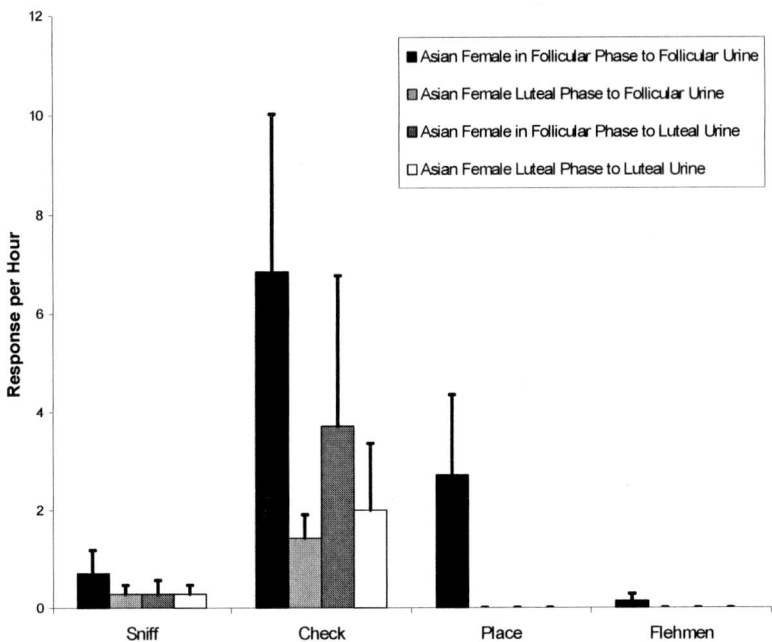

**Figure 2.** Main chemosensory responses (see Figure 1) by captive female Asian elephants in their luteal and follicular stages of the estrous cycle to luteal and follicular urine from a conspecific female (see Slade et al., 2003).

signal detection (dependent variables) are likely to reveal even more about signal meaning for particular types of receivers (independent variable).

## 4. ADDITIONAL BEHAVIORS: EXPANDING THE ETHOGRAM

The flehmen trunk movement is easy to differentiate from other trunk movements irrespective of elephant species or sex. Variation in the duration and frequency of flehmen may be dependent on the chemical sample or perhaps on the particular style of an individual elephant (Rasmussen, unpubl. data). Check and place responses also are distinct. Because African elephants have two trunk tip fingers, compared to one for the Asian species, checks can be more variable for the African species (i.e., anterior, posterior or simultaneous checks are possible). Behaviorally, the multiple positions in which the trunk moves for sniffing are perhaps the most intriguing types of variation. As defined for a bioassay scenario, a sniff occurs when the trunk hovers over a sample that we place on the ground (Figure 1). This is the most direct means by which an elephant can intake headspace over the sample; however, air intake can occur using other trunk positions.

The posture of the trunk may reflect the most efficient means of detecting odors from different locations. Chemical signals may emanate from urine, feces, temporal gland secretion or other bodily fluid on a horizontal or vertical substrate, as well as from visible conspecifics or distant individuals. We describe a number of trunk positions and discuss sample intake and potential functional significance beyond airflow to and from the lungs. An elephant may reach directly out toward the nearby object of interest (reach). The elephant looks larger as it thrusts out its trunk, a formidable weapon as well as a passageway to the main olfactory system. To sample air currents, elephants commonly raise their trunk in a periscopic fashion (Poole, 1999; pers. obs.). This position places the trunk above the body height of nearby elephants, potentially sampling cues from a distance. The posture almost shows a disinterest in the immediate surroundings and appears non-threatening. When elephants are within tens of meters of objects, their trunk position becomes enigmatic. The trunk may curl over (overpass), under (underpass), sideways (side door), or around in a helical fashion (roundabout). The purpose of these various postures is unclear. The elephant may be investigating itself (e.g., during an underpass the trunk tip is pointed up toward the elephant's own head), but typically the trunk tip points toward an object such as an observer or another elephant. The curvature exhibited by the trunk may be beneficial for retaining volatile compounds by allowing longer association /disassociation times with the proteinaceous compounds of the truncal mucus. The trunk position also may convey a behavioral message to a receiver. For instance, a direct reach posture may imply a potential threat in addition to improving the efficiency of air sampling. The curled postures may be less hostile or even provide a dishonest signal of pretended friendliness. Elephants may appear to leave the area only to move around behind the object of interest and approach from a vantage position that is potentially more informative. We are tracking the use of different trunk postures to assess potential communicative function to their use. Not only are behaviors after signal detection important, but receiver psychology may be decoded by careful scrutiny of the means of signal reception.

Events related to sensory reception involve both the intake of putative chemical signals and other intricate processes. We have documented at least ten distinct, additional behaviors that occur during sample investigation (Table 1). For some of these behaviors, the purpose seems apparent but not quantitatively proven. The trunk tip may dig into, blow on, suck up or rub around in urine soaked dirt or dried feces to locate liquid bound chemicals or release volatile chemicals. The tip can pinch shut as if to close off additional odor intake and to isolate odors already acquired into the trunk. The trunk may shake back and forth after touching a sample and this may precede flehmen (Scott, 2002; Rasmussen and Greenwood, 2003). Some of these same behaviors also may serve to clear the trunk of odor or particles. Blow, flick, and shake are all behaviors that release material from the trunk. Audible behaviors such as blow might communicate to others, either intentionally or through eavesdropping (Bradbury and Vehrencamp, 1998). Vocalizations are more likely to be purposeful responses to chemical message reception. We have observed elephants dusting themselves with chemical samples, such as dilute acetic acid or musth urine. Such anointing occurs in other species such as moose to facilitate mate attraction (Miquelle and Van Ballenberghe, 1985), but the purpose in

**Table 1.** Possible function and description of elephant trunk behaviors related to chemical signal reception.

| Possible function | Behavior | Operational definition |
|---|---|---|
| Increase sample detection | Dig | Trunk tip moves substrate, creating a hole |
| | Blow | Forceful audible exhalation often accompanied by particle release and sometimes substrate movement |
| | Pinch | End of trunk closes in or above sample |
| | Rub or Scrub | End of trunk flattens and moves around in sample |
| | Shake | Trunk is off substrate and lower half or more of the trunk swings up and often around in a corkscrew pattern |
| | Suck | Strong inhalation with trunk on substrate accompanied by observable, muscular, truncal contractions |
| Clearance | Blow | See above |
| | Flick | End of trunk (< ca. 25 cm) moves back and forth similar to a human wrist motion |
| | Shake | See above |
| Anoint | Dust | Trunk grabs substrate containing chemical sample and distributes substrate on elephant |
| Respond to signaler or others | Blow | See above |
| | Vocalize | Audible or infrasonic sound production while investigating sample |
| Lose interest or distracted | Aborted flehmen | Trunk exhibits a check or place response (Figure 1) and curls to chin level and then is dropped before entering palate |

elephants is uncertain. At times, an elephant appears to be performing a flehmen but the trunk tip stops at about chin level. In some cases, the animal seemed to be distracted but in other instances, a loss of interest would be a more appropriate explanation. In the latter circumstances, processing of olfactory reception may break off the chemical transfer to the VNO openings in the palate.

Numerous other behaviors can vary among elephants upon detecting a particular chemical signal. Males exhibit several penis-involved behaviors such as unsheathing, erection, belly hitting, and trunk touching. Changes in body positions include anterior elevations in stature, pinna of ears assuming a 90-degree position to main axis of body, and ear(s) flapping or waving. The directional movement of the elephant also may change. An individual may back up, become motionless, depart at an angle to its original direction, or go around and then forward. Each behavior potentially reveals something about signal meaning to the individual receiver.

The challenge lies in documenting the occurrence of these various additional behaviors to understand better the sender-receiver interaction. We are doing this through field observations and more tightly controlled biological assays with elephants. We can select elephants with specific attributes and in particular conditional states as bioassay subjects and for donors of chemical samples, such as urine. Chemical samples from the same sender can vary in numerous ways, including amount of deposition, substrate, time of deposition, and condition of the sender. Because chemical signals can be received in the absence of the sender, variation of response behaviors to the same chemical sample reflects attributes of the receiver, alterations in environmental conditions or a combination of the two.

## 5. USING VARIATION TO UNDERSTAND RECEIVER PSYCHOLOGY

The ability to detect chemical signals clearly varies with environmental conditions such as temperature, relative humidity, wind velocity and direction. We are recording such important environmental variables to correlate them with measures of response. Currently, we are most interested in variation in receiver responses that cannot be explained by abiotic environmental factors. Of special interest is the influence of group composition, sex, and learning on response patterns. We are examining the development of chemosensory responses to particular types of chemical samples, namely urine from female elephants at different periods in their reproductive cycle (e.g., luteal and follicular phase of estrous). We have documented conspecific adult patterns of behavior to urine from Asian elephants (Rasmussen et al., 1982, 1996; Schulte and Rasmussen, 1999; Scott, 2002; Slade et al., 2003) and are collecting data on responses to African elephant urine. At some point in maturation, elephants respond in an appropriate fashion to a specific signal. The type and degree of response shown by adults has an ontogenetic history that we are trying to map.

Our expanded ethogram provides greater information on the developmental pattern for understanding variation among adults to a specific chemical signal, such as an estrus pheromone. Male and female elephants mature at similar ages physiologically, but females are able to mate at a much younger age than most males because of social factors (Laws, 1969; Eisenberg, 1980; Hildebrandt et al., 1998; Poole, 1999; Moss, 2001). Males enter their first full musth during their early twenties and are not generally able to mate females until this age. Male Asian elephants experience a moda musth during their teenage subadult years (Rasmussen et al., 2002), but until they are larger cannot compete with older males for access to females. We predict that direct assessment of follicular urine and an estrus pheromone through our main bioassay behaviors (Fig. 1) will vary with maturation and social status (e.g., see Rasmussen et al., 1997). In addition, we are examining if additional behaviors will show a complementary pattern that elucidates receiver psychology. We present one hypothetical example to illustrate.

Male elephants with the greatest ability to mate with females should exhibit the greatest interest in urine from reproductively ready females (Fig. 3). If we suppose that male elephants reach their reproductive peak during their forties (Poole, 1987, 1989a,b; Moss, 2001), then these males would show the greatest rates of flehmen. Inexperience and a lower social status would inhibit males at a younger age from fully responding to the urinary signal. High response rates could be especially dangerous if the female's signal has a priming type effect, making the male more amorous and likely to be confronted by larger males in musth. The reduced condition of extremely old males could dampen their responsiveness to the urinary signal, or lengthy experience could quicken inspection.

Rates of response for the main chemosensory behaviors may be a somewhat large-grain measure of variation by age. By incorporating additional behaviors, we might gain insight on finer grain developmental patterns (Figure 3). For instance, younger elephants, especially those pre-musth, would be predicted to move away from pre-ovulatory urine sooner, vocalize more, and blow more readily to vent the odor. These behaviors would curtail as males mature and have greater opportunity to mate. Other behaviors may increase with age, such as remaining motionless at the urine, possibly with ears erect in a listening stance, as well as digging and sucking at the sample to enhance uptake. By developing general patterns of response, we can better understand the status and social history of particular individuals. Comparisons across populations could be revealing for conservation purposes. If the hypothetical pattern illustrated in Figure 3 were documented, then the relative age for peaks in behaviors may show differences in demographics and breeding success among populations. For example, a heavily poached population might show a strong leftward shift in the pattern. The benefit of introducing older males could be assessed by the shift in behaviors over time.

The primary benefit of including more behaviors in a standard bioassay, especially those done in the field, is to determine the relevant sender-receiver complex for chemical signals and their meaning. A particular chemical may have evolved under a singular

**Figure 3.** Hypothetical responses of male elephants at different ages to pre-ovulatory urine that contains an estrous pheromone. The main response of signal detection is flehmen. Additional behaviors may reveal aspects of receiver condition, status, and experience. Behavior one might include "moves away", "vocalize", and "blow". Behavior two might include motionless with ears erect, dig, and suck. Ages are estimates and would vary with populations and their histories of human intrusion and poaching.

selective pressure and thus have a singular meaning to one type of receiver (i.e., the intended receiver). Yet, for many animals, this meaning may change with age, reproductive state, social status, and the like. The suites of behaviors that accompany the primary receptive responses are likely to expose the role of such signals across a wide range of receiver types. In Chaos Theory, slight variation in initial conditions, such as the flapping of a butterfly's wings, can have a profound effect on the observed outcome, such as weather patterns a thousand kilometers away (hence, known as the butterfly effect, Lorenz, 1963). The release of a chemical signal may signify the flapping butterfly of communication in elephant and other animal societies.

## 6. ACKNOWLEDGEMENTS

We thank our respective colleges and universities for support of our scholarly activities and Oregon State University for hosting the tenth meeting. Funding is provided from the Division of Integrative Biology and Neuroscience of the National Science Foundation (Award Number IBN-0217068), Biospherics Research Corporation, Georgia Southern University (Faculty Research Grant), and Hendrix College. Numerous facilities housing captive elephants have provided assistance and expertise throughout our studies. We especially thank Riddle's Elephant Sanctuary, the Oregon Zoo, Roger Williams Park Zoo, and Ringling's CEC. Additional thanks to the following zoos: Bowmanville, Cameron Park, Indianapolis, Jacksonville, Knoxville, Louisville, Miami Metro, Nashville (R. and C. Pankow), North Carolina, Sedgwick County, Seneca Park, Six Flags Marine World, West Palm Beach, and Wildlife Safari Park. We have benefited from research on African elephants at Addo Elephant National Park with assistance from SANP and TERU and at the Save the Elephants facility in Samburu, Kenya. Studies with Asian elephants outside of North America have occurred through cooperation with the Union of Myanmar and the Asian Elephant Research and Conservation Center, India. Dr. Mimi Halpern provided valuable comments on flehmen and the VNO. Our families have given us unbounded support throughout our studies.

## 7. REFERENCES

Bradbury, J. W., and Vehrencamp, S.L., 1998, *Principles of Animal Communication*, Sinauer Associates, Massachusetts, pp. 737-738.
Doty, R., and Ferguson-Segal, M., 1989, Influence of castration on the odor detection performance of male rats, *Behav. Neurosci.* **103**:691-693.
Eisenberg, J. F., 1980, Recent research on the biology of the Asiatic elephant (*Elephas m. maximus*) on Sri Lanka, *Spolia Zeglavica* **35**:213-218.
Estes, R.D., 1972, The role of the vomeronasal organ in mammalian reproduction, *Mammalia* **36**:315–341.
Firestein, S., 2001, How the olfactory system makes sense of scents, *Nature* **413**:211-218.
Guilford, T. and Dawkins, M. S., 1991, Receiver psychology and the evolution of animal signals, *Anim. Behav.* **42**:1-14.
Hildebrand, J. G., 1995, Analysis of chemical signals by nervous systems, in: *Chemical Ecology: The Chemistry of Biotic Interaction*, T. Eisner and J. Meinwald, eds., National Academy Press, Washington, D.C., pp. 161-181.
Hildebrandt, T. B., Göritz, F., Pratt, N. C., Schmitt, D. L., Quandt, S., Raath, J., and Hofmann, R. R., 1998, Reproductive assessment of male elephants (*Loxodonta africana* and *Elephas maximus*) by ultrasonography, *J. Zoo Wildl. Med.* **29**:114-128.

Jainudeen, M. R., Katongole, C. B., and Short, R. V., 1972a., Plasma testosterone levels in relation to musth and sexual activity in the male Asiatic elephant, *Elephas maximus, J. Reprod. Fertil.* **29**:99-103.
Jainudeen, M .R., McKay, G. M., and Eisenberg, J. F., 1972b, Observations on musth in the domesticated Asiatic elephant, *Mammalia* **36**:247-261.
Johnston, R. E., 1999, Neural mechanism of communication: from pheromones to mosaic signals, in: *Chemical Signals in Vertebrates 9*, A. Marchlewska-Koj, J.J. Lepri and D. Müller-Schwarze, eds., Kluwer Academic/Plenum Publishers, New York, pp. 61-68.
Karlson, P., and Lüscher, M., 1959, 'Pheromones': a new term for a class of biologically active substances, *Nature* **183**:55-56.
Keverne, E. B., 1999, The vomeronasal organ, *Science* **286**:716-720.
Kinzer, G. W., Fentiman, A. F., Jr, Page, T. F., Jr, Foltz, R. L., Vitė, J. P., and Pitman, G. B., 1969, Bark beetle attractants: identification, synthesis and field bioassay of a new compound isolated from Dendroctonus, *Nature* **221**:447-448.
Langbauer, W. R., Jr, 2000, Elephant communication, *Zoo Biol.* **19**:425-446.
Laws, R. M., 1969, Aspects of reproduction in the African elephant, *Loxodonta africana, J. Reprod. Fertil.* **6**:193-217.
Lazar, J., Greenwood, D. R., Rasmussen, L. E. L., and Prestwich, G. D., 2002, Molecular and functional characterization of an odorant binding protein of the Asian elephant, *Elephas maximus*: implications for the role of lipocalins in mammalian olfaction, *Biochemistry* **41**:11786-11794.
Lorenz, E.N., 1963, Deterministic nonperiodic flow, *Journal of the Atmospheric Sciences* **20**:130-141.
Mackintosh, J.H., 1985, The bioassay of mammalian olfactory signals, *Mammal Rev.* **15**:57-70.
Miquelle, D.G., and Van Ballenberghe, V., 1985, The moose bell: a visual or olfactory communicator? *Alces* **21**:191-213.
Moss, C.J., 2001, The demography of an African elephant (*Loxodonta africana*) population in Amboseli, Kenya, *J. Zool. London* **255**:145-156.
Poole, J.H., 1987, Rutting behavior in African elephants: the phenomenon of musth, ***Behaviour* 102:283-316.**
Poole, J.H., 1989a, Announcing intent: the aggressive state of musth in African elephants., *Anim. Behav.* **37**:140-152.
Poole, J.H., 1989b, Mate guarding, reproductive success and female choice in African elephants, *Anim. Behav.* **37**:842-849.
Poole, J.H., 1999, Signals and assessment in African elephants: evidence from playback experiments, *Anim. Behav.* **58**:185-194.
Rasmussen, L.E.L., 1988, Chemosensory responses in two species of elephants to constituents of temporal gland secretion and musth urine, *J. Chem. Ecol.* **14**:1687-1711.
Rasmussen, L.E.L., 1998, Chemical communication: an integral part of functional Asian elephant (*Elephas maximus*) society, *Ecoscience* **5**:410-426.
Rasmussen, L.E.L., and Greenwood, D.R., 2003, Frontalin: a chemical message of musth in Asian elephants, *Elephas maximus, Chemical Senses* **28**:433-446.
Rasmussen, L.E.L., and Munger, B., 1996, The sensorineural specializations of the trunk tip (finger) of the Asian elephant (*Elephas maximus*), *Anat. Rec.* **246**:127-134.
Rasmussen, L.E.L., and Schulte, B.A., 1998, The importance of chemical signals in the reproduction of Asian and African elephants, *Anim. Reprod. Science* **53**:19-34.
Rasmussen, L.E.L., Schmidt, M.J., Henneous, R., Groves, D., and Daves, G.D., Jr, 1982, Asian bull elephants: flehmen-like responses to extractable components in female elephant estrous urine, *Science* **217**:159-162.
Rasmussen, L.E.L., Schmidt, M.J., and Daves, G.D., Jr, 1986, Chemical communication among Asian elephants, in: *Chemical Signals in Vertebrates IV*, D. Duvall, D. Müller-Schwarze, and R. Silverstein, eds., Plenum, NY, pp. 627-645.
Rasmussen, L.E.L., Lee, T.D., Roelofs, W.L., Zhang, A., and Daves, G.D., Jr, 1996, Asian elephants and Lepidoptera have a common sex pheromone, *Nature* **379**:684.
Rasmussen, L.E.L., Lee, T.D., Zhang, A., Roelofs, W.L., and Daves, Jr., G.D., 1997, Purification, identification, concentration and bioactivity of (Z)-7-dodecen-1-yl acetate: sex pheromone of the female Asian elephant, *Elephas maximus, Chemical Senses* **22**:417-437.
Rasmussen, L.E.L., Riddle, H.S., and Krishnamurthy, V., 2002, Mellifluous matures to malodorous in musth, *Nature* **415**:975-976.
Rasmussen, L.E.L., Lazar, J., and Greenwood, D.R., 2003, Olfactory adventures of elephantine pheromones, *Biochemical Society Transactions* **31**:137-141.
Rowe, C., 1999, Receiver psychology and the evolution of multicomponent signals, *Anim. Behav.* **58**:921-931.
Schneider, K.M., 1930, Das Flehmen, *Zool. Gart.* **43**:183-198.

Schulte, B.A., and Rasmussen, L.E.L., 1999, Signal-receiver interplay in the communication of male condition by Asian elephants, *Anim. Behav.* **57**:1265-1274.
Scott, N.L., 2002, Chemical communication and musth in captive male elephants, M.Sc. thesis, Portland State University, Portland, OR, 71 pp.
Slade, B.E., Schulte, B.A. and Rasmussen, L.E.L., 2003, Oestrous state dynamics in chemical communication by captive female Asian elephants, *Anim. Behav.* **65**:813-819.
Wyatt, T.D., 2003, Pheromones and Animal Behaviour: Communication by Smell and Taste, Cambridge University Press, U.K., pp. 23-36.

# THE GLAND AND THE SAC – THE PREORBITAL APPARATUS OF MUNTJACS

Susan J. Rehorek, Willem J. Hillenius, John Kennaugh, and Norma Chapman[*]

## 1. INTRODUCTION

A preorbital fossa is a bilateral depression in the skull, located anterior to the eye. Such depressions occur in several ungulate mammals, but are most well known in bovid and cervid artiodactyls. These depressions, and the "glands" that are typically associated with them, are known by a variety of names, including antorbital, preorbital, suborbital fossa and gland, and even maxillary gland or organ (see Schaffer, 1940 for review). Usually, little attention is paid to the degree of glandular development, and the term "gland" is typically applied quite indiscriminately. However, the glandular nature of the integumentary epithelium lining the preorbital fossa varies considerably among taxa. In antilopine and cephalophine bovids the fossa contains a single, well-developed, and encapsulated gland, drained by 1-2 ducts (Schaffer, 1940; Richter, 1971, 1973; Kuhn, 1976; Mainoya, 1978). In contrast, in other bovids (e.g, sheep), and in deer, such a specialized gland is not present. In these cases the lining of the fossa is broadly open to the exterior, and possesses only a few more glandular elements than the surrounding skin (Schaffer, 1940; Quay, 1955; Quay and Müller-Schwarze, 1970; Lincoln, 1971; Mossing and Källquist, 1981; Gray et al., 1988). It is inappropriate to use the term "gland" in this context, and we propose the term preorbital sac for these cases.

In many cervids and bovids, a pigmented furrow connects the anterior aspect of the orbit to the posterior surface of the preorbital sac (Schaffer, 1940). It has been proposed that orbital secretions pass from the orbit to the preorbital sac via this furrow in deer and thus perhaps contribute to the exudate of the preorbital sac (Harder, 1694; Schaffer, 1940; Quay and Müller-Schwarze, 1970). The nearest orbital gland is the Harderian gland, which is located in the anterior aspect of the orbit, and the ducts of which open on the

---

[*] Susan J. Rehorek, Slippery Rock University, Slippery Rock, Pennsylvania, 16257.   Willem J. Hillenius, College of Charleston, Charleston, South Carolina, 29424. John Kennaugh, Prestwich, United Kingdom, M25 3DR. Norma Chapman, Barton Mills, Suffolk, United Kingdom, 1P28 6AA.

medial surface of the nictitating membrane, very near the pigmented furrow. The chemical nature of the secretant from this gland is largely unknown for artiodactyls, with only a few studies thus far published (Harder, 1694; Miessner, 1900; Sakai, 1981; Rehorek et al., in prep).

Sexual dimorphism of both the size of the preorbital fossa and the development of the preorbital sac have been observed among several cervids and bovids (Schaffer, 1940; Chapman and Chapman, 1982). Comparatively little is known about the histological nature of this difference, especially its effect upon the glandular epithelial lining of the fossa (Schaffer, 1940; Lincoln, 1971; Quay and Müller-Schwarze, 1971; Mossing and Källquist, 1981). There have been no reports of sexual dimorphism in deer Harderian glands.

Muntjacs (Cervidae: Muntiacinae) are very primitive members of the deer family, which nevertheless possess the largest pre-orbital fossa and sac of any cervid (Geist, 1998). Though the anatomy and histology of the pre-orbital fossa of muntjacs (*Muntiacus muntjac* and *M. reevesi*) has been described previously (see Schaffer, 1940 for review), the marked size difference of this feature between sexes (described by Chapman and Chapman, 1982) was not discussed. The focus of this study was to examine the histological nature of the sexual dimorphism of the preorbital sac and to describe the structure of the Harderian gland of the Chinese muntjac, *Muntiacus reevesi*.

## 2. MATERIAL AND METHODS

Specimens of *M. reevesi* were obtained from the collection of Norma Chapman, which is derived from feral populations in the U.K. Thirty adult males and females were examined. Preorbital sacs and Harderian glands were dissected out, fixed in 10% formalin, processed into paraffin and sectioned at 7 – 10 µm. Alternate slides of both the preorbital sac and the Harderian gland were stained with either haematoxylin and eosin or Mallory's stain. The remaining Harderian gland slides were treated with histochemical stains: mucosubstances were detected by the periodic acid-Schiffs (PAS) method, whereas bromphenol blue (BPB) was used to detect protein (Barka and Anderson, 1965). For lipid detection, frozen Harderian gland sections (15 µm) were stained with the supersaturated isopropanol method (Oil Red O: Lillie, 1961), using exposure to 70% alcohol as a control.

## 3. RESULTS

There was marked sexual dimorphism in both the preorbital sac and the Harderian gland in Muntjac. For both structures, the glandular material in males was better developed than in the females.

### 3.1 Preorbital Sac

The preorbital sac of Muntjac lies rostral to the anterior corner of the eye, in the pre-orbital fossa (which is formed by the maxillary, ethmoid and lacrimal bones). The opening of the sac is a long, sigmoidal slit, measuring 20 – 25 mm in fixed material. A pigmented furrow, extending posteriorly from this slit, crosses over the orbital rim and pigmented furrow, extending posteriorly from this slit, crosses over the orbital rim and connects the preorbital sac to the orbit. This furrow appears to be positioned to allow fluid from the anterior region of the eye (at the region of the nictitating membrane) to pass into the preorbital sac.

The fossa and the sac lining it are roughly circular in shape, although the sac is subdivided into two pockets by a ridge, which runs from the posterodorsal to the anteroventral aspects of the sac (Figure 1). Within this ridge, comparatively close to the exterior, runs the vein of the preorbital fossa, a tributary of the sphenopalatine venous plexus. The two pockets of the preorbital sac are lined by debris comprised of shed keratinous cells and various secretions, forming a white mass. Sparse white hair lines the epithelium of the pockets. Histologically, both pockets (anterior and posterior) contain both sebaceous and apocrine sweat glands. Myoepithelial cells surround the tubules of the apocrine sweat glands.

These pockets are not only markedly different from each other, but also exhibit sexual dimorphism. The posterior pocket of both sexes exhibits a similar feature: the upper layers of the epidermis, which are high in lipid content, appear to be sloughing off readily and form a pasty layer covering in the pocket (Figures 1, 2 and 3). In the females, the wall of the posterior pocket is lined by fine hairs (20 µm in diameter), which are associated with small sebaceous glands, few apocrine sweat glands and no apparent arrector pili (Figure 2). However, in the males, not only was more hair present, but also the sebaceous glands were very large, multi-lobed structures and apocrine sweat glands were readily identifiable (Figure 3). The hairs in the posterior pocket of males could furthermore be put into two categories based on thickness: 25 µm and 50 µm. Only the latter category is associated with the sebaceous glands.

The lining of the anterior pocket is thinner, with less evidence of sloughing cells. In comparison to the posterior pocket, the anterior pocket possesses smaller sebaceous glands, and larger, more convoluted sweat glands. In females, the sweat glands were smaller, thin-walled tubules filled with an eosinophilic material (Figure 4). In males, the sweat glands were larger, thicker walled tubules which produced a more globular secretion (Figure 5).

## 3.2 Harderian Gland

The Harderian gland of *M. reevesi* is a large structure, associated with the nictitating membrane. It is composed of two lobes (cf. Rehorek et al., in prep): a white, anterior lobe (formerly regarded as a separate "nictitans" gland) and a posterior, reddish-brown lobe. There is notable sexual dimorphism in the Harderian gland: the posterior lobe of males (average mass $7.62 \pm 1.34$ g, n = 25) is significantly larger ($P<0.0001$) than that of the female (average mass $1.42 \pm 0.4$ g, n = 24). There is no nictitating membrane muscle.

Histologically, both lobes are suffused with much connective tissue, forming septae and organizing the lobes into distinct lobules. There is much interstitial adipose tissue present, though more prominent in the anterior part of the red lobe and its junction with the posterior end of the white lobe. A summary of the histochemistry of the Harderian gland is presented in Table 1. The anterior white lobe is a mucous secreting structure, whereas the posterior lobe appears to produce serous as well as lipid secretions. The posterior lobe is highly vascular, with numerous blood vessels passing within the septae, which accounts for the distinct color.

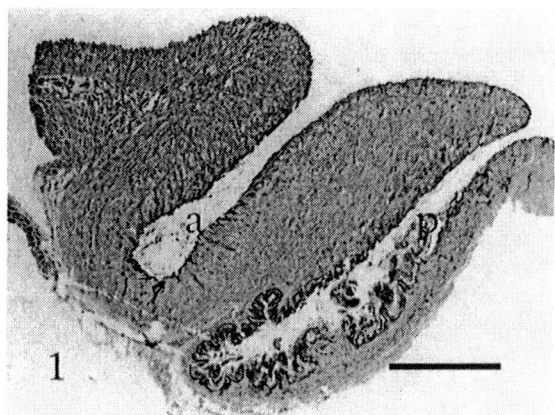

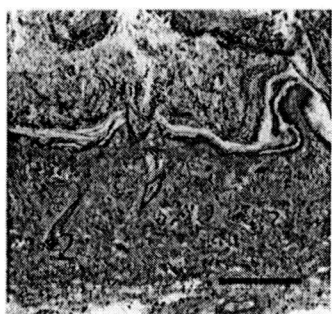 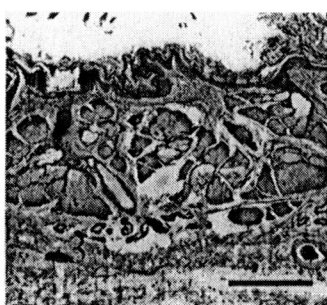

**Figures 1 – 3.** Light micrographs of the preorbital sacs and Harderian glands of *M. reevesi*. Oblique section through preorbital sac of muntjac, showing anterior (a) and posterior (p) pockets (**Figure 1**). Higher magnification of posterior pocket of female (**Figure 2**) and male (**Figure 3**) muntjacs. Scale bars = 3.5 mm (Figure 1) and 0.6 mm (Figures 2 and 3).

**Table 1.** Histochemistry of the Harderian glands of *Muntiacus reevesi*.

| Stain | White lobe | Red lobe |
|---|---|---|
| PAS for mucus | ++ | + |
| BPB for protein | +/- | +/- |
| Oil Red O for lipids | + | ++ |

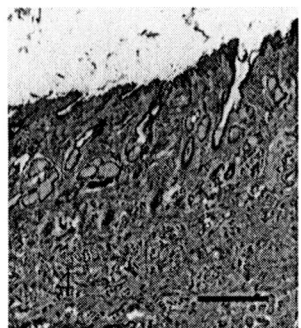
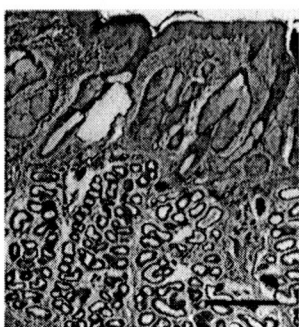

Figures 4 and 5. Light micrographs of the pre-orbital sacs of *M. reevesi*. Higher magnification of the anterior pocket of female (**Figure 4**) and male (**Figure 5**) muntjacs. Scale bars = 0.5 mm (Figure 4) and 0.57 mm (Figure 5).

## 4. DISCUSSION

The preorbital sac of Muntjac share many common features with other deer. Schaffer (1940) noted that the preorbital sac of ungulates is lined by modified skin, whose integumentary structures (especially sebaceous glands, sweat glands and hair) exhibit varying levels of hypertrophy or atrophy among species. In most respects, our observations of the preorbital sac of *M. reevesi* agree closely with Schaffer's (1940) summary on both *M. muntjak* and *M. reevesi*. However, Schaffer (1940) reported neither the presence of two pockets nor any sexual dimorphism. He did, however, cite the Harderian gland as a potential contributor to the secretions found in the preorbital sac.

To date, there has been no other description of a preorbital sac with two pockets in artiodactyls. This condition appears to be unique to *M. reevesi* (or at least in the introduced British populations). Though numerous authors have noted the sloughing of the lining epithelium (Lincoln, 1971; Quay and Müller-Schwarze, 1971), this appears to be solely limited to the posterior pocket in *M. reevesi*; the anterior pocket is free of such detritus. The two pockets are separated by a small ridge, wherein runs the preorbital vein. Though this vein has been described in numerous other ungulates (Rehorek et al., 2001), this is the first known instance of a ridge superimposed upon the path of this vein.

The preorbital sac of *M. reevesi* exhibits sexual dimorphism in both the development of sweat glands in the anterior pocket (large amount in male, very few in female) and sebaceous glands (well developed in both pockets of the male, with little development in the female pockets). The only other cervids known thus far to exhibit sexual dimorphism in the preorbital sac are reindeer (*Rangifer tarandus*: Odocoileinae) wherein the apocrine sweat glands are larger in males than females (Mossing and Källquist, 1981), and *Cervus elaphus* (Cervinae), of which only behavioral observations indicate a greater activity in the male preorbital sac during rutting (Lincoln, 1971). No sexual dimorphism was noted in *Odocoileus* (Odocoileinae: Quay and Müller-Schwarze, 1970). In contrast, sexual dimorphism was described in three bovid subfamilies (Cephalophinae, Antilopinae and Caprinae), including more highly developed, more pigmented and larger preorbital sacs/glands in males than in females (Schaffer, 1940). Histologically, sexual dimorphism has only been described in *Cephalophus* (Bovidae: Cephalophinae) where males have more developed sebaceous glands than females (Schaffer, 1940). Thus, muntjac exhibits a unique combination of features, combining attributes seen in *Rangifer* (the hypertrophied nature of the apocrine sweat glands in the male) with others seen in *Cephalophus* (larger sebaceous glands in males).

The presence of the pigmented furrow, connecting the orbital region to the preorbital sac has been noted on numerous occasions (Harder, 1694; Schaffer, 1940). It has been proposed that secretions from the nearest orbital gland, the Harderian gland, may pass into the preorbital sac via this route (Harder, 1694; Schaffer, 1940; Quay and Müller-Schwarze, 1970, 1971). Our examination showed that the Harderian gland of *M. reevesi* is a bilobed structure, comprised of an anterior, white, mucous-secreting lobe and a posterior, red, lipid/serous-secreting lobe. The anterior lobe (often cited as a separate nictitans gland) has been documented in several deer species, but the posterior lobe has thus far only been reported in a few cervids, including fallow deer *(Dama dama)*, red deer *(Cervus elaphus*: Cervinae: Harder, 1694; Miessner, 1900), reindeer *(Rangifer tarandus*: Odocoileinae: pers. obs.) and muntjac. This red lobe is absent in numerous other odocoileine deer, including roe deer (*Capreolus capreolus*: Miessner, 1900), white-tailed deer (*Odocoileus virginianus:* pers. obs.) and mule deer (*O. hemionus hemionus*: pers. obs.). Little is known about the morphology of the Harderian gland in bovids, though in sheep the structure is a single-lobed mucous secreting structure (Sakai, 1981) and in the cow it appears to be a bilobed structure (pers. obs.). From these few comparative observations it may be deduced that, at least for the Cervidae, the presence of the posterior, red lobe appears to be correlated with the presence of a relatively well-developed preorbital sac (wherein the epithelial lining is slightly/somewhat more glandular than surrounding skin). Further comparative analyses are required, especially in the cephalophines and antilopines, to see if this correlation exists in other artiodactyls.

In conclusion, the epithelial lining of the preorbital sac in *M. reevesi* is more convoluted than previously described, as it is subdivided into two pockets. The marked sexual dimorphism of the glandular epithelium together with that of the Harderian gland supports the contention that these two disparate structures may have a functional connection. Future research on the preorbital apparatus in ungulates should therefore consider also the potential input from the Harderian gland.

## 5. ACKNOWLEDGMENTS

The authors wish to thank Janet Sanjur (for technical assistance) and Dean DeNicola (for statistical support). This work was made possible by a SRU international travel grant and travel funding from C of C. Light-microscopy photography was made possible by a National Science Foundation equipment grant, application 9970189. All animals were obtained through either routine culling practice in the UK or road kill. No permit was required to collect these specimens.

## 6. REFERENCES

Barka, T., and Anderson, P. J., 1965, *Histochemistry: Theory, Practice and Bibliography*, Harper and Row, New York, pp. 63-64.
Chapman, D. I., and Chapman, N. G., 1982, The taxonomic status of feral muntjac deer (*Muntiacus sp.*) in England, *J. Nat. Hist.* **16**:381-387.
Geist, V., 1998, *Deer of the World: Their Evolution, Behaviour, and Ecology*, Stackpole Books, Mechanicsburg.
Gray, D. R., Flood, P. F., and Rowell, J. E., 1988, The structure and function of musk ox preorbital glands, *Can. J. Zool.* **67**:1134-1142.
Harder, J. J., 1694, Glandulae nova lachrymalis una cum ductu excretorio in cervis et dama, *Acta-Eruditorium-Lipsiae.* **1694**:49-52.
Kuhn, H. J., 1976, Antorbitaldrüse und Tränennasengang von *Neotragus pygmaeus*, *Z. Säugetier.* **41**:369-380.
Lillie, R. D., 1961, *Histopathogenic Technic and Practical Histochemistry.* Blakiston Co., New York, pp. 120.
Lincoln, G. A., 1971, The seasonal reproductive changes in the red deer stag (*Cervus elaphus*), *J. Zool, Lond.* **163**:105-123.
Mainoya, J. R., 1978, Histological aspects of the preorbital and interdigital glands of the red duiker (*Cephalophus natalensis*), *E.Afr. Wildl. J.* **16**:265-272.
Miessner, H., 1900, Die Drüsen des dritten Augenlides einiger Säugethiere, *Arch. Wiss. Prakt. Tierheilkd.* **26**:122-154.
Mossing, T., and Källquist, L., 1981, Variation in cutaneous glandular structures in reindeer (*Rangifer tarandus*), *J. Mamm.* **62**(3):606-612.
Müller-Schwarze, D., 1970, Pheromones in black-tailed deer (*Odocoileus hemionus columbianus*), *Anim. Behav.* **19**:141-152.
Quay, W. B., 1955, Histology and cytochemistry of skin gland areas in the caribou, *Rangifer, J. Mamm.* **36**(2):187-201.
Quay, W. B., 1959, Microscopic structure and variation in the cutaneous glands of the deer, *Odocoileus virginianus*, *J. Mamm.* **40**(1):115-128.
Quay, W. B., and Müller-Schwarze, D., 1970, Functional histology of integumentary glandular regions in the rocky mountain mule deer (*Odocoileus hemionus hemionus*), *J. Mamm.* **51**(4):675-694.
Rehorek, S. J., Hillenius, W. J., and Solounias, N., 2001, Pre-orbital fossae in mammals: a re-examination of this unusual structure, *J. Morph.* **248**(3):274.
Rehorek, S. J., Hillenius, W. J., Sanjur, J., and Chapman, N., One gland, two lobes: organogensis of the "Harderian" and "nictitans" glands of two deer, (*in prep*).
Richter, J., 1971, Untersuchungen an Antorbitaldrüsen von *Madoqua* (Bovidae, Mammalia), *Z. Säugetier.* **36**:334-342.
Richter, J., 1973, Zur Kenntnis der Antorbitaldrüsen der Cephalophinae (Bovidae: Mammalia), *Z. Säugetier.* **38**:303-313.
Sakai, T., 1981, The mammalian Harderian gland: morphology, histochemistry, function and phylogeny. *Arch. Histol. Jap.* **44**:299-333.
Schaffer, J., 1940, *Die Hautdrüsenorgane der Säugetiere*, Urban and Schwarzenberg, Berlin, Germany.
Thiessen, D., and Rice, M., 1976, Mammalian scent gland marking and social behaviour, *Psych. Bull.* **83**(4):505-539.

# THE CHEMISTRY OF SCENT MARKING IN TWO LEMURS:

## *Lemur catta* and *Propithecus verreauxi coquereli*

R. Andrew Hayes, Toni-Lyn Morelli, and Patricia C. Wright[*]

## 1. INTRODUCTION

Prosimians are generally acknowledged to be the most olfaction-oriented of primates, retaining olfactory complexity in addition to developing other sensory modalities (Jolly, 1966). In spite of this, surprisingly little is known about the specifics of how they use scent for communication. The majority of earlier studies of olfaction in prosimians have investigated the behavioural response to scent marks, but have not attempted to determine the nature of the scent chemically, or to measure chemical differences between individuals (e.g. Epple et al., 1987; Evans, 1980; Harrington, 1974, 1977; Price and Feistner, 1994; Ramsay and Giller, 1996; Vick and Conley, 1976).

The present study is a preliminary attempt to begin to counter this deficit. It concentrates on two species of lemur, the ring-tailed lemur (*Lemur catta*) and Coquerel's sifaka (*Propithecus verrauxi coquereli*). Both of these species are classified in the IUCN Red List of Threatened Species as Vulnerable (*L.* catta - VU A1c, *P. v. coquereli* - VU A2cd).

Coquerel's sifaka, *P. v. coquereli*, is a diurnal, primarily folivorous primate, inhabiting the forests of northwestern Madagascar. The ring-tailed lemur, *L. catta*, is also diurnal, but rather than eating leaves, it is principally frugivorous, is smaller than the sifaka, and is much more terrestrial than other lemurs. Ring-tailed lemurs are found in the forests and arid, open areas of southern and southwestern Madagascar. Neither species displays significant sexual dimorphism in appearance (Kappeler, 1990b).

---

[*] R. Andrew Hayes, School of Natural Resource Sciences, Queensland University of Technology, Brisbane, Queensland, 4001, Australia. Toni-Lyn Morelli, Department of Ecology and Evolution, State University of New York at Stony Brook, Stony Brook, New York, 11794-4364, USA. Patricia C. Wright, Department of Anthropology, State University of New York at Stony Brook, Stony Brook, New York, 11794-4364, USA.

The mean group size of *P. v. coquereli* is smaller than that of *L. catta*, but both live in multi-male, multi-female groups (Richard et al., 2002), which are dominated by females (Jolly, 1966; Kubzdela et al., 1992; Pereira et al., 1990). Both *L. catta* and *P. v. coquereli* social groups have overlapping home ranges (Gould and Overdorff, 2002; Richard et al., 2001). However, ring-tailed lemur females defend territory, with some groups having frequent inter-group encounters, whereas sifaka inter-group encounters appear to be rare and non-confrontational (Jolly 1966; Kubzdela et al., 1992; Richard 1974; Richard et al., 2002). Both species are strongly seasonal breeders (Rasmussen, 1985; Wright, 1999).

The ring-tailed lemur is probably the best studied species of prosimian, and much has been published on scent-marking behaviour in this species, especially with respect to the anogenital gland. Little is known about the olfactory behavior of *P. v. coquereli* (Brockman and Whitten, 1996),

Like many mammals, both *L. catta* and *P. verreauxi*, have specialized glands for scent marking their surroundings. Both species place marks in the environment with their anogenital region (Mertl, 1977; Mertl-Millhollen, 1979). In addition to this, male ring-tailed lemurs have specialized glands, located on the flexor surface of each forearm near a keratinized spur (referred to as antebrachial glands), and brachial organs, found just above each clavicle (Montagna and Soonyun, 1962). The spur is used to gouge the surface of a tree, and scent is then deposited by the antebrachial organ. The secretion from the antebrachial organ is often mixed with that from the brachial organ and the combined secretion is used in the mark. During "stinkfights," males rub their tails with these two secretions and wave them at other males (Mertl, 1976). Similarly, male sifakas possess specialized throat glands that they use to mark surrounding branches and tree trunks.

The anogenital marks placed by male ring-tailed lemurs seem to be deposited for the information of other males (Kappeler, 1990a, 1998; Oda, 1999), although the female's mark is believed to convey information about her reproductive status. Male ring-tailed lemurs mark more frequently (Gould and Overdorff, 2002) and are more interested in female secretion in the breeding as opposed to the non-breeding season (Dugmore et al., 1984). Also, female marking rate shows a high point during and after estrus (Jolly, 1972; Kappeler, 1990a, 1998)

Boundaries of group territories are marked by anogenital secretion of both male and female *L. catta*, as well as other glands in the male (Mertl-Millhollen, 1988). Male ring-tailed lemurs respond differently to the scent marks from male and female donors (Dugmore et al., 1984; Ramsay and Giller, 1996), although frequency of anogenital marking is the same in males and females (Oda, 1999).

In addition to the anogenital gland, researchers have also studied the antebrachial gland in *L. catta*. This gland is used by ring-tailed males to mark both objects in the surrounding territory and their own tails, and is commonly associated with agonistic interactions between males (Jolly, 1972). Oda (1999) suggested that antebrachial marking is used for intra- rather than inter-group communication. Antebrachial marking occurs at more than twice the frequency of anogenital marking (Dugmore et al., 1984; Oda, 1999), and high-ranking males antebrachial mark at a higher rate than low-ranking males (Oda, 1999). Season does not affect the rate of antebrachial marking (Dugmore *et al.*, 1984). Male lemurs can distinguish between the antebrachial secretions of other males (Dugmore et al., 1984; Mertl, 1975).

A chemical analysis of the marks from the scent glands of both of these species of lemur will begin to allow the answering of several interesting questions relating to the biology of these species. These include: what is it about this secretion that functions to communicate so much information? What are the components of the secretion, and how do they vary among individuals? Finally, what, if any, are the inter-specific differences of the scent mark, and can we detect them chemically?

## 2. METHODS

Swabs were taken of the glandular regions of captive lemurs held at the Duke University Primate Center (DUPC), Durham, NC. All but one of the animals from which samples were collected was captive-born. *L. catta* at DUPC, including the test subjects, are housed in semi-free-range enclosures (9.1 hectares) during the warmer months (approximately April to October), while during the winter they are housed in indoor/outdoor enclosures with other *L. catta*. *P. v. coquereli* are housed in a separate building in open-air enclosures that still admit natural lighting when closed in for the winter months. At the DUPC, *L. catta* are fed vegetables and commercial primate pellets (PMI Feeds Inc. (Purina) Lab Diet Product 5038 Monkey Diet), whereas *P. v. coquereli* are fed vegetables, nuts, leaves, and a different type of commercial primate pellet (Mazuri Leaf Eater Primate Diet Mini-biscuit #5672).

Swabs were taken from the antebrachial and brachial glands of the five male *L. catta* and from the throat glands of the five male *P. v. coquereli*. Swabs were also collected from the anogenital region of males and females of both species, the results of which are discussed elsewhere (Hayes et al, in press). The swab was collected by rubbing the area of the scent gland of a hand-captured lemur with 1 cm × 4 cm glass filter paper (after Salamon, 1994). The swabs were then sealed in an airtight vial and stored at a reduced temperature (of approximately -75°C) until being shipped to Australia for analysis.

In the laboratory, the vial containing the swab was opened and the filter paper inserted into the liner of the cooled injector port (45°C) of a gas chromatograph (GC; Hewlett Packard 5890 Series II) coupled to a mass spectrometer (MS; Hewlett Packard 5971A) and fitted with a silica capillary column (J&W, model DB5-HT, no. 122-5731, 30 m long × 0.25 mm ID × 0.1 µm film thickness). The sample was cryofocused onto the front of the GC column by maintaining the column at low temperature (2°C), and ramping up the injector to 150°C at 27°C/min.

Data were acquired under the following GC conditions: inlet temperature: 150°C, carrier gas: helium at 1.1 mL/min, split ratio 1:2, detector temperature: 280°C, temperature program: initial temperature 2°C, initial time 4 min, rate 20°C/min to 50°C then 5°C/min, final temperature 250°C, final time 2 min. The MS was held at 190°C in the ion source using ionization energy of 70 eV and a scan rate of 0.9 scans/s. Tentative identities were assigned to peaks with respect to the Wiley mass spectral library.

Non-parametric statistical analysis (Bray-Curtis cluster analysis and multidimensional scaling ordination (MDS)) was performed on the GC-MS output to ascertain if any differences could be detected between the animals. MDS has been shown to be a useful technique for analysing chromatographic data, which can be difficult to analyse statistically (see Hayes et al., 2002, 2003). Each point in the MDS plot represents an individual lemur and points that are close together (clumped) correspond to individuals with similar peak composition (presence and abundance). As they represent relative differences between samples, the axes of an MDS plot are dimensionless.

Analysis of Similarity (ANOSIM) tests were used to determine if clusters of individuals are significantly different. These tests are a range of Mantel-type permutations of randomization procedures, making no distributional assumptions; they depend only upon rank similarities, and thus are appropriate for this type of data. Software used for the multivariate analysis was Primer 5 for Windows (V 5.2.4, 2001).

## 3. RESULTS

Several of the compounds found on the swabs from the lemurs were identified by mass-spectrometry (Table 1). These compounds were primarily identified as straight and branched long-chain hydrocarbons with alcohol and aldehyde components. In addition, several of the components were oxygenated compounds, particularly esters. The percentage of individuals in each group from which the component was identified is also shown (Table 1). It should be noted that there is a contaminant peak at approximately 51 minutes that appears in all chromatgrams, including blank samples of paper without secretion. This peak acted as an internal standard, and was thus invaluable in allowing comparisons of retention times to be made between chromatograms

In *L. catta*, these swabs were from the antebrachial and the brachial glands (Figure 1a and b respectively). In *P. v. coquereli* they were from the throat gland (Figure 1c). The differentiation between swabs of these glands is not as clear as that described previously for the anogenital swabs (Hayes et al., in press). For *L. catta*, the anogenital swabs are not distinguishable from the antebrachial swabs (ANOSIM: $R = 0.456$, $P = 0.071$); the difference between anogenital and brachial swabs were not highly significant (ANOSIM: $R = 0.491$, $P = 0.048$). The swabs of the antebrachial and brachial glands are indistinguishable according to the present methods (ANOSIM: $R = -0.167$, $P = 0.80$).

The throat glands from *P. v. coquereli*, on the other hand, are different from the anogenital swabs of males of both species (ANOSIM: $R = 0.736$, $P = 0.016$). However, the pairwise comparison test between the throat gland and both the antebrachial and brachial glands showed no difference, likely due to the small number of replicates.

The multidimensional scaling (MDS) output of the male glands is also shown (Figure 2). This gives a visual representation of the data, described by the ANOSIM (above). Each point on the figure represents an individual lemur swab. Points that are close together are more similar, and those further away are more different. Anogenital swabs are different from all other groups of swabs with the strongest difference between the anogenital and throat swabs.

## 4. DISCUSSION

Results to date have indicated that *L. catta* anogenital scent marks may communicate detailed information to the recipient, including the sex of the marker (Hayes et al., in press). However, male and female sifaka anogenital scent marks are not distinguishable by our tests. We have previously suggested that this lack of distinguishibility may be because a message defining sex is encoded by some other information besides the volatile

**Table 1.** Tentative identity and approximate retention times of compounds identified from swabs of the glandular regions of males of the two species of lemur. Number indicates percentage of individuals sampled in which the given compound was identified.

| Name | Retention time (min.) | *L. catta* antebrachial | *L. catta* brachial | *P. v. coquereli* throat |
|---|---|---|---|---|
| cyclooctane | 20.0 | 20 | 0 | 0 |
| 1-dodecanol | 21.6 | 20 | 0 | 0 |
| methyl-cyclodecane | 22.5 | 20 | 0 | 0 |
| (Z)-2-decene | 24.2 | 20 | 0 | 0 |
| 4-decene | 25.0 | 20 | 0 | 20 |
| 1butyl-1-methyl-2-propyl-cyclopropane | 26.6 | 60 | 40 | 80 |
| 6-tridecene | 27.4 | 20 | 0 | 0 |
| hydrocarbon | 28.9 | 20 | 0 | 20 |
| cyclododecane | 29.8 | 60 | 40 | 100 |
| 3-methyl-6-octen-1-ol | 30.0 | 20 | 0 | 0 |
| hydrocarbon | 30.2 | 20 | 0 | 0 |
| long chain alcohol | 30.7 | 20 | 0 | 0 |
| 1-tridecene | 31.2 | 20 | 40 | 40 |
| 1-pentadecanol | 32.0 | 20 | 40 | 20 |
| 1-hexadecanol | 33.3 | 20 | 40 | 20 |
| long chain alcohol | 34.1 | 40 | 40 | 20 |
| long chain alcohol | 35.3 | 40 | 20 | 40 |
| 1-heptadecanol | 36.0 | 20 | 40 | 100 |
| (Z)-7-hexadecene | 37.2 | 40 | 20 | 20 |
| long chain ester | 37.5 | 20 | 20 | 20 |
| 1-octadecanol | 38.0 | 20 | 40 | 100 |
| long chain alcohol | 38.2 | 20 | 0 | 0 |
| long chain alcohol | 39.3 | 20 | 0 | 0 |
| long chain alcohol | 39.8 | 20 | 0 | 60 |
| hydrocarbon | 41.0 | 20 | 0 | 20 |
| (E)-β-farnesene | 41.4 | 20 | 0 | 0 |
| long chain ester | 42.1 | 40 | 0 | 0 |
| long chain ester | 42.7 | 20 | 20 | 40 |
| octanoic acid, hexadecyl ester | 44.3 | 0 | 40 | 80 |
| hydrocarbon | 46.2 | 20 | 0 | 0 |
| long chain ester | 47.7 | 20 | 0 | 0 |
| 1-hexadecanal, acetate | 48.5 | 40 | 40 | 40 |
| long chain alcohol | 53.2 | 0 | 20 | 60 |
| long chain alcohol | 55.3 | 0 | 20 | 0 |
| long chain alcohol | 57.9 | 0 | 0 | 20 |
| long chain alcohol | 58.6 | 0 | 20 | 0 |
| long chain alcohol | 60.4 | 0 | 20 | 0 |

chemicals in the anogenital mark. For example, the secretion from the throat gland sampled in the *P. v. coquereli* males was distinguishable from the anogenital secretion of both sexes in both species. Perhaps males can use secretions of the throat glands in contexts in which sex differentiation is important, as they are distinguishable from anogenital marks and are not present in females. If so, and if *L. catta* secretions from different glands are truly indistinguishable, these species may have evolved different mechanisms for communicating messages in which the sex of the sender is important. In the future it would be interesting and worthwhile to use habituation tests to determine whether *P. v. coquereli* can in practice differentiate between the scent marks of males and females from the different glands (after Epple, 1986).

(a) *L. catta* antebrachial gland

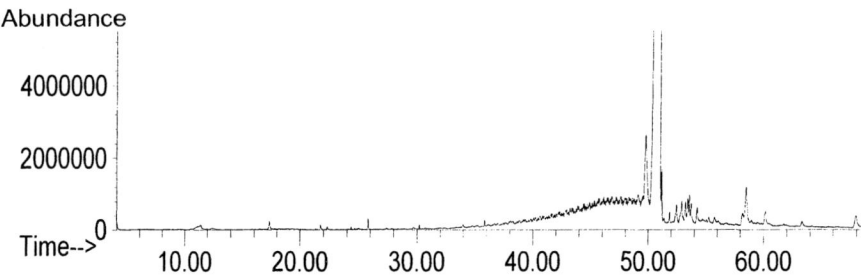

(b) *L. catta* brachial gland

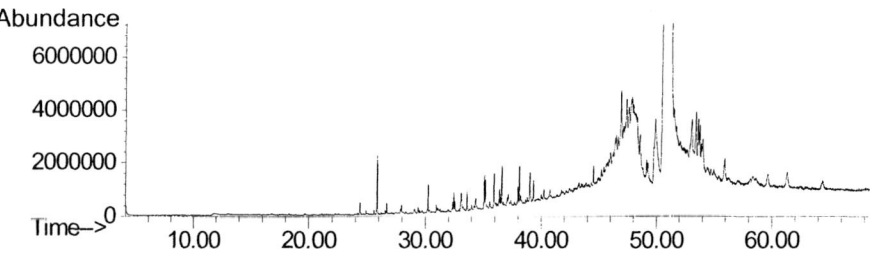

(c) *P. v. coquereli* throat gland

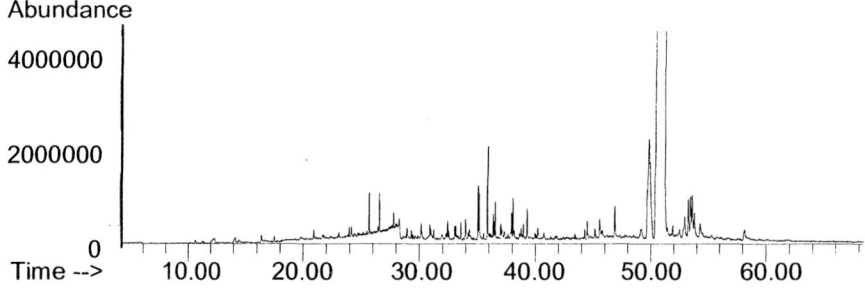

**Figure 1.** Chromatograms produced by typical swabs from the (a) antebrachial and (b) brachial glands of *L. catta* males, and the (c) throat gland of *P. v. coquereli* males. The peak at approximately 51 minutes is an internal standard.

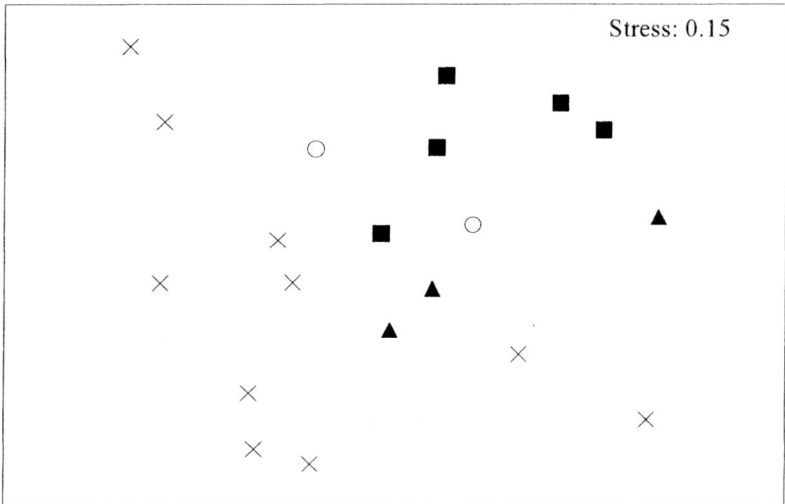

**Figure 2.** The 2-dimensional MDS ordination of the samples from the four gland types from male lemurs. The plot is based on untransformed abundances and a Bray-Curtis similarity matrix. Note that the brachial (*L. catta*) and throat (*P. v. coquereli*) glands are separate from the anogenital glands (both species); all other groupings are indistinguishable. Symbols represent: anogenital gland - ×, antebrachial gland - ▲, brachial gland - o, throat gland - ■.

In *L. catta* it is not possible to differentiate antebrachial and brachial swabs. This may be due to the manner in which these glands are used by the animals. Before marking with the antebrachial gland, male *L. catta* typically mix this secretion with that from their brachial gland (Mertl, 1976). Thus, there may be some contamination of swabs from each of these locations with secretion from the other gland. Therefore, our inability to distinguish between the swabs may be more a reflection of the collection technique than true correspondence between the glands. The research described in the present chapter, together with that published previously (Hayes et al., in press) demonstrates that it is possible to detect, both chemically and statistically, information in lemur scent marks relating to sex, reproductive status, and individuality, as well as species. Future habituation experiments such as those described above may help to clarify the context in which these scent marks are used by these species and the importance of the differences between these and other lemur species, as well as lead to a better understanding of the evolution of olfactory communication in prosimians.

## 5. ACKNOWLEDGEMENTS

The authors would like to thank Ratelolahy Jean Felix, Safia Salimo and Ravalison of ICTE-MICET and Bill Hess and David Haring of Duke University Primate Center (DUPC) for their advice and help in the sample collections. This research was carried out under DUPC IACUC Registry No. A026-02-01. Samples were imported into Australia under AQIS permit No. 200205460. Funding for this research was provided by DUPC

Director's Research Incentive Grant, the Wenner Gren Foundation for Anthropological Research Inc. to PCW, and a Sokal/Slobodkin travel award to T-LM. This is DUPC publication #775.

## 6. REFERENCES

Brockman, D. K., and Whitten, P. L., 1996, Reproduction in free-ranging *Propithecus verreauxi* - estrus and the relationship between multiple partner matings and fertilization, *Am. J. Phys. Anthropol.* **100**:57-69.
Dugmore, S. J., Bailey, K., and Evans, C. S., 1984, Discrimination by male ring-tailed lemurs (*Lemur catta*) between the scent marks of males and those of female conspecifics, *Int. J. Primatol.* **5**:235-245.
Epple, G., 1986, Communication by chemical signals, in: *Comparative Primate Biology*, G. Mitchell, and J. Erwin, eds., A. R. Liss, New York, pp. 531-580.
Epple, G., Alveario, M. C., Belcher, A. M., and Smith, A. B. III, 1987, Species and subspecies specificity in urine and scent marks of saddle-back tamarins (*Saguinus fuscicollis*), *Int. J. Primatol.* **8**:663-681.
Evans, C. S., 1980, Diosmic response to scent-signals in *Lemur catta*, in: *Chemical Signals. Vertebrates and Aquatic Invertebrates*, D. Müller-Schwarze, and R. M. Silverstein, eds., Plenum Press, New York, pp. 417-420.
Gould, L., and Overdorff, D. J., 2002, Adult male scent-marking in *Lemur catta* and *Eulemur fulvus rufus*, *Int. J. Primatol.* **23**:575-586.
Harrington, J. E., 1974, Olfactory communication in *Lemur fulvus*, in: *Prosimian Biology*, R. D. Martin, G. A. Doyle, and A. C. Walker, eds., Duckworth, London, pp. 331-346.
Harrington, J. E., 1977, Discrimination between males and females by scent in *Lemur fulvus*, *Anim. Behav.* **25**:147-151.
Hayes, R. A., Morelli, T-L., and Wright, P. C., in press, Scent marking in *Lemur catta* and *Propithecus verreauxi coquereli*: a preliminary chemical examination, *Am. J. Primatol.*
Hayes, R. A., Richardson, B. J., Claus, S. C., and Wyllie, S. G., 2002, Semiochemicals and social signalling in the wild European rabbit in Australia. II. Variations in chemical composition of chin gland secretion across sampling sites, *J. Chem. Ecol.* **28**:2613-2625.
Hayes, R. A., Richardson, B. J., and Wyllie, S. G., 2003, To fix or not to fix: the role of 2-phenoxyethanol in rabbit, *Oryctolagus cuniculus*, chin gland secretion, *J. Chem. Ecol.* **29**:1051-1064.
Jolly, A., 1966, *Lemur Behavior: A Madagascar Field Study*, University of Chicago Press, Chicago.
Jolly, A., 1972, Troop continuity and troop spacing in *Propithecus verreauxi* and *Lemur catta* at Berenty (Madagascar), *Folia Primatol.* **17**:335-362.
Kappeler, P. M., 1990a, Social status and scent marking behaviour in *Lemur catta*, *Anim. Behav.* **40**:774-776.
Kappeler, P. M., 1990b, The evolution of sexual size dimorphism in prosimian primates, *Am. J. Primatol.* **21**:201-214.
Kappeler, P. M., 1998, To whom it may concern: the transmission and function of chemical signals in *Lemur catta*, *Behav. Ecol. Sociobiol.* **42**:411-421.
Kubzdela, K. S., Richard, A. F., and Pereira, M. E., 1992, Social relations in semi-free-ranging sifakas (*Propithecus verreauxi coquereli*) and the question of female dominance, *Am. J. Primatol.* **28**:139-145.
Mertl, A. S., 1975, Discrimination of Individuals by Scent in a Primate, *Behav. Biol.* **14**:505-509.
Mertl, A. S., 1976, Olfactory and visual cues in social interactions of *Lemur catta*, *Folia Primatol.* **26**:151-161.
Mertl, A. S., 1977, Habituation to territorial scent marks in the field by *Lemur catta*, *Behav. Biol.* **21**:500-507.
Mertl-Millhollen, A. S., 1979, Olfactory demarcation of territorial boundaries by a primate-*Propithecus verreauxi*, *Folia Primatol.* **32**:35-42.
Mertl-Millhollen, A. S., 1988, Olfactory discrimination of territorial but not range boundaries by *Lemur catta*, *Folia Primatol.* **50**:175-187.
Montagna, W., and Soonyun, J., 1962, Skin of Primates 10. Skin of ring-tailed lemur (*Lemur catta*), *Am. J. Phys. Anthropol.* **20**:95.
Oda, R., 1999, Scent marking and contact call production in ring-tailed lemurs *(Lemur catta)*, *Folia Primatol.* **70**:121-124.
Pereira, M. E., Kaufman, R., Kappeler, P. M., and Overdorff, D. J., 1990, Female dominance does not characterize all of the Lemuridae, *Folia Primatol.* **55**:96-103.
Price, E. C., and Feistner, T. C., 1994, Responses of captive aye-ayes (*Daubentonia madagascarensis*) to the scent of conspecifics: a preliminary investigation, *Folia Primatol.* **62**:170-174.
Ramsay, N. F., and Giller, P. S., 1996, Scent-marking in ring-tailed lemurs: responses to the introduction of "foreign" scent in the home range, *Primates* **37**:13-23.

Rasmussen, D. T., 1985, A comparative study of breeding seasonality and litter size in eleven taxa of captive lemurs (*Lemur* and *Varecia*), *Int. J. Primatol.* **6**:501-511.
Richard, A. F., 1974, Patterns of mating in *Propithecus verreauxi*, in: *Prosimian Biology*, R. D. Martin, A. C. Walker, and G. Doyle, eds., Duckworth, London, pp 49-74.
Richard, A. F., Dewar, R. E., Schwatrz, M., and Ratsirason, J., 2001, Mass change, environmental variability and female fertility in wild *Propithecus verreauxi*, *J. Hum. Evol.* **39**:381-391.
Richard, A. F., Dewar, R. E., Schwatrz, M., and Ratsirason, J., 2002, Life in the slow lane? Demography and life histories of male and female sifaka (*Propithecus verreauxi verreauxi*), *J. Zool.* **256**:421-436.
Salamon, M., 1994, Seasonal, sexual and dietary induced variations in the sternal scent secretion in the brushtail possum (*Trichosurus vulpecula*), in: *Chemical Signals in Vertebrates VII*, R. Apfelbach, D. Müller-Schwarze, K. Reutier, and E. Weiber, eds., Pergamon Press, Oxford, pp. 211-222.
Vick, L. G., and Conley, J. M., 1976, An ethogram for *Lemur fulvus*, *Primates* **17**:125-144.
Wright, P. C., 1999, Lemur traits and Madagascar ecology: Coping with an island environment, *Yearb. Phys. Anthropol.* **42**:31-72.

# SOILED BEDDING FROM GROUP-HOUSED FEMALES EXERTS STRONG INFLUENCE ON MALE REPRODUCTIVE CONDITION

Sachiko Koyama[1, 2] and Shinji Kamimura[2]

## 1. INTRODUCTION

Odors are one of the major means of communication for house mice. They influence the behavior and the physiological states of other individuals (Cavaggioni et al., 1999; Vandenbergh, 1994). Examples for influences of odors of a mouse on the physiological states of others have been found from early in 1950s. Odors of female mice kept in groups have been known to lengthen females' estrous cycle (Lee and van der Boot, 1955; 1956) and odors of males were found to induce estrous and to synchronize estrous cycles in females (Whitten, 1956, 1958). Male odors were also found to accelerate puberty in female pups (Vandenbergh, 1969). Odors of strange males were found to block the establishment of pregnancy in females (Bruce, 1959; Parkes and Bruce, 1961), which suggested the role of cognitive processes in some of those influences. The studies especially of the primer effects, as such, tended to be biased towards female mice until these days. Although there have been studies on primer effects in male mice, most of these have concentrated on hormone levels and there have been only a few studies investigating parameters that directly indicate the reproductive activities of males (e.g. sperm density, sperm motility, or sperm velocity).

In one of our previous studies, we found that female odors have induced increased sperm density in males (Koyama and Kamimura, 2000). Females were all kept in isolated housing conditions in that study. We chose that housing condition because it had been known that singly housed females come into estrus frequently (Lee and van der Boot, 1955; 1956) and because we thought that odors of estrous females might have stronger influence on the reproductive condition of males compared to odors of diestrous

---

[1] Present address: Dept. Chemistry, Indiana Univ., Bloomington, Indiana 47405-4001, U.S.A.
[2] Div. Biol., Dept. Life Sciences, Grad. Sch. Arts & Sci., Univ. of Tokyo, 3-8-1 Komaba, Meguro-ku, Tokyo, 153-8902, Japan.

females. In order to investigate whether such differences in the influence of female odors on males exist depending on the housing conditions of females, we kept females in groups and under isolated condition and provided males the soiled bedding of those females. As the density of a female group has been found to be positively correlated with the magnitude of the influence of female odors on lengthening estrous cycle lengths (Coppola and Vandenbergh, 1985), we kept group-housed females in 2 different densities, i.e., in high density and in low density. As properties of female odors have been shown to vary depending on their housing conditions (Mugford and Nowell, 1970, 1971), we expected the influences of females' odors should be modified by their housing conditions.

## 2. METHODS

### 2.1. Animals

Male mice (192 ddY strain; offspring of the breeding colony kept at the lab) were kept in male-male pairs from 5 weeks of age. All pairs were randomly assigned to 4 experimental groups (24 pairs each) that were provided with 4 different types of odor stimulation as below. Odor donor female mice were kept under one of the following 3 different conditions from 3 weeks of age; isolated (24 females), in groups of low density (12 females, 4 mice per group), or of high density (12 females, 12 mice per group). These housing conditions resulted in significant differences of numbers of estrous females, i.e. 54, 31 and 24% in isolated females, in female groups in low density, and in a female group in high density, respectively (*Chi-square* = 22.614, df = 2, $P < 0.001$). The average length of estrous cycle of isolated females was significantly shorter than that of grouped females (4.3, 6.1 and 7.2 days in isolated females, in female groups of low density and a high-density female group, respectively. ANOVA, $F_{2,40} = 27.676$, $P < 0.001$). 25% of the females kept in high density did not come into estrus during the study.

The females were provided with 4 g of shredded filter papers per each mouse from 5 weeks of age. The filter papers were collected 2 or 3 days later to be provided to males. In providing those soiled filter papers to males, we carefully prepared them to be the same weight (4 g) collected from the same number of females that were non-littermates of the males that received them. Half of the 24 isolated females were used to provide males with different odor stimulation condition: that is, in order to investigate the effects of female numbers, another group of males were provided with the soiled filter papers of one fixed isolated female throughout the experiment. Therefore, there were 4 experimental groups of males that were provided with 4 different types of odor stimuli as follows: (1) odors of 12 females kept in isolation, (2) odors of 12 females kept in groups of low density, (3) odors of 12 females kept in a group of high density, and (4) odors of one fixed female that was kept in isolation.

### 2.2. Observation of Sperm

To examine the density, velocity, and motility of the sperm of males, mice were anaesthetized with Nembutal (0.07 ml for each mouse) and sperm was collected from the cauda epididymis. Details of the procedures to determine sperm activity have been described previously (Koyama and Kamimura, 1999, 2000). In short, in determining the velocity and motility, sperm was collected from the right cauda epididymis, diluted in Biggers, Whitten, and Whittingham (BWW) medium (Biggers et al., 1971; Koyama and Kamimura, 1999), observed using phase-contrast microscope (Optiphoto, Nikon Co., Ltd., Tokyo, Japan, with SPlan x 20NH, Olympus Optical Co., Ltd. Tokyo, Japan) and recorded on video tapes with a CCD camera (DXC-151, Sony Co., Ltd., Tokyo, Japan). Thermo Plate (Model MATS-SS, Tokai Hit, Shizuoka, Japan) was used to keep the specimen at 37 °C under the microscope during observation. The percentage motility and velocity of spermatozoa were determined from the recorded images.

To determine the density of sperm, the left cauda epididymis was used. After removing the epithelium of the isolated cauda epididymis, 1 cm of the tubular duct was dissected. Sperm was then squeezed out on a glass slide using an edge of a cover slip. The sperm was diluted with 0.5 ml of 0.9 % NaCl medium, and was observed on a blood cell counter (Thoma). The number of spermatozoa was determined under the phase-contrast microscope.

After sperm activity was observed, blood was collected from the heart and kept at -20 °C until use. Testosterone EIA kit (Cayman Chemical Co., Ann Arbor, USA) was used to determine plasma concentration of testosterone.

## 3. RESULTS AND DISCUSSION

The most striking result was found in the sperm density of male mice (Table 1; ANOVA, $F$ group$_{3,79}$ = 5.993, $P$ = 0.001). Sperm density became significantly high when males were provided with soiled bedding of females kept in groups (Table 1). The effects were not dependent on the number but on the density of females, since the sperm density showed no significant difference between that of one and 12 isolated females (Table 1). Some odor substances found in females kept in groups are suggested to have strong influence on males, changing their reproductive conditions to increase sperm density. As one of the factors that possibly affected sperm density in males, we measured the plasma concentration of testosterone (Table 1), which is known to be related to spermatogenesis (McLachlan et al., 1995; Sharpe, 1994; Sun et al., 1989). However, we could not find any positive correlation with the sperm density (Table 1). Sperm motility and swimming velocity of spermatozoa showed no difference among experimental groups (Table 1).

Table 1. Influence of female odors on sperm and testosterone concentration

| | Housing conditions of odor donor females | | | |
|---|---|---|---|---|
| | 12 in high density | 12 in low density | 12 Isolated | 1 Isolated |
| Sperm density* | 35 ± 15$^a$ | 30 ± 17$^a$ | 22 ± 14$^b$ | 16 ± 11$^b$ |
| Sperm motility** | 62 ± 23 | 64 ± 19 | 65 ± 17 | 63 ± 14 |
| Sperm velocity*** | 163 ± 25 | 163 ± 27 | 177 ±28 | 164 ±34 |
| Testosterone Concentration**** | 18 ± 24 | 17 ± 27 | 16 ± 25 | 9 ± 13 |

\* arbitrary unit; ** percentage of motile spermatozoa; *** micrometer/sec; **** pg/ml a > b, $P < 0.05$

In the 1970s, there have been studies in which male mice were exposed to urine of females. Female urine was found to trigger luteinizing hormone (LH) in males (Maruniak and Bronson, 1976). And it was shown that urine from females in a diestrous state, proestrous state, and even ovariectomized females increased LH levels in males. These results indicate that the influence of female odors on males is not dependent on ovary steroids at least in increasing the LH. There are also studies that showed males exposed to females (not to the urine of females) increased androgen production (Batty, 1978; Macrides et al., 1975). Such studies suggest that there might be a stimuli-response pathway by means of LH, which in turn increases the androgen level and enhances spermatogenesis, and results in the increased sperm density as in the present study.

In the present study, however, the plasma testosterone concentration did not show positive correlation with sperm density. Several reasons can be considered to explain such lack of correlation. For example, testosterone concentration may correlate to the activity of spermatogenesis when it takes place in the testes. This means that in order to investigate the role of testosterone, it might be necessary to investigate the concentration of it 40 to 50 days preceding the date sperm density was measured. As another possibility, it may be necessary to investigate the role of other hormones, such as estrogen and prolactin, on increasing sperm density. Estrogen is involved in the reabsorption of fluid at efferent ductules (Hess et al., 1997). Fluid reabsorption increases the concentration of sperm in the corpus and cauda epididymis. Recently, prolactin was found to play an important role in males. It increases both the number of LH receptors in the testis and the binding of human chorionic gonadotropin (hCG) to LH receptors, and it also enhances testosterone synthesis and secretion (Hussein and Zipf, 1988; Zipf et al., 1978). Prolactin concentration in the testis increases in adulthood compared to the juvenile and pubertal period (Imaoka et al., 1998) and prolactin can be detected in all parts of the epididymis (Brumlow and Adams, 1990).

Studies on the influence of odors on the reproductive activities of males are still at the starting point. The questions on the odor substances that are related to such phenomena, e.g., the neuroendocrinological changes induced by those key odors and the hormonal changes induced by such neuroendocrinological changes, are not solved yet. The clarification of these mechanisms may not only shed light on an aspect of the system of chemical communications in mice, but may also benefit the areas of infertility in human medical studies and reproductive biology of endangered species.

## 4. REFERENCES

Batty, J., 1978, Acute changes in plasma testosterone levels and their relation to measures of sexual behaviour in the male house mouse (*Mus musculus*), *Anim. Behav.* **26**:349-357.
Biggers, J. D., Whitten, W. K., and Whittingham D.G., 1971, in: *Methods in Mammalian Embryology*, 1st ed., J.C. Daniel, ed., W.H. Freeman, San Francisco, pp.86-116.
Bruce, H. M., 1959, An exteroceptive block to pregnancy in the mouse, *Nature* **184**:105.
Brumlow, W. B., and Adams, C.S., 1990, Immunocytochemical detection of prolactin or prolactin-like immunoreactivity in epididymis of mature male mouse, *Histochemistry* **93**:299-304.
Cavaggioni, A., Mucignat, C., and Tirindelli, R, 1999, Pheromone signaling in the mouse: role of urinary proteins and vomeronasal organ, *Arch. Ital. Biol.* **137**:193-200 (1999).
Coppola, D. M., and Vandenbergh, J. G., 1985, Effect of density, duration of grouping and age of urine stimulus on the puberty delay pheromone in female mice, *J. Reprod. Fertil.* **73**:517-522.
Hess, R. A., Bunick, D., Lee, K.-H., Bahr, J., Taylor, J. A., Korach, K. S., and Lubahn, D. B., 1997, A role for oestrogens in the male reproductive system, *Nature* **390**:509-512.

Hussein, M. O., and Zipf, W. B., 1988, Temporal relationship of the prolactin-dependent LH-induced LH receptor to the LH stimulus, *J. Cell. Physiol.* **134**:137-142.

Imaoka, T., Matsuda, M., and Mori, T., 1998, Expression of prolactin messenger ribonucleic acid in the mouse gonads during sexual maturation, *Life Sci.* **63**:2251-2258.

Koyama, S., and Kamimura, S., 1999, Lowered sperm motility in mice of subordinate social status, *Physiol. Behav.* **65**:665-669.

Koyama, S., and Kamimura, S., 2000, Influence of social dominance and female odor on the sperm activity of male mice, *Physiol. Behav.* **71**:415-422.

Lee, S., and van der Boot, L. M., 1955, Spontaneous pseudopregnancy in mice, *Acta Physiol. Pharmaco. Neerl.* **4**: 442-443.

Lee S., and van der Boot, L. M., 1956, Spontaneous pseudopregnancy in mice II, *Acta Physiol.Pharmaco. Neerl.* **5**: 213-214.

Macrides, F., Bartke, A., and Dalterio, S., 1975, Strange females increase plasma testosterone levels in male mice, *Science* **189**:1104-1106.

Maruniak, J. A., and Bronson, F. H., 1976, Gonadotropic responses of male mice to female urine. *Endocrinology* **99**:963-969.

McLachlan, R. I., Wreford, N. G., Robertson, D. M., and de Kretser, D. M., 1995, Hormonal control of spermatogenesis, *TEM* **6**:95-101.

Mugford, R. A., and Nowell, N. W., 1970, Pheromones and their effect on aggression in mice, *Nature* **226**:967-968.

Mugford, R. A., and Nowell, N. W., 1971, Endocrine control over production and activity of the anti-aggression pheromone from female mice, *J. Endocrinol.* **49**, 225-232.

Parkes, A. S., and Bruce, H. M., 1961, Olfactory stimuli in mammalian reproduction, *Science* **134**:1049-1054.

Sharpe, R.M., 1994, ???, in: *The Physiology of Reproduction*, 2$^{nd}$ ed., E. Knobil, and J.D. Neill, eds., Raven Press, Ltd., New York, pp. 1363-1433.

Sun, Y-T., Irby, D. C., Robertson, D. M., and de Kretser, D. M., 1989, The effects of exogenously administered testosterone on spermatogenesis in intact and hypophysectomized rats, *Endocrinology* **125**:1000-1010.

Vandenbergh, J. G., 1969, Male odor accelerates female sexual maturation in mice, *Endocrinology* **84**:658-660.

Vandenbergh, J., 1994, in: *The Physiology of Reproduction*, 2$^{nd}$ ed., E. Knobil, and J.D. Neill, eds., Raven Press, Ltd., New York, pp. 343-359.

Whitten, W. K., 1956, Modification of the oestrus cycle of the mouse by external stimuli associated with the male, *J. Endocrinol.* **13**:399-404.

Whitten, W. K., 1958, Modification of the oestrus cycle of the mouse by external stimuli associated with the male: changes in the oestrus cycle determined by vaginal smears, *J. Endocrinol.* **17**:307-313.

Zipf, W. B., Payne, A. H., and Kelch, R. P., 1978, Prolactin, growth hormone and luteinizing hormone in the maintenance of testicular luteininzing hormone receptors, *Endocrinology* **103**:595-600.

# THE ROLE OF THE MAJOR HISTOCOMPATIBILITY COMPLEX IN SCENT COMMUNICATION

Michael D. Thom[1], Robert J. Beynon[2], and Jane L. Hurst[1]

## 1. INTRODUCTION

The major histocompatibility complex (MHC), which encodes proteins involved in immunological self-nonself recognition, shows an extraordinarily high degree of genetic polymorphism. This enables the MHC to fulfil its vital role in the recognition of a wide range of novel antigens. Two nonexclusive mechanisms for the maintenance of MHC variability have been proposed. Firstly, a rapidly evolving pathogen and parasite burden could favour rare alleles, and thus maintain diversity through negative frequency-dependent selection (Penn and Potts, 1999; Penn, 2002). Secondly, both males and females might gain fitness benefits by selecting complementary genotypes, and dissimilar MHC types in particular. This could be mediated indirectly by the selection of mating partners which maximise overall genotypic dissimilarity, or through direct assessment of MHC type. The latter mechanism would require evaluation of the MHC haplotype by potential mates. There is considerable evidence for an association between MHC and scent in a number of species that could be used for this purpose (e.g. rodents: Singh, 2001; fish: Aeschlimann et al., 2003; Olsen et al., 2002; lizards: Olsson et al., 2003; birds: Zelano and Edwards, 2002; humans: Porter and Moore, 1981; Wedekind et al., 1995). The considerable importance of disassortative mating in maintaining MHC diversity and disease resistance suggests that odours linked to MHC might have a specific role in the maintenance of the MHC's own polymorphism. This would involve individuals recognizing classes of animals which share or do not share their own MHC type. The observation that these odours can be discriminated under a variety of conditions has also led to the idea that they might function in other areas of social communication. As an extension of the hypothesis that animals can recognize MHC classes in a mate choice context, they might also use this ability to distinguish kin from non-kin. Finally,

---

[1] Animal Behaviour Group and [2]Protein Function Group, Faculty of Veterinary Science, University of Liverpool, Neston CH64 7TE, UK

the extreme polymorphism of MHC-associated odours could be sufficient that each individual has an essentially unique signature – in this case, they could be implicated in mediating individual recognition (Yamazaki et al., 1999; see also Hurst et al., this volume). This would require a more intricate mechanism than that involved in either mate choice or kin recognition, since each individual would need to be recognized as a unique entity, rather than merely assigned to the classes "similar" or "dissimilar". Here we assess the evidence for the possible roles for MHC-associated odours; are they involved only in discriminating classes of MHC, or has the polymorphism of the MHC also led to it being seconded to a role in individual recognition? We begin by discussing the evidence for MHC-associated odours and the different experimental designs that have revealed their potential functions.

## 2. EXPERIMENTAL EVIDENCE

Four main experimental paradigms have been used in the study of MHC-associated odours (Eggert et al., 1999). These are: mate choice; pregnancy block (the Bruce effect); trained discrimination; and habituation-dishabituation.

### 2.1. Mate Choice and Kin Recognition

The first empirical evidence for a behavioural effect of MHC-associated odours came from mate choice studies in mice. These demonstrated that both males and females exhibit a bias toward mating partners from an MHC-congenic strain over partners of their own strain (Yamazaki et al., 1976, 1978). There is also evidence for an MHC-disassortative mating bias in semi-natural populations (Potts et al., 1991). The presumed mechanism underlying the effect is that MHC-associated odours mediate the observed mating bias, and some preference studies using urinary odours alone support this (Egid and Brown, 1989; Ninomiya and Brown, 1995 but see Ehman and Scott, 2001). The direction of preference can be reversed by cross-fostering (Beauchamp et al., 1988; Penn and Potts, 1998), which implies that mice compare the odour of potential mating partners to those of their parents, rather than to their own scent. Although there is considerable evidence for a tendency to disassortative mating in mice, this trend has not been observed in every MHC-congenic pair tested (see reviews in Eklund, 1999; Penn and Potts, 1999; Yamazaki et al 1999), and the reasons for this remain unresolved. Other factors, such as marking rate, can override the importance of MHC compatibility in some circumstances (Roberts and Gosling, 2003), while territory ownership was of overriding importance in the semi-natural populations studied by Potts et al. (1991). Furthermore, Manning et al. (1992a) noted that female mice from inbred strains are less choosy than wild females, perhaps as a result of artificial selection on incestuous matings, and as a result they may be unsuitable for use in mate-choice experiments.

Functional tests of behaviour have also demonstrated the importance of the MHC in kin recognition. For example, Manning et al., (1992b) established that female mice in a semi-natural enclosure preferred nesting partners which shared their MHC alleles. Female mice preferentially retrieve offspring of the same MHC type as their own over MHC-congenic pups, while pups orient towards odours sharing their mother's MHC type even after cross-fostering (e.g. Yamazaki et al., 2000).

Both mate choice and kinship studies suggest that MHC-associated odours are likely to have a role in mediating the identification of close relatives or those sharing similar genotypes, a function that could be maintained through direct or inclusive fitness. These paradigms demonstrate that mice are able to discriminate a familiar odour (own strain) from that of other individuals which differ at the MHC. None of these tests provide evidence for a separate or additional role for MHC-associated odours in individual recognition. While it remains entirely possible that in a natural context mice individually identify alternative mating partners, there is no necessity for them to do so in order to select an MHC-dissimilar individual in mate choice and kin recognition tests. The only requirement is that mice can correctly discriminate classes of "familiar" (similar to self or to parents) from "unfamiliar" odours (Halpin, 1986; Penn, 2002).

Importantly, the role of background genes linked to MHC type has not been explored in the context of mate choice and kin selection, and linked genes rather than MHC genes themselves could control observed preferences. Nevertheless, there is good evidence that mice can recognize MHC classes, and that recognition of these classes has significant influences on behaviour. In particular, they reveal two potentially crucial functional roles for MHC-associated odours, in mate selection and in kin recognition. While this type of discrimination may be less precise than that required for individual recognition, it is clear that MHC-associated odours could be of considerable functional importance. Two non-functional paradigms, trained discrimination and habituation-dishabituation, have been used to further identify the role and specificity of odours in the discrimination of MHC classes; these are discussed in turn below.

## 2.2. Trained Discrimination

This type of experiment uses the repeated presentation of an odour together with positive or negative reinforcement to determine whether subjects can be trained to discriminate between two odours. The experiment usually begins with a training regime, which together with the final testing takes place in a Y-shaped maze. The end of each arm of the maze contains a different odour source, and air is drawn over these sources and into each arm. Subjects are usually water-deprived, and rewarded with a water drop for correctly approaching one arbitrarily chosen rewarded odour. Training often begins with non-social odours (e.g. cinnamon and juniper), then moves on to the urine types under test. If mice can successfully be trained to discriminate between two odour sources (with success usually defined as approximately 80% correct responses) then it is assumed that they can discriminate the odours. Generalization trials usually follow, in order to test whether the subject can discriminate odours of the same type, but derived from different sources and in the absence of reinforcement. Using this technique, mice can be trained to discriminate between urine from two MHC-congenic strains, but not between urine samples originating from individuals of the same inbred strain (Yamaguchi et al., 1981). Subjects could even discriminate between odours originating from mice which exhibited a single-point mutation in the H-2K region (Yamazaki et al., 1983a). Thus, even single-locus mutations can elicit detectable changes in odour profile.

An advantage over some other test paradigms is that both test urine samples are equally familiar to the subject, and thus responses cannot be due to a simple discrimination between familiar and unfamiliar (Halpin, 1986). These tests thus demonstrate that mice are able to detect differences in the MHC odours of two unfamiliar samples with sufficient resolution to discriminate between conspecifics which differ only subtly at the MHC, which further supports the hypothesis that MHC disassortative mate choice and kin recognition are based on odour differences. The responses obtained from trained discrimination tests are also consistent with the second possible function of MHC-associated odours, that of individual recognition. However, trained discrimination methods do not provide conclusive evidence of a functional individual recognition system based on MHC-associated odours. Mice can clearly recognize that the two test odours differ from each other, but this does not amount to individual recognition. Subjects can correctly discriminate juniper from cinnamon, as is demonstrated by the training regime; however this does not imply that these odours have functional significance. Similarly, rats can discriminate between the odours of inbred mouse strains differing at MHC loci (e.g. Beauchamp, 1985), again an ability that is unlikely to have any natural function. There is also a limitation for testing the functional importance of odours in that the studies operate under entirely artificial conditions. Subjects are severely water deprived (often for 23 hours) and are rewarded with water for making a "correct" choice. Although animals will often learn to discriminate between the test odours, this provides no indication of how they react to such odours under normal conditions (Halpin, 1986). In fact the observation that several hundred training trials are often required to achieve the desired success rate (e.g. Yamaguchi et al., 1981; Yamazaki et al.1990, 1994; Bard et al., 2000) suggests that mice may find the task inherently difficult, and that intensive training is a prerequisite for discrimination of an odour difference. It is known from other studies that mice can be trained to recognize odours they do not normally detect (Wang et al., 1993).

### 2.3. Habituation-dishabituation

The habituation-dishabituation test relies on an explicitly predicted difference in the spontaneous response to novel and familiar odours (Halpin, 1986). A subject is presented repeatedly with the same odour type, during which time its investigation response – time sniffing the scent or the number of sniffing bouts – is expected to decline (habituation). When presented with a second, test odour, the investigation response generally increases (dishabituation). As a control, test animals are presented with an odour from a source genetically identical to the source of the first odour, to control for non-genetic odour differences. This method has been used to demonstrate that wild mice are very sensitive to differences in MHC-associated odours, and can discriminate the odours of mice that differ at only a single locus (dm2; Penn and Potts, 1998). This acuity remains even after randomising background genetic differences by back-crossing the mutant strains (Carroll et al., 2002).

These habituation-dishabituation studies avoid the problems of intensive subject training, but re-introduce the problem of differing odour familiarity (Halpin, 1986). Although the response is spontaneous, its functional significance is unclear. While the dishabituation response could be interpreted as evidence of the ability to recognize the scents of different individuals, it could equally be explained by the ability to discriminate between MHC classes for the purposes of mate choice or kin recognition. Thus these tests do not currently allow distinction between the use of MHC-associated odours only for mate selection, and their broader use in individual recognition. Habituation-dishabituation tests also demonstrate that odour sources besides MHC can have a strong influence on odour, since subjects dishabituate to the control scent, even when this is taken from animals genetically identical to those providing the habituation odours (Penn and Potts, 1998; Carroll et al., 2002). While a variety of genetic and non-genetic sources are likely to influence individual odour, the signal for true individuality signalling is likely to have a genetic basis (see Hurst et al., this volume). Controlling for background non-genetic odours by pooling the urine from several individuals is essential for identifying the source of the underlying genetic component signalling identity in these tests. However pooling samples to mask non-genetic variation also removes the opportunity for subjects to demonstrate individual recognition, since they are not being offered individuals to identify. An additional difficulty with the paradigm was highlighted by Mayeaux and Johnston (2002), who demonstrated that the strongest dishabituation response is observed when the novel odour also occupies a novel location. It is thus important to distinguish between a location effect and any novel scent effect. This is likely to be a particular problem in the modified habituation-discrimination paradigm, where both the control and test odours are presented simultaneously in the final trial (e.g. Johnston and Jernigan, 1994; Johnston and Bullock, 2001). In this type of test, at least one odour must always occupy a novel position.

The habituation-dishabituation method suffers from some of the problems associated with the trained discrimination procedure; principally that it lacks a known functional context. However there is another important fundamental limitation in the habituation-dishabituation paradigm, which is that it relies entirely on an investigation response. As we discuss in Section 3, investigation alone can never be a definitive measure of individual recognition.

## 2.4. Pregnancy Block

The mate choice and kin recognition paradigms are consistent with a role for MHC-associated odours in maintaining allelic diversity in the MHC region. A further array of non-functional tests has been used to focus our understanding of these odours. However a separate spontaneous physiological response, pregnancy block, might also provide evidence for true individual recognition. Pregnancy block (or the Bruce effect) refers to the termination of pregnancy at the pre-implantation stage by the presence of an unfamiliar male (Bruce, 1960). This can occur when unfamiliar male odours are presented alone (Bruce and Parrott, 1960), indicating that the difference between familiar and unfamiliar males is based on chemical cues. These observations were adapted into an experimental assay by Yamazaki et al. (1983b), who demonstrated that the rate of pregnancy block was significantly greater when the stud and unfamiliar males differed only at the MHC, than when both were from the same strain. This effect has been shown to occur when the novel male differs from the stud at only a single MHC locus (Yamazaki et al., 1986). Significantly, the effect of male MHC type on the rate of pregnancy block has been demonstrated using urine alone, suggesting that the effect is mediated by MHC-associated odours.

The pregnancy block paradigm demonstrates that females can detect a difference between two males which differ at MHC loci. Does this represent a functional response? One functional interpretation of the Bruce effect is that females recognize the presence of a new individual male, and pregnancy is terminated in response to this recognition. However the functional significance of the pregnancy block response remains controversial. Importantly, it has not been tested in wild mice in a natural context, where females must regularly encounter many males. Other findings suggest that the response could occur in a variety of stressful conditions (Weir and de Fries, 1963), in which case exposure to a new and unfamiliar male could invoke a non-functional stress response. It would be interesting to determine whether other differences that alter odour familiarity have a similar effect, for example food type, metabolic changes and microbial flora. If so, the response is likely to be one of novelty, rather than specifically related to the MHC or compatibility of the male. In the context of this last point, we note that the response is not specific to male odours, since females of a different MHC type to the subject can also induce pregnancy block (Yamazaki et al., 1983b).

In fact it may be impossible to detect true individual recognition with this paradigm. Because the experimental design does not require a behavioural response by females, it cannot reveal whether recognition occurs at a physiological or a cognitive level. If the effect does rely on cognitive discrimination between male odours, females are nevertheless not required to individually identify each of the males, simply to recognize that an unfamiliar male differs in MHC-associated odours from a familiar male. The pregnancy block paradigm may thus provide evidence for female discrimination of classes of animals (grouped according to MHC type), but does not tell us whether familiar animals can be individually identified.

## 3. INVESTIGATION AND PREFERENCE

In both the mate choice studies (where mating is prevented) and the habituation-dishabituation test, investigation of an odour source is measured as the behavioural response. However, little attention is usually paid to the functional meaning of investigation in these contexts. We believe that investigation is of limited value in studies of individual recognition, because it is not a specific functional response.

In some mate choice studies, urine is presented in place of the whole animal in order to demonstrate that odours are, or can be, mediators of disassortative mating (e.g. Ninomiya and Brown, 1995; Ehman and Scott, 2001). In the absence of more direct measures, such as asking human subjects which odour they prefer (e.g. Wedekind et al., 1995), the duration of investigation or time spent near an odour are generally assumed to equate directly to preference. This can be misleading when the assumption is not explicitly tested. For instance, female voles have been shown to exhibit an odour preference for unparasitized males which does not correlate with mate selection (Klein et al., 1999).

Investigation is not always taken to mean preference. Nevertheless, we must address the interpretation of investigation behaviour in each context. For example, in the habituation-dishabituation paradigm, subjects typically investigate the scents for only a few seconds even at first exposure. Furthermore the mean difference in investigation time between the test and control odours is sometimes only a second or less (see figures in Penn and Potts, 1998; Carroll et al., 2002). When the novel scent comes from a pool with a different MHC type to the habituated scent, the mice spend barely any more time investigating this than they would if it was genetically identical to the new odour. There appear to be at least two opposing interpretations of this behaviour: firstly, that the scent contains virtually no additional information and the mice are scarcely able to detect a difference (i.e. discrimination is difficult), or secondly, that the odour difference is so striking that only a tiny amount of time is required to acquire the information contained in the scent (i.e. discrimination is easy). Further, do differences in investigation time measure the ease with which odours can be distinguished, or their relative interest or importance to the subject? Because of these uncertainties regarding the meaning of investigation behaviour, interpretation of the results of habituation-dishabituation tests is complicated.

There is an even more serious problem when using the habituation-dishabituation paradigm to examine individual recognition. The underlying logic is that an increased investigation response to the final odour demonstrates recognition of a difference from the odour used in the habituation presentations. Although we do not disagree with this interpretation, the extended assumption that detection of a difference implies individual recognition may be an over-interpretation. It is impossible to determine from a subject's investigation response whether it has recognized a different individual, or merely a change in the same individual's scent. If we consider a human example, someone shown a photograph of a friend taken many years ago may scan the picture for a few moments before deciding they recognize the person, while a recent photo will take almost no time to identify. But the amount of time they take to examine the picture is not a good indication of whether or not they eventually identify the subject correctly, merely of how difficult the task is. So in a test where a mouse is presented repeatedly with the scent of one individual and then the scent of a completely different individual, the subject may spend a long time investigating the final scent, but still arrive at the (erroneous) conclusion that it is, after all, from the same individual as the habituation scent. Examining investigation duration alone can never tell us what information the subject gained from the scent, as it does not involve a functional response.

## 4. WHAT IS THE SIGNIFICANCE OF MHC-ASSOCIATED ODOURS?

It is well established that many genetic and non-genetic factors contribute to the overall scent signature of an individual (see also Hurst et al., this volume), and among these odours are those associated with MHC type. Do these odours have functional importance, or are they merely one component contributing to the complex mixture of volatile odorants in urine? Understanding the mechanism by which MHC-associated odours might be delivered can help elucidate the information they may carry.

The carrier hypothesis proposes that fragments of MHC class I proteins bind volatiles in an allele-specific manner. This results in a unique combination of volatiles being drawn from the set of available metabolites, and excreted in the urine to produce a distinctive, MHC-specific odour (Singh, 1999, 2001; Pearse-Pratt et al., 1999). This mechanism would supply sufficient variability in odour signatures to enable recognition of individual MHC types. If there is sufficient MHC polymorphism, it might also provide enough diversity in genetically determined odours to enable individual recognition.

Alternatively, the MHC might influence individual odours indirectly through developmental and other effects that influence the mixture of volatile metabolites excreted in urine (Boyse et al., 1987). Although this hypothesis has previously been dismissed as being too difficult to test (Singh, 2001), it remains an important alternative to the carrier hypothesis. Indeed there is evidence that MHC type is associated with a range of physiological traits (Iványi, 1978), which are likely to influence the complex of volatiles that are known to bind to urinary proteins such as MUPs. In this case the MHC would have a more indirect role, along with other genetic differences between individuals that influence the range of volatiles contributing to the overall individual odortype.

The MHC carrier model could elicit odour signatures that would be characteristic of specific MHC types and recognizable even against a variable genetic background, due to allele-specific binding. This could result in an odour signal that communicates MHC type specifically. However the second mechanism is more likely to produce an odour signature that reflects the whole set of genes influencing volatile metabolites. Thus although MHC types might be discriminable against a similar genetic (and environmental) background, specific MHC types are likely to be masked by other genetic and non-genetic differences that also influence the general mixture of volatile metabolites. In this case, the apparent ability of mice and other species to discriminate MHC odours derived from inbred strains might be an artificial response, arising from the absence of background genetic "chemical noise" masking the MHC-specific odour signal. In this case, recognizing MHC type might be important only when individuals are genetically very similar, such as in a highly inbred population.

## 5. CONCLUSIONS

There is incontrovertible evidence that the MHC region is associated with type-specific odours in a number of species. Existing experimental paradigms have demonstrated that animals can discriminate between classes of MHC-associated odour with great acuity. Furthermore, experiments on mate choice and kin selection provide convincing evidence of a functional role for these odours. At the moment, evidence for genuine individual recognition or discrimination, meaning the "learned discrimination among conspecific individuals" (Halpin, 1986; p. 44), is lacking. This is partly because the four main experimental paradigms have not been used explicitly to address the question of individual recognition, but rather to test whether broader subgroups of MHC-associated odours exist and are distinguishable. These types of experiment can certainly be adapted to test individual recognition. For example Gheusi et al., (1997) found evidence for individual discrimination by rats using operant training, although this study did not examine MHC-associated odours specifically. The experimental paradigms described here have been vital in narrowing down the search for potential sources of individual odours, and testing odour discrimination abilities. Ultimately however, they can only allow an assessment of olfactory acuity, rather than whether this acuity is put to any use. In order to demonstrate that MHC-associated odours have an adaptive role in individual recognition, functional tests in a realistic behavioural context are required. Such functional tests exist (e.g. see Rich and Hurst, 1999; Hurst et al., 2001; Lai and Johnston, 2002) and now need to be applied to the unresolved issue of MHC-associated odours and individual recognition.

## 6. ACKNOWLEDGEMENTS

Thanks to members of the Animal Behaviour Group for comments and discussion, and to the BBSRC for financial support.

## 7. REFERENCES

Aeschlimann, P. B., Häberli, M. A., Reusch, T. B. H., Boehm, T., and Milinski, M., 2003, Female sticklebacks *Gasterosteus aculeatus* use self-reference to optimize MHC allele number during mate selection, *Behav. Ecol. Sociobiol.* **54**:119-126.
Bard, J., Yamazaki, K., Curran, M., Boyse, E. A., and Beauchamp, G. K., 2000, Effect of B2m gene disruption on MHC-determined odortypes, *Immunogenetics* **51**:514-518.
Beauchamp, G. K., Yamazaki, K., Bard, J., and Boyse, E. A., 1988, Preweaning experience in the control of mating preferences by genes in the major histocompatibility complex of the mouse, *Behav. Genetics* **18**:537-547.
Beauchamp, G. K., Yamazaki, K., Wysocki, C. J., Slotnick, B. M., Thomas, L., and Boyse, E. A., 1985, Chemosensory recognition of mouse major histocompatibility types by another species, *Proc. Natl. Acad. Sci. USA* **82**:4186-4188.
Boyse, E. A., Beauchamp, G. K., and Yamazaki, K., 1987, The genetics of body scent, *Trends Genet.* **3**: 97-102.
Bruce, H. M., 1960, A block to pregnancy in the mouse caused by proximity of strange males, *J. Reprod. Fert.* **1**:96-103.
Bruce, H. M., and Parrott, D. M. V., 1960, Role of olfactory sense in pregnancy block by strange males, *Science* **131**:1526.
Carroll, L. S., Penn, D. J., and Potts, W. K., 2002, Discrimination of MHC-derived odors by untrained mice is consistent with divergence in peptide-binding region residues, *Proc. Natl. Acad. Sci. USA* **99**:2187-2192.
Eggert, F., Muller-Ruchholtz, W., and Ferstl, R., 1999, Olfactory cues associated with the major histocompatibility complex, *Genetica* **104**:191-197.
Egid, K., and Brown, J. L., 1989, The major histocompatibility complex and female mating preferences in mice, *Anim. Behav.* **38**:548-549.
Ehman, K. D., and Scott, M. E., 2001, Urinary odour preferences of MHC congenic female mice, *Mus domesticus*: implications for kin recognition and detection of parasitized males, *Anim. Behav.* **62**:781-789.
Eklund, A. C., 1999, Use of the MHC for mate choice in wild house mice (*Mus domesticus*), *Genetica* **104**:245-248.
Gheusi, G., Goodall, G., and Dantzer, R., 1997, Individually distinctive odours represent individual conspecifics in rats, *Anim. Behav.* **53**:935-944.
Halpin, Z. T., 1986, Individual odors among mammals: origins and functions, *Adv. Stud. Behav.* **16**:39-70.
Hurst, J. L., Payne, C. E., Nevison, C. M., Marie, A. D., Humphries, R. E., Robertson, D. H. L., Cavaggioni, A., and Beynon, R. J., 2001, Individual recognition in mice mediated by major urinary proteins, *Nature* **414**:631-634.
Iványi, P., 1978, Some aspects of the H-2 system, the major histocompatibility system in the mouse, *Proc. Roy. Soc. Lond. B.* **202**:117-158.
Johnston, R. E., and Bullock, T. A., 2001, Individual recognition by use of odours in golden hamsters: the nature of individual representations, *Anim. Behav.* **61**:545-557.
Johnston, R. E., and Jernigan, P., 1994, Golden hamsters recognize individuals, not just individual scents, *Anim. Behav.* **48**:129-136.
Klein, S. L., Gamble, H. R., and Nelson, R. J., 1999, *Trichinella spiralis* infection in voles alters female odor preference but not partner preference, *Behav. Ecol. Sociobiol.* **45**:323-329.
Lai, W. S., and Johnston, R. E., 2002, Individual recognition after fighting by golden hamsters: a new method, *Physiol. Behav.* **76**:225-239.
Manning, C. J., Potts W. K., Wakeland, E. K., and Dewsbury, D.A., 1992a, What's wrong with MHC mate choice experiments?, in: *Chemical Signals in Vertebrates VI*, R. L. Doty and D. Müller-Schwarze, eds., Plenum Press, New York, pp. 229-235.
Manning, C. J., Wakeland, E. K., and Potts, W. K., 1992b, Communal nesting patterns in mice implicate MHC genes in kin recognition, *Nature* **360**:581-583.
Mayeaux, D. J., and Johnston, R. E., 2002, Discrimination of individual odours by hamsters (*Mesocricetus auratus*) varies with the location of those odours, *Anim. Behav.* **64**:269-281.
Ninomiya, K., and Brown, R.E., 1995, Removal of the preputial glands alters the individual odors of MHC-congenic mice and the preferences of females for these odors, *Physiol. Behav.* **58**:191-194.
Olsen, K. H., Grahn, M., and Lohm, J., 2002, Influence of MHC on sibling discrimination in Arctic char, *Salvelinus alpinus* (L.), *J. Chem. Ecol.* **28**:783-795.
Olsson, M., Madsen, T., Nordby, J., Wapstra, E., Ujvari, B., and Wittsell, H., 2003, Major histocompatibility complex and mate choice in sand lizards, *Proc. R. Soc. Lond. B. (Supplement)* doi: **10.1098/rsbl.2003.0079**.
Pearse-Pratt, R., Schellinck, H. M., Brown, R. E., Singh, P. B., and Roser, B., 1999, Soluble MHC antigens and olfactory recognition of genetic individuality: the mechanism, *Genetica*, **104**:223-230.

Penn, D. J., 2002, The scent of genetic compatibility: sexual selection and the major histocompatibility complex, *Ethology* **108**:1-21.
Penn, D. J., and Potts, W. K., 1998, MHC-disassortative mating preferences reversed by cross-fostering, *Proc. R. Soc. Lond. B.* **265**:1299-1306.
Penn, D. J., and Potts, W. K., 1998, Untrained mice discriminate MHC-determined odors, *Physiol. Behav.* **63**:235-243.
Penn, D. J., and Potts, W. K., 1999, The evolution of mating preferences and major histocompatibility complex genes, *Am. Nat.* **153**:145-164.
Porter, R. H., and Moore, D., 1981, Human kin recognition by olfactory cues, *Physiol. Behav.*, **27**:493-495.
Potts, W. K., Manning, C.J., and Wakeland, E.K., 1991, Mating patterns in seminatural populations of mice influenced by MHC genotype, *Nature* **352**:619-621.
Rich, T. J., and Hurst, J. L., 1999, The competing countermarks hypothesis: reliable assessment of competitive ability by potential mates, *Anim. Behav.* **58**:1027-1037.
Roberts, S. C., and Gosling, L. M., 2003, Genetica similarity and quality interact in mate choice decisions by female mice, *Nat. Genet.* **35**:103-106.
Singh, P. B., 2001, Chemosensation and genetic individuality, *Reproduction* **121**:529-539.
Singh, P. B., 1999, The present status of the 'carrier hypothesis' for chemosensory recognition of genetic individuality, *Genetica* **104**:231-233.
Wang, H., Wysocki, C. J., and Gold, G. H., 1993, Induction of olfactory receptor sensitivity in mice, *Science* **260**:998-1000.
Wedekind, C., Seebeck, T., Bettens, F., and Paepke, A., 1995, MHC-dependent mate preferences in humans, *Proc. Roy Soc. Lond. B.* **260**:245-249.
Weir, M. W., and de Fries, J. C., 1963, Blocking of pregnancy in mice as a function of stress, *Psych. Rep.* **13**:365-366.
Yamaguchi, M., Yamazaki, K., Beauchamp, G. K., Bard, J., Thomas, L., and Boyse, E. A., 1981, Distinctive urinary odors governed by the major histocompatibility locus of the mouse, *Proc. Natl. Acad. Sci. USA* **78**:5817-5820.
Yamazaki, K., Beauchamp, G. K., Bard, J., and Boyse, E. A., 1990, Chemosensory identity and the Y chromosome, *Behav. Genet.* **20**:157-165.
Yamazaki, K., Beauchamp, G. K., Curran, M., Bard, J., and Boyse, E. A., 2000, Parent-progeny recognition as a function of MHC odortype identity, *Proc. Natl. Acad. Sci. USA* **97**:10500-10502.
Yamazaki, K., Beauchamp, G. K., Egorov, I. K., Bard, J., Thomas, L., and Boyse, E. A., 1983a, Sensory distinction between $H-2^b$ and $H-2^{bm1}$ mutant mice, *Proc. Natl. Acad. Sci. USA* **80**:5685-5688.
Yamazaki, K., Beauchamp, G. K., Matsuzaki, O., Kupniewski, D., Bard, J., Thomas, L., and Boyse, E. A., 1986, Influence of a genetic difference confined to a mutation of H-2K on the incidence of pregnancy block in mice, *Proc. Natl. Acad. Sci. USA* **83**:740-741.
Yamazaki, K., Beauchamp, G. K., Shen, F.-W., Bard, J., and Boyse, E. A., 1994, Discrimination of odortypes determined by the major histocompatibility complex among outbred mice, *Proc. Natl. Acad. Sci. USA* **91**:3735-3738.
Yamazaki, K., Beauchamp, G. K., Singer, A., Bard, J., and Boyse, E. A., 1999, Odortypes: their origin and composition, *Proc. Natl. Acad. Sci. USA* **96**:1522-1525.
Yamazaki, K., Beauchamp, G. K., Wysocki, C. J., Bard, J., Thomas, L., and Boyse, E. A., 1983b, Recognition of H-2 types in relation to the blocking of pregnancy in mice, *Science* **221**:186-188.
Yamazaki, K., Boyse, E. A., Miké, V., Thaler, H. T., Mathieson, B. J., Abbott, J., Boyse, J., Zayas, Z. A., and Thomas, L., 1976, Control of mating preferences in mice by genes in the major histocompatibility complex, *J. Expt. Med.* **144**:1324-1335.
Yamazaki, K., Yamaguchi, M., Andrews, P. W., Peake, B., and Boyse, E. A., 1978, Mating preferences of F2 segregants of crosses between MHC-congenic mouse strains, *Immunogenetics* **6**:253-259.
Zelano, B., and Edwards, S. V., 2002, An MHC component to kin recognition and mate choice in birds: predictions, progress, and prospects, *Am. Nat.* **160**:S225-S237.

# CHARACTERISATION OF PROTEINS IN SCENT MARKS: PROTEOMICS MEETS SEMIOCHEMISTRY

Duncan H.L. Robertson[*], Sarah Cheetham[*,§], Stuart Armstrong[*], Jane L. Hurst[§] and Robert J Beynon[*,†]

## 1. INTRODUCTION

The wide range of semiochemical secretions used by animals has a correspondingly diverse mixture of chemical constituents. Many classes of molecule have been associated with scents and scent marks, predominantly small volatile molecules. However, it is only relatively recently that we have come to appreciate the multiple roles of high molecular weight components, especially proteins and peptides. Early research hinted at this possibility - the observation that the high molecular weight fraction of male mouse urine accelerated the onset of female puberty (Vandenberg et al., 1975, 1976) recalled the well-known observation that mouse urine contained large quantities of protein (Parfentjev and Perlzweig, 1933; Finlayson and Baumann, 1958).

It is now known that proteins play many different roles in scent communication. Proteins are involved in the biosynthesis of low molecular weight scent components (e.g. Seybold and Tittiger, 2003) (although, by and large, the enzymology of scent biosynthesis is unknown), as precursors of proteolytically cleaved water-borne peptides (Kikuyama and Toyoda, 1999), as mediators of secretion and maintenance of scent marks in the environment (Hurst et al., 1998), as concentrators or buffers of scent signals at the first phase of the reception process (Tegoni et al., 2000), as G-protein coupled receptors (Buck and Axel, 1991), or in the subsequent signal transduction processes (Ronnett and Moon, 2002).

Techniques for the analysis of low molecular weight scent components are well evolved, but the emergent interest in the protein components of scent marks requires new

---

[*] Protein Function Group, Faculty of Veterinary Science, University of Liverpool, Liverpool, L69 3BX, United Kingdom.
[§] Animal Behaviour Group, Faculty of Veterinary Science, University of Liverpool, Liverpool, L69 3BX, United Kingdom.
[†] To whom all correspondence should be addressed

experimental approaches to characterise these macromolecular components. Proteomics, the analysis of all the proteins within a cell, tissue or secretion, is a new discipline that has developed in response to the publication of entire genomic DNA sequences from a variety of species (Aebersold and Mann, 2003; Gershon, 2003; Lin et al., 2003). The demands of proteomics have stimulated development of a raft of new techniques for the identification and characterisation of large numbers of proteins in complex mixtures - techniques that are readily applied to the characterisation of proteins that play a role in some aspect of the semiochemical information transduction process. In this paper, we document the currently available methods and strategies for identification and characterisation of proteins in scent secretions, illustrated with data from our own research programmes.

## 2. ASSESSING THE PROTEIN COMPLEXITY OF A SECRETION

As might be expected from the variety of scent secretions used in chemical communication, the protein complexity can vary greatly. The optimal strategy to identify and characterise the proteins within a scent will largely depend on the amount of protein present and the proportion of the protein of interest, relative to other proteins that 'contaminate' the sample. Consequently, assessing the variety and amount of protein in a secretion is an essential first step. The optimal strategy for resolution and identification of proteins in a scent secretion will depend on the abundance of the sample, the complexity of the sample and the overall chemical composition and nature of the scent. An abundant, aqueous sample such as urine presents a much more amenable challenge than, for example, a waxy sebaceous secretion that is limited in quantity.

### 2.1 Electrophoretic Techniques

Polyacrylamide gel electrophoresis (PAGE) is a well established technique that separates proteins on the basis of size. A porous, highly hydrated polyacrylamide gel provides a stabilising matrix in which the electrophoresis is conducted, limiting diffusion that would otherwise degrade the quality of the separation. The most common variant of this technique treats proteins with the anionic detergent sodium dodecyl sulphate (SDS) prior to their separation in an electric field (SDS-PAGE). SDS binds to proteins at a constant weight ratio. This gives the protein-SDS complex a uniform charge density, unfolds the protein structure, and separates the proteins as a set of highly charged, largely unstructured polymers, suppressing charge differences between individual proteins and 'filtering' the proteins through the acrylamide gel, the potential field providing the driving force. Low cost and ease of use, along with its high resolving power make this technique ideal for an initial appraisal of protein complexity (Figure 1).

A second technique used widely for analysis of protein heterogeneity is isoelectric focussing (IEF) (Nguyen et al., 1977). A pH gradient is established in a gel matrix, and application of an electric field to the gel causes proteins to migrate to the point in the pH gradient equal to their iso-electric point (where they have no net charge). Manipulation of the range and slope of the pH gradient in IEF provides scope for separation of proteins that differ by slight variation in overall charge. Because IEF is a focussing technique, the longer the gel is run, the higher the resolution of the mixture, and IEF is probably the most effective method for separation of complex mixtures of proteins (Figure 1).

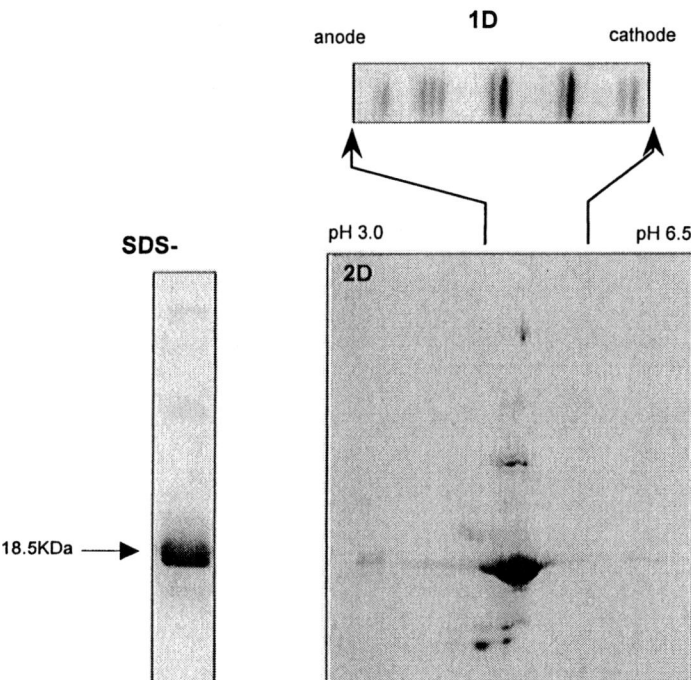

Figure 1. Separation of protein by one and two-dimensional electrophoresis. A sample of mouse urine was separated by SDS-PAGE (size-based separation), isoelectric focusing (charge based separation) and by a two dimensional (charge, then size) separation. Proteins were visualised by staining with Coomassie blue protein stain or silver.

For particularly complex samples, the two techniques can be combined into a two-dimensional separation (2DGE or 2D PAGE). Proteins are initially separated by IEF in a thin strip of gel and at the end of the separation, SDS is added to the strip. This strip is subsequently laid onto the top of a slab of an acrylamide gel, in which the proteins undergo a second, orthogonal separation according to their size (the SDS-PAGE separation). Although relatively protracted, the resolving power of this technique is unsurpassed, allowing the separation of thousands of proteins in a single gel (e.g., Simpson et al., 2004).

## 2.2 Chromatographic techniques

Although chromatographic techniques have, for practical purposes, lower resolution than electrophoresis, proteins separated by column chromatography can be retained in their native state and are easily recovered for use in further analyses or bioassays. The two methods most commonly used are size exclusion chromatography (SEC) and ion exchange chromatography (IEC). Because these techniques capitalise on different molecular properties to achieve separation (molecular mass and net charge respectively) they are also amenable to orthogonal combination.

SEC is also known as gel filtration, and operates by exploiting the relative ability of different-sized proteins to penetrate porous beads. Larger proteins cannot enter the pores, circumvent the beads and are eluted first in the eluate. Smaller molecules enter the pores and take a correspondingly longer path through the column, eluting later. SEC is particularly useful as it can separate low molecular weight scent components from macromolecules. The first indications of the primer pheromonal properties of MUPs were discovered by bioassay of urine fractionated by SEC (Vandenberg et al., 1975). More recently, we used SEC to implicate MUPs as the main component in male house mouse urine that stimulates males to countermark another male's scent (Humphries et al., 1999).

Variations in the composition of charged amino acids in different proteins means that many proteins carry distinct net charges. In IEC, proteins interact differentially with a charged chromatographic medium. Proteins of similar charge to the medium pass through the column, but oppositely charged proteins adhere to the column medium by electrostatic interactions. Adhered proteins are then selectively desorbed by progressive weakening of the electrostatic interactions by introduction of competing salt ions. IEC has been used extensively for characterising MUPs (Robertson et al., 1995, Veggerby et al., 2002, Mechref et al., 2000), where proteins can differ by as little as one charged amino acid residue. For particularly complex secretions a combined chromatographic approach is possible, where the secretion is initially separated by SEC and the emerging fractions are loaded onto an IEC column for further analysis. Analogous to 2D-PAGE, this type of orthogonal separation has the potential to resolve many more proteins within a complex mixture.

It is customary to monitor the elution of proteins by virtue of an increase in absorbance of the eluate, either at 280nm or, for greater sensitivity, at 214nm. Some caution is advised here, as many organic compounds absorb light at these wavelengths, and an absorbance peak at 280nm cannot be assumed to reflect a peak of proteins. Further assays are required to confirm the presence of protein and the complexity of the mixture.

## 2.3 Mass Spectrometry

Mass measurement of proteins and peptides has become a routine procedure, made possible by the development of methods to bring involatile proteins and peptides into the gas phase without incurring damage. The two most common 'soft' ionisation modes are electrospray ionisation (ESI) and matrix associated laser desorption ionisation (MALDI) (Gershon, 2003). Once in the gas phase, the peptide and proteins can be mass-measured by a range of mass analysers that differ in mass range, resolution and accuracy. ESI is an effective method of nebulising a solution of protein molecules, but is relatively intolerant of contaminants such as detergents and salts. MALDI, which employs a laser to volatilise and charge proteins and peptides in the solid state, is relatively tolerant to contaminants. A detailed discussion of different mass analysers is beyond the scope of this chapter, but see Ashcroft (2003) and Chalmers and Gaskell (2000) for recent reviews.

## 2.4 Combined Strategies

It is perfectly feasible to combine one or more of these methods into a multi-dimensional characterisation of a protein mixture. Because the methodologies are orthogonal in the information they produce, they can often provide additional insights into the complexity of a mixture. A good example of this approach is provided by our early work on MUP characterisation in inbred mouse strains (Robertson et al., 1995). An ESI-MS survey of a total protein extract of BALB/cJ mouse urine revealed three proteins of masses 18643Da, 18694Da and 18706Da. However, when the same protein sample was resolved by high resolution ion exchange chromatography, four discrete peaks could be discriminated. Each of these four fractions were then analysed again by ESI-MS. From this analysis, proteins of five discrete masses could be resolved. Two of the proteins were isobaric (having the same mass of 18693Da at the resolution of ESI-MS) but because they possessed different overall charges, were separable by ion exchange chromatography. A lower abundance, fifth protein at 18698Da was unmasked because the purification process simplified the set of analytes submitted to ESI-MS and subsequent data processing.

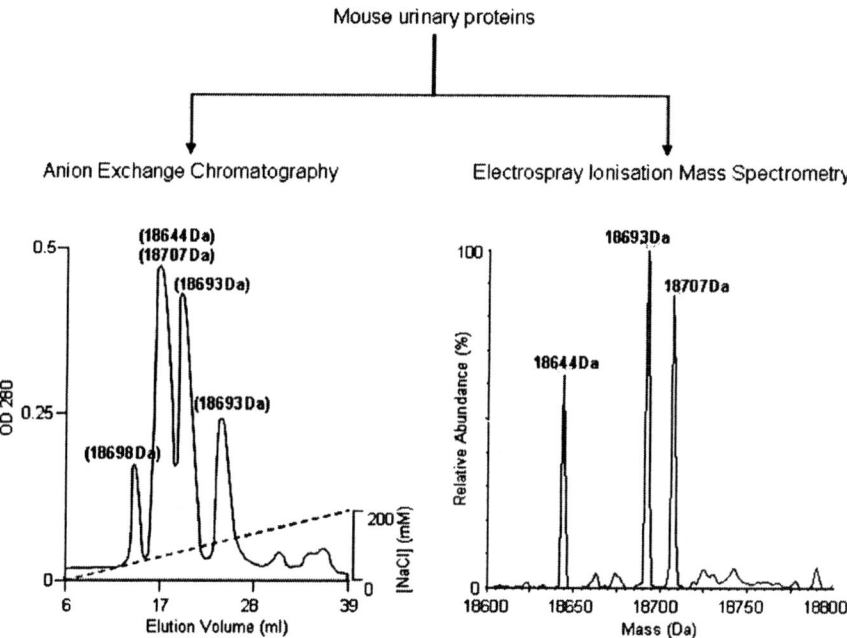

Figure 2. Assessment of the complexity of a protein mixture by combined strategies. A protein mixture from mouse urine was analysed by ion exchange chromatography (charge based separation) and by mass spectrometry of the intact proteins. Both methods allude to the complexity of the mixture. However, when the proteins separated by chromatography were analysed by mass spectrometry, the largest peak was demonstrated to contain two proteins. Moreover, the chromatographic separation indicated that there were two protein peaks that contained proteins of mass 18693Da. The techniques give orthogonal and complementary views of the complexity of the analyte.

## 3. IDENTIFICATION OF PROTEINS

### 3.1 Strategies for Identification

It is worth considering what we mean by 'identification' of a protein. The term has different meanings, depending on the perspective of the researcher. To most molecular biologists, it could embrace a full cDNA and gene sequence; to a structural biologist it might be extended to encompass the precise three dimensional structure of the protein. However, we would advocate a more pragmatic approach. In some circumstances, it may be enough to know that a protein is a member of a class of proteins such as pheromone binding proteins. In other situations, the entire protein sequence or cDNA sequence may be a necessary prelude to heterologous expression of a recombinant protein. In the first instance, a valid question is "does my protein of interest have an identity or resemblance to any known protein in the sequence databases?" To put this in perspective, there are well over one million known protein sequences in the current protein databases, and to be able to match an unknown protein to one of those sequences is a major gain in information.

If the genome sequence of the animal is known, or databases of cDNA or EST sequences (expressed sequence tags derived from mRNA) are available, then the process of protein identification becomes significantly easier, but most species in which scent marking is studied are not primary candidates for genome sequencing programmes. For example, the bank vole (*Clethrionomys glareolus*) has single figure entries that represent true proteins, contrasting dramatically with the deposition of approx. 150,000 mouse (*Mus* spp) entries in the same database. In the absence of genome sequence data, identification is still feasible, although one has to be prepared for the possibility of discovering an entirely novel protein that had not hitherto been identified in the genomes of yeast, a nematode worm, fruit fly, mouse, or man. Further, proteins involved in species-specific communication are likely to have been subject to diversifying positive selection pressure (Emes et al., 2003) increasing the possibility of an exciting discovery of a novel protein.

A general strategy for identifying proteins in scent marks is difficult to generalize; individual applications may require specific modifications and embellishments. In practical terms, extraction of proteins from a scent mark or secretion is the first step in an identification strategy and the complexity of the extraction process will reflect the nature of the secretion. Prevention of enzymatic degradation of the proteins is routine, usually through treatment of the sample with a cocktail of protease inhibitors (Beynon and Bond, 1999). Extraction of proteins from aqueous secretions is easily achieved with SEC or centrifugal filtration, which partitions the proteins from the many other smaller contaminant molecules. Other more complex or lipid-containing samples will need further treatment to solubilise the protein and remove contaminating lipid or other material before the proteins can be studied. The protein solution should be assessed for overall complexity by 1D or 2D gel electrophoresis, routine methods that are readily adopted by most laboratories. Correlation of protein patterns with behavioural parameters such as status or seasonality may help to locate proteins that are involved in mediation of scents, as does sexual dimorphism in protein expression patterns.

## 3.2 Identification Techniques for Previously Characterised Proteins or Genomes

Where the amino acid sequence of the protein is already known, or the complete genome of the producer organism has been sequenced (e.g. *Mus domesticus*), identification is relatively straightforward.

If the protein is soluble in aqueous media, mass spectrometric methods can measure the average mass of the protein. Typically, this mass can be acquired to an accuracy of 1Da in 10,000Da. Although very expensive, high performance instruments can exceed this level of accuracy by an order of magnitude. Average mass cannot be used to search entire databases of proteins, because this single parameter carries insufficient information for unambiguous identification, but can be used to confirm the expression of a protein for which a predicted mass can be calculated from the inferred protein sequence. However, even calculation of a predicted mass carries some difficulties, notably the discrepancy between predicted and actual mass caused by post-translational modifications such as proteolysis, disulphide bond formation, acetylation and so forth. Given good knowledge of the sequences and likely modifications of the candidate proteins, mass measurement of proteins can be of sufficiently high quality to identify proteins in mixtures (Hayter et al., 2002). Our initial mass spectrometric analysis of MUPs from BALB/c urine yielded measured masses that agreed closely with the theoretical masses predicted from cDNA sequences. On the basis of this measurement we confirmed the amino acid sequences of the MUPs expressed by inbred BALB/c mice. Subsequent analysis of MUPs from wild mice (Robertson et al., 1997, Beynon et al., 2002) demonstrated that in addition to expressing the same proteins, wild mice also expressed others with masses that were not predicted by any known cDNA sequence.

Identification of the proteins in scent marks is enhanced if additional information about the protein can be recovered. Peptide mass fingerprinting (PMF) is a method that can be applied to proteins that have been purified to near homogeneity, whether by column chromatography or gel electrophoresis. The purified protein is exposed to a proteolytic enzyme, such as trypsin, that cleaves the protein at specific amino acid residues. Because each protein molecule is cleaved in the same way, the outcome of this digestion is a set of 'limit peptides' that defines the positions of the cleavage sites in the protein sequence. The masses of these limit peptides are then determined by mass spectrometry. A typical protein of around 300 amino acids would be expected to be cleaved into about 30 fragments by trypsin, and we might expect to be able to measure the mass of about half of these peptides. This is the 'peptide mass fingerprint'. Search algorithms (e.g. Perkins et al., 1999; Zhang and Chait, 2000) match the experimental fingerprint with a succession of theoretical fingerprints generated from every known protein sequence, and the best matches, provided they exceed a probabilistic threshold, can be considered as candidate identities for the protein. Identification of putative proteins in pig saliva (Figure 3) illustrates the application of PMF.

PMF is most effective when the precise sequence of the protein already exists in the protein databases. It works effectively with proteins isolated by either chromatographic or electrophoretic methods, and with a sensitivity that is commensurate with such separation techniques applied to microgram or nanogram quantities of protein. For PMF to work optimally, the protein for which a fingerprint has been obtained must be present in the sequence database. A significant limitation in PMF is the difficulty of matching a fingerprint derived from species other than those that are represented in the sequence database. Even small sequence changes (such as substitution of a leucine residue with the chemically similar residue valine) elicit a change in mass (14Da) that far exceeds the tolerance of any search engine. If such changes are present in several of the peptides in the fingerprint, there is a much reduced probability of a corresponding match. New tools are being developed that compensate for the possibility of such conservative amino acid changes, but PMF works best when the source protein has a sequence that is very close to the target database entry. Of course, the ability to match a fingerprint across a species boundary depends on the sequence similarity between the novel and known sequences. In many cases, 'cross-species matching' can yield high probability identifications, and certainly justifies an attempt at PMF as a first pass strategy.

### 3.3 Identification Techniques for Uncharacterised Proteins or Genomes

Identification of proteins from those species where few protein or cDNA sequences are available presents a greater challenge, particularly when a good quality MALDI-TOF fingerprint has failed to hint at an identity. Under these circumstances, there is a need to recover additional information, the most accessible of which is partial amino acid

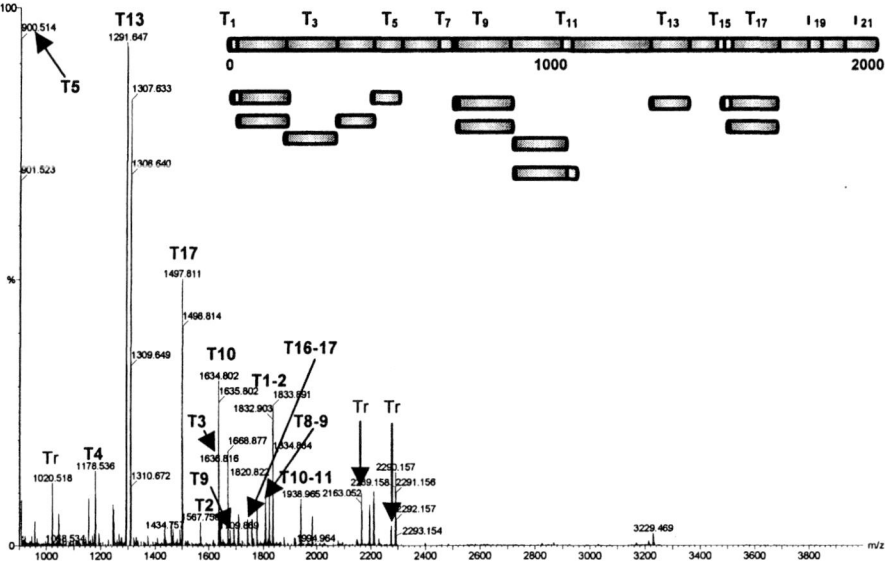

**Figure 3.** Identification of a protein by peptide mass fingerprinting. The protein constituents of pig saliva were separated by SD-PAGE and a protein band was digested with trypsin. The resultant tryptic peptides were mass-measured using MALDI-ToF mass spectrometry. The peptides in the mass spectrum were either derived from trypsin self-digestion (T) or were derived from the protein in the gel. Database searching with the masses of these peptides led to an unequivocal identification of the protein as SAL (salivary lipocalin). The inset map shows the theoretical tryptic digestion map of this protein, and underneath are the peptides that were observed. In many instances, smaller peptides were visible as partial digestion products.

sequence data. The accurate mass of a peptide does not give information on either the amino acids that constitute that peptide or their sequence, so recovery of the sequence of even a few amino acids is a substantial gain in information content. The sequence is then used as a search term against the entire database of known protein sequences. Because the search algorithms are tolerant of amino acid changes, and especially those that are structurally conservative (such as leucine / isoleucine, for example) matches can more readily cross species boundaries.

The two most common means of determining amino acid sequences are Edman degradation and tandem mass spectrometry (MS/MS or $MS^2$). The technique of Edman degradation determines the amino terminal sequence of proteins or peptides by labelling the N-terminal amino acid, cleaving it from the protein and identifying the amino acid from its chromatographic behaviour. The cycle is then repeated to identify a contiguous stretch of amino acids. Although Edman degradation is not trivial and requires a significant quantity of sample, it has the potential to generate a long stretch of amino acid sequence.

Tandem mass spectrometry also has the potential to recover sequence data, but with substantially less starting material. Most commonly, an isolated protein is treated with a proteinase to produce a mixture of smaller peptides, and it is these peptides, usually in a mixture, that are subject to mass spectrometric sequencing. A tandem mass spectrometer, and specifically, a "tandem in space" instrument, as the name suggests defines a complex instrument comprising two physically linked mass spectrometers. (There are also "tandem in time" mass spectrometers, where a peptide ion can be stored in the mass spectrometer, fragmented, and then the same mass spectrometer can be used to measure the masses of the fragment ions, but these are beyond the scope of this article.) For peptide sequencing, each peptide is subject to pseudo-random fragmentation of the polypeptide chain, at the peptide bond. The resulting population of product ions differ from each other by one amino acid residue and are transferred to the second mass spectrometer that measures their mass accurately. As the majority of amino acids have unique signature masses, the amino acid sequence is deduced from the mass differences of the peaks within the product ion spectrum.

Tandem mass spectrometry offers a number of potential advantages over Edman degradation. As with PMF, it can be applied to peptides extracted from SDS or 2D-PAGE gels. It is also fast, easily automated and unlike Edman degradation, it produces sequence information from peptides recovered along the length of a protein chain. Amongst the drawbacks to the techniques are the inability to resolve leucine and isoleucine (these are isobaric, having the same mass) and the inability to control the fragmentation to yield information that always covers the entire peptide.

A sequence "tag" from an unknown protein allows a number of further options for characterisation. Even a short stretch of amino acid sequence provides a powerful means of interrogating a protein database, and may provide a useful alternative to PMF in poor quality samples that may have peptides derived from more than one protein. More advanced database searching systems will find proteins with homologous sequences to that of the peptide tag. A good example of this type of database searching system is BLAST (Basic Local Alignment Sequence Tool), (Altschul et al., 1997). A modified form of this (MS BLAST) was used in conjunction with PMF to characterise the proteome of the yeast *Pichia pastoris*, whose genome has not been fully sequenced provide strong evidence for the identity or class membership, such as the lipocalins, for example. Some sequence tags are more informative than others. For example, the lipocalins as a class is typified by several short sequence motifs, which if found in the sequenced peptides enhances the likelihood of the unknown protein also being a lipocalin. Finally, short tracts of sequence are often enough to encourage chemical synthesis of the equivalent peptide, in order to raise antibodies which will, under the best circumstances, act as a specific reagent for identification, localisation, characterisation and quantification of the parent protein. They can additionally be used to direct the design of degenerate primers for cDNA cloning and subsequent over-expression of the protein in a heterogeneous host (Figure 4).

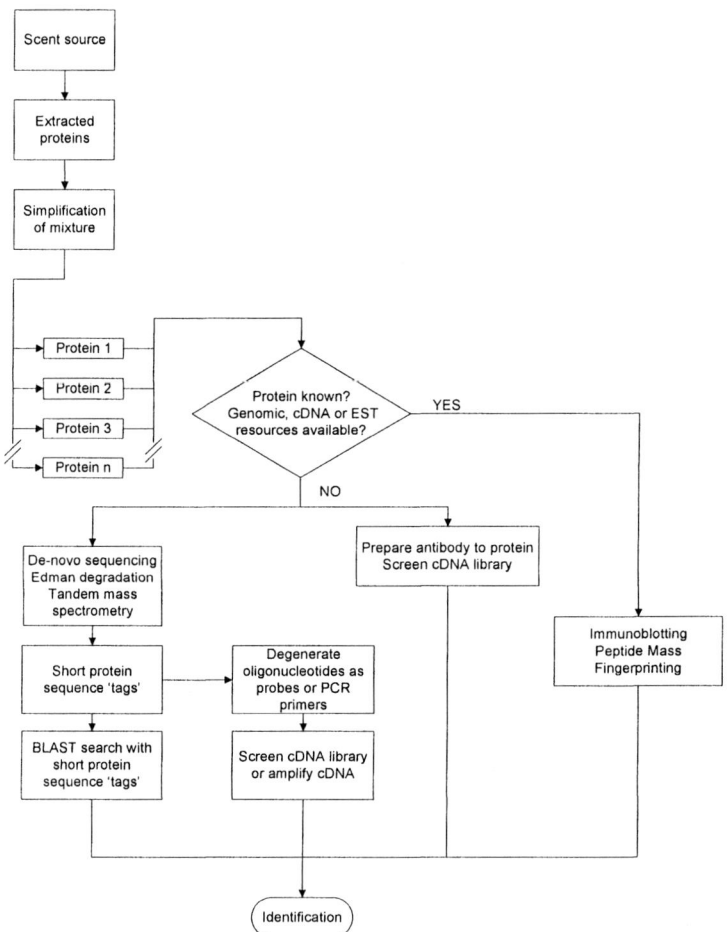

**Figure 4.** Strategy for analysis of protein in scent sources.

## 4. CHARACTERISATION OF PROTEIN POLYMORPHISMS

Proteins often exist in a variety of closely related forms. MUPs are a particularly good example of this (Robertson et al., 1995 and 1997; Beynon et al., 2002) but the same phenomenon has also been observed in OBPs (Pes and Pelosi, 1995), and rat alpha-2U proteins (Mertens et al., 1983). The proteins of the Major Histocompatibility Complex (MHC) are also highly heterogenous (Klein, 1979). The heterogeneity of MUPs and the MHC has functional significance, (Hurst et al., 2001; Cunningham, 1977) but equivalent roles of heterogeneity in other systems are not well understood. In order to attribute functions to protein heterogeneity, it must first be characterised. The following case study of MUP heterogeneity in the grassland mouse *Mus spretus* (Figure 5) illustrates how this can be achieved and how the identification strategy outlined above can be applied.

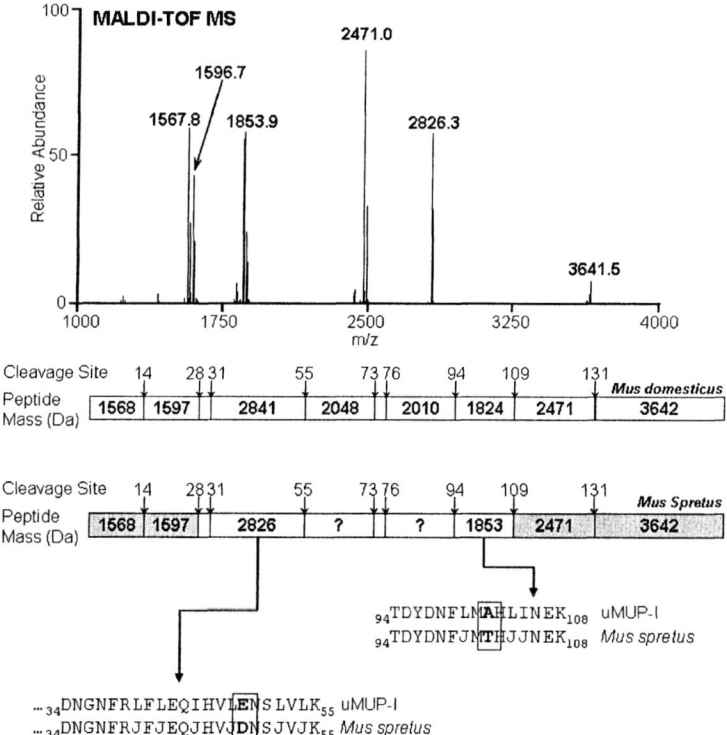

**Figure 5.** Mapping of protein polymorphisms by mass spectrometry. A Major Urinary Protein from *Mus spretus* was purified and Lys-C digestion fragments were mass measured by MALDI-ToF mass spectrometry. Four of the peptides had the same masses as the equivalent protein from *Mus domesticus*. However, for two of the remainder, of novel mass, tandem mass spectrometry led to the identification of an amino acid change that was consistent with the new peptide mass. (J is used to define leucine or isoleucine, which cannot be distinguished by this type of mass spectrometry).

Proteins were extracted from small samples (10-50μl) of *Mus spretus* urine using SEC run in centrifuged columns (Robertson et al., 1993). An initial anion exchange chromatography separation resolved the proteins into one large and two small peaks. By ESI-TOF MS the large peak consisted of a single protein of molecular mass 18759Da which was unique when compared to the theoretical masses of all the MUP sequences lodged in the SwissProt database. Such theoretical masses were determined using an experimentally determined N-terminal motif (EEASS) to separate the mature sequence from the signal peptide known to exist in MUPs prior to export from the liver cell (Clark et al., 1985).

To establish the identity of this protein, a preparation was digested with endoproteinase Lys-C (Lys-C), which cleaves the peptide chain at each lysine residue. The masses of the resulting peptides were measured using MALDI-TOF mass spectrometry and compared to a set of theoretical masses derived from reference *Mus domesticus* MUP sequences. Four of the six peptide masses measured in this experiment matched those from the reference MUP sequence, confirming that the *Mus spretus* protein was a MUP. Two peptides had masses that were apparently unique to the *Mus spretus* protein. These two peptides were analysed further by tandem mass spectrometry, which permitted recovery of the complete amino acid sequence of both peptides (Figure 6). When these two sequences were aligned to the corresponding peptides from the reference MUP, they were each found to contain one amino acid change that could explain the differences in mass by MALDI-TOF mass spectrometry.

## 5. ANALYSIS OF PROTEIN LIGANDS

Many of the proteins in scent marks are members of the lipocalin superfamily, characterised by their distinctive barrel shaped structure and central ligand binding pocket. Given the well established role of volatile compounds in olfactory communication (Buck and Axel, 1991), the function of semiochemical proteins may be to modulate scent signals by interaction with volatile components. Thus, a further step in characterising a novel scent secretion protein may be to identify and characterise any associated ligands.

### 5.1 Identification of Protein-associated Molecules

Prior to analysis of protein-associated ligands, it is essential to eliminate any contaminating molecules that may exist freely in the secretion. This is achieved simply on small volumes by size exclusion separation ('spun columns') whereby free volatile molecules are retained by the column whilst proteins and their ligands pass through. Centrifugal filters can also be used for this purpose and have the added advantages that the protein fraction can be concentrated, and that free ligands and other small molecules are captured. Dialysis through semi-permeable membranes is effective, if slow.

For volatile ligands, capillary gas chromatography (GC) provides a ready means of separating mixtures of protein associated volatile molecules. The separations achieved by GC are of high quality, and direct coupling of the gas phase eluent to a mass spectrometer (GC-MS) generates a fragmentation mass spectrum of the volatile, which can be compared to databases of the fragment ions of known compounds. Volatile molecules can

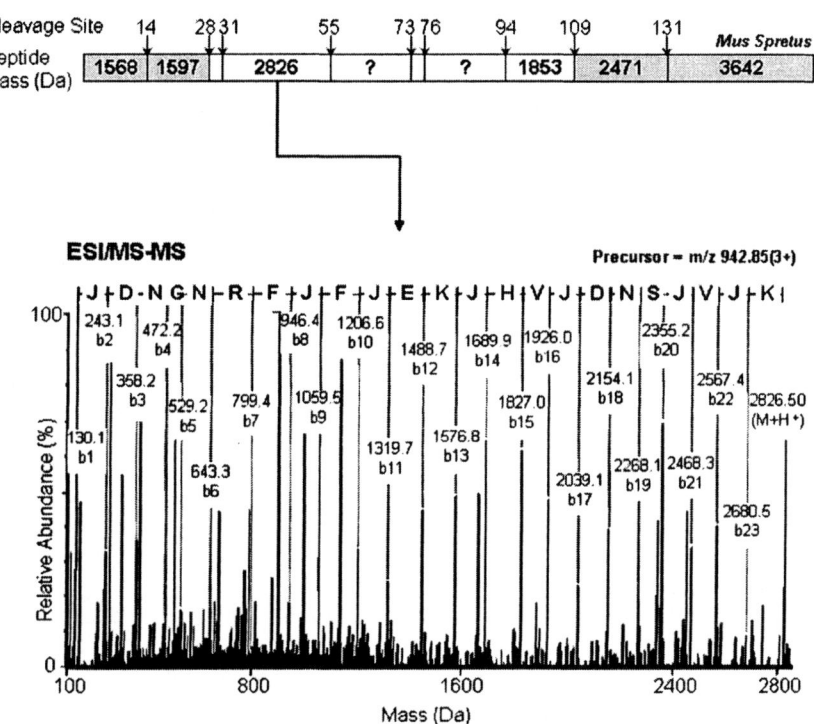

Figure 6. Tandem mass spectrometry for protein sequencing. One peptide (2826Da) from the *Mus spretus* peptide mixture (Figure 5) was selected for sequencing by mass spectrometry. When the peptide ion is fragmented in the mass spectrometer, a complex pattern of ions are obtained, corresponding to the successive loss of amino acids from one end or the other of the peptide. From the mass differences between the successive ions, it is possible to reconstruct the peptide sequence of the protein. (J = leucine / isoleucine)

be extracted from secretions in three ways; solvent extraction, headspace sampling and solid phase micro-extraction (SPME). GC and GC-MS are widely used techniques and a detailed coverage is not attempted in this text. Semiochemical applications of GC-MS with each extraction technique are illustrated in Schwende et al. (1986), Robertson et al. (1993), Rasmussen (2001), and Humphries et al. (1999).

In scents that are mediated by aqueous media, the semiochemical might be the protein itself, a peptide derived there from by enzymatic fragmentation, or the active component might be bound to a carrier protein. Such scents are analysed completely in the aqueous phase in the first instance, using size exclusion chromatography, and electrophoretic and low molecular weight analytical methods as well as protein and peptide mass spectrometry. Aqueous denaturants (urea, guanidine hydrochloride, detergents, heat) can be used to separate a putative ligand from a protein, and in most instances, recover the ligand in the aqueous phase.

## 5.2 Characterisation of Protein-ligand Interactions

A number of methods have been used to investigate both the manner in which ligands are bound and released and the effect that changes in the protein structure can have on this process. Displacement assays, such as that used to characterise the binding affinities for a range of odorants for OBP (Pevsner et al., 1990) equilibrate a preparation of the semiochemical protein with a radioactive ligand. The protein is then incubated with varying concentrations of an unlabelled test ligand, the binding affinity of which is calculated from the ease with which the radioactive ligand is displaced. Isothermal Titration Calorimetry accurately measures the temperature changes that occur when a protein and ligand are mixed. From these measurements, it is possible to characterise a number of thermodynamic and kinetic parameters for the interaction, including the dissociation constant (Kd) and the stoichiometry. This technique was used to characterise the interaction between MUP and one of its ligands (thiazole) and compare the interactions with a number of thiazole analogues (Sharrow et al., 2002, 2003). Of course, analysis of the role of proteins in ligand release should attempt to mimic natural conditions, such as humidity and temperature. Further, scent marks dry out, and the changes within the scent mark can influence the rate of release of bound semiochemical. To illustrate this, the role of MUPs in scent mark ageing was studied by following the release of MUP ligands after deposition of urine samples onto inert glass fibre filters. These were left open to the air for differing times, after which residual MUP ligands (thiazole and brevicomin) were determined by GC/MS from static headspace extractions (Robertson et al., 2001).

Non-natural ligands can be used to probe ligand binding cavities, for example. An exogenous, fluorescent reporter ligand has been employed to investigate the effect of amino acid changes on the MUP ligand binding pocket (Marie et al., 2001). The reporter ligand (NPN) has a markedly enhanced fluorescence spectrum when it resides in the hydrophobic environment of the MUP binding pocket. Incubation of different MUP preparations with increasing concentrations of NPN allowed the kinetic parameters of the MUP-NPN complex to be determined. The study concluded that amino acid changes in three of the four heterogeneous positions (50, 136 and 140) had no effect on the binding of NPN. A F/V substitution at the fourth position (56) did however have a significant effect on NPN binding, and by implication altered the properties of the MUP ligand binding pocket.

## 6. CONCLUSIONS

The complex and intriguing roles played by proteins and peptides in scent communication is only now beginning to emerge. New experimental strategies are required to discover the nature and role of these proteins, but the growing awareness of proteins in scents is timely. A whole range of techniques are evolving in the field of proteomics that can be directly applied to the study of scent marks and secretions. The precise way that such techniques are applied will depend on the nature of the secretion, the biological function of the protein and the aims of the study. However, these methodologies are increasingly accessible, and we can anticipate a dramatic growth in information on the role of proteins in chemical communication.

## 7. REFERENCES

Aebersold, R., and Mann, M., 2003, Mass spectrometry-based proteomics, *Nature* **422**:198-207.
Altschul, S. F., Madden, T. L., Schäffer, A. A., Zhang, J., Zhang, Z., Miller, W., and Lipman, D. J., 1997, Gapped BLAST and PSI-BLAST: a new generation of protein database search programs, *Nucleic Acids Res.* **25**:3389-3402.
Ashcroft, A. E., 2003, Protein and peptide identification: the role of mass spectrometry in proteomics, *Nat. Prod. Rep.* **20**:202-215.
Beynon, R. J., and Bond, J. S., 1999, *Proteolytic Enzymes a Practical Approach*, IRL press, Oxford.
Beynon, R. J., Veggerby, C., Payne, C. E., Robertson, D. H., Gaskell, S. J., Humphries, R. E., and Hurst, J. L., 2002, Polymorphism in major urinary proteins: molecular heterogeneity in a wild mouse population, *J. Chem. Ecol.* **28**:1429-1446.
Buck, L., and Axel, R., 1991, A novel multigene family may encode odorant receptors: a molecular basis for odour recognition, *Cell* **65**:175-187.
Chalmers, M. J., and Gaskell, S. J., 2000, Advances in mass spectrometry for proteome analysis, *Curr. Opin. Biotechnol.* **11**:384-390.
Clark, A. J., Ghazal, P., Bingham, R. W., Barrett, D., and Bishop, J. O., 1985, Sequence structures of a mouse major urinary protein gene and pseudogene compared, *EMBO J.* **4**:3159-3165.
Cunningham, B. A., 1977, The structure and function of histocompatibility antigens, *Sci. Am.* **237**:96-107.
Emes, R. D., Goodstadt, L., Winter, E. E., and Ponting, C. P., 2003, Comparison of the genomes of human and mouse lays the foundation of genome zoology. *Hum. Mol. Genet.* **12**:701-709.
Finlayson, J. S., and Baumann, C. A., 1958, Mouse proteinuria, *Amer. J. Physiol.* **192**:69-72.
Gershon, D., 2003, Proteomics technologies: Probing the proteome, *Nature* **424**:581-587.
Hayter, J. R., Robertson D. H., Gaskell, S. J., and Beynon R. J., 2003, Proteome analysis of intact proteins in complex mixtures, *Mol. Cell. Proteom.* **2**:85-95.
Humphries, R., Hurst, J. L., Robertson, D. H. L., and Beynon R. J., 1999, Unravelling the chemical basis of competitive scent marking in house mice, *Anim. Behav.* **58**:1177-1190.
Hurst, J. L., Robertson, D. H. L., Tolladay, U., and Beynon, R. J., 1998, Lipocalins in mouse urine provide a slow release mechanism for olfactory signals, *Anim. Behav.* **55**:1289-1297.
Hurst, J. L., Payne, C. E., Nevison, C. M., Marie, A. D., Humphries, R. E., Robertson D. H. L., Cavaggioni, A., and Beynon, R. J., 2001, Major urinary proteins play a key role in the mechanism of individual recognition in mice, *Nature* **414**:631-634.
Kikuyama, S., and Toyoda, F., 1999, Sodefrin: a novel sex pheromone in a newt, *Rev. Reprod.* **4**:1-4.
Klein, J., 1979, The major histocompatibility complex of the mouse, *Science* **213**:516-521.
Laemmlli, U. K., 1970, Cleavage of structural proteins during the assembly of the head of bacteriophage T4, *Nature* **227**:680-685.
Lin, D., Tabb, D. L. and Yates, J. R., 2003, Large-scale protein identification using mass spectrometry, *Biochim. Biophys. Acta.* **1646**: 1-10.
Marie A. D., Veggerby C., Robertson D. H., Gaskell S. J., Hubbard S. J., Martinsen L., Hurst J. L., and Beynon R. J., 2001, Effect of polymorphisms on ligand binding by mouse major urinary proteins, *Protein Sci.* **10**: 411-417.
Mechref, Y., Zidek, L., Ma, W., and Novotny, M. V., 2000, Glycosylated major urinary protein of the house mouse: characterization of its N-linked oligosaccharides, *Glycobiology* **10**:231-235.
Mertens, B., Vandoren, G., Opdenakker, G., Volckaert, G., and Verhoeven, G., 1983, Heterogeneity of alpha 2u-globulin gene products in different translational systems, plasma and urine, *FEBS Lett.* **162**:296-299.
Nguyen, N. Y., Salokangas, A., and Chrambach, A., 1977, Electrofocusing in natural pH gradients formed by buffers: gradient modification, *Anal. Biochem.* **78**:287-294.
Parfentjev, J. A., and Perlzweig, W. A., 1933, The comparison of the urine of white mice, *J. Biol. Chem.* **100**:551-555.
Perkins, D. N., Pappin, D. J., Creasy, D.M., and Cottrell, J.S., 1999, Probability-based protein identification by searching sequence databases using mass spectrometry data, *Electrophoresis* **20**:3551-3567.
Pevsner, J., Hou, V., Snowman, A. M., and Snyder, S.H., 1990, Odorant binding protein: characterisation of ligand binding, *J. Biol. Chem.* **265**:6118-6125.
Rasmussen, L. E., 2001, Source and cyclic release pattern of (Z)-7-dodecenyl acetate, the pre-ovulatory pheromone of the female Asian elephant, *Chem. Senses* **26**:611-623.
Robertson, D. H. L., Beynon, R. J., and Evershed, R. P., 1993, Extraction, characterisation and binding analysis of two pheromonally active ligands associated with the major urinary protein of the house mouse (*Mus musculus*), *J. Chem. Ecol.* **19**:1405-1416.

Robertson, D. H. L., Evershed, R. P., Cox, K., Gaskell, S., and Beynon, R.J., 1996, Molecular heterogeneity of urinary proteins of the house mouse *Mus musculus*. *Biochem. J.* **316**: 265-272.

Robertson, D. H. L., Hurst, J. L., Bolgar, M. S., Gaskell, S. J., and Beynon, R.J., 1997, Molecular heterogeneity of urinary proteins in wild house mouse populations, *Rap. Comm. Mass Spec.* **11**:786-790.

Robertson, D. H. L. Marie, A. D., Veggerby, C., Hurst J. L., and Beynon, R.J., 2001, Characteristics of ligand binding and release by major urinary proteins, in: *Chemical Signals in Vertebrates 9* A. Marchlewska-Koj, J.J. Lepri and D. Muller-Schwarze, eds., Kluwer Academic/Plenum Publishers, New York, pp. 169-176.

Ronnett, G. V. and Moon, C., 2002, G proteins and olfactory signal transduction, *Annu. Rev. Physiol.* **64**:189-222.

Schwende, F. J., Weisler, D., Jorgenson, J. W., Carmack, M., and Novotny, M., 1986, Urinary volatile constituents of the house mouse Mus musculus and their endocrine dependency, *J. Chem. Ecol.* **12**:277-296.

Seybold, S. J., and Tittiger, C., 2003, Biochemistry and molecular biology of de novo isoprenoid pheromone production in the Scolytidae, *Ann. Rev. Entomol.* **48**:425-453.

Sharrow, S. D., Vaughn, J. L., Zidek, L., Novotny, M. V., and Stone, M. J., 2002, Pheromone binding by polymorphic mouse major urinary proteins, *Protein Sci.* **11**:2247-2256.

Sharrow, S. D., Novotny, M. V. and Stone, M. J., 2003, Thermodynamic analysis of binding between mouse major urinary protein-I and the pheromone 2-sec-butyl-4,5-dihydrothiazole, *Biochemistry* **42**:6302-6309.

Shevchenko, A., Sunyaev, S., Loboda, A., Shevchenko, A., Bork, P., Ens, W., and Standing, K.G., 2001, Charting the proteomes of organisms with unsequenced genomes by MALDI-quadrupole time-of-flight mass spectrometry and BLAST homology searching, *Anal Chem.* **73**:1917-1926.

Simpson, D. M., Beynon, R. J., Robertson, D. H. L., Loughran, M., and Haywood, S., 2004, Copper associated liver disease: a proteomics study of copper challenge in a sheep model, *Proteomics* **4** (in press).

Singh, P. B., Brown, R. E., and Roser, B., 1987, MHC antigens in urine as olfactory recognition cues, *Nature* **327**:161-164.

Tegoni, M., Pelosi, P., Vincent, F., Spinelli, S., Campanacci, V., Grolli, S., Ramoni, R., and Cambillau, C., 2000, Mammalian odorant binding proteins, *Biochim. Biophys. Acta.* **1482**:229-240.

Vandenbergh, J. G., Whitsett, J. M., and Lombardi, J. R., 1975, Partial isolation of a pheromone accelerating puberty in female mice, *J. Reprod. Fert.* **45**:515-523.

Vandenberg, J. G., Finlayson, J. S., Dobrogosz, W. J., Dills, S. S., and Kost, T. A., 1976, Chromatographic separation of puberty accelerating pheromone from male mouse urine, *Biol. Reprod.* **15**:260-265.

Zhang, W., and Chait, B. T., 2000, ProFound: an expert system for protein identification using mass spectrometric peptide mapping information, *Anal. Chem.* **72**:2482-2489.

# THE "SCENTS" OF OWNERSHIP

Jane L. Hurst[1], Michael D. Thom[1], Charlotte M. Nevison[1], Richard E. Humphries[1] and Robert J. Beynon[2]

## 1. INTRODUCTION

The ability to recognise individuals, rather than different classes of conspecifics (such as male/female, familiar/unfamiliar, dominant/subordinate, kin/unrelated), allows animals to learn information about the abilities or behaviour of specific individuals and modulate their behaviour in subsequent interactions. Such information may be gained through direct interaction with a particular animal, through eavesdropping on the behaviour or interactions of the individual with others, or simply through picking up signals or cues that provide information about the individual's status or location. The ability to learn information about specific individuals has many advantages in a range of social situations, including competitive interactions, cooperation between conspecifics, and in mate selection and recognition. Animals similarly can gain considerable advantages from being recognised, particularly when signalling information about their status or intentions to others. We might thus expect strong selection both for the ability to recognise specific individuals and for signals of individual identity, among vertebrate species that have the mental capacity to remember information about particular individuals.

During direct interactions, or when eavesdropping on interactions between others, recognition may involve a wide range of cues such as scent, visual and/or vocal characteristics, together with more complex information such as individual-specific behavioural patterns and location. In addition, many mammals use scent marks deposited in the environment to provide information in their absence. These scent marks can only inform about a specific individual if the scent encodes a reliable signal of individual ownership. Here, we discuss the characteristics required of ownership signals in scent marks and review recent studies from our laboratory that have attempted to ascertain the

---

[1] Animal Behaviour Group, Faculty of Veterinary Science, University of Liverpool, Neston CH64 7TE, UK
[2] Protein Function Group, Faculty of Veterinary Science, University of Liverpool, Neston CH64 7TE, UK

## 2. INDIVIDUALITY AND OWNERSHIP

Many factors contribute to the scents produced by individuals. Genetic variation at a variety of loci is known to produce discriminable differences in scents among laboratory rodents (e.g. Boyse et al., 1987; Beauchamp et al., 1990; Eggert et al., 1996; Carroll et al., 2002). Scents also vary according to an individual's current status, including its social class, reproductive condition, health status, and food sources (see reviews in Brown and Macdonald 1985; Brown 1995; Penn and Potts 1998; also Schellinck et al., 1997; Ehman and Scott 2001; Yamazaki et al., 2002). Other incidental environmental factors such as the microbial flora that animals pick up from their environment can contribute (e.g. Singh et al., 1990). Every individual therefore produces a complex and unique mixture of scents that have been termed individuality odours or "olfactory fingerprints" (Brown, 1979, 1995), while odours specifically determined by polymorphic genes are termed odortypes (Yamazaki, 1999). Although other conspecifics are able to discriminate differences in scents that derive from such a wide range of sources, it is not yet clear which of these factors determine the ownership signals in scent marks, the scents used to recognise an individual when animals meet, or the chemical basis of such recognition scents. Simple discrimination between scents does not assess individual recognition. To understand which cues are used to recognise an individual scent owner requires a functional test of individual recognition (Halpin, 1986).

From a theoretical viewpoint, the ownership signals in scent marks need to be stable and persistent once deposited, since it is essential that the signal of ownership does not change as the scent mark ages. Further, ownership signals should not be altered by metabolic or environmental fluctuations that could disrupt an identity signal. Ideally, identity signals should be genetically determined so that they are a "hard-wired" characteristic of the individual. The genes involved need to be sufficiently polymorphic to provide each individual in the local population with a unique signature.

Two highly polymorphic gene complexes have been shown to contribute to discriminable differences in scents and show such high variability between even closely related individuals that they could provide the main basis of individual ownership scent signatures. The major histocompatibility complex (MHC) is involved in self-nonself recognition at the cellular level in the immune system. This gene complex also affects the scents produced by different individuals in a wide range of species including rodents (see Singh, 2001 for review), fish (Olsen et al., 1998; Reusch et al., 2001) and humans (Wedekind and Furi, 1997; Jacob et al., 2002). These scents play an important role in the selection of genetically compatible mates and in kin recognition, although their role in individual scent signatures remains to be fully established (see Thom et al., this volume). Although the molecular mechanism underlying MHC-specific scents are incompletely understood, in rodents MHC-associated odours appear to be a complex mixture of volatiles bound and release by urinary proteins (Singer et al., 1993, 1997).

A second highly polymorphic system that contributes to the urinary scents of mice is the major urinary protein (MUP) complex (Beynon and Hurst, 2003). These small lipocalin proteins have a central cavity that binds and releases semiochemicals. Unlike MHC, the only known function of these proteins is in chemical signalling and they are present at considerably higher concentrations in mouse urine than are MHC peptides. MUPs and their ligands play an important role in status signalling through scent marks among male mice (see Beynon and Hurst, 2003; Malone et al., this volume). Individual wild mice typically express at least 7-16 different urinary MUPs, with considerable combinatorial diversity between individual mice (Payne et al., 2001, see also Robertson et al., this volume). The diversity of profiles among wild-caught animals suggests that this might be sufficient to provide an individual identity signature even in isolated populations (Beynon et al., 2002) provided that other mice can discriminate between different MUP profiles.

## 3. ASSESSMENT OF SCENT OWNSHIP SIGNATURES

While studies assessing the influence of MHC genes on scent cues have utilized MHC-congenic strains of laboratory rodents, similar congenic strains are not available to investigate the influence of MUP type. However, in the wild, animals must be able to identify individuals against a very heterogeneous genetic and environmental background. We thus took a different approach to examine the contribution of MUPs to scent mark ownership signals by examining behavioural responses to different MUP types against the variable genetic background of wild-derived mice (Hurst et al., 2001a). To ensure that subjects were genetically heterogeneous at many different loci, and thus differences between individuals should easily be recognised, we crossed unrelated animals captured from different populations. Although MUP and MHC gene complexes are highly polymorphic in mice, some siblings will inherit the same haplotypes from their parents. Within each litter, males expressed the same MUP pattern as some of their brothers but a different pattern to other brothers, while unrelated males always expressed different MUP patterns. To examine whether MUP type played a role in scent ownership recognition, we tested whether males showed functional recognition of their own scent marks versus those from other males.

Male mice use scent marks to advertise their territory ownership and competitive ability (see Hurst et al., 2001b for review; also Malone et al., this volume). These scent marks are attractive to females which are able to recognise the individual owners of scent marks (Rich and Hurst, 1998, 1999). When males encounter competing scents from other males within their territory, they countermark by increasing their own rate of scent marking in the vicinity of the intruder's scent, a response expected because fresh marks from a competitor reduce the territory owner's attractiveness to females (Hurst and Rich, 1999). We used this important functional response to assess whether males recognised scent marks introduced into their individual territories as their own or from another male. We measured time spent in the vicinity of the scent stimulus together with the number of scent marks deposited by the territory owner in the vicinity of the introduced scent. Mice showed little interest when own scent was introduced, responding no more than to a control water stimulus. By contrast, males spent much more time close to urine from an unrelated male and increased their scent marking to countermark as expected. Males showed the same response to urine from a brother of different MUP type to own, clearly recognising that the urine was not own but from a competing male. Although brothers are close kin, adult male mice are highly territorial and compete strongly, regardless of relatedness. However, urine from a brother of same MUP type as own stimulated no more interest than own urine or water, and males failed to increase countermarking (Hurst et al., 2001a). This suggests that despite the many genetic differences that would have been apparent between outbred brothers that were not genetically linked to MUP type, males did not recognise scent marks of the same MUP type as different from own. We confirmed that response was due to MUP type rather than genes linked to MUP by showing that scent marking increased in response to own urine when a recombinant MUP was added to change the individual's MUP profile while controlling for total urinary protein concentration (Hurst et al., 2001a).

A limitation of this study was that we did not measure the time mice spent directly sniffing at scent stimuli to gather information as we were unable to see the precise location of stimulus urine during the test, so instead we measured time spent in the vicinity of the stimulus. It was thus not clear whether the wild-derived mice failed to discriminate any difference between own scents and those sharing the same MUP type, or whether differences in scents were discriminated (as evidenced by investigation) but had no functional meaning for ownership recognition and thus failed to stimulate a functional response. To distinguish between these two possibilities, we conducted a second experiment in which stimulus urine was covered by a small square of metal mesh through which mice could push their noses to investigate the scent closely. This allowed us to measure direct sniffing of the stimulus unambiguously, as well as measuring the total time spent in the vicinity of the stimulus and number of scent marks deposited as before. Animals were also given much more social experience prior to tests in case the unnaturally low social experience of laboratory reared mice resulted in poor recognition abilities.

Despite the different genetic background and experience of mice used in this second study, which showed increased responsiveness to introduced stimuli compared to those in the first study, the same pattern of response was apparent. Mice spent more time in the vicinity of scent stimuli of different MUP pattern to own, whether from an unrelated male or from a brother, and scent countermarking was greater than in response to own urine or to a water control. This was not the case in response to urine of same MUP type as own. However, the duration of sniffing revealed that mice clearly discriminated a difference in the scent of a brother of same MUP type compared to own urine. Urine from other males induced prolonged close investigation from the territory owner on the first 2-3 visits. Urine from a brother of same MUP type stimulated as much sniffing as urine of different MUP type from a brother or an unrelated male, while own urine induced no more close investigation than water. Thus mice detected a difference between own and same MUP type urine and were stimulated immediately to gather further information from all non-self urine stimuli by close contact investigation. However, having investigated a scent stimulus closely, males returned repeatedly to spend prolonged periods only near to scent of different MUP type, as if waiting for the owner to reappear, and only elevated their scent marking to countermark a different MUP type scent regardless of any relatedness to the owner. After investigating scent from a brother of same MUP type closely, males did not show this typical response to an intruder's scent mark. Although males detected non-MUP related differences in scents, as predicted from laboratory tests of scent discrimination, these differences did not appear to have functional meaning in terms of ownership identity, at least with respect to discriminating between own scent marks and those of other males (Hurst et al., unpublished data).

Recognizing whether scent marks are own or from another equivalent male is only one aspect of individual ownership recognition. It is also important to establish whether and how animals recognize the owners of scents from different individuals that are equal in all other respects (thus animals are not distinguishing particular classes such as self versus non-self, familiar versus unfamiliar, dominant versus subordinate, etc). We have recently developed a new assay to show that mice can identify the owner of scent marks from one of two familiar neighbours. When scent marks from different equivalent neighbours were introduced into a male's territory, preliminary results suggest that mice showed functional recognition of the appropriate individual neighbour. Further, mice again appeared to use MUP type to recognize the owner of the scent marks even against a variable genetic background (Hurst et al., unpublished data).

## 4. MECHANISM UNDERLYING MUP OWNERSHIP SIGNAL

How might differences in MUP type be perceived? First, ownership may be signalled by the pattern of volatiles differentially bound and released by MUPs (model I). If MUP isoforms have different ligand binding affinities, the MUP pattern expressed by an individual may define a mixture of volatiles that is specific to the MUP-type (independent of other genetic or status differences between animals). Since we already know that other genetic and status factors contribute to the volatiles emanating from mouse urine (see earlier), animals would need to recognize MUP-associated volatile patterns against this complex and dynamic background. Alternatively, MUP isoforms may differentially bind small molecular weight ligands as above, but mice detect the pattern of non-volatile MUP-ligand complexes (model II) rather than MUP-associated volatiles alone. Scents are actively pumped to the vomeronasal organ in mammals such as mice, allowing for the transport of involatile as well as volatile molecules (Halpern and Martinez-Marcos, 2003; Luo et al., 2003). Thus, MUPs may transport and present a specific pattern of ligands to receptors in the vomeronasal organ, which might help to eliminate 'noise' from other factors that affect volatile odorants. A third possibility is that the ownership signal is held by the involatile MUPs themselves, although animals are stimulated to investigate scents closely by differences in the complex mixture of volatiles (model III). This model requires vomeronasal receptors that are sensitive to the small differences in MUP isoforms. Most polymorphisms occur on the surface of the MUP and do not appear to affect ligand binding affinities, but as yet the existence of specific receptors for MUPs remains controversial (Beynon and Hurst, submitted).

To establish whether the ownership signal is held in the pattern of volatiles released by MUPs (model I), we compared responses to scent marks from another male when the territory owner could contact the scent or when contact was prevented by a porous sheet of nitrocellulose membrane that binds proteins (Nevison et al., 2003). This prevented mice gaining access to the involatile MUPs but allowed volatiles to pass through. Although mice spent just as long investigating scents whether they could contact the scent source or not, confirming that they could detect volatiles emanating from the scent source, they only recognized scent from another male and increased scent marking to countermark when they could contact the scent source. The ownership signal in scent marks thus does not appear to be volatile and animals need to make contact with the scent source in order to assess individual ownership. This suggests that the ownership signal is either held in the involatile MUP-ligand complex (model II) or is signalled by the MUPs themselves (model III). Recent evidence from a study of laboratory mice investigating anaesthetized conspecifics suggests that scent ownership recognition may occur through the vomeronasal system (Luo et al., 2003). Nasal contact during investigation of a conspecific appears to be essential for activation of vomeronasal system neurons, with individual neurons sensitive only to specific combinations of the strain and sex of the mouse investigated.

## 5. Individual recognition

The fact that mice require contact with a scent source to assess individual ownership does not mean that volatiles are unimportant for scent ownership or individual recognition. Indeed, volatiles play an essential role in drawing attention to a scent source and stimulating animals to investigate closely, and for advertising scent age. An unfamiliar pattern of volatiles is clearly recognized and stimulates much more prolonged investigation than a familiar scent source. However, when animals such as mice detect volatiles emanating from a new scent source, their response is usually to approach and attempt to contact the scent source during investigation (Humphries et al., 1999). While close contact will maximize sensitivity to components present at low concentration, it also allows access to non-volatile compounds. The fact that the vomeronasal system of intact and conscious mice is only activated when mice make nasal contact with a scent source suggests that either scents are only pumped to the vomeronasal organ on contact, or the vomeronasal system is only activated by involatile components of scents such as MUPs or MUP-ligand complexes. Notably, Guo et al. (1997) found no increase in c-fos mRNA expression in the accessory olfactory bulb when male pheromones normally bound to MUPs were presented alone, but considerable activity when these pheromones were presented with MUPs. Urinary proteins also increase the likelihood that low molecular weight components in the urine of unfamiliar males will induce pregnancy block (Peele et al., 2003).

Although making contact with a scent mark to detect involatile components is relatively easy, making close contact with a conspecific during direct interaction may be considerably more dangerous. Further, a requirement always to approach and contact a scent source to gain involatile information about ownership would have additional time demands that would interrupt and compete with other activities. It is likely that animals learn to recognize the complex of volatile scents associated with familiar individuals, particularly those of familiar aggressors that would be dangerous to approach closely. However, an individual's volatile profile is probably unstable, influenced by many non-genetic factors and therefore reflecting its current status and environment. When animals detect an unfamiliar mixture of volatiles (or other novelty such as an apparently familiar scent in an unfamiliar site – Mayeaux and Johnston, 2002), they approach the scent source to investigate closely. This close contact would provide animals with the opportunity to form an association between an involatile ownership signal and the volatile profile of the scent (Figure 1a). If the same, now familiar, mixture of volatiles is encountered subsequently, animals may recall the associated involatile ownership signal, obviating the need for contact with the scent source (Figure 1b). However, if any novelty is encountered (e.g. either a new or altered mixture of volatiles), close contact investigation is again induced (assuming that volatiles do not indicate that the scent source would be dangerous to approach). On close investigation, animals might encounter a new ownership signal from an unfamiliar individual which will then be associated with the new volatile profile (Figure 1c). Alternatively, animals might encounter an ownership signal from a familiar individual whose volatile profile has changed (e.g. due to metabolic changes or other change in status). Close investigation would provide an opportunity to update the association between the individual's stable involatile ownership signal with its altered volatile profile (Figure 1d).

This model of association between volatile (or varying) and involatile (or constant) profiles appears to correspond to behaviour when animals interact. When unfamiliar animals first meet, the usual response is mutual close contact investigation, often involving many different areas of the body. Close investigation between unfamiliar competitors often ends when one or both animals attack. When competitors meet subsequently, they will often sniff towards each other from a distance, and then flee or attack based on information gained only from volatile scents (e.g. Hurst 1993). However, animals also have the opportunity for close investigation of scent marks in the local environment. Scent marks may thus play an important role in allowing animals to update

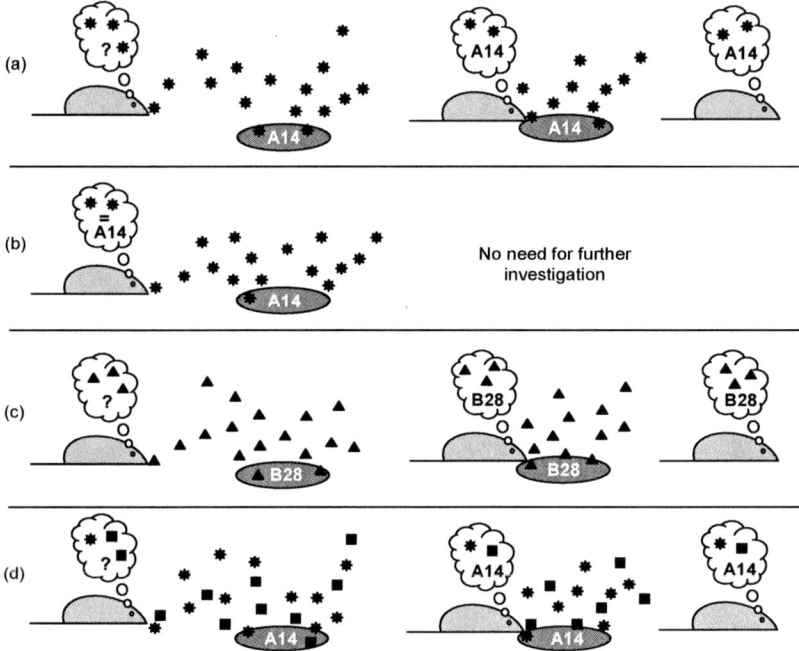

**Figure 1.** Model of association between volatile scent (symbols) and involatile ownership signal (text code) when animals encounter scents. (a) Animal encounters unfamiliar volatile scents and contacts the scent source to investigate involatile components that provide a stable scent ownership signal; volatile and involatile profiles become associated in memory. (b) Animal encounters familiar volatile profile associated with an involatile ownership signal in memory; no further investigation required. (c) Unfamiliar volatile profile stimulates contact with scent source containing new ownership signal; volatile and involatile profiles become associated in memory. (d) Unfamiliar components of volatile profile stimulate contact with scent source containing familiar ownership signal; association between involatile ownership signal and altered volatile profile updated in memory.

the association between stable involatile profiles and changing volatile profiles. By contrast, animals usually show little close investigation of familiar individuals living within their own social group. However, animals that have been separated for a period typically show increased close contact investigation when they meet, while animals that have picked up scents from elsewhere can induce considerable close contact investigation from familiar group members. Such close investigation may fulfill the dual function of checking information concerning the current status of the individual in addition to reconfirming their identity. A learnt association between involatile and volatile components may be a general feature involved in recognition, not only of the individual but also of the opposite sex. Although naïve female mice show an innate attraction to male scents when they can contact involatile components, they are only attracted to male volatiles alone once they have learnt an association between the involatile and volatile components through contact with male scent (Moncho-Bogani et al., 2002). Male mice that have encountered females artificially odorized with perfume subsequently emit ultrasonic courtship vocalizations to the perfume itself, apparently associating the perfume with recognition of a female mouse (Nyby et al., 1978). Contact investigation of urine scents activates both the vomeronasal and main olfactory systems (Guo et al., 1997), allowing the possibility of associative learning between involatile components (the complex of MUPs and odorants) that activate the vomeronasal organ with the pattern of volatile odorants that activate the main olfactory epithelium (Moncho-Bogani et al., 2002).

## 6. CONCLUSIONS

Scent marks need to encode a reliable signal of ownership to inform about the specific owner. Although tests of discrimination have told us much about the abilities of animals to detect differences in conspecific scents deriving from a wide range of sources, we know little about the components involved in scent ownership recognition or individual recognition when animals meet because this requires functional tests of individual recognition. Ownership signals in scent marks need to be stable and persistent, ideally genetically determined and sufficiently polymorphic. Recent work from our laboratory, using functional tests of scent mark recognition, suggest that the pattern of polymorphic MUPs in the urinary scent marks of male house mice provides an ownership signal. The ownership signal is involatile, requiring investigatory contact with the scent source, and involves either involatile complexes between MUPs and their bound odorants or the MUPs themselves, probably detected through the vomeronasal system. However, mice also detect non-MUP related differences in urinary volatiles. We propose a model of learnt association between involatile and volatile components that would allow mice to recognize previously encountered volatile profiles from familiar individuals or animals of the same sex without requiring close contact investigation. Investigation of fresh scent marks deposited around the environment would allow animals to update the proposed association between an individual's stable involatile profile with any changes in its volatile profile. Further research is required to test this model and to establish its generality in other mammalian species.

## 7. ACKNOWLEDGEMENTS

This work was supported by BBSRC research grants to JLH and RJB.

## 8. REFERENCES

Beynon, R. J., and Hurst, J. L., 2003, Multiple roles of major urinary proteins in the house mouse, *Mus domesticus, Biochem. Soc. Trans.* **31**:142-6.
Beynon, R. J., Veggerby, C., Payne, C. E., Robertson, D. H., Gaskell, S. J., Humphries, R. E., and Hurst, J. L., 2002, Polymorphism in major urinary proteins: molecular heterogeneity in a wild mouse population, *J. Chem. Ecol.* **28**:1429-46.
Boyse, E. A., Beauchamp, G. K., and Yamazaki, K., 1987, The genetics of body scent, *Trends Genet.* **3**:97-102.
Brown, R. E., 1979, Mammalian social odors: a critical review, *Adv. Study Behav.* **10**:103-162.
Brown, R. E., 1995, What is the role of the immune system in determining individually distinct body odours? *Int. J. Immunopharmacol.* **17**:655-61.
Brown, R. E., and McDonald, D. W., 1985, *Social Odours in Mammals, Vol 1 and 2*, Oxford, Clarendon Press.
Carroll, L. S., Penn, D. J., and Potts, W. K., 2002, Discrimination of MHC-derived odors by untrained mice is consistent with divergence in peptide-binding region residues, *Proc. Nat.l Acad. Sci. USA* **99**:2187-92.
Eggert, F., Holler, C., Luszyk, D., Muller-Ruchholtz, W., and Ferstl, R., 1996, MHC-associated and MHC-independent urinary chemosignals in mice, *Physiol. Behav.* **59**:57-62.
Guo, J., Zhou, A., and Moss, R. L., 1997, Urine and urine-derived compounds induce *c-fos* mRNA expression in accessory olfactory bulb, *Neuroreport* **8**:1679-1683.
Halpern, M., and Martinez-Marcos, A., 2003, Structure and function of the vomeronasal system: an update, *Prog. Neurobiol.* **70**:245-318.
Halpin, Z. T., 1986, Individual odours among mammals: origins and functions, *Adv. Study Behav.* **16**:39-71.
Humphries, R. E., Robertson, D. H. L., Beynon, R. J., and Hurst, J. L., 1999, Unravelling the chemical basis of competitive scent marking in house mice, *Anim. Behav.* **58**:1177-1190.
Hurst, J. L., 1993, The priming effects of urine substrate marks on interactions between male house mice, *Mus musculus domesticus* Schwarz and Schwarz, *Anim. Behav.* **45**:55-81.
Hurst, J. L., Beynon, R. J., Humphries, R. E., Malone, N., Nevison, C. M., Payne, C. E., Robertson, D. H. L., and Veggerby, C., 2001, Information in scent signals of competitive social status: the interface between behaviour and chemistry, in: *Chemical Signals in Vertebrates*, A. Marchelewska-Koj, D. Muller-Schwarze, and J. Lepri, eds., Plenum Press, New York, pp. 43-52.
Hurst, J. L., Payne, C. E., Nevison, C. M., Marie, A. D., Humphries, R. E., Robertson, D. H., Cavaggioni, A., and Beynon, R. J., 2001, Individual recognition in mice mediated by major urinary proteins, *Nature* **414**:631-4.
Hurst, J. L., and Rich, T. J., 1999, Scent marks as competitive signals of mate quality, in: *Advances in Chemical Communication in Vertebrates*, R. E. Johnson, D. Muller-Schwarze and P. Sorensen, eds., Plenum Press, New York, pp. 209-226.
Jacob, S., McClintock, M. K., Zelano, B., and Ober, C., 2002, Paternally inherited HLA alleles are associated with women's choice of male odor, *Nat. Genet.* **30**:175-9.
Luo, M., Fee, M. S., and Katz, L. C., 2003, Encoding pheromonal signals in the accessory olfactory bulb of behaving mice, *Science* **299**:1196-201.
Malone, N., Armstrong, S. D., Humphries, R. E., Beynon, R. J., and Hurst, J. L., submitted, Scent mark investment and the communication of competitive ability, this volume.
Mayeaux, D. J., and Johnston, R. E., 2002, Discrimination of individual odours by hamsters (*Mesocricetus auratus*) varies with the location of those odours, *Anim. Behav.* **64**:269-281.
Moncho-Bogani, J., Lanuza, E., Hernandez, A., Novejarque, A., and Martinez-Garcia, F., 2002, Attractive properties of sexual pheromones in mice: innate or learned? *Physiol. Behav.* **77**:167-176.
Nevison, C. M., Armstrong, S., Beynon, R. J., Humphries, R. E., and Hurst, J. L., 2003, The ownership signature in mouse scent marks is involatile, *Proc. R. Soc. Lond. B* **270**:1957-1963.
Nyby, J., Whitney, G., Schmitz, S., and Dizinno, G., 1978, Postpubertal experience establishes signal value of mammalian sex odor, *Behav. Biol.* **22**:545-552.
Olsen, K. H., Grahn, M., Lohm, J., and Langefors, A., 1998, MHC and kin discrimination in juvenile Arctic charr, *Salvelinus alpinus* (L.), *Anim. Behav.* **56**:319-327.
Payne, C. E., Malone, N., Humphries, R. E., Bradbrook, C., Veggerby, C., Beynon, R. J., and Hurst, J. L., 2001, Heterogeneity of major urinary proteins in house mice: population and sex differences, in: *Chemical*

*Signals in Vertebrates*, A. Marchelewska-Koj, D. Muller-Schwarze and J. Lepri, eds., Plenum Press, New York, pp. 233-240.

Peele, P., Salazar, I., Mimmack, M., Keverne, E. B., and Brennan, P. A., 2003, Low molecular weight constituents of male mouse urine mediate the pregnancy block effect and convey information about the identity of the mating male, *Eur. J. Neurosci.* **18**:622-628.

Penn, D., and Potts, W. K., 1998, Chemical signals and parasite-mediated sexual selection, *Trends Ecol. Evol.* **13**:391-396.

Reusch, T. B., Haberli, M. A., Aeschlimann, P. B., and Milinski, M., 2001, Female sticklebacks count alleles in a strategy of sexual selection explaining MHC polymorphism, *Nature* **414**:300-2.

Rich, T. J., and Hurst, J. L., 1998, Scent marks as reliable signals of the competitive ability of mates, *Anim. Behav.* **56**:727-735.

Rich, T. J., and Hurst, J. L., 1999, The competing countermarks hypothesis: reliable assessment of competitive ability by potential mates, *Anim. Behav.* **58**:1027-1037.

Robertson, D. H. L., Cheetham, S. A., Armstrong, S. D., Hurst, J. L., and Beynon, R. J., submitted, Characterisation of proteins in scent marks: proteomics meets semiochemistry, this volume.

Schellinck, H. M., Slotnick, B. M., and Brown, R. E., 1997, Odors of individuality originating from the major histocompatibility complex are masked by diet cues in the urine of rats, *Anim. Learn Behav.* **25**:193-199.

Singer, A. G., Beauchamp, G. K., and Yamazaki, K., 1997, Volatile signals of the major histocompatibility complex in male mouse urine, *Proc. Natl. Acad. Sci. USA* **94**:2210-4.

Singer, A. G., Tsuchiya, H., Wellington, J. L., Beauchamp, G. K., and Yamazaki, K., 1993, Chemistry of odortypes in mice - fractionation and bioassay, *J. Chem. Ecol.* **19**:569-579.

Singh, P. B., 2001, Chemosensation and genetic individuality, *Reproduction* **121**:529-39.

Singh, P. B., Herbert, J., Roser, B., Arnott, L., Tucker, D. K., and Brown, R. E., 1990, Rearing rats in a germ-free environment eliminates their odors of individuality, *J. Chem. Ecol.* **16**:1667-1682.

Thom, M. D., Beynon, R. J., and Hurst, J. L., submitted, The role of the major histocompatibility complex in scent communication, this volume.

Wedekind, C., and Furi, S., 1997, Body odour preferences in men and women: do they aim for specific MHC combinations or simply heterozygosity? *Proc. R. Soc. Lond. B Biol. Sci.* **264**:1471-9.

Yamazaki, K., Beauchamp, G. K., Bard, J., and Boyse, E. A., 1990, Chemosensory identity and the Y chromosome. *Behav. Genet.* **20**:157-65.

Yamazaki, K., Boyse, E. A., Bard, J., Curran, M., Kim, D., Ross, S. R., and Beauchamp, G. K., 2002, Presence of mouse mammary tumor virus specifically alters the body odor of mice. *Proc. Natl. Acad. Sci. USA* **99**:5612-5.

Yamazaki, K., Yamaguchi, M., Baranoski, L., Bard, J., Boyse, E. A., and Thomas, L., 1979, Recognition among mice. Evidence from the use of a Y-maze differentially scented by congenic mice of different major histocompatibility types, *J. Exp. Med.* **150**:755-60.

# THE ROLE OF SCENT IN INTER-MALE AGGRESSION IN HOUSE MICE & LABORATORY MICE

Julia C. Lacey and Jane L. Hurst[*]

## 1. INTRODUCTION

Understanding the factors that control competitive behaviour is important not only for addressing questions concerning the evolution and significance of different social strategies, but knowledge of these factors may also be applied to allow for the appropriate management and husbandry of captive animals. House mice (*Mus domesticus*) are a gregarious species capable of forming complex social relationships and readily seek social contact (Van Loo et al., 2001). It has been frequently documented that house mice use urinary scent marks to signal social status and regulate social interactions according, primarily, to differences in scent mark quality. With contributions from more than one species/subspecies of wild mouse, more than 450 inbred laboratory strains of mice have been developed over the past century for genetic and other studies including olfaction (Beck et al., 2000). However, fundamental differences in competitive behaviour and responses to scents exist between laboratory and wild strains of male mice, which is likely to be a consequence of severe inbreeding and abnormal environment in the laboratory. Here, we review how scent controls competitive behaviour in wild male house mice and discuss how captive management impinges on this in the laboratory.

## 2. SIGNALLING COMPETITIVE ABILITY

House mice reside together in territorial family groups, consisting of a dominant male with one or more breeding females and subordinate males (Hurst, 1987). Dominant males scent mark their territories extensively with small spots and streaks of urine and will further increase their scent marking rate around competing urine marks found within their territory (Desjardins et al., 1973; Hurst et al., 2001). Dominant males will also

---

[*] Julia C. Lacey and Jane L. Hurst, Animal Behaviour Group, Faculty of Veterinary Science, University of Liverpool, Leahurst, Neston CH64 7TE, UK

attack and chase subordinate residents that attempt to compete (Hurst, 1990). Subordinate males advertise their non-competitive status by depositing much fewer but larger pools of urine, of a different quality, away from highly valued resource sites (Desjardins et al., 1973; Harvey et al., 1989).

There is growing evidence that the protein fraction of urine, mostly consisting of the major urinary proteins (MUPs), is a significant mediator of chemical messages between wild mice (Hurst et al., 2001). MUPs are lipocalins known to bind and slowly release volatile, low-molecular-mass ligands (Hurst et al. 1998; Beynon et al., 2001). These ligands are semiochemicals with specific behavioural effects and their concentration may vary according to social status. For example, 2-(sec-butyl)-4,5-dihydrothiazole (thiazole) and α- and β-farnesene are elevated in dominant male urine compared to subordinate males (Harvey et al., 1989; Novotny et al., 1990). Thiazole and 2,3-dehydro-*exo*-brevicomin (brevicomin) are androgen-dependent volatile semiochemicals held in the central calyx of MUPs (Robertson et al., 1993) that act synergistically in stimulating aggressive behaviour in male mice when spiked into the urine of castrated males (Novotny et al., 1985). Farnesenes are also associated with aggression and high social status in mice. They are produced by the preputial glands, which are heavier in dominant mice compared to their subordinate counterparts when a clear dominant-subordinate relationship exists (Barnett et al., 1980). The application of farnesenes alone is sufficient to stimulate aggression between males, although the preputial gland secretion is more effective when added to urine (Jones and Nowell, 1973c). Thus the quality of a male scent mark, in terms of the concentration of different semiochemicals it contains, is a significant mediator of inter-male aggression.

## 3. FUNCTION OF MALE SCENT SIGNALS IN COMPETITION

Early studies described the function of male mouse urinary odour cues as aggression-promoting or aversive to other males. However, the response to a scent depends on a wide range of different factors and scent marks are generally deposited as a broadcast signal that will be encountered by many different individuals (Hurst, 2004). Thus it is more appropriate to consider the information provided by male scents as signals that advertise specific aspects of a male's quality or competitive ability.

We have described how the quality of male scent marks differs between dominants and subordinates; accordingly the response of the male receiver differs depending on the type of male mark encountered. If a male competitor attempts to put down urine marks in a pattern that resembles that of a dominant male, or the urinary scent mark is of a similar quality to that of a dominant rather than a subordinate male, then the resident dominant male is more likely to increase his level of aggression; by contrast, a subordinate male is likely to flee on encountering the scent mark owner (Jones and Nowell, 1973a; Hurst, 1993).

The ability to discriminate between individual conspecifics is important in facilitating the development and maintenance of stable social groups (Hurst et al., 1993, 1994). The urinary scents of wild house mice provide each individual with a unique olfactory signature. Two highly polymorphic gene clusters appear to provide the main basis for individuality scents in mice: the major histocompatibility complex (MHC) and the major urinary proteins (MUPs) (see Thom et al. and Hurst et al., this volume). These individual odour signatures within a group of mice may contribute to a familiar group odour due to scent transfer between group members that may occur, for example, at times of allo-grooming (Barnard et al., 1991). Familiarity of scent cues is a major factor that influences aggression between male mice. Wild adult male house mice can maintain stable groups with relatively little aggression shown between them so long as each member of the group continues to contribute to the shared substrate odour in order to maintain familiarity and tolerance with each other (Hurst et al., 1993). Thus, familiarity is more than just being about kinship in terms of sharing an odour that is familiar to self; it is about sharing an odour that is familiar among all group members (Hurst et al., 1994). Familiarity of individual odour signatures also plays a role in signalling competitive ability in that mice can associate an individual odour with a previous aggressor and will avoid an area containing that particular odour (Jones and Nowell, 1973b; Hurst, 1993).

Context also has an important influence on inter-male aggression. Male mice are much more likely to aggressively challenge opponents when surrounded by their own scent (Gosling and McKay, 1990). Hence it is the scent cues present that provide the context upon which a male may base his decision to attack or not. However, the resident dominant male needs to successfully demonstrate his aggressive dominance by making sure that his urinary scent marks are the most prominent and recent throughout the territory if they are to avoid aggressive challenges by competitors (Hurst, 1993; Humphries et al., 1999). When in an area where the substrate is soiled by male scents that are unfamiliar to both opponents, inter-male aggression is more likely to be observed due to the nature of other urinary scent marks that may be present, than it is in a completely clean cage (Jones and Nowell, 1975; Gray and Hurst, 1995). This makes comparisons between studies that examine aggression in scent marked home cages or clean test arenas difficult.

## 4. AGGRESSION IN THE LABORATORY

Most laboratory animal environments have been designed primarily according to experimental and practical imperatives rather than considering the animals needs (Chamove, 1989). This means that laboratory mice are often housed in same-age, single-sex groups of four to five in relatively small cages. Housing adult male mice in small groups with no means of escape from aggressive cagemates can lead to sudden outbreaks of escalating aggression, sometimes resulting in serious injury and fatalities within previously harmonious groups (Van Loo et al., 2003). Consequently, many aggressive strains of laboratory mice are housed individually, which may increase levels of stress for a species with a strong motivation for social contact (Brain, 1975). The reasons underlying such aggressive outbreaks have yet to be fully established. We have seen how under natural conditions, a sophisticated system of scent signalling regulates competitive interactions between males, leading to a stable social system. However, the laboratory environment is far removed from the natural environment of wild house mice and several factors impinge on this well-adapted system, which we will go on to discuss.

## 4.1. The Role of Inbreeding

Severe inbreeding of mice may play a part in destabilising social relationships within groups. Hundreds of inbred strains have been established whereby mice within strains show homozygosity at most loci and are hence virtually genetically identical (Cohen, 1999). This may result in a reduction in data variation and improve data reproducibility (Guénet and Bonhomme, 2003). However, such severe inbreeding is likely to have an impact on individual recognition and the formation of social relationships. Inbred mice of the same strain and sex share the same genetically determined individuality scent signature (Boyse et al., 1987; Nevison et al., 2000), resulting in inbred laboratory mice being unable to discriminate between familiar and unfamiliar individuals of their own strain (Nevison et al., 2003). Moreover, they do not recognize the urine of another mouse of the same strain and sex as being any different to their own and consequently show no competitive countermarking in response to the scent marks of other males of their own strain (Nevison et al., 2000). If a dominant male is unable to detect competing urinary scent marks from cagemates on the home-cage substrate, then potential aggression may be reduced since a subordinate's scent is not recognized as being different from the dominant's scent (Nevison et al., 2000). Moreover, if cagemates all have a similar scent then they may perceive each other as close kin and will thus show reduced levels of aggression towards each other due to kin bias (Kareem and Barnard, 1982). However, if a subordinate male is not able to clearly advertise his subordinate status, and cagemates have difficulty in recognizing the dominant male because inbred mice are unable to link social status with individuality through substrate scent cues, then social relationships may become unstable.

As yet the impact of the loss of variability in individual identity cues on social relationships and inter-male aggression is poorly understood. The level of aggressiveness between cagemates varies considerably between strains, with males of most outbred strains being almost invariably aggressive whilst most inbred strains show comparatively little aggression (Miczek et al., 2001). It remains to be established whether genetic similarity among cagemates contributes to the low levels of aggression observed in many inbred strains and what role artificial selection plays in inter-male aggression.

## 4.2. Environmental Disturbances

Social stability within groups of laboratory mice is likely to be vulnerable to disruption by external or internal changes in olfactory signals. For example, experimental procedures that change the odour of one or more of the cagemates might cause mice to be recognized as suddenly unfamiliar among a group of previously familiar mice, thus increasing the risk of attack (Nevison et al., 2003). Additionally, handling the mice and removing substrate odours without cleaning the cage base and grill results in a disruption of odours that induces aggression between male cagemates (Gray and Hurst, 1995). Further important questions remain to be answered, such as whether mice housed in open cage racks are aware of the scent of others in neighbouring cages, how much information they can gain about neighbours, how far the volatile odorants travel, and whether scents can stimulate aggression in neighbouring cages.

## 4.3. Lack of Olfactory Experience

A further consequence of extreme inbreeding is that inbred mice have no experience of differences in individual scent signatures or other variation in environmental odours. However, normal neuronal development depends on the complexity of an animal's environment (Bateson, 1981). Olfactory deprivation by closure of the nostrils induces dramatic anatomical and neurochemical changes in the olfactory bulbs (Najbauer and Leon, 1995; Philpot et al., 1997). There is a growing need to study the effects of olfactory experience and stimulation on aggression due to the increasing use of individually ventilated cages (IVCs). IVCs allow for better disease control and a reduction in the spread of laboratory animal allergens, but reduce the complexity of the olfactory environment. IVCs may reduce the occurrence of unpredictable outbreaks of aggression if exposure to scent cues from animals housed in the same room disrupts social stability within groups. However, mice are a gregarious species with a strong motivation to gain access to social contact and use scent as the primary source of gaining social information. The reduction in olfactory stimulation might thus result in increased abnormal behaviour and stress responses that would have both experimental and welfare implications.

## 5. CONCLUSION

Mice maintain stable social groups by assessing the gender, relatedness, familiarity, social and reproductive status of other individuals in the group and modify their own behaviour accordingly. Among territorial wild house mice, males need to recognise the male that dominates a particular territory to avoid potentially damaging aggressive encounters. Males advertise social status via the quality and the spatial and temporal pattern of urinary scent marks, and other males modulate their aggressive or aversive response to the signaller according to social status, familiarity, age of scent mark and any other odours that may be found in the vicinity of the scent mark. However, aggression in laboratory mice is very much a different story with artificial selection, genetic inbreeding, a simple olfactory environment and artificial social groups impinging on the role of scent in inter-male aggression in laboratory mice. Much work remains to be done in investigating the roles of experience, development, neurophysiology, and genetics in the aggressive behaviour of laboratory mice if problems of animal welfare and experimental variability are to be overcome.

## 6. ACKNOWLEDGEMENTS

We would like to thank members of the Animal Behaviour Group and Rob Beynon for comments, suggestions and encouragement, and the BBSRC for financial support.

## 7. REFERENCES

Barnard, C. J., Hurst, J. L., and Aldhous, P., 1991, Of mice and kin – the functional significance of kin bias in social behaviour, *Biol. Rev.* **66**:379-430.

Barnett, S. A., Dickson, R. G., and Warth, K. G., 1980, Social status, activity and preputial glands of wild and domestic house mice, *Zool. J. Linn. Soc-Lond.* **70**:421-430.
Bateson, P., 1981, Ontogeny of behaviour, *Brit. Med. Bull.* **37**:159-164.
Beck, J. A., Lloyd, S., Hafezparast, M., Lennon-Pierce, M., Eppig, J. T., Festing, M. F. W., and Fisher, M. C., 2000, Genealogies of mouse inbred strains, *Nature Genetics*, **24**:23-25.
Beynon, R. J., Hurst, J. L., Gaskell, S. J., Hubbard, S. J., Humphries, R. E., Malone, N., Marie, A. D., Martinsen, L., Nevison, C. M., Payne, C. E., Robertson, D. H. L., and Veggerby, C., 2001, Mice, MUPs and myths: Structure-function relationships of the major urinary proteins, in: *Chemical Signals in Vertebrates Vol. 9*, A. Marchlewska-Koj, J. J. Lepri, and D. Müller-Schwarze, ed., Kluwer Academic/Plenum Publishers, New York, pp. 149-156.
Boyse, E. A., Beauchamp, G. K., and Yamazaki, K., 1987, The genetics of body scent, *Trends Genet.* **3**:97-102.
Brain, P., 1975, What does individual housing mean to a mouse? *Life Sci.* **16**:187-200.
Brain, P. F. and Nowell, N. W., 1970, The effects of differential grouping on endocrine function of mature male albino mice, *Physiol. Behav.* **5**:907-910.
Chamove, A. S., 1989, Cage design reduces emotionality in mice, *Lab. Anim.* **23**:215-219.
Cohen, J. J., 1999, Individual variability and immunity, *Brain. Behav. Immun.* **13**:76-79.
Desjardins, C., Maruniak, J. A., and Bronson, F. H., 1973, Social rank in the house mouse: differentiation revealed by ultra-violet visualisation of urinary marking patterns, *Science*, **182**:939-941.
Gosling, L. M. and McKay, H. V., 1990, Competitor assessment by scent matching: an experimental test, *Behav. Ecol. Sociobiol.* **26**:415-420.
Gray, S. and Hurst, J. L., 1995, The effects of cage cleaning on aggression within groups of male laboratory mice, *Anim. Behav.* **49**:821-826.
Guénet, J-L. & Bonhomme, F., 2003, Wild mice: an ever-increasing contribution to a popular mammalian model, *Trends. Genet.* **19**:24-31.
Harvey, S., Jemiolo, B., and Novotny, M., 1989, Pattern of volatile compounds in dominant and subordinate male mouse urine, *J. Chem. Ecol.* **15**:2061-2072.
Humphries, R. E., Robertson, D. H. L., Beynon, R. J., and Hurst, J. L., 1999, Unravelling the chemical basis of competitive scent marking in house mice, *Anim. Behav.* **58**:1177-1190.
Hurst, J. L., 1987, The functions of urine marking in a free-living population of house mice *Mus domesticus* Rutty, *Anim. Behav.* **35**:1433-1442.
Hurst, J. L., 1990, Urine marking in populations of wild house mice *Mus domesticus* Rutty. I. Communication between males, *Anim. Behav.* **40**:209-222.
Hurst, J. L., 1993, The priming effects of urine substrate marks on interactions between male house mice, *Mus musculus domesticus* Schwarz and Schwarz, *Anim. Behav.* **45**:55-81.
Hurst, J. L., 2004, Scent marking and social communication, in: *Animal Communication Networks*, P. K. McGregor, ed., Cambridge University Press, Cambridge, in press.
Hurst, J. L., Beynon, R. J., Humphries, R. E., Malone, N., Nevison, C. M., Payne, C. E., Robertson, D. L. H., and Veggerby, C., 2001, Information in scent signals competitive social status: the interface between behaviour and chemistry, in: *Chemical Signals in Vertebrates Vol. 9*, A. Marchlewska-Koj, J. J. Lepri, and D. Müller-Schwarze, ed., Kluwer Academic/Plenum Publishers, New York, pp. 43-52.
Hurst, J. L., Fang, J., and Barnard, C., 1993, The role of substrate odours in maintaining social tolerance between male house mice, *Mus musculus domesticus, Anim. Behav.* **45**:997-1006.
Hurst, J. L., Fang, J., and Barnard, C., 1994, The role of substrate odours in maintaining social tolerance between male house mice, *Mus musculus domesticus*: relatedness, incidental kinship effects and the establishment of social status, *Anim. Behav.* **48**:157-167.
Hurst, J. L., Robertson, D. H. L., Tolladay, U., and Beynon R. J., 1998, Proteins in urine scent marks of male house mice extend the longevity of olfactory signals, *Anim. Behav.* **55**:1289-1297.
Jones, R. B. and Nowell, N. W., 1973a, Aversive and aggression-promoting properties of urine from dominant and subordinate male mice, *Anim. Learn. Behav.* **1**:207-210.
Jones, R. B. and Nowell, N. W., 1973b, Aversive effects of the urine of a dominant male mouse upon the investigatory behaviour of its defeated opponent, *Anim. Behav.* **21**:707-710.
Jones, R. B. and Nowell, N. W., 1973c, Effects of preputial and coagulating gland secretions upon aggressive behaviour in male mice: a confirmation, *J. Endocrinol.* **59**:203-204.
Jones, R. B. and Nowell, N. W., 1975, Effects of clean and soiled sawdust substrates and of different urine types upon aggressive behaviour in male mice, *Aggressive. Behav.* **1**:111-121.
Kareem, A. M. and Barnard, C. J., 1982, The importance of kinship and familiarity in social interactions between mice, *Anim. Behav.* **30**:594-601.
Mackintosh, J. H. and Grant, E. C., 1966, The effect of olfactory stimuli on the agonistic behaviour of laboratory mice, *Z. Tierpschol.* **23**:584-587.

Miczek, K. A., Maxson, S. C., Fish, E. W., and Faccidomo, S., 2001, Aggressive behavioural phenotypes in mice, *Behav. Brain. Res.* **125**:167-181.

Najbauer, J., and Leon, M., 1995, Olfactory experience modulates apoptosis in the developing olfactory bulb, *Brain Res.* **674**:245-251.

Nevison, C. M., Barnard, C. J., Beynon, R. J., and Hurst, J. L., 2000, The consequences of inbreeding for recognizing competitors, *P. Roy. Soc. Lond. B. Bio.* **267**:687-694.

Nevison, C. M., Barnard, C. J., and Hurst, J. L., 2003, The consequence of inbreeding for modulating social relationships between competitors, *Appl. Anim. Behav. Sci.* **2018**:1-12.

Novotny, M., Harvey, S., Jemiolo, B., and Alberto, J., 1985, Synthetic pheromones that promote inter-male aggression in mice, *P. Natl. Acad. Sci. USA* **82**:2059-2061.

Novotny, M., Harvey, S., Jemiolo, B., 1990, Chemistry of male dominance in the house mouse, *Mus domesticus, Experientia*, **46**:109-113.

Philpot, B. D., Foster, T. C., and Brunjes, P. C., 1997, Mitral/tufted cell activity is attenuated and becomes uncoupled from respiration following naris closure, *J. Neurbiol.* **33**:374-386.

Robertson, D. H. L., Evershed, R. P., and Beynon, R. J., 1993, Extraction, characterization and binding analysis of two pheromonally active ligands associated with the major urinary protein of the house mouse, *J. Chem. Ecol.* **19**:1405-1415.

Smith, J., Hurst, J. L., and Barnard, C. J., 1994, Comparing behaviour in wild and laboratory strains of the house mouse: levels of comparison and functional interference, *Behav. Process.* **32**:79-86.

Van Loo, P. L. P., de Groot, A. C., Van Zupten, B. F. M., and Baumans, V., 2001, Do male mice prefer or avoid each other's company? Influence of hierarchy, kinship, and familiarity, *J. Appl. Anim. Welf. Sci.* **4**:91-103.

Van Loo, P. L. P., Van de Meer, E., Kruitwagen, C. L. J. J., Koolhaas, J. M., Van Zutphen, L. F. M., and Baumans, V., 2003, Strain-specific aggressive behaviour of male mice submitted to different husbandry procedures, *Aggressive. Behav.* **29**:69-80.

# CHEMICAL SIGNALS AND VOMERONASAL SYSTEM FUNCTION IN AXOLOTLS (*AMBYSTOMA MEXICANUM*)

Heather L. Eisthen[1] and Daesik Park[1,2]

## 1. INTRODUCTION

Most tetrapods, the group of vertebrates that includes amphibians, reptiles, and mammals, possess separate olfactory and vomeronasal systems. Why? What are the functions of these two anatomically distinct chemosensory systems? Although the vomeronasal system is often presumed to be specialized for mediating responses to pheromones and the olfactory system assumed to respond to "general odorants" that are not pheromones, this dichotomy does not hold up for any group of vertebrates studied to date. In mammals, some pheromonal effects are mediated by the olfactory system (Dorries et al., 1997; Cohen-Tannoudji et al., 1989), and the vomeronasal organ responds to some general odorants (Sam et al., 2001). In squamate reptiles the vomeronasal system is critically involved in both foraging and in responding to pheromones (Halpern and Kubie, 1984). In salamanders, the vomeronasal organ has recently been suggested to play a role in foraging (Placyk and Graves, 2002), and both the olfactory and vomeronasal organs respond to pheromones (Toyoda et al., 1999; Park and Propper, 2002). Thus, the vomeronasal system cannot be considered to be a labeled line for pheromone processing (Shepherd, 1985), although certainly many pheromonal effects are mediated by the vomeronasal system (see Halpern and Martínez-Marcos, 2003, for a recent review).

An alternative hypothesis suggests that the vomeronasal system mediates responses to large, high-molecular-weight molecules that are not volatile in air, and that the olfactory system responds to molecules that have a sufficiently low molecular weight to become airborne and contact the olfactory epithelium (Halpern and Kubie, 1980; Wysocki et al., 1980). This hypothesis is attractive, as it might explain why the vomeronasal system has been lost repeatedly in aquatic and arboreal tetrapods, including in crocodilians and birds (Parsons, 1959, 1967, 1970), some cetaceans (Lowell and Flanigan, 1980; Oelschläger and Buhl, 1985), manatees (Mackay-Sim et al., 1985),

---

[1] Department of Zoology, Michigan State University, East Lansing, MI 48824.
[2] Department of Science Education, Kangwon National University, Chuncheon, Kangwon 200-701, Korea.

several species of bat (Bhatnagar, 1980), and some primates (Smith et al., 2001, 2002). Further, if mammalian pheromones tend to be high-molecular-weight molecules (Singer, 1991; Nevison et al., 2003), this hypothesis might also explain why many pheromone effects in mammals are mediated by the vomeronasal system.

The vomeronasal system is generally present in aquatic amphibians (Eisthen et al., 1994; Eisthen, 2000), an observation that seems to contradict the molecular-weight hypothesis, as volatility is not relevant in water. To date, the pheromones that have been isolated from amphibians are all peptides and proteins (Kikuyama et al., 1995; Rollmann et al., 1999; Wabnitz et al., 1999; Yamamoto et al., 2000), and the only study demonstrating selective stimulation of the vomeronasal organ used stimuli that were 3 kDa and larger (Wirsig-Wiechmann et al., 2002). Perhaps the vomeronasal system of amphibians detects large molecules, and the olfactory system responds to smaller molecules.

Nevertheless, some data concerning the presence of the vomeronasal system in aquatic vertebrates remain difficult to explain. For example, the vomeronasal system has been lost in proteids, a family of aquatic salamanders (Seydel, 1895; Anton, 1911; Farbman and Gesteland, 1974; Eisthen, 2000). Without a firm understanding of the relative functions of the olfactory and vomeronasal systems, the causes of the loss of the system in this group of animals remain mysterious. Another anomalous detail concerns the presence of the vomeronasal system in teleost fishes: although distinct vomeronasal organs and accessory olfactory bulbs are clearly absent in teleosts (Eisthen, 1992, 1997), members of the family of genes encoding mammalian vomeronasal odorant receptors are expressed in the olfactory epithelium of goldfish and pufferfish (Cao et al., 1998; Asano-Miyoshi et al., 2000). Thus, we are far from a solid understanding of the organization and function of the vomeronasal system in vertebrates.

In our laboratory, we are studying chemosensory-guided behavior and the neurobiology of chemosensory systems in axolotls (*Ambystoma mexicanum*), a species of salamander that is aquatic throughout life. Through our studies, we hope to determine the relative functions of the olfactory and vomeronasal systems in axolotls, and in aquatic salamanders in general. We expect that this work will help us understand the function of the vomeronasal system in amphibians and, more generally, in aquatic vertebrates.

We have chosen to study axolotls for several reasons. First, axolotls are hardy, and are easily maintained and bred in the laboratory (Armstrong et al., 1989). Unlike many salamanders, axolotls will court and breed year-round in the laboratory. Axolotls are larger than many salamanders, making the surgeries and dissections for tract-tracing and recording from live cells fairly simple; they also have large cells, simplifying experiments involving single-cell physiology. Finally, axolotls are essentially a subspecies of tiger salamander (Shaffer, 1993), and much of the available data concerning neuroanatomy and olfactory system physiology in tiger salamanders (e.g., Herrick, 1927, 1948; Firestein and Werblin, 1987, 1989; Kauer, 2002) can be applied directly to axolotls.

## 2. ANATOMY OF NASAL CHEMOSENSORY SYSTEMS IN AXOLOTLS

In axolotls, as in most other salamanders, the vomeronasal organ consists of a pouch that protrudes from the lateral edge of the nasal cavity (Figure 1). The vomeronasal organ

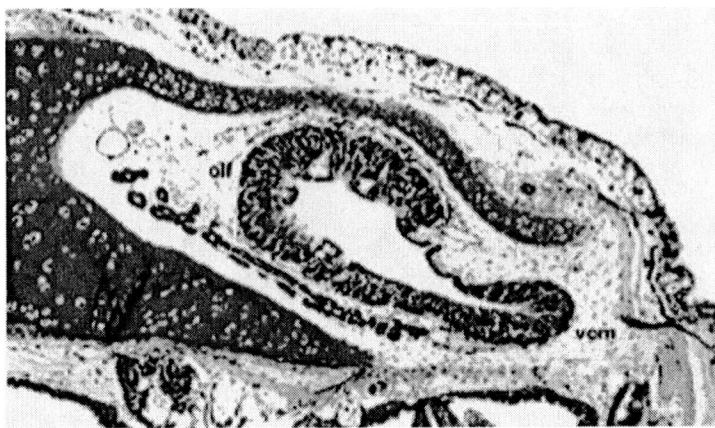

Figure 1. Cross section through the snout of an axolotl, decalcified and stained with cresylecht violet. The midline of the snout is shown on the left, lateral is to the right, and the dorsal surface of the snout is toward the top. This section is taken through the point at which the vomeronasal organ connects with the nasal cavity. The vomeronasal organ is lined with vomeronasal sensory epithelium (vom), and the medial portions of the nasal cavity are lined with olfactory epithelium (olf).

of axolotls is relatively small, extending approximately 10% of the total length of the nasal cavity. In cross-section, the organ is wider than it is tall.

The structure of the olfactory and vomeronasal epithelia differ considerably (Eisthen et al., 1994). In axolotls, as in other vertebrates, the nasal sensory epithelia contain sensory neurons, supporting (sustentacular) cells, and basal cells. The olfactory epithelium of axolotls contains both ciliated and microvillar receptor neurons, and the two types of neurons are found in homogeneous clusters throughout the epithelium. In contrast, vomeronasal receptor neurons terminate in microvilli, although a rare ciliated dendrite can be found in the vomeronasal epithelium. In both the olfactory and vomeronasal epithelia, sustentacular cells terminate in short processes; the vomeronasal epithelium contains a second class of supporting cells with a large surface that is covered with highly active, motile cilia. These cells may function to move mucus and odorants into the relatively small vomeronasal organ and across the sensory epithelium.

As illustrated in Figure 2, the axons of most olfactory receptor neurons project to the main olfactory bulb, at the rostral pole of the telencephalon; the axons of the vomeronasal neurons project to a distinct accessory olfactory bulb on the lateral edge of the telencephalon, immediately caudal to the main olfactory bulb. The central projections of the main and accessory olfactory bulbs have been described in tiger salamanders by Herrick (1927, 1948) and by Kokoros and Northcutt (1977), and we assume that the projections are the same in axolotls; similar projections have also been described in other salamanders (Herrick, 1933; Schmidt and Roth, 1990). Briefly, as illustrated in Figures 2A and B, the afferents from the main olfactory bulb project via medial and lateral tracts to the septum, striatum, and medial amygdala, and to the dorsal, lateral, and medial pallia (the homologue of the mammalian hippocampus). The accessory olfactory bulb projects to the lateral amygdala (Figure 2D); as in mammals and reptiles, different portions of the

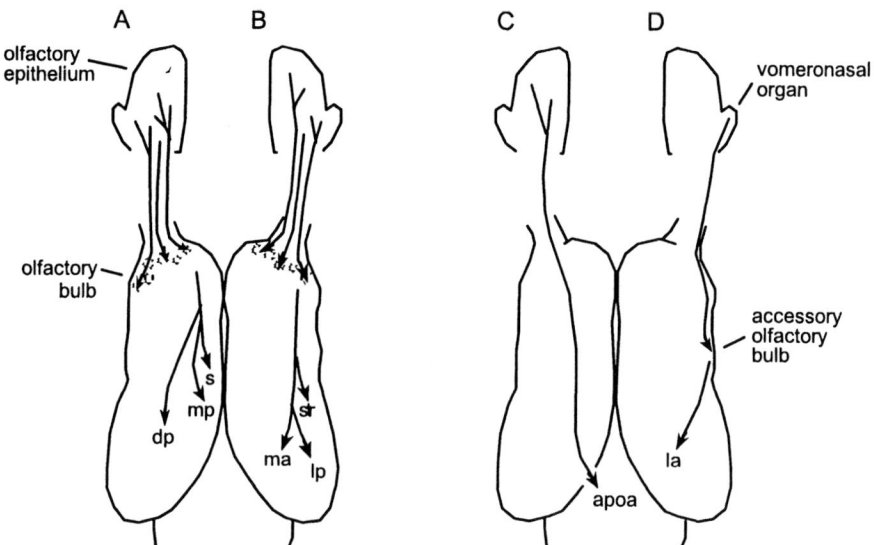

**Figure 2.** Schematic diagram of the nasal cavities and forebrain of a salamander, illustrating the central projections of the olfactory and vomeronasal systems in dorsal view. Anterior is toward the top of the figure, and only ipsilateral projections are shown. The medial (A) and lateral (B) olfactory tracts arise from the olfactory bulb. (C) The extra-bulbar olfactory pathway bypasses the olfactory bulb and projects directly to the anterior preoptic area. (D) The accessory olfactory bulb, which receives input from the vomeronasal organ, projects to the lateral amygdala (la). Other abbreviations: apoa = anterior preoptic area; dp = dorsal pallium; lp = lateral pallium; mp = medial pallium; ma = medial amygdala; s = septum; str = striatum. Based on descriptions in Herrick, 1927, 1933, 1948; Kokoros and Northcutt, 1977; and Schmidt and Roth, 1990.

amygdala receive input from the main and accessory olfactory bulbs (reviewed in Halpern and Martínez-Marcos, 2003). Finally, axolotls appear to possess an extra-bulbar olfactory pathway (Szabo et al., 1991; Hofmann and Meyer, 1992), in which axons of olfactory receptor neurons bypass the olfactory bulb and project directly to the anterior preoptic area (Fig. 2C; unpublished observations).

## 3. CHEMOSENSORY-GUIDED BEHAVIOR IN AXOLOTLS

Unfortunately, axolotls are almost extinct in the wild, and little known is about their behavior and ecology in natural settings; nevertheless, it seems reasonable to assume that many aspects of their behavior are similar to those of other neotenic populations of tiger salamanders. We are investigating the roles of chemosensory cues in guiding reproductive and foraging behavior in axolotls in the laboratory, with the goal of using these cues in neurobiological experiments to examine the relative functions of the olfactory and vomeronasal systems.

We have observed axolotls courting in the laboratory, and the basic sequence of courtship behaviors in axolotls appears to be similar to that which has been described for other ambystomids (e.g., *A. talpoideum*, Shoop, 1960; *A. tigrinum*, Arnold, 1976). The initial stage in courtship consists of the male nudging the female with its snout, first along the flank and then in the cloacal region. This leads to the "waltz" behavior described by Shoop (1960) for *A. talpoideum*, in which each animal holds its snout near the other's cloaca as the pair moves in a circle. After several minutes of waltzing, the male moves away from the female, which follows closely behind, often touching his tail with her snout. During this stage of courtship, the male slowly shakes the posterior parts of the body and tail from side to side with an undulating motion. Arnold (1976) calls this behavior "vent shuffling", and it resembles the "hula dance" performed by male red-spotted newts (*Notophthalmus viridescens*; Verrell, 1982). Unlike newts, however, the female also shakes her tail while following, although not as consistently as does the male. After the female has followed the male for a brief period, the male pauses and lowers his cloacal area while vigorously shaking his abdomen and tail. He deposits a spermatophore on the substrate, and then moves forward approximately one body length. The female follows. Once she is positioned with her cloaca over the spermatophore, the female lowers her cloaca to pick up the spermatophore while rapidly shaking her tail. At this point her behavior is almost indistinguishable from that of a male depositing a spermatophore. After the female has retrieved the sperm cap, the two animals move independently. Eventually, nudging and waltzing may be re-established. This stage of courtship is usually briefer during the second and subsequent encounters, and the female shortly begins following the male again. The entire process of courtship and sperm transfer is generally repeated several times during the course of an encounter.

Females fertilize their eggs internally, then lay most or all of their eggs in a spawn that may contain several hundred eggs. Given that in natural conditions females probably spawn only once in a year, males may be under considerable pressure to discriminate gravid females, which are full of eggs and ready to spawn, from those that have recently spawned or are otherwise not in breeding condition. In contrast, males can mate successfully at 2- to 3-week intervals (Armstrong et al., 1989), and females may not need to discriminate between males that have spawned recently and those that have not.

As a first step toward understanding the role of chemical cues in courtship behavior, we have conducted experiments to determine whether sexually-mature male and female axolotls respond differently to odorants from conspecifics in different reproductive conditions (Park et al., 2004). We presented both male and female axolotls with whole-body odorants from same-sex conspecifics and from opposite-sex conspecifics that had spawned recently or had not spawned recently, as well as a clean-water control, and measured both general activity and the frequency and duration of hula displays by subjects. We found that in males, exposure to whole-body odorants from females led to an increase in general activity, regardless of the reproductive condition of the odorant donor. Intriguingly, males that had no prior experience with adult females were more active than experienced males when exposed to odorants from females; because of our housing regime, males that had not been paired with a female prior to testing had no experience with odorants from sexually-mature females, so we do not know whether the effect we observed requires behavioral interactions with females, or simply exposure to chemical cues from females. We found no differences in the frequency or duration of hula displays by males in any testing condition. Among female subjects, exposure to odorants from conspecifics did not affect general activity, but hula displays were performed only in response to exposure to odorants from males. No difference in the frequency or duration of hula displays was observed in response to odorants from males that had spawned recently or not. We found that both sexually experienced and inexperienced females performed hula displays when exposed to whole-body odorants from males, suggesting that odorants from male axolotls could function to increase sexual receptivity in female axolotls, as has been demonstrated in other salamanders (Houck and Reagan, 1990; Rollmann et al., 1999). Overall, our results demonstrate that both male and female axolotls can use odorant cues to discriminate the sex of conspecifics, but do not indicate whether either males or females can distinguish the reproductive status of opposite-sex conspecifics based on chemical cues.

The role of chemosensory systems in foraging behavior in aquatic salamanders has been a topic of little research, although chemical cues have been shown to play a role in prey-catching in a terrestrial species (David and Jaeger, 1981; Lindquist and Bachmann, 1982; Placyk and Graves, 2002). Axolotls have well-developed lateral line systems that include both mechanosensory and electrosensory organs (Münz et al., 1984). Given that salamanders eat live prey, the lateral lines presumably play a major role in helping axolotls locate and capture prey. Nevertheless, axolotls raised in the laboratory on a diet of food pellets appear to use chemosensory cues to locate pellets on the substrate. When searching for food pellets, axolotls engage in a stereotypical behavior in which the snout is held at a sharp angle to the substrate, presumably to bring the dorsally-located nostrils

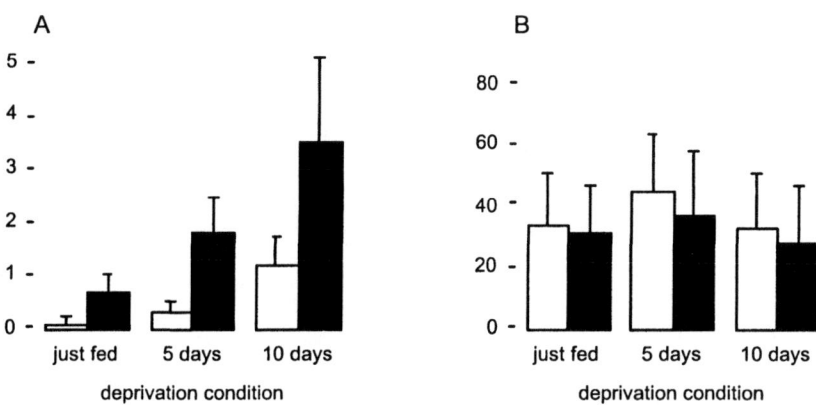

**Figure 3.** Chemosensory cues play a role in foraging in axolotls. Ten adult axolotls of both sexes were tested immediately after feeding, after 5 days, and after 10 days of food deprivation, in an aquarium containing clean Holtfreter's solution, the standard medium for maintaining axolotls (white bars), or in Holtfreter's solution containing food odors (black bars). Prior to the first test, each subject was fed commercial salmon pellets (Rangen Inc., Buhl, ID) *ad lib*. Thirty to ninety min later, the subject was placed in clean 10-gal aquarium containing Holtfreter's solution and its activity recorded. Each animal was tested again after 5 and 10 days, and then retested later using the same procedure, except that 250 ml food suspension (3 g food pellets mixed into 250 ml $dH_2O$ and filtered) was added to the aquarium. (A) The frequency of searching the substrate for food pellets increased in the presence of food odors (two-way repeated measures ANOVA, $F_{1,54} = 5.91$, $P = 0.018$), and with food deprivation ($F_{2,54} = 3.61$, $P = 0.034$). No interaction between these factors was observed ($F_{2,54} = 0.66$, $P = 0.52$). Searching was defined as holding the snout such that the angle between the lower jaw and aquarium floor was more than 30° for at least 5 sec. (B) Overall activity, measured as the number of times a line down the middle of the aquarium was crossed, was not affected by the presence of food odors ($F_{1,54} = 0.11$, $P = 0.75$) or by food deprivation. ($F_{2,54} = 0.20$, $P = 0.82$). Again, no interaction was observed ($F_{2,54} = 0.01$, $P = 0.99$).

closer to the substrate. As demonstrated in Figure 3, axolotls engage in this behavior more when food deprived than when recently fed, and more in the presence of the odorants from food pellets than in a clean aquarium. We plan to use similar odorants to examine the roles of the olfactory and vomeronasal systems in foraging in axolotls.

## 4. PHYSIOLOGY OF THE OLFACTORY AND VOMERONASAL EPITHELIA IN AXOLOTLS

Although tiger salamanders have long been used as a model for studies of odorant processing in the olfactory epithelium and bulb (e.g., Kauer and Moulton, 1974; Mackay-Sim and Kubie, 1981), few researchers have studied the physiology of the vomeronasal organ and accessory olfactory bulb in ambystomids. We have examined voltage-activated currents in olfactory and vomeronasal receptor neurons in axolotls, and find that the currents in olfactory receptor neurons are like those described in larval tiger salamanders (Firestein and Werblin, 1987). In general, vomeronasal receptor neurons have the same complement of currents, although, as in mice (Liman and Corey, 1996), the kinetics of the calcium current differ between olfactory and vomeronasal neurons. Specifically, vomeronasal neurons possess a rapidly-inactivating calcium current (Figure 4), but a sustained calcium current is found in olfactory receptor neurons. Thus, the vomeronasal receptor neurons in axolotls appear to be both morphologically and physiologically similar to those described in other vertebrates.

We have used electro-olfactogram (EOG) recording techniques to examine odorant responses in the olfactory and vomeronasal epithelia of adult axolotls (Park and Eisthen, 2003; Park et al., 2003, 2004). Overall, we have found the odorant responsiveness of the two epithelia to be remarkably similar: for example, both epithelia respond vigorously to amino acids, such as L-methionine (Figure 5). In conjunction with the study of behavioral responses to conspecific odorants described above, we also examined EOG responses to the same stimuli (Park et al., 2004). Because the magnitude of odorant responses varies across the sensory epithelium due to such factors as differential distribution of odorant receptors (Mackay-Sim et al., 1982), we recorded from eight sites on the olfactory epithelium and two on the vomeronasal epithelium. As shown in Figure 5, in both males and females, olfactory and vomeronasal EOG responses were larger in response to whole-body odorants from individuals of the opposite sex than from individuals of the same sex as the subject. In males, odorants from gravid and recently-spawned females elicited differential responses from one location each in the olfactory and vomeronasal epithelia, but no differences in response magnitude were recorded at the other eight locations examined. Among females, we found a curious effect of social experience on EOG responses: at some locations in the olfactory epithelium, EOG responses to odorants from both male and female conspecifics were larger in females that had no experience with adult males than in those that did. Responses in the vomeronasal organ also appeared to differ between experienced and inexperienced females, but our data set was small and this difference did not prove to be statistically significant. Overall, these results indicate that in axolotls both the olfactory and vomeronasal systems may be involved in discriminating the sex and reproductive condition of conspecifics, and that social or sexual experience alters responsiveness to some chemical cues from conspecifics.

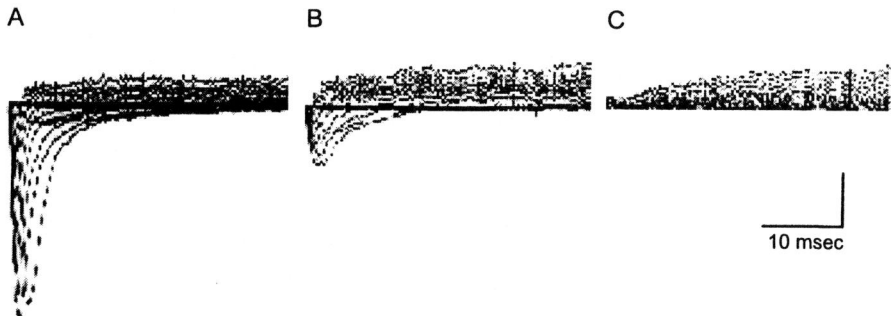

**Figure 4.** Whole-cell currents recorded from an axolotl vomeronasal receptor neuron in an epithelial slice. Currents were elicited by applying a series of voltage pulses, from -80 mV to +60 mV, in 10 mV increments. $V_{hold}$ = -80 mV. (A) Currents recorded in standard amphibian physiological saline, before any channel blockers were washed onto the slice. (B) Currents recorded 3 min after 10 μM tetrodotoxin (TTX) was washed onto the cell. TTX blocks voltage-gated sodium currents, revealing a second inward current. (C) Currents recorded 9 min after a solution containing both 10 μM TTX and 1 mM $CoCl_2$ was washed onto the slice. Cobalt blocks calcium currents, demonstrating that the rapidly-inactivating inward current shown in panel B was carried by calcium.

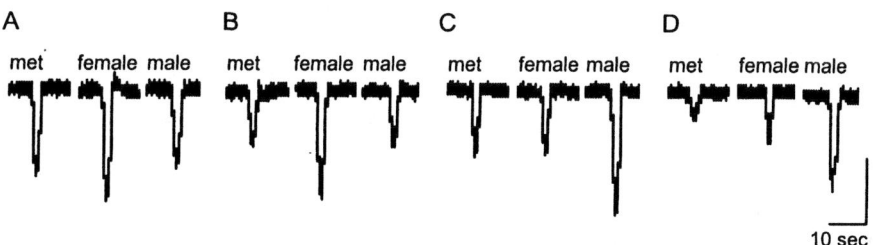

**Figure 5.** Electro-olfactogram (EOG) responses recorded from the olfactory and vomeronasal epithelia in axolotls. Odorants consisted of 100 μl of 1 mM L-methionine (met), of water containing whole-body odorants from sexually-mature adult females (female), or of water containing whole-body odorants from size-matched, sexually-mature adult males (male). (A) Responses recorded from the olfactory epithelium of an adult male; (B) responses recorded from the vomeronasal organ of the same individual. (C) Responses recorded from the olfactory epithelium of an adult female; (D) responses recorded from the vomeronasal organ of the same female. Note that in all cases, responses elicited by odorants from opposite-sex individuals are larger than those elicited by odorants from same-sex individuals. Adapted from Park et al., 2004.

Centrifugal feedback has been shown to shape responses in the retina and inner ear in vertebrates (Akopian, 2000; Ashmore et al., 2000). Similarly, we have found that peptides, presumably released by the terminal nerve, modulate electrophysiological activity and odorant responses in the olfactory epithelium of axolotls (Park and Eisthen, 2003; Park et al., 2003). Given that the cells of the terminal nerve receive input from olfactory areas in the forebrain (Yamamoto and Ito, 2000), it seems possible that the terminal nerve functions to specifically enhance and suppress responding to various olfactory cues; for example, perhaps the terminal nerve increases responding to conspecific chemical signals during courtship and mating behavior. A complete understanding of peripheral processing of chemical signals will have to include a consideration of the role of the terminal nerve in shaping responses to these signals.

## 5. CONCLUSIONS

We are studying the anatomy and physiology of the olfactory and vomeronasal systems in axolotls with the goal of determining the behavioral functions of these two chemosensory systems in aquatic amphibians. Our anatomical studies demonstrate that the vomeronasal epithelium of axolotls is much like that of other tetrapods. Other studies indicate that the projections from the olfactory and vomeronasal epithelia into the central nervous system are separate through several synapses, suggesting that these chemosensory systems serve different functions. Our electrophysiological experiments have not revealed striking differences in odorant responsivity between the olfactory and vomeronasal epithelia, but we are just beginning to work in this area and cannot draw yet strong conclusions about the relative quality or strength of odorant responses in these sensory epithelia. We have begun to show that odorant cues play a role in both foraging and in social behavior in axolotls; we hope that by combining neurobiological and behavior studies, we will be able to fully understand the ways in which chemosensory stimuli are processed to mediate behavior in axolotls.

Given that separate olfactory and vomeronasal systems are present in amphibians and in amniotes, the vomeronasal system must have been present in the last common ancestor of these two groups, and this animal is now thought to have been fully aquatic (Panchen, 1991; Lebedev and Coates, 1995). An understanding of the function of the vomeronasal system in aquatic amphibians may help shed light on the factors that led to the evolutionary origin of the vomeronasal system.

## 6. ACKNOWLEDGMENTS

We thank Sandra Borland, Susan Duhon, and Jill Gresens of the Indiana University Axolotl Colony, without whose help this work would not have been possible. Thanks to Vince Dionne, in whose laboratory the data presented in Figure 4 were collected, and to Kosha Baxi for comments on the manuscript. We are grateful for support from the National Science Foundation (IBN 9982934) and the National Institutes of Health (DC05366).

## 7. REFERENCES

Akopian, A., 2000, Neuromodulation of ligand- and voltage-gated channels in the amphibian retina, *Microsc. Res. Tech.* **50**:403-410.

Anton, W., 1911, Die Nasenhöhle der Perennibranchiaten, *Morphol. Jahrb.* **44**:179-199.

Armstrong, J. B., Duhon, S. T., and Malacinski, G. M., 1989, Raising the axolotl in captivity, in: *Developmental Biology of the Axolotl*, J. B. Armstrong and G. M. Malacinski, eds., Oxford University Press, New York, pp. 220-227.

Arnold, S. J., 1976, Sexual behavior, sexual interference and sexual defense in the salamanders *Ambystoma maculatum*, *Ambystoma tigrinum* and *Plethodon jordani*, *Z. Tierpsychol.* **42**:247-300.
Asano-Miyoshi, M., Suda, T., Yasuoka, A., Osima, S., Yamashita, S., Abe, K., and Emori, Y., 2000, Random expression of main and vomeronasal olfactory receptor genes in immature and mature olfactory epithelia of *Fugu rubripes*, *J. Biochem.* **127**:915-924.
Ashmore, J. F., Geleoc, G. S., and Harbott, L., 2000, Molecular mechanisms of sound amplification in the mammalian cochlea, *Proc. Natl. Acad. Sci. U. S. A.* **97**:11759-11764.
Bhatnagar, K. P., 1980, The chiropteran vomeronasal organ: Its relevance to the phylogeny of bats, in: *Proceedings of the Fifth International Bat Research Conference*, D. E. Wilson and A. L. Gardner, eds., Texas Tech Press, Lubbock, TX, pp. 289-315.
Cao, Y., Oh, B. C., and Stryer, L., 1998, Cloning and localization of two multigene receptor families in goldfish olfactory epithelium, *Proc. Natl. Acad. Sci. U. S. A.* **95**:11987-11992.
Cohen-Tannoudji, J., Lavenet, C., Locatelli, A., Tillet, Y., and Signoret, J., 1989, Non-involvement of the accessory olfactory system in the LH response of anoestrous ewes to male odour, *J. Reprod. Fert.* **86**:135-144.
David, R. S., and Jaeger, R. G., 1981, Prey location through chemical cues by a terrestrial salamander, *Copeia* **1981**:435-440.
Dorries, K. M., Adkins-Regan, E., and Halpern, B. P., 1997, Sensitivity and behavioral responses to the pheromone, androstenone, are not mediated by the vomeronasal organ in the pig, *Brain Behav. Evol.* **49**:52-63.
Eisthen, H. L., 1992, Phylogeny of the vomeronasal system and of receptor cell types in the olfactory and vomeronasal epithelia of vertebrates, *Micr. Res. Tech.* **23**:1-21.
Eisthen, H. L., 1997, Evolution of vertebrate olfactory systems, *Brain Behav. Evol.* **50**:222-233.
Eisthen, H. L., 2000, Presence of the vomeronasal system in aquatic salamanders, *Phil. Trans. Roy. Soc. (London)* **355**:1209-1213.
Eisthen, H. L., Sengelaub, D. R., Schroeder, D. M., and Alberts, J. R., 1994, Anatomy and forebrain projections of the olfactory and vomeronasal organs in axolotls (*Ambystoma mexicanum*), *Brain Behav. Evol.* **44**:108-124.
Farbman, A. I., and Gesteland, R. C., 1974, Fine structure of the olfactory epithelium in the mud puppy, *Necturus maculosus*, *Am. J. Anat.* **139**:227-244.
Firestein, S., and Werblin, F., 1987, Gated currents in isolated olfactory receptor neurons of the larval tiger salamander, *Proc. Natl. Acad. Sci. U. S. A.* **84**:6292-6296.
Firestein, S., and Werblin, F., 1989, Odor-induced membrane currents in vertebrate olfactory receptor neurons, *Science* **244**:79-82.
Halpern, M., and Kubie, J. L., 1980, Chemical access to the vomeronasal organs of garter snakes, *Physiol. Behav.* **24**:367-371.
Halpern, M., and Kubie, J. L., 1984, The role of the ophidian vomeronasal system in species-typical behavior, *Trends Neurosci.* **7**:472-477.
Halpern, M., and Martínez-Marcos, A., 2003, Structure and function of the vomeronasal system: An update, *Prog. Neurobiol.* **70**:245-318.
Herrick, C. J., 1927, The amphibian forebrain IV: The cerebral hemispheres of *Amblystoma*, *J. Comp. Neurol. Psychol.* **43**:231-325.
Herrick, C. J., 1933, The amphibian forebrain IV: *Necturus*, *J. Comp. Neurol.* **58**:1-288.
Herrick, C. J., 1948, *The Brain of the Tiger Salamander*, University of Chicago Press, Chicago.
Hofmann, M. H., and Meyer, D. L., 1992, Peripheral origin of olfactory nerve fibers by-passing the olfactory bulb in *Xenopus laevis*, *Brain Res.* **589**:161-163.
Houck, L. D., and Reagan, N. L., 1990, Male courtship pheromones increase female receptivity in a plethodontid salamander, *Anim. Behav.* **39**:729-734.
Kauer, J. S., 2002, On the scents of smell in the salamander, *Nature* **417**:336-342.
Kauer, J. S., and Moulton, D. G., 1974, Responses of olfactory bulb neurones to odour stimulation of small nasal areas in the salamander, *J. Physiol.* **243**:717-737.
Kikuyama, S., Toyoda, F., Ohmiya, Y., Matsuda, K., Tanaka, S., and Hayashi, H., 1995, Sodefrin: A female-attracting peptide pheromone in newt cloacal glands, *Science* **267**:1643-1645.
Kokoros, J. J., and Northcutt, R. G., 1977, Telencephalic afferents of the tiger salamander *Ambystoma tigrinum* (Green), *J. Comp. Neurol.* **173**:613-628.
Lebedev, O. A., and Coates, M. I., 1995, The postcranial skeleton of the Devonian tetrapod *Tulerpeton curtum* Lebedev, *Zool. J. Linn. Soc.* **114**:307-348.
Liman, E. R., and Corey, D. P., 1996, Electrophysiological characterization of chemosensory neurons from the mouse vomeronasal organ, *J. Neurosci.* **16**:4625-4637.

Lindquist, S. B., and Bachmann, M. D., 1982, The role of visual and olfactory cues in the prey catching behavior of the tiger salamander, *Ambystoma tigrinum*, *Copeia* 1982:81–90.
Lowell, W. R., and Flanigan, W. F., Jr., 1980, Marine mammal chemoreception, *Mamm. Rev.* 10:53-59.
Mackay-Sim, A., Duvall, D., and Graves, B. M., 1985, The West Indian manatee (*Trichechus manatus*) lacks a vomeronasal organ, *Brain Behav. Evol.* 27:186-194.
Mackay-Sim, A., and Kubie, J. L., 1981, The salamander nose: A model system for the study for the study of spatial coding of olfactory quality, *Chem. Senses* 6:249-257.
Mackay-Sim, A., Shaman, P., and Moulton, D. G., 1982, Topographic coding of olfactory quality: Odorant-specific patterns of epithelial responsivity in the salamander, *J. Neurophysiol.* 48:584-596.
Münz, H., Claas, B., and Fritzsch, B., 1984, Electroreceptive and mechanoreceptive units in the lateral line of the axolotl *Ambystoma mexicanum*, *J. Comp. Physiol. A* 154:33-44.
Nevison, C. M., Armstrong, S., Beynon, R. J., Humphries, R. E., and Hurst, J. L., 2003, The ownership signature in mouse scent marks is involatile, *Proc. R. Soc. Lond. B Biol. Sci.* 270:1957-1963.
Oelschläger, H. A., and Buhl, E. H., 1985, Development and rudimentation of the peripheral olfactory system in the harbor porpoise *Phocoena phocoena* (Mammalia: Cetacea), *J. Morph.* 184:351-360.
Panchen, A. L., 1991, The early tetrapods: Classification and the shapes of cladograms, in: *Origins of the Higher Groups of Tetrapods: Controversy and Consensus*, H.-P. Schultze and L. Trueb, eds., Cornell University Press, Ithaca, NY, pp. 100-144.
Park, D., and Eisthen, H. L., 2003, Gonadotropin releasing hormone (GnRH) modulates odorant responses in the peripheral olfactory system of axolotls, *J. Neurophysiol.* 90:731-738.
Park, D., McGuire, J. M., Majchrzak, A. L., Ziobro, J. M., and Eisthen, H. L., 2004, Discrimination of conspecific sex and reproductive condition using chemical cues in axolotls (*Ambystoma mexicanum*), *J. Comp. Physiol. A., in press.*
Park, D., and Propper, C. R., 2002, The olfactory organ is activated by a repelling pheromone in the red-spotted newt *Notophthalmus viridescens*, *Korean J. Biol. Sci.* 6:233-237.
Park, D., Zawacki, S. R., and Eisthen, H. L., 2003, Olfactory signal modulation by molluscan cardioexcitatory tetrapeptide (FMRFamide) in axolotls (*Ambystoma mexicanum*), *Chem. Senses* 28:339-348.
Parsons, T. S., 1959, Nasal anatomy and the phylogeny of reptiles, *Evolution* 13:175-187.
Parsons, T. S., 1967, Evolution of the nasal structures in the lower tetrapods, *Am. Zool.* 7:397-413.
Parsons, T. S., 1970, The origin of Jacobson's organ. *Forma Functio* 3:105-111.
Placyk, J. S., Jr., and Graves, B. M., 2002, Prey detection by vomeronasal chemoreception in a plethodontid salamander, *J. Chem. Ecol.* 28:1017-1036.
Rollmann, S. M., Houck, L. D., and Feldhoff, R. C., 1999, Proteinaceous pheromone affecting female receptivity in a terrestrial salamander, *Science* 285:1907-1909.
Sam, M., Vora, S., Malnic, B., Ma, W., Novotny, M. V., and Buck, L.B., 2001, Odorants may arouse instinctive behaviors, *Nature* 412:142.
Schmidt, A., and Roth, G. 1990, Central olfactory and vomeronasal pathways in salamanders, *J. Hirnforsch.* 31:543-553.
Seydel, O., 1895, Über die Nasenhöhle und das Jacobson'sche Organ der Amphibien: Eine vergleichend-anatomische Untersuchung, *Morphol. Jahrb.* 23:453-543.
Shaffer, H. B., 1993, Phylogenetics of model organisms: The laboratory axolotl, *Ambystoma mexicanum*, *Syst. Biol.* 42:508-522.
Shepherd, G. M., 1985, Are there labeled lines in the olfactory pathway?, in: *Taste, Olfaction, and the Central Nervous System*, D. W. Pfaff, Rockefeller University Press, New York, pp. 307-321.
Shoop, C. R., 1960, The breeding habits of the mole salamander, *Ambystoma talpoideum* (Holbrook), in Southeastern Louisiana, *Tulane Stud. Zool. Bot.* 8:65-82.
Singer, A. G., 1991, A chemistry of mammalian pheromones, *J. Steroid Biochem. Mol. Biol.* 39:627-632.
Smith, T. D., Bhatnagar, K. P., Shimp, K. L., Kinzinger, J. H., Bonar, C. J., Burrows, A. M., Mooney, M. P., and Siegel, M. I., 2002, Histological definition of the vomeronasal organ in humans and chimpanzees, with a comparison to other primates, *Anat. Rec.* 267:166-176.
Smith, T. D., Siegel, M. I., and Bhatnagar, K. P., 2001, Reappraisal of the vomeronasal system of catarrhine primates: Ontogeny, morphology, functionality, and persisting questions, *Anat. Rec.* 265:176-192.
Szabo, T., Blähser, S., Denizot, J. -P., and Ravaille-Véron, M., 1991, Projection olfactive primaire extrabulbaire chez certaines poissons téléostéens, *C. R. Acad. Sci. Paris* 312:555-560.
Toyoda, F., Hayakawa, Y., Ichikawa, M., and Kikuyama, S., 1999, Olfactory responses to a female-attracting pheromone in the newt, *Cynops pyrrhogaster*, in: R. E. Johnston, D. Müller-Schwarze, and P. W. Sorensen, eds., *Advances in Chemical Signals in Vertebrates*, Kluwer Academic / Plenum, New York, pp. 607-615.
Verrell, P. A., 1982, The sexual behavior of the red-spotted newt, *Notophthalmus viridescens* (Amphibia: Urodela: Salamandridae), *Anim. Behav.* 30:1224-1236.

Wabnitz, P. A., Bowie, J. H., Tyler, M. J., Wallace, J. C., and Smith, B. P., 1999, Aquatic sex pheromone from a male tree frog, *Nature* **401**:444-445.

Wirsig-Wiechmann, C. R., Houck, L. D., Feldhoff, P. W., and Feldhoff, R. C., 2002, Pheromonal activation of vomeronasal neurons in plethodontid salamanders. *Brain Res.* **952**:335-344.

Wysocki, C. J., Wellington, J. L., and Beauchamp, G. K., 1980, Access of urinary nonvolatiles to the mammalian vomeronasal organ, *Science* **207**:781-783.

Yamamoto, N., and Ito, H., 2000, Afferent sources to the ganglion of the terminal nerve in teleosts, *J. Comp. Neurol.* **428**:355-375.

Yamamoto, K., Kawai, Y., Hayashi, T., Ohe, Y., Hayashi, H., Toyoda, F., Kawahara, G., Iwata, T., and Kikuyama, S., 2000, Silefrin, a sodefrin-like pheromone in the abdominal gland of the sword-tailed newt, *Cynops ensicauda*, *FEBS Lett.* **472**:267-270.

# FROM THE EYE TO THE NOSE: ANCIENT ORBITAL TO VOMERONASAL COMMUNICATION IN TETRAPODS?

Willem J. Hillenius and Susan J. Rehorek*

## 1. INTRODUCTION

Despite its early discovery (Harder, 1694), the precise physiological role of the Harderian gland has remained elusive; it is one of the last vertebrate structures for which a confirmed function cannot yet be confidently ascribed (Webb et al., 1992; Payne, 1994). Found in most tetrapods, the Harderian gland (HG) is an orbitoconjunctival gland that was traditionally assumed to function principally in the lubrication of the cornea and eyelids (e.g., Löwenthal, 1896; Davis, 1929; Walls, 1942; Cohn, 1955). However, a growing body of recent literature suggests that in many tetrapod groups, the HG has a diverse array of extraocular functions as well. In rodents, these include pheromone production (e.g., Thiessen et al., 1976; Payne, 1977, 1979; Harriman and Thiessen, 1985; cf. Thiessen, 1992; Payne, 1994), extraretinal photoreception (e.g., Wetterberg et al., 1972; Pevet et al., 1984; Hugo et al., 1987; Spike et al., 1992), and thermoregulation (e.g., Thiessen, 1977, 1992; Thiessen and Kittrell, 1980; Harriman and Thiessen, 1983; Harlow, 1984). In birds, the HG is implicated in the regional immune response of the upper respiratory tract (e.g., Bang and Bang, 1968; Burns, 1992; Montgomery and Maslin, 1992; Shirama et al., 1996), and in some turtles it contributes to extrarenal salt excretion (Chieffi-Baccari et al., 1992). The HG of caecilians and squamates appears to be involved in the vomeronasal sense (Broman, 1920; Bellairs, 1970; Heller, 1982; Wake, 1985; Rehorek et al., 2000c), and it may also play a role in digestion in some snakes (e.g., Gans, 1974; Saint-Girons, 1982, 1989).

The HG is a compound acinar structure located deep in the anterior portion of the orbit, although in some cases it has expanded medial to or even posterior to the eye (Gans, 1974; Sakai, 1989; Wake, 1985; Saint-Girons, 1988). It opens by one or more ducts into the anterior angle of the orbital conjunctiva (Figure 1). In mammals, the HG

---

* Willem J. Hillenius, College of Charleston, Charleston, South Carolina, 29424. Susan J. Rehorek, Slippery Rock University, Slippery Rock, Pennsylvania, 16257.

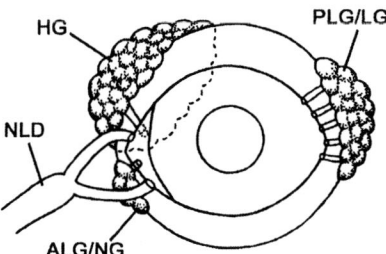

**Figure 1.** Schematic diagram of the orbital glands of a generalized tetrapod. *ALG/NG* anterior lacrimal gland (sauropsids) or "nictitans" gland (mammals), *HG* Harderian gland, *NLD* nasolacrimal duct, *PLG/LG* posterior lacrimal gland (sauropsids) or lacrimal gland (mammals).

was long considered distinct from the so-called nictitans gland (e.g., Miessner, 1900; Sakai, 1981, 1992), but recent embryological studies indicate that these are merely different lobes of the same glandular body (Rehorek et al., submitted).

The histochemical nature of HG secretions varies among taxa, and the gland is often heterogeneous within individuals (Löwenthal, 1892b, 1896; Miessner, 1900; Paule, 1957; Weaker, 1981; Sakai and Van Lennep, 1984; Rehorek, 1997; Payne, 1994). HG secretions are generally serous or mucoserous, but in many mammals, a posterior lobe of the gland is partly or wholly lipoidal (e.g., Cohn, 1955; Paule, 1957; Watanabe, 1980; Weaker, 1981; Sakai, 1981, 1992; Payne, 1994). Small amounts of lipids have also been documented in anurans (Minucci et al., 1989), geckonoid lizards (Rehorek et al., 1997, 2000a; Chieffi-Baccari et al., 2000), and birds (see Burns, 1992; Payne, 1994).

The literature on the HG is burgeoning, and current understanding of HG biology is compiled in several comprehensive reviews (e.g., Sakai, 1981, 1992; Olcese and Wesche, 1989; Webb et al., 1992; Payne, 1994; Rehorek, 1997). However, few studies place these observations in a broader evolutionary context. For example, there has been no attempt to determine which functions might be primitive or derived, or whether similar functions in different taxa represent shared ancestral or independently evolved states. This is particularly unfortunate, since the HG is probably the oldest of the tetrapod orbital glands: it is the only orbital gland common to all major groups of tetrapods (e.g., Walls, 1942; Webb et al., 1992; Payne, 1994), and may thus have been the only one present in ancestral tetrapods. Analysis of possible primitive functions of the HG may therefore provide novel insights into the evolutionary history of the tetrapod lacrimal apparatus.

Moreover, many of these studies also fail to consider HG function in context with other orbital and extraorbital structures, and as a result potential interactions of this gland with such structures have been overlooked. In particular, a peculiar aspect of many of the proposed functions is that the HG exudates are thought to achieve their putative roles only after passing through the nasolacrimal duct (e.g., Broman, 1920; Bellairs and Boyd, 1950; Thiessen, 1992; Schmidt and Wake, 1990; Burns, 1992; Rehorek, 1997, 2000c). Typically, the HG duct opens onto the conjunctiva in the anterior canthus (cf. Figure 1), and in most cases, its fluids are first spread over the cornea, eyelids and nictitating membrane (and are thus usually exposed to the air). In many tetrapods, these fluids then pass into the nasolacrimal duct (NLD) for delivery to their putative targets. For example, in rodents the HG exudates are conveyed to the nostril, and from there are spread over the fur during grooming behavior (Thiessen, 1977, 1992; Coulson and Pinter, 1996). In birds, the NLD is thought to deposit the HG secretions in the nasopharynx, from where they reach the upper respiratory tract (Bang and Bang, 1968; Mueller et al., 1971; Burns, 1992; Shirama et al., 1996). The NLD of squamates and caecilians leads to the vomeronasal organ (e.g., Broman, 1920; Bellairs and Boyd, 1950; Schmidt and Wake, 1990; Rehorek et al., 2000c). In all these cases, delivery of the HG exudates thus appears to involve two distinct duct systems, which are typically separated by an exposure to the ambient environment. Such an arrangement is rare, if not unique in the vertebrate exocrine system, yet it nevertheless appears to be present in many different taxa, from lissamphibians to birds and mammals.

In this paper, we review the morphological and empirical evidence that suggests a potential interaction between the HG and structures downstream of the NLD. In particular, we point out that the vomeronasal organ (VNO) is typically located at or very near the distal terminus of the NLD, and may be much more commonly a recipient of HG secretions than is generally acknowledged. An interaction between HG and VNO has been previously suggested for both squamates and caecilian lissamphibians, but we suggest that the morphological arrangements of the HG, NLD and VNO in other lissamphibians and in at least some mammals are such that communication between these organs is also possible in these taxa. It is therefore conceivable that such an interaction may represent a primitive aspect of HG biology.

## 2. HG - VNO: A POTENTIALLY ANCIENT INTERACTION?

Current evidence for a possible functional interaction between HG and VNO consists of numerous morphological observations and some anecdotal empirical data. This evidence is strongest in squamates (especially snakes) and in caecilian lissamphibians, but the morphological arrangements of the HG, NLD, and VNO of anurans, urodeles, *Sphenodon*, and even of many mammals suggest that a possible interaction between these structures cannot be ruled out. These data are reviewed here.

### 2.1. Squamates

In squamates, the HG is typically the largest of the orbital glands, and in some cases (e.g., snakes) exceeds the eye itself in size (Bellairs and Boyd, 1947; Gans, 1974; Saint-Girons, 1982, 1988; Payne, 1994; Rehorek, 1997, 2003). It is principally a serous structure (Saint-Girons, 1982, 1988, 1989; Rehorek, 1997), although small amounts of lipids have been described in all geckos studied thus far, including pygopodids (Rehorek et al., 1997, 2000a; Chieffi-Baccari et al., 2000). The communication of the HG with the orbit varies among squamates, but in all cases HG fluids are eventually drained via the NLD. Usually, the HG duct retains its connection to the medial canthus, and HG secretions pass into the NLD only after passing over the cornea (Bellairs and Boyd, 1947; Saint-Girons, 1982; Rehorek, 1997). However, in both pygododid geckos and in snakes (in which the eyelids have fused to form a transparent spectacle) the HG duct opens directly into the NLD and bypasses the conjunctival space altogether. The NLD itself still opens into the conjunctival space of pygopods and primitive snakes, but in colubroid snakes this latter connection is vestigial and the NLD conveys almost exclusively HG

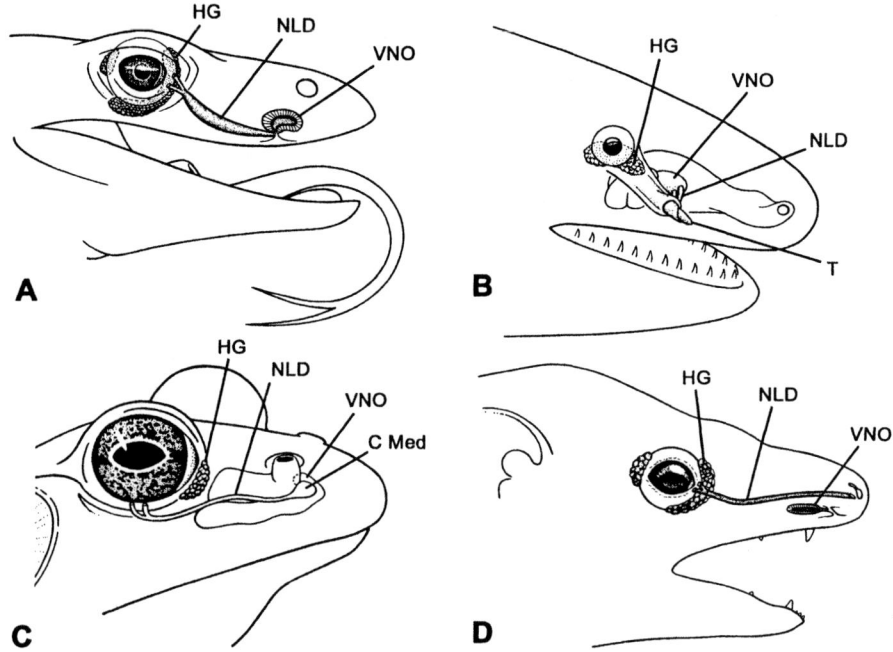

Figure 2. The Harderian gland, nasolacrimal duct, and vomeronasal organ of (A) squamates, (B) caecilians, (C) anurans, and (D) mammals. *C Med* cavum medium of nasal cavity *HG* Harderian gland, *NLD* nasolacrimal duct, *T* tentacle, *VNO* vomeronasal organ. After Hillenius 2000 (A, B, D) and Hillenius et al. 2001 (C).

fluids (Bellairs and Boyd, 1947; Saint-Girons, 1982; Rehorek et al., 2000a, 2003).

In most squamates, the NLD passes to the choanal region, where it opens into the anterior end of the choanal groove, immediately adjacent to the opening of the VNO duct; in snakes, the NLD opens directly in the VNO duct (Figure 2a; Bellairs, 1941; Pratt, 1948; Bellairs and Boyd, 1950). Several empirical studies confirm earlier suggestions that HG secretions reach the VNO via the NLD. For example, fluorescent chemicals injected into the HG of garter snakes appeared shortly afterwards on the tongue (Heller, 1982). In many squamates, the tongue is implicated in collecting and transmitting odor compounds to the VNO lumen (Halpern, 1992; Schwenk, 1993, 1994). Serous compounds observed in the VNO lumen of snakes, geckos and skinks most likely came from the HG (Rehorek et al., 2000c), as the squamate VNO typically lacks intrinsic serous-secreting structures. Finally, radioactive protein tracer compound injected into the HG of snakes appeared shortly afterwards in all crevices of the VNO (Rehorek et al., 2000d).

It is currently not known what function, if any, HG fluids have in the vomeronasal sense. Both Broman (1920) and Bellairs (1970) have suggested that the HG secretions may act as a solvent for scent particles deposited in the VNO by the tongue. Unlike the main (nasal) olfactory epithelium, the VNO lacks Bowman's glands (Halpern, 1987; Eisthen, 1992; Rehorek et al., 2000b, 2000c). The serous Bowman's glands are presumed to provide the protein component of the mucus lining of the nasal olfactory epithelium (Halpern, 1992), and the HG secretions may serve a similar function in the VNO (Rehorek et al., 2000c). Bellairs (1970) also suggested that HG secretions might serve to flush odors from the VNO. However, so far no experiments have been performed to verify whether a functional interaction between these structures exists, or to determine the nature of this interaction.

### 2.2. Caecilians

The arrangement of the orbital conjunctiva, the NLD, and the VNO of caecilian lissamphibians is strikingly different from that of squamates, but these structures nevertheless appear to serve very similar functions. The eyes of caecilians are reduced, and the orbital conjunctiva and eyelids have been modified into a highly specialized tentacular organ (Figure 2b; Sarasin and Sarasin, 1889; Jurgens, 1971; Badenhorst, 1978; Billo and Wake, 1987; Schmidt and Wake, 1990). The HG is the only orbital gland in caecilians (Wake, 1985; Billo and Wake, 1987; Schmidt and Wake, 1990). As in snakes, the gland is comparatively enormous in caecilians; it typically fills the orbit, surrounding the eyeball and in some cases extending into the tentacular region (Wake, 1985). The histochemical character of the caecilian HG has not been reported. HG secretions are released into the base of the tentacle sac, and lubricate the tentacle organ (Sarasin and Sarasin, 1889; Walls, 1942; Badenhorst, 1978; Wake, 1985; Billo and Wake, 1987; Schmidt and Wake, 1990). The caecilian tentacle is considered a tactile and chemosensory organ that samples ambient odor molecules in a manner analogous to that of the squamate tongue. The tentacle is connected to the VNO by means of the tentacle duct, which represents the modified NLD (Sarasin and Sarasin, 1889; Jurgens, 1971; Badenhorst, 1978; Billo and Wake, 1987). It is widely presumed that odor particles are dissolved into the HG secretions that lubricate the tentacle organ, and that these particles are subsequently conveyed to the VNO via the NLD (Parsons, 1967; Jurgens, 1971; Badenhorst, 1978; Wake, 1985; Duellman and Trueb, 1986; Billo and Wake, 1987; Schmidt and Wake, 1990). Chemicals experimentally injected into the tentacle sheath were subsequently found in the lumen of the VNO, and could only have been transported there via the tentacle duct (Schmidt and Wake, 1990). This presumed function of the caecilian HG thus closely resembles that postulated for the gland of squamates.

### 2.3. Anurans

The HG is the only orbital gland in anurans. It is located deep in the medial angle of the orbit, anterior to the eyeball, and opens by a single duct medial to the nictitating membrane (Walls, 1942; Di Matteo et al., 1989; Chieffi et al., 1992). It is principally mucoserous (Di Matteo et al., 1989; Minucci et al., 1990; Chieffi et al., 1992), although Minucci et al. (1989) also report seasonal secretion of small quantities of lipids in female *Bufo viridis*. No empirical data is available about HG function in anurans, but most authors presume that it principally serves in orbital lubrication (Di Matteo et al., 1989; Minucci et al., 1990; Chieffi et al., 1992). The HG secretions are released onto the cornea, and are subsequently drained through the NLD. Distally, the NLD of anurans terminates in the cavum medium of the nasal cavity, between the main cavum principale and the cavum inferior (Figure 2c; Helling, 1938; Trueb, 1970; Jurgens, 1971; Trueb and Cannatella, 1982; Duellman and Trueb, 1986). Droplets of india ink placed on the cornea

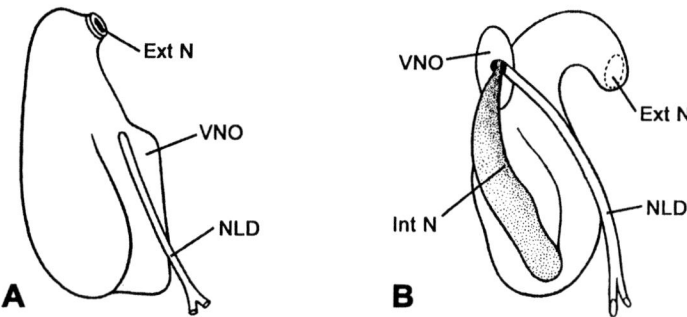

**Figure 3.** Diagrammatic representation of the nasal cavities of (**A**) urodeles (dorsal view) and (**B**) *Sphenodon* (ventral view). *Ext N* external naris, *Int N* internal naris, *NLD* nasolacrimal duct, *VNO* vomeronasal organ. Modified from Jurgens 1971 (**A**), and Bellairs and Boyd 1950 (**B**).

of *Rana* were found to reach the VNO (i.e. the medial diverticulum of the cavum inferior) via the NLD (Hillenius et al., 2001). Consequently, a potential role of these fluids in vomeronasal function can therefore not be excluded.

### 2.4. Urodeles

In contrast to other tetrapods, the orbital glands of urodeles are comparatively poorly differentiated. Typically, a row of glandular acini occurs along the lower eyelid, of which the anterior and posterior ends are often the best developed (Walls, 1942; Chieffi et al., 1992); these are generally considered to represent the Harderian and (posterior) lacrimal glands of other tetrapods, respectively (Walls, 1942; Sakai, 1981; Duellman and Trueb, 1986; Romer and Parsons, 1986). To date, neither the histochemical nature of these glands nor their function has been studied in detail. In most urodeles, the NLD terminates in the anterolateral portion of the lateral diverticulum of the nasal cavity, where it opens into the area of sensory VNO epithelium (Figure 3a; Parsons, 1967; Jurgens, 1971; Duellman and Trueb, 1986; Dawley and Bass, 1988, 1989). Consequently, secretions of the orbital glands can therefore potentially reach the VNO in these animals. However, no empirical studies of the trajectory of these fluids are available. A NLD is absent in *Amphiuma, Siren* and proteid salamanders; notably, the VNO is poorly developed or absent in these same taxa (Jurgens, 1971; Duellman and Trueb, 1986; Eisthen, 2000), which suggests that development of these structures are perhaps correlated.

### 2.5. Sphenodon

The Tuataras, *Sphenodon spp*, are the sister taxon to squamates (Gauthier et al., 1988a), and are therefore very significant to studies on the evolutionary biology of the squamate Harderian gland and VNO. The HG is the only orbital gland in *Sphenodon*, and produces mucoserous secretions (Walls, 1942; Saint-Girons, 1982, 1985, 1988, 1989; Chieffi et al., 1992). Its duct opens medial to the nictitating membrane, and the HG secretions presumably pass over the cornea before draining into the NLD; in this respect, *Sphenodon* resembles typical squamates (Bellairs and Boyd, 1947). Rostrally, the NLD opens in the anterior region of the choana, immediately opposite the VNO, and is connected to the opening of the VNO duct by a semicircular groove along the anterior rim of the choana (Figure 3b; Fuchs, 1908; Schauinsland, 1900; Hoppe, 1934; Malan, 1946; Pratt, 1948). To our knowledge, no empirical information is available for *Sphenodon* on the destination or function of the HG fluids, on the destination of fluids from the NLD or, for that matter, on any aspect of VNO function or odor sampling. Nonetheless, the arrangement of these structures is consistent with a potential interaction between HG and VNO, mediated via the NLD.

## 2.6. Mammals

Mammals typically retain a HG, although this gland is absent in chiropterans, higher primates, and scattered members of other orders (Paule, 1957; Olcese and Wesche, 1989; Payne, 1994). Histochemically, HG secretions are highly variable, ranging from seromucous to lipoidal-porphyrin (Paule, 1957; Krause and McMenamin, 1992; Bodyak and Stepanova, 1994; Payne, 1994). In some cases, different lobes of the same gland may have different secretory products (Löwenthal, 1892a; Weaker, 1981; Sakai and Van Lennep, 1984). Sakai (1981, 1989, 1992) and Sakai and Van Lennep (1984) differentiate between the HG and a "nictitans gland," based only on the histochemical nature of their secretions, defining the HG as an exclusively lipoidal structure. However, this histochemical definition is inconsistent with numerous anatomical considerations, and should be avoided (cf. Bodyak and Stepanova, 1994; Payne, 1994; Buzzell, 1996). Moreover, recent embryological observations indicate that these two structures are merely different lobes of the same glandular body (Rehorek et al., submitted). The HG ducts open in the anterior part of the orbit, either medial or lateral to the nictitating membrane (Paule, 1957; Johnston et al., 1983), and HG exudates usually drain into the NLD after passing over the cornea (Thiessen, 1992; Coulson and Pinter, 1996).

HG function has been examined primarily in rodents, and in these, generally only the lipoidal and porphyrin secretions have been studied. In rodents, the HG has been implicated in a wide range of extraocular functions, including thermoregulation, pheromone secretion and photoreception. HG secretions are delivered to the rostrum via the NLD, and are subsequently spread across the fur by autogrooming, in a manner analogous to the preening behavior of birds (Thiessen, 1977, 1992; Coulson and Pinter, 1996). In at least some rodents, the lipoidal secretions of the HG affect insulatory capacity of the fur coat (Thiessen, 1977; Thiessen and Kittrell, 1980; Kittrell, 1981; Harriman and Thiessen, 1983; Harlow, 1984; Pevet et al., 1984; see Thiessen, 1992; Payne, 1994 for reviews), and also reduce cutaneous water loss (Harriman and Thiessen, 1983; Harlow, 1984). HG secretions are also thought to contain pheromones that affect intraspecific interactions, such as female proceptive behavior, intermale aggression and parent/offspring interactions (e.g., Harriman and Thiessen, 1985; Thiessen and Harriman, 1986; Olcese and Wesche, 1989; Thiessen, 1992; Coulson and Pinter, 1996). Porphyrin compounds in rodent HG secretions are thought to influence photoperiod perception (e.g., Wetterberg et al., 1972; Pevet et al., 1984; Hugo et al., 1987; Spike et al., 1992). Little is known about HG function in other mammals.

Thus far, the mammalian HG has not been implicated in VNO function, but a potential interaction cannot be ruled out. A direct connection between the NLD and the

VNO is absent: unlike the condition in lepidosaurs and lissamphibians, the NLD of most mammals continues rostrally beyond the level of the VNO, and typically opens in the floor of the narial vestibule (Figure 2d; Born, 1879a, b, 1883; De Beer, 1937; Bang and Bang, 1959; Evans and Christensen, 1979; Zeller, 1989). NLD-conveyed fluids are generally discharged onto the rhinarium, the moist exterior of the nose, and, as described above, these fluids are subsequently spread over the fur, at least in rodents. However, in many mammals, the rhinarium is also the primary site for collection of VNO-sensitive odor molecules: odor samples are dissolved in moisture on the rhinarium when the snout is rubbed against the substrate during "nuzzling" behavior, and these samples are then thought to travel to the VNO via the philtrum of the upper lip and the nasopalatine duct (Maier, 1980; Wöhrmann-Repenning, 1980, 1984a-c; Poran et al., 1993a, b; Asher, 1998). It is therefore possible that the HG exudates of these mammals may, as a component of the rhinarial fluids, be involved in the acquisition of odor samples for the VNO, in a manner analogous to that postulated for the HG of squamates and caecilians.

## 2.7. The others: Turtles, Crocodilians and Birds

In these remaining taxa, the NLD and the VNO are apomorphically modified or lost. The HG is typically present, but a VNO-associated function for this gland is not possible.

In turtles, a functional VNO is present (McCotter, 1917; Seydel, 1896; Matthes, 1934; Parsons, 1959, 1970; Okamoto et al., 1996), but unlike that of other tetrapods, it is not enclosed in a distinct diverticulum. Instead, VNO epithelium is spread diffusely in shallow sulci in the floor and ventral walls of the nasal passage (Parsons, 1959, 1970; Gauthier et al., 1988b). This is considered a highly derived feature of turtles (Gauthier et al., 1988b). The NLD is absent (Parsons, 1959; Bellairs and Kamal, 1981), and the secretions of the orbital glands consequently remain in the orbit or flow out over the cheeks (Payne, 1994). Turtles have two orbital glands, one each in the anterior and posterior corners of the orbit. Recently, there has been some confusion with regards to the nomenclature of these glands (e.g., Chieffi-Baccari et al., 1992), but most workers consider the anterior gland to be the HG (Cowan, 1969; Saint-Girons, 1985, 1988, 1989; Rehorek and Hillenius, submitted). This anterior gland produces both serous and mucous secretions (Cowan, 1969; Saint-Girons, 1985, 1988, 1989), but according to Chieffi-Baccari et al. (1992) it also includes some salt-secreting cells.

In contrast, the VNO is absent in both crocodilians and birds, although a transient primordium of the organ appears briefly in early embryonic stages of crocodilians (Matthes, 1934; Parsons, 1959, 1970, 1971; Portmann, 1961; Gauthier et al., 1988b). Accordingly, the NLD fails to reach the rostral portion of the nasal cavity in both taxa, opening instead to the nasopharyngeal region (Matthes, 1934; Parsons, 1959, 1970, 1971; Witmer, 1995). In both taxa, the HG is well developed (Walls, 1942; Slonaker, 1918; Saint-Girons, 1985; Burns, 1992; Payne, 1994). It produces seromucous secretions in crocodilians (Saint-Girons, 1985, 1988, 1989), whereas in birds, these are mucolipoidal rather than serous (Wight et al., 1971; Burns, 1992; Payne, 1994). In addition, the avian HG also contains lymphoid tissues (Bang and Bang, 1968; Mueller et al., 1971; Schramm, 1980; Montgomery and Maslin, 1992; Shirama et al., 1996), and it has been implicated in the immune response of the eye and, via the NLD, of the upper respiratory tract (Bang and Bang, 1968; Montgomery and Maslin, 1992; see Olcese and Wesche, 1989; Burns, 1992; and Payne, 1994 for reviews). HG function in crocodilians is poorly understood.

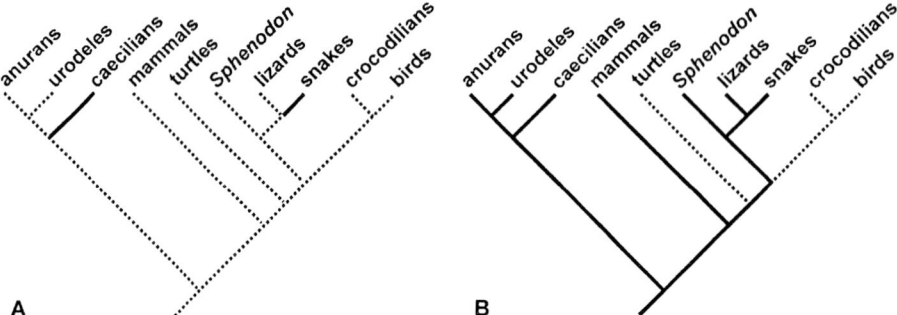

Figure 4. Two scenarios for the origin of the orbital-to-vomeronasal communication. In **A**, this function represents a highly derived condition, independently evolved in snakes and caecilians, presumably in concert with VNO specializations (solid lines). In **B**, at least some aspects of HG – VNO interaction are a synapomorphic attribute of lissamphibians and amniotes, and may date back to the very origin of these structures in tetrapods (solid lines). Cladograms after Gauthier et al. (1988b).

## 3. SUMMARY, WORKING HYPOTHESES, AND CONCLUSIONS

It has long been suspected that HG secretions may play a role in the vomeronasal sense in squamates (esp. snakes) and caecilian lissamphibians. In both cases, the HG is the only possible source of orbital fluids in the NLD, which leads from the orbit directly into the VNO. The specific role of the HG secretions in VNO function remains unknown for both taxa, but it has been suggested in each case that they may act as a solvent for VNO-sensitive odor particles, possibly as an analog to the secretions of the Bowman's glands of the nasal olfactory organ.

The VNO complexes of both snakes and caecilians clearly represent independently derived, apomorphic specialized conditions. The VNO of snakes is characterized by a suite of features (such as the deeply bifurcate tongue, tongue-flicking behavior, and the relative size of the VNO and accessory olfactory bulb) that are more highly developed than in other squamates (e.g., Halpern, 1992; Schwenk, 1993, 1994), while caecilians uniquely possess the highly derived tentacle organ (Wake, 1985; Schmidt and Wake, 1990). Therefore, one possible hypothesis is that the functional association of the HG and the VNO in these taxa represents an apomorphic feature, correlated with the specializations of the VNO apparatus, and evolved independently in these unrelated taxa (Figure 4a).

However, the morphological arrangements of HG, NLD and VNO of more generalized squamates (lizards), *Sphenodon*, anurans, urodeles and mammals, although less direct than those of snakes and caecilians, nevertheless suggest that a functional interaction between HG and VNO may also be present in each of these taxa. Therefore, a possible alternative hypothesis is that at least some aspects of a HG-VNO interaction may represent a symplesiomorphic attribute of lissamphibians and amniotes (Figure 4b). It is worth noting that all three structures, HG, NLD, and VNO, are synapomorphic at the tetrapod level, and all three structures probably first appeared in early tetrapods (Walls, 1942; Parsons, 1967, 1971; Eisthen, 1992; Webb et al., 1992). Therefore, a functional interaction between these structures that dates back to the early evolution of tetrapods is perhaps not so far-fetched. If, for example, HG secretions indeed act as a solvent for odor molecules, such a role may have arisen from inadvertent entrapment of odor particles on a cornea and conjunctiva lubricated by the HG in early tetrapods, perhaps especially as the eyes emerged above the water surface. Once a NLD became established, linking the orbit to the nasal region (e.g., see Schmalhausen, 1958, 1968; Parsons, 1967, 1971), the VNO may have become specialized, at least in part, to analyze this influx of odor compounds. From such an ancestral state, the diverse specializations of the VNO and its odor acquisition mechanisms seen in extant tetrapods could readily have been derived. For instance, the modification of the orbital conjunctiva into a specialized tentacle complex in caecilians would represent a comparatively simple adaptation of this ancestral corneal/conjunctival odor collection mechanism. And the different methods for VNO access among squamates and mammals, including tongue flicking, nuzzling, and flehmen behaviors, could represent independent adaptations of this mechanism to odor collection in a fully terrestrial environment. Presumably, when the corneal/conjunctival mechanism became obsolete in these cases, HG secretions were simply transferred to another surface (tongue or rhinarium) to facilitate odor pick-up. In this scenario, the absence of this feature in turtles, crocodilians and birds constitutes a secondary loss, most likely correlated with modification of the VNO and loss of the NLD in turtles, and with complete loss of the VNO in archosaurs.

These are but two opposite poles of a range of possible scenarios. Current information remains inadequate for more definite statements about the nature or evolutionary history of the interaction between orbital glands and vomeronasal sensory organ. However, future studies of both HG and VNO function should consider the possibility of such an interaction, even in tetrapods such as anurans, urodeles, lizards, and mammals, in which the anatomical arrangements between these organs are more subtle than those of snakes and caecilians.

## 4. REFERENCES

Asher, R. J., 1998, Morphological diversity of anatomical strepsirrhinism and the evolution of the lemuriform toothcomb, *Am. J. Phys. Anthr.* **105**:355-367.
Badenhorst, A., 1978, The development and the phylogeny of the Organ of Jacobson and the tentacular apparatus of *Ichthyophis glutinosus* (Linne), *Ann. Univ. Stellenbosch. Serie 2A* **1**:1-26.
Bang, B. G., and Bang, F. B., 1959, A comparative study of the vertebrate nasal chamber in relation to upper respiratory infections, *Bull. Johns Hopkins Univ.* **104**:107-149.
Bang, B. G., and Bang, F. B., 1968, Localized lymphoid tissues and plasma cells in paraocular and paranasal organ systems in chickens, *Am. J. Path.* **53**:735-751.
Bellairs, A., 1941, Observations on Jacobson's organ and its innervation in *Vipera berus*, *J. Anat.* **76**:167-177.
Bellairs, A., 1970, *The Life of Reptiles*, Universe Books, New York, 590p.
Bellairs, A., and Boyd, J. D., 1947, The lachrymal apparatus in lizards and snakes - I. The brille, the orbital glands, lachrymal canaliculi and origin of the lachrymal duct, *Proc. Zool. Soc. (Lond.)* **117**:81-108.
Bellairs, A., and Boyd, J. D., 1950, The lachrymal apparatus in lizards and snakes - II. The anterior part of the lachrymal duct and its relationship with the palate and with the nasal and vomeronasal organs, *Proc. Zool. Soc. (Lond.)* **120**:167-310.
Bellairs, A., and Kamal, A. M., 1981, The chondrocranium and the development of the skull in Recent reptiles, in: *Biology of the Reptilia*, Vol. 11, C. Gans and T. S. Parsons, eds., Acad, Press, London, pp. 1-263.
Billo, R., and Wake, M. H., 1987, Tentacle development in *Dermophis mexicana* (Amphibia, Gymnophiona) with an hypothesis of tentacle origin, *J. Morph.* **192**:101-111.
Bodyak, N. D., and Stepanova, L. V., 1994, Harderian gland ultrastructure of the Black Sea Bottlenose dolphin (*Tursiops truncatus ponticus*), *J. Morph.* **220**:207-221.

Born, G., 1879a, Die Nasenhöhlen und der Thränennasengang der amnioten Wirbeltiere I, *Morph. Jb.* **5**:62-137.
Born, G., 1879b, Die Nasenhöhlen und der Thränennasengang der amnioten Wirbeltiere II, *Morph. Jb.* **5**:401-429.
Born, G., 1883, Die Nasenhöhlen und der Thränennasengang der amnioten Wirbeltiere III, *Morph. Jb.* **8**:188-232.
Broman, I., 1920, Das Organon Vomero-Nasale Jacobsoni - ein Wassergeruchsorgan! *Arb. anat. Inst. Wiesbaden (Anat. H. Abt. I)* **58**:137-191.
Burns, R. B., 1992, The harderian gland in birds: histology and immunology, in: *Harderian Glands: Porphyrin Metabolism, Behavioral and Endocrine Effects*, S.M. Webb, R.A. Hoffman, M.L. Puig-Domingo, and R.J. Reiter, eds., Springer Verlag, Berlin, pp. 155-163.
Buzzell, G. R., 1996, The Harderian gland: perspectives, *Micr. Res. Tech.* **34**:2-5.
Chieffi, G., Chieffi-Baccari, G., Di Matteo, L., d'Istria, M., Marmorino, C., Minucci, S., and Varriale, B., 1992, The harderian gland of amphibians and reptiles, in: *Harderian Glands: Porphyrin Metabolism, Behavioral and Endocrine Effects*, S.M. Webb, R.A. Hoffman, M.L. Puig-Domingo, and R.J. Reiter, eds., Springer Verlag, Berlin, pp. 91-108.
Chieffi-Baccari, G., Di Matteo, L., and Minucci, S., 1992, The orbital glands of the chelonians *Pseudemys scripta* and *Testudo graeca*: comparative histological, histochemical and ultrastructural investigations, *J. Anat.* **180**:1-13.
Chieffi-Baccari, G., Chieffi, G., Di Matteo, L., Dafnis, D., De Rienzo, G., and Minucci, S., 2000, Morphology of the Harderian gland of the gecko *Tarentola mauritanica*, *J. Morph.* **244**:137-142.
Cohn, S. A., 1955, Histochemical observations on the Harderian gland of the albino mouse, *J. Hist. Cyt.* **3**:342-353.
Coulson, J. O., and Pinter, A. J., 1996, The harderian gland of the northern grasshopper mouse, *Onychomys leucogaster*, *Can. J. Zool.* **74**:1220-1228.
Cowan, F. B. M., 1969, Gross and microscopic anatomy of the orbital glands of *Malaclemmys* and other emydine turtles, *Can. J. Zool.* **47**:723-729.
Davis, F. A., 1929, The anatomy and histology of the eye and orbit of the rabbit, *Trans. Am. Ophth. Soc.* **27**:401-441.
Dawley, E. M., and Bass, A. H., 1988, Organization of the vomeronasal organ in a plethodontid salamander, *J. Morph.* **198**:243-255.
Dawley, E. M., and Bass, A. H., 1989, Chemical access to the vomeronasal organs of a plethodontid salamander, *J. Morph.* **200**:163-174.
De Beer, G. R., 1937, *The Development of the Vertebrate Skull*, Oxford Univ. Press, Oxford, 554p.
Di Matteo, L., Minucci, S., Chieffi-Baccari, G., Pellicciari, C., d'Istria, M., and Chieffi, G., 1989, The Harderian gland of the frog, *Rana esculenta*, during the annual cycle: histology, histochemistry and ultrastructure, *Bas. Appl. Histochem.* **33**:93-112.
Duellman, W. E., and Trueb, L., 1986, *Biology of Amphibians*, McGraw-Hill, New York, 670 p.
Eisthen, H. L., 1992, Phylogeny of the vomeronasal system and of receptor cell types in the olfactory and vomeronasal epithelia of vertebrates, *Micr. Res. Tech.* **23**:1-21.
Eisthen, H. L., 2000, Presence of the vomeronasal system in aquatic salamanders, *Phil. Trans. R. Soc. Lond.* B **355**:1209-1213.
Evans, H. E., and Christensen, G. C., 1979, *Miller's Anatomy of the Dog*, W.B. Saunders, Philadelphia, 1181p.
Fuchs, H., 1908, Untersuchungen über Ontogenie und Phylogenie der Gaumenbildung bei den Wirbeltieren, *Z. Morph. Anthr.* **11**:153-248.
Gans, C., 1974, *Biomechanics: an Approach to Vertebrate Biology*, J.B. Lippincott, Philadelphia, 261p.
Gauthier, J. A., Estes, R., and de Queiroz, K., 1988a, A phylogenetic analysis of Lepidosauromorpha, in: *Phylogenetic Relationships of the Lizard Families*, R. Estes and G. Pregill, eds., Stanford Univ. Press, Stanford, pp. 15-99.
Gauthier, J. A., Kluge, A. G., and Rowe, T., 1988b, Amniote phylogeny and the importance of fossils. *Cladistics* **4**:105-209.
Halpern, M., 1987, The organization and function of the vomeronasal system, *Ann. Rev. Neurosci.* **10**:325-362.
Halpern, M., 1992, Nasal chemical senses in reptiles: structure and function, in: *Biology of the Reptilia*, Vol. 18, C. Gans and D. Crew, eds., Univ. Chicago Press, Chicago, pp. 424-532.
Harder, J. J., 1694, Glandula nova lachrymalis una cum ductu excretorio in cervis et damis, *Acta Erud. Lipsiae* 49-52.
Harlow, H. J., 1984, The influence of harderian gland removal and fur lipid removal on heat loss and water flux to and from the skin of muskrats (*Ondatra zibeticus*), *Physiol. Zool.* **57**:349-356.
Harriman, A. E., and Thiessen, D. D., 1983, Removal of Harderian exudates by sandbathing contributes to osmotic balance in mongolian gerbils, *Physiol. Behav.* **31**:317-323.

Harriman, A. E., and Thiessen, D. D., 1985, Harderian letdown in male Mongolian gerbils (*Meriones unguiculatus*) contributes to proceptive behavior, *Horm. Behav.* **19**:213-219.

Heller, S. B., 1982, *The Role of the Vomeronasal Organ in Garter Snake Aggregation and Shelter Selection*, PhD Diss., SUNY Brooklyn, NY, 143p.

Helling, H., 1938, Das Geruchsorgan der Anuren, vergleichend-morphologisch betrachtet, *Z. ges. Anat.* **108**:587-643.

Hillenius, W. J., 2000, Septomaxilla of nonmammalian synapsids: soft-tissue correlates and a new functional interpretation, *J. Morph.* **245**:29-50.

Hillenius, W. J., Watrobski, L. K., and Rehorek, S. J., 2001. Passage of tear duct fluids through the nasal cavity of frogs, *J. Herp.* **35**:701-704.

Hoppe, G., 1934, Das Geruchsorgan von *Hatteria punctata*, *Z. Anat. EntwGesch.* **102**:434-461.

Hugo, J., Krijt, J., Vokura, M., and Janousek, V., 1987, Secretory response to light in rat Harderian gland: possible photoreceptive role of Harderian porphyrin, *Gen. Physiol. Biophysiol.* **6**:401-404.

Johnston, H. S., McGadey, J., Thompson, G. G., Moore, M. R., and Payne, A. P., 1983, The Harderian gland, its secretory duct and porphyrin content in the mongolian gerbil (*Meriones unguiculatus*), *J. Anat.* **137**:615-630.

Jurgens, J. D., 1971, The morphology of the nasal region of Amphibia and its bearing on the phylogeny of the group, *Ann. Univ. Stellenbosch, Serie A (Sool.)* **46**:1-146.

Kittrell, E. M. W., 1981, *The Harderian Gland and Thermoregulation*, PhD Diss., UT Austin, TX, 156p.

Krause, W. J., and McMenamin, P. G., 1992, Morphological observations on the Harderian gland of the North American opossum (*Didelphis virginiana*), *Anat. Embr.* **186**:145-152.

Löwenthal, N., 1892a, Notiz über die Harder'sche Drüse des Igels, *Anat. Anz.* **7**:48-54.

Löwenthal, N., 1892b, Beitrag zur Kenntnis der Harder'schen Drüse bei den Säugetieren, *Anat. Anz.* **7**:546-556.

Löwenthal, N., 1896, Drüsenstudien, *Int. Monats. Anat. Physiol.* **13**:27-65.

Maier, W., 1980, Nasal structures in Old and New World primates, in: *Evolutionary Biology of the New World Monkeys and Continental Drift*, R. Ciochon, L. and A.B. Chiarelli, eds., Plenum, New York, pp. 219-241.

Malan, M. E., 1946, Contributions to the comparative anatomy of the nasal capsule and the organ of Jacobson of the Lacertilia, *Ann. Univ. Stellenbosch* **24**:69-137.

Matthes, E., 1934, Geruchsorgan, in: *Handbuch der vergleichenden Anatomie der Wirbeltiere*, L. Bolk, E. Göppert, E. Kallius, and W. Lubosch, eds., Urban & Schwarzenberg, Berlin, pp. 879-948.

McCotter, R. E., 1917, The vomero-nasal apparatus in *Chrysemys punctata* and *Rana catesbiana*, *Anat. Rec.* **13**:51-67.

Miessner, H., 1900, Die Drüsen des dritten Augenlides einiger Säugethiere, *Arch. wiss. prakt. Tierheilk.* **26**:122-154.

Minucci, S., Chieffi-Baccari, G., Di Matteo, L., and Chieffi, G., 1989, A sexual dimorphism of the Harderian gland of toad, *Bufo viridis*, *Bas. Appl. Histochem.* **33**:299-310.

Minucci, S., Chieffi-Baccari, G., Di Matteo, L., Marmorino, C., d'Istria, M., and Chieffi, G., 1990, Influence of light and temperature on the secretory activity of the Harderian gland of the green frog, *Rana esculenta*, *Comp. Biochem. Physiol.* **95A**:249-252.

Montgomery, R. D., and Maslin, W. R., 1992, A comparison of the gland of Harder response and head-associated Lymphoid Tissue (HALT) Morphology in chickens and turkeys, *Avian Dis.* **36**:755-759.

Mueller, A. P., Sato, K., and Glick, B., 1971, The chicken lacrimal gland, gland of Harder, caecal tonsil and accessory spleen as sources of antibody-producing cells, *Cell. Imm.* **2**:140-152.

Okamoto, K., Tokumitsu, Y., and Kashiwayanagi, M., 1996, Adenyl Cyclase activity in turtle vomeronasal and olfactory epithelium, *Biochem. Biophys. Res. Comm.* **220**:98-101.

Olcese, J., and Wesche A., 1989, The Harderian gland, *Comp. Biochem. Physiol.* **93A**:655-665.

Parsons, T. S., 1959, Studies on the comparative embryology of the reptilian nose, *Bull. MCZ* **120**:104-277.

Parsons, T. S., 1967, Evolution of the nasal structure in the lower tetrapods, *Am. Zool.* **7**:397-413.

Parsons, T. S., 1970, The nose and Jacobson's organ, in *Biology of the Reptilia*, Vol. 2, C. Gans, ed., Acad. Press, New York, pp. 99-191.

Parsons, T. S., 1971, Anatomy of nasal structures from a comparative viewpoint, in: *Handbook of Sensory Physiology*, L. M. Beidler, ed., Springer Verl., Berlin, pp. 1-28.

Paule, W. J., 1957, *The Comparative Histochemistry of the Harderian Gland*, PhD Diss., Ohio St. Univ., Columbus, 106p.

Payne, A. P., 1977, Pheromonal effects of Harderian gland homogenates on aggressive behaviour in the hamster, *J. Endocr.* **73**:191-192.

Payne, A. P., 1979, The attractiveness of Harderian gland smears to sexually naive and experienced male golden hamsters, *Anim. Beh.* **27**:897-904.

Payne, A. P., 1994, The Harderian gland, a tercentennial review, *J. Anat.* **185**:1-49.

Pevet, P., Heth, G., Hiam, A., and Nevo, E., 1984, Photoperiod reception in the blind mole rat (*Spalax ehrenbergi* Nehring): involvement of the harderian gland, atrofied eyes and melatonin, *J. Exp. Zool.* **232**:41-50.

Poran, N. S., Tripoli, R., and Halpern, M., 1993a, Nuzzling in the gray short-tailed opossum II: familiarity and individual recognition, *Physiol. Behav.* **53**:969-973.

Poran, N. S., Vandoros, A., and Halpern, M., 1993b, Nuzzling in the gray short-tailed opossum I: delivery of odors to vomeronasal organ, *Physiol. Behav.* **53**:959-967.

Portmann, A., 1961, Sensory organs: skin, taste and olfaction, in: *Biology and Comparative Physiology of Birds*, A. J. Marshall, ed., Acad. Press, New York, pp. 37-48.

Pratt, C. W. M., 1948, The morphology of the ethmoidal region of *Sphenodon* and lizards, *Proc. Zool. Soc. Lond.* **118**:171-201.

Rehorek, S. J., 1997, Squamate harderian gland: an overview, *Anat. Rec.* **248**:301-306.

Rehorek, S. J., Firth, B.T., and Hutchison, M.N., 1997, Morphology of the harderian gland of some australian geckos, *J. Morph.* **231**:253-259.

Rehorek, S. J., Firth, B. T., and Hutchison, M. N., 2000a, Can an orbital gland function in the vomeronasal sense? A study of the pygopodid Harderian gland, *Can. J. Zool.* **78**:648-654.

Rehorek, S. J., Firth, B. T., and Hutchison, M. N., 2000b, The structure of the nasal chemosensory system in squamate reptiles. 1. The olfactory organ, with special reference to olfaction in geckos, *J. Biosci.* **25**:173-179.

Rehorek, S. J., Firth, B. T., and Hutchison, M. N., 2000c, The structure of the nasal chemosensory system in squamate reptiles. 2. Lubricatory capacity of the vomeronasal organ, *J. Biosci.* **25**:181-190.

Rehorek, S. J., Hillenius, W. J., Quan, W. and Halpern, M., 2000d, Passage of the Harderian gland secretions to the vomeronasal organ of *Thamnophis spp*. (Serpentes, Colubridae), *Can. J. Zool* **78**:1284-1288.

Rehorek, S. J., Halpern, M., Firth, B. T., and Hutchison, M. N., 2003, The Harderian gland of two species of snakes: *Pseudonaja textilis* (Elapidae) and *Thamnophis sirtalis* (Colubridae), *Can. J. Zool* **81**:357-363.

Rehorek, S. J. and Hillenius, W. J., submitted, Homology of the tetrapod Harderian gland, *J Morph.*, abstracts for ICVM 7.

Rehorek, S. J., Hillenius, W. J., Sanjur, J. and Chapman, N., submitted, One gland, two lobes: organogenesis of the "Harderian" and "nictitans" glands of two deer, *Anat. Embr.*

Romer, A. S., and Parsons, T. S., 1986, *The Vertebrate Body*, 6th ed., Saunders College, Philadelphia. 678p.

Saint-Girons, H, 1982, Histologie comparée des glandes orbitaires de lépidosauriens, *Ann. Sci. Nat. Zool. Paris, 12e série* **4**:171-191.

Saint-Girons, H., 1985, Histologie des glandes orbitaires des crocodiles et des tortues, et comparison avec les lépidosauriens, *Ann. Sci. Nat. Zool. Paris, 13e série* **7**:249-264.

Saint-Girons, H., 1988, Les glandes céphaliques exocrines des reptiles I - Données anatomiques et histologiques, *Ann. Sci. Nat. Zool. Paris, 13e série* **9**:221-255.

Saint-Girons, H., 1989, Les glandes céphaliques exocrines des reptiles II - Considérations fonctionelles et évolutives, *Ann. Sci. Nat. Zool. Paris, 13e série* **10**:1-17.

Sakai, T., 1981, The mammalian harderian gland: morphology, biochemistry function and phylogeny, *Arch. Histol. Japan* **44**:299-333.

Sakai, T., 1989, Major ocular glands (harderian gland and lacrimal gland) of the musk shrew (*Suncus murinus*) with a review on the comparative anatomy and histology of the mammalian lacrimal glands, *J. Morph.* **201**:39-57.

Sakai, T., 1992, Comparative anatomy of mammalian harderian glands, in: *Harderian Glands: Porphyrin Metabolism, Behavioral and Endocrine Effects*, S.M. Webb, R.A. Hoffman, M.L. Puig-Domingo, and R.J. Reiter, eds., Springer Verl., Berlin, pp. 7-23.

Sakai, T., and Van Lennep, E. W., 1984, The harderian gland in australian marsupials, *J. Mamm.* **65**:159-162.

Sarasin, P., and Sarasin, F., 1889, Zur Entwicklungsgeschichte und Anatomie der Ceylonesischen Blindwühle *Ichthyophis glutinosus* (*Epicrium glutinosum* Aut.), in: *Ergebnisse Naturwissenschaftlicher Forschungen auf Ceylon in den Jahren 1884-1886. Zweiter Band*, Kreidel's Verlag, Wiesbaden pp. 153-263.

Schauinsland, H., 1900, Weitere Beiträge zur Entwicklungsgeschichte der Hatteria. Skelettsystem, schalleitender Apparat, Hirnnerven etc., *Arch. mikr. Anat. Entw.* **56**:747-867.

Schmalhausen, I. I., 1958, Slezno-nosovoy protok y septomaxillare chvostatich amphibiy (the nasolacrimal duct and septomaxillare of urodele amphibians), *Zool. Zh.* **37**:570-583 (In Russian).

Schmalhausen, I. I., 1968, *The Origin of Terrestrial Vertebrates*, Acad. Press, New York, 314p.

Schmidt, A., and Wake, M. H., 1990, Olfactory and vomeronasal systems of Caecilia, *J. Morph.* **205**:255-268.

Schramm, U., 1980, Lymphoid cells in the Harderian gland of birds, *Cell Tissue Res.* **205**:85-94.

Schwenk, K., 1993, The evolution of chemoreception in squamate reptiles: a phylogenetic approach, *Brain Behav. Evol.* **41**:124-137.

Schwenk, K., 1994, Why snakes have forked tongues, *Science* **263**:1573-1577.

Seydel, O. 1896. Über die Nasenhöhle und das Jacobson'sche Organ der Land- und Sumpfschildkröten, *Fests. 70. Geb. Gegenb.* **2**:385-486.
Shirama, K., Satoh, T., Kitamura, T., and Yamada, J., 1996, The avian Harderian gland: morphology and immunology, *Micr. Res. Tech.* **34**:16-27.
Slonaker, J. R., 1918, A physiological study of the anatomy of the eye and its accessory parts of the English sparrow (*Passer domesticus*), *J. Morph.* **31**:351-459.
Spike, R. C., Payne, A. P., and Moore, M. R., 1992, Porphyrins and their possible significance in Harderian glands, in: *Harderian Glands: Porphyrin Metabolism, Behavioral and Endocrine Effects*, S.M. Webb, R.A. Hoffmann, M.L. Puig-Domingo, and R.J. Reiter, eds., Springer Verl., Berlin, pp. 165-193.
Thiessen, D. D., 1977, Thermoenergetics and the evolution of pheromone communication, in: *Progress in Psychobiology and Physiological Psychology*, J. M. Sprague and A. N. Epstein, eds., Acad. Press, New York, pp. 91-191.
Thiessen, D. D., 1992, The function of the harderian gland in the Mongolian gerbil, *Meriones unguiculatus*, in: *Harderian Glands: Porphyrin Metabolism, Behavioral and Endocrine Effects*, S.M. Webb, R.A. Hoffmann, M.L. Puig-Domingo, and R.J. Reiter, eds., Springer Verl., Berlin, pp. 127-140.
Thiessen, D.D., Clancy, A., and Goodwin, M., 1976, Harderian gland pheromone in the Mongolian Gerbil, *Meriones unguiculatus*, *J. Chem. Ecol.* **2**:231-238.
Thiessen, D. D., and Harriman, A. E., 1986, Harderian gland exudates in the male *Meriones unguiculatus* regulate female proceptive behavior, aggression, and investigation, *J. Comp. Psych.* **100**:85-87.
Thiessen, D. D., and Kittrell, E. M. W., 1980, The Harderian gland and thermoregulation in the gerbil (*Meriones unguiculatus*), *Physiol. Behav.* **24**:417-427.
Trueb, L., 1970, Evolutionary relationships of casque-headed tree frogs with co-ossified skulls (family Hylidae), *Univ. Kansas Publ. Mus. Nat. Hist.* **18**:547-716.
Trueb, L., and Cannatella, D. C., 1982, The cranial osteology and hyolaryngeal apparatus of *Rhinophrynus dorsalis* (Anura: Rhynophrynidae) with comparisons to recent pipid frogs, *J. Morph.* **171**:11-40.
Wake, M. H., 1985, The comparative morphology and evolution of the eyes of caecilians (Amphibia: Gymnophiona). *Zoomorph.* **105**:277-295.
Walls, G. L., 1942, The vertebrate eye and its adaptive radiation, *Bull. Cranbrook Inst. Sci.* **19**:1-785.
Watanabe, M., 1980, An autoradiographic, biochemical, and morphological study of the Harderian gland of the mouse, *J. Morph.* **163**:349-365.
Weaker, F. J. 1981, Light microscopic and ultrastructural features of the Harderian gland of the nine-banded armadillo, *J. Anat.* **133**:49-65.
Webb, S. M., Hoffman, R.A., Puig-Domingo, M. L., and Reiter, R. J., eds., 1992, *Harderian Glands: Porphyrin Metabolism, Behavioral and Endocrine Effects*, Springer Verl., Berlin, 325p.
Wetterberg, L., Ulrich, R., and Yuwiler. A., 1972, Light, the Harderian gland and the rodent pineal, in: *Proceedings of the International Congress of Endocrinology*, Exerpta Medica, Amsterdam, pp. 268-272.
Wight, P. A. L., Mackenzie, G. M., Rothwell, B., and Burns, R. B., 1971, The Harderian glands of the domestic fowl, *J. Anat.* **110**:323-333.
Witmer, L. M., 1995, Homology of facial structures in extant archosaurs (birds and crocodilians), with special reference to paranasal pneumaticity and nasal conchae, *J. Morph.* **225**:269-327.
Wöhrmann-Repenning, A., 1980, The relationship between Jacobson's organ and the oral cavity in a rodent, *Zool. Anz.* **204**:391-399.
Wöhrmann-Repenning, A., 1984a, Phylogenetische Aspekte zur Topographie der Jacobsonschen Organe und der Ductus nasopalatini bei Insectivora, Primates, *Tupaia* und *Didelphis*, *Anat. Anz.* **157**:137-149.
Wöhrmann-Repenning, A., 1984b, Vergleichend anatomische Untersuchungen am Vomeronasalkomplex und am rostralen Gaumen verschiedener Mammalia. Teil I. *Gegenb. morph. Jb* **130**:501-530.
Wöhrmann-Repenning, A., 1984c, Vergleichend anatomische Untersuchungen am Vomeronasalkomplex und am rostralen Gaumen verschiedener Mammalia. Teil II, *Gegenb. morph. Jb.* **130**:609-637.
Zeller, U., 1989, Die Entwicklung und Morphologie des Schädels von *Ornithorhynchus anatinus* (Mammalia: Prototheria: Monotremata), *Abh. Senckenb. naturf. Ges.* **545**:1-188.

# PREY CHEMICAL SIGNAL TRANSDUCTION IN THE VOMERONASAL SYSTEM OF GARTER SNAKES

Mimi Halpern, Angel R. Cinelli, and Dalton Wang[*]

## 1. INTRODUCTION

The olfactory and vomeronasal (VN) systems are the major nasal chemosensory systems involved in prey detection in terrestrial vertebrates. Arguably, the garter snake vomeronasal organ prey chemical recognition system is the most completely characterized signal transduction pathway in the nasal chemical senses. Garter snakes have proven to be particularly good subjects for these studies because they have exceptionally well developed VN systems, their prey preferences are present at birth, prior to any postnatal feeding experience, and they typically have restricted diets that are difficult to modify. In addition, snakes use their tongues to deliver odorants to their VN organs (Halpern and Kubie, 1980), thereby providing the researcher with an observable correlate to VN activation.

The peripheral organ for the olfactory system is the olfactory sensory epithelium, located in the dorsal-posterior portion of the nasal cavity. This epithelium, composed of several different types of cells, contains bipolar neurons (also known as olfactory receptor cells) whose axons extend to the brain and terminate in the main olfactory bulb. The VN organ is also situated in the periphery, and similarly contains several different types of cells, the most numerous being bipolar neurons whose axons terminate in the accessory olfactory bulb. The VN organ opens into the roof of the mouth via a very narrow channel, the VN duct. In snakes the tongue delivers odorants to the opening of the VN duct and, by a mechanism at present not understood, these odorants reach the dendritic tips of the bipolar neurons (Halpern and Kubie, 1980).

The studies described herein used garter snakes belonging to the species *Thamnophis sirtalis*. Most of these snakes have a diet that includes primarily earthworms and fish,

---

[*] SUNY Downstate Medical Center, 450 Clarkson Avenue, Brooklyn NY 11203

although some eat small amphibians as well. The studies described in this paper all employ earthworm products.

## 2. BEHAVIORAL AND ANATOMICAL STUDIES

Early studies demonstrated that the vomeronasal system was critical for snakes to make discriminated responses to earthworm wash. Using the technique developed by Wilde and Burghardt, snakes were presented with cotton swabs dipped in earthworm extract or distilled water. The snakes responded to the earthworm-soaked swabs by attacking the swab, but did not attack swabs soaked in distilled water. Complete, bilateral vomeronasal nerve lesions resulted in a loss of this discriminated response, but complete bilateral olfactory nerve lesions or sham lesions were without deleterious effect (Halpern and Frumin, 1979).

Garter snakes can be trained to correctly follow trails of earthworm extract in a four-choice maze. Snakes are most accurate following high concentration trails and increase their tongue-flick rate as a function of earthworm extract concentration (Kubie and Halpern, 1978). Vomeronasal nerve lesions, but not olfactory nerve lesions, result in a loss of accurate trail following and a loss of increased tongue-flicking in response to trail odor concentrations (Kubie and Halpern, 1979).

Snakes require direct lingual contact with the source of the odor to accurately follow prey trails. They are able to accurately follow a dry trail, but are unable to follow a trail removed from direct lingual contact (Kubie and Halpern, 1978). However, airborne odors are discriminable (Begun et al., 1988) and result in increased tongue-flick rates (Halpern and Kubie, 1983). This increase in tongue-flicking to airborne odorants depends primarily on a functional olfactory system (Halpern et al., 1997). Thus, both vomeronasal and olfactory systems of snakes subserve the function of increasing tongue-flick rates in response to nonvolatile and volatile prey chemicals.

The anatomical substrate for tongue flicking in response to prey odors has been demonstrated using tract-tracing techniques. The neurons of the accessory olfactory bulb project their axons to the nucleus sphericus, which projects to the medial amygdala. The neurons of the main olfactory bulb project to the lateral pallium, which projects to the medial amygdala. The latter projects, in turn, to the lateral hypothalamus. The lateral hypothalamus projects to the hypoglossal nucleus, which controls the tongue musculature. Thus, both olfactory and vomeronasal information are relayed to the medial amygdala which provides access to the control circuitry for tongue flicking (Martínez-Marcos et al., 2001; Martínez-Marcos et al., 2002).

## 3. MODEL OF SIGNAL TRANSDUCTION

Once inside the VNO, how is a chemical signal transduced into a neural message? A working model of the transduction process, based on currently available data is presented in Figure 1.

In order to test the accuracy of this model, each element in the pathway needed to be demonstrated: ligand identification, characterization and purification; specific ligand binding; sequencing, cloning and activity of the recombinant ligand, the presence of G-

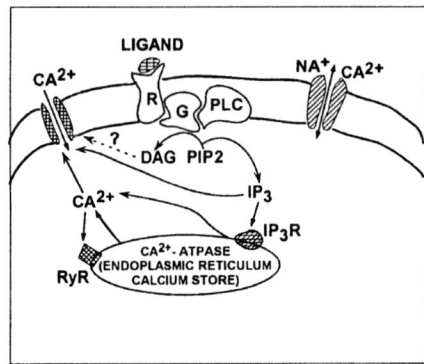

Figure 1. Model of signal transduction in the snake VN system. Ligand binding to a G-protein (G)-coupled receptor (R) activates a phosphatidyl inositol-specific phospholipase C (PLC) which, in turn, hydrolyzes phosphatidyl inositol 4,5-bisphosphate ($PIP_2$) producing diacylglyserol (DAG) and 1,4,5-inositol trisphosphate ($IP_3$). $IP_3$ acts directly on $IP_3$ receptors on the smooth endoplasmic reticulum to release calcium ($Ca^{2+}$) from intracellular stores and on an $IP_3$- sensitive $Ca^{2+}$ channel in the cell membrane allowing calcium influx from the extracellular space. The elevated levels of intracellular $Ca^{2+}$ results in activation of the ryanodine receptor (RyR) on the membrane of the endoplasmic reticulum, resulting in additional $Ca^{2+}$ release from intracellular stores, a phenomenon known as calcium-induced calcium release (CICR). Intracellular $Ca^{2+}$ levels return to prestimulation levels by efflux of $Ca^{2+}$ and influx of sodium ($NA^+$) through a $NA^+/ Ca^{2+}$ exchanger.

proteins and their coupling to the receptor; identification of second messengers, ligand-induced increases in intracellular $Ca^{2+}$, which may give rise to the receptor potential, and their possible correlation.

## 4. PURIFICATION OF CHEMOATTRACTANT

A 20 kDa protein was isolated from earthworm electric shock secretion (ESS), by use of gel filtration on a Bio-gel P-2 column (Figure 2) and polyacrylamide slab gel electrophoresis under nondenaturing conditions (Jiang et al., 1990; Figure 3). The bioassay used to identify the active component in ESS involved placing 50-100 µl of the fractionated material on an artificial earthworm and observing the snake attack the worm. Control stimuli were the elutants used in the separation. Of the multiple protein bands resolved on the polyacrylamide slab gel one, band 7, contained virtually all of the chemoattractive activity. The protein in this band was found to have an isoelectric point of approximately 4.0, and a molecular mass of 20 kDa. The purified protein, ES20, had approximately 100 times more specific chemoattractant activity than crude ESS. This activity was abolished following proteolytic digestionb (Jiang et al., 1990). The purified ES20 protein was partially sequenced from the N-terminal. The resulting 15 amino acid residues were used to design degenerative oligodeoxynucleotide probes and used in RT-PCR and in screening an earthworm cDNA library that was constructed in sense orientation. A gene was cloned from a polymerase chain reaction as well as from the cDNA library screening and expressed in an expression host, *E. coli*. The recombinant protein was fully active in the bioassay (Liu et al., 1997).

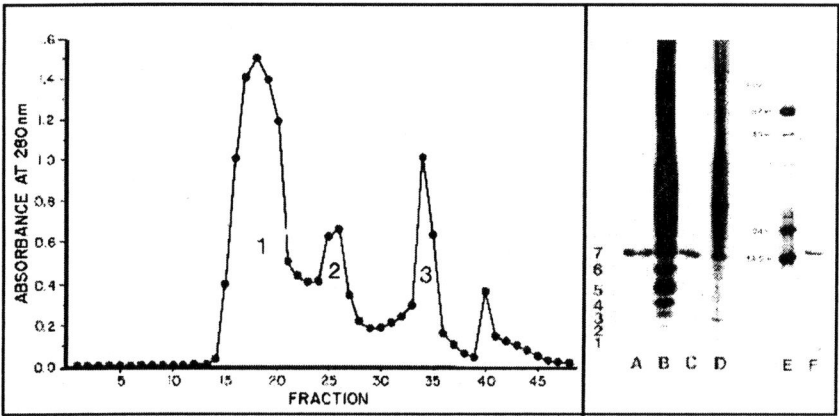

**Figure 2** (left). Permeation chromatography of ESS on Bio-Gel P-2 column. Three major peaks were observed. All of the chemoattractive activity for snakes was in Peak 1, none in peaks 2 or 3. Peak 3 contained alarm pheromone for earthworms. **Figure 3** (right). Polyacrylamide gel electrophoresis of ESS (B and D), purified chemoattractant (A) on non-denaturing gel; protein markers (E) and purified chemoattractant (F) on SDS-denaturing gel. From Jiang et al. (1990) with permission.

## 5. CHEMOATTRACTANT RECEPTORS ARE G-PROTEIN COUPLED RECEPTORS

ES20 bound specifically to VN sensory epithelial membranes in a reversible and saturable manner with an apparent dissociation constant of 0.3μm and a $B_{max}$ of 0.4 nmol/mg protein (Figure 4; Jiang et al., 1990). Immunowestern blots revealed that the snake VN epithelium contained the α subunits of $G_i$, $G_o$, $G_s$ and immunohistochemistry of the epithelium confirmed the presence of $G_i$ and $G_o$ subunits[12]. The affinity of the ES20 receptor was decreased by GTPγS (Figure 5), which is known to activate G-

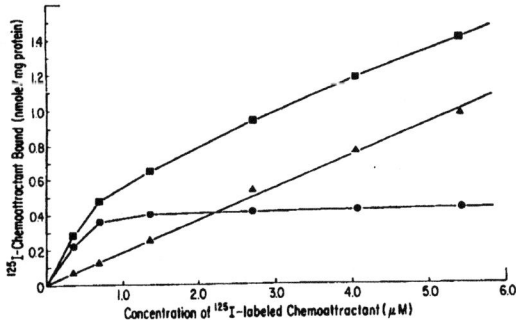

**Figure 4.** Binding curves for $^{125}$I-ES20 to its receptor on VN sensory epithelial membrane. ▲___▲, nonsaturable binding, ■___■, total binding, ●___●, saturable binding. From Jiang et al. (1990) with permission.

proteins, suggesting that the ES20 receptors are coupled to G-proteins. This interpretation was supported by a reduction of ADP-ribosylation of G-proteins. As shown in Figure 6, ES20 binding to its receptor caused an ES20 concentration-dependent reduction in ADP-ribosylation of G-proteins catalyzed by pertussis toxin (PTX). Thus, the chemoattractant receptors are coupled to $G_i/G_o$ proteins (Luo et al., 1994).

## 6. GENERATION OF SECOND MESSENGERS BY CHEMOATTRACTANT RECEPTOR BINDING

ES20-receptor binding increased intracellular levels of $IP_3$ (Figure 7) and decreased intracellular levels of cAMP (Figure 8; Luo et al., 1994), suggesting that the ES20-induced signal transduction pathways involved both PTX-sensitive and insensitive G-proteins. ES20-receptor binding activates PTX-sensitive G-proteins which in turn activate phospholipase C (PLC). PLC hydrolyzes $PIP_2$ to generate the second messengers DAG and $IP_3$. $IP_3$ mobilizes cytosolic calcium (see below). On the other hand, the activated PTX-insensitive G-proteins activate adenylate cyclase (AC) which regulates the levels of cAMP in VN sensory epithelium. The activity of VN AC is sensitive to $Ca^{2+}$ (Wang et al., 1997). A gene encoding adenylate cyclase was cloned and referred to as $AC_{VN}$ (Liu et al., 1998). It showed a high degree of homology to the $Ca^{2+}$-sensitive type VI AC.

## 7. CHEMOATTRACTANT EVOKES ELECTICAL POTENTIONALS IN VN MEMBRANES AND THE AOB

When applied to the VN epithelium of a garter snake from which activity was recorded in the AOB, ES20 resulted in a concentration-dependent increase in firing of AOB mitral cells (Jiang et al., 1990). Dialysis of $IP_3$ into patch-clamped VN receptor cells resulted in an inward, depolarizing current similar to that observed with ESS, which was blocked by ruthenium red (Figure 9; Taniguchi et al., 2000). Dialysis of cAMP into VN receptor cells had no effect (Figure 10; Taniguchi et al., 2000).

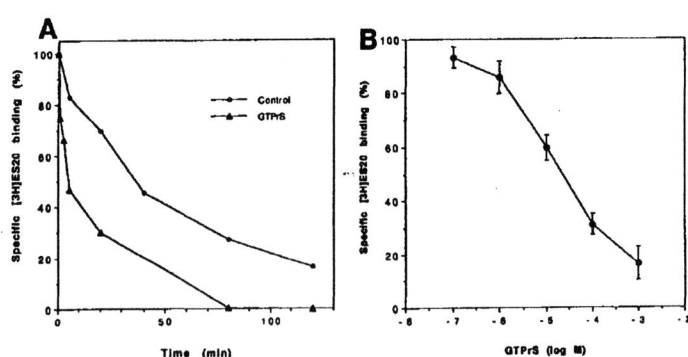

Figure 5. Effect of GTPγS on the binding of [$^3$H]ES20 to its receptors in VN membranes. GTPγS reduced the affinity of [$^3$H]ES20 to its receptor (A) in a concentration-dependent manner (B). From Luo et al. (1994) with permission.

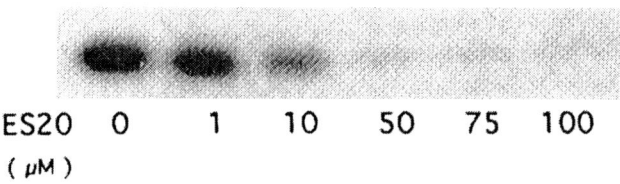

**Figure 6.** Effect of ES20 on ADP-ribosylation of 41-kDA G-protein in VN membranes by pertussis toxin. From Luo et al. (1994) with permission.

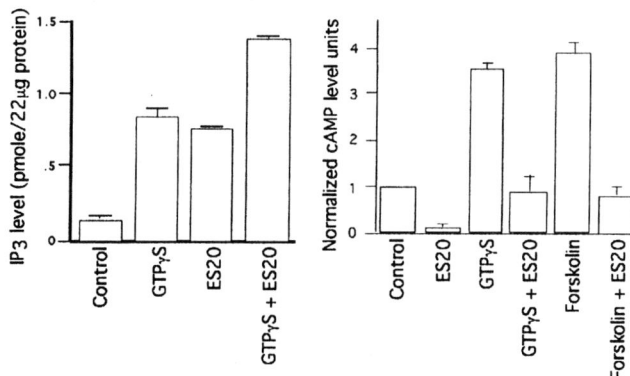

**Figure 7** (left). Effect of ES20 and GTPγS on intracellular $IP_3$ in VN sensory epithelium. **Figure 8** (right). Effect of ES20, GTPγS and Forskolin on generation of cAMP in VN membranes. Drawn from data in Luo et al. (1994).

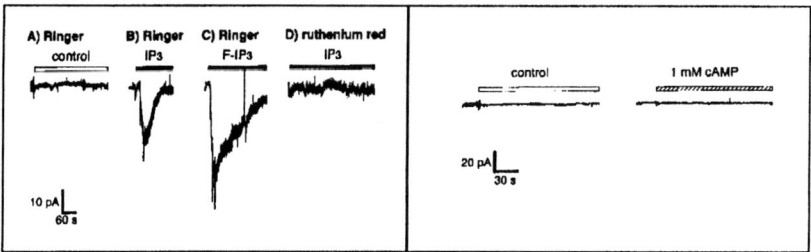

**Figure 9** (left). Response induced by intracellular application of $IP_3$ and its derivative. This response is blocked by ruthenium red, an $IP_3$-channel blocker. **Figure 10** (right). No response is induced by intracellular application of cAMP. From Taniguchi et al. (2000) with permission.

## 8. CALCIUM IMAGING

Since one of the major functions of $IP_3$ is to release $Ca^{2+}$ from intracellular stores, the issue of whether chemoattractant binding to its receptor increased intracellular calcium was addressed (Cinelli et al., 2002). A calcium sensitive dye, Calcium green 1, was injected into the accessory olfactory bulb of snakes. The animals survived 4-6 days to permit retrograde transport of the dye to the cell bodies of receptor cells in VN sensory epithelium. Thus, in VN sensory epithelial slices (~240 μm thick) only mature VN neurons were stained with the calcium-sensitive dye and this population gave rise to the $Ca^{2+}$ increases in response to ESS, bradykinin (a stimulator of phosphoinositide turnover) and forskolin (an adenylate cyclase activator).

ESS (Cinelli et al., 2002) and ES20 (unpublished observations) applied to VN sensory epithelial slices produced a transient cytosolic accumulation of $Ca^{2+}$. This $Ca^{2+}$ transient was first observed in the dendritic region of the epithelium, reaching peak amplitude within 1 second (Figure 11). The increase in cytosolic $Ca^{2+}$ was concentration-dependent, in the concentration range of 0.5-4.5 mg/ml protein. Actin, used as a control, produced no increase in cytosolic $Ca^{2+}$. Bradykinin produced $Ca^{2+}$ transients similar in shape and size to ESS-produced $Ca^{2+}$ transients (Figure 12).

To determine whether the cytosolic increase in $Ca^{2+}$ originated from release of $Ca^{2+}$ from intracellular stores or resulted from influx of $Ca^{2+}$ from the extracellular space, both ESS stimulation and bradykinin stimulation were performed under conditions of 0 $Ca^{2+}$ in the extracellular space. Under these conditions, in most areas of the epithelium, there was no change in the stimulated $Ca^{2+}$ transient; however, in the apical regions there was a marked diminution in the peak of the $Ca^{2+}$ transient. This was true for both ESS- and bradykinin-stimulations (Figure 12). Furthermore, with repeated stimulations, there was a gradual decline in the magnitude of the $Ca^{2+}$ transient, until it disappeared altogether. These results were interpreted as demonstrating that most of the $Ca^{2+}$ response results from release of $Ca^{2+}$ from intracellular stores, whereas the initial drop in the $Ca^{2+}$ transient in the apical region results from a rather restricted loss of $Ca^{2+}$ influx from the extracellular compartment. The gradual loss of the $Ca^{2+}$ transient upon repeated stimulation appears to be a manifestation of the inability to replenish intracellular stores when the extacellular fluid is $Ca^{2+}$-free.

This issue was further explored using two reagents that deplete intracellular $Ca^{2+}$ stores: thapsigargin and ryanodine. Thapsigargin is an inhibitor of the intracellular $Ca^{2+}$ pump and prevents $Ca^{2+}$ reuptake into intracellular $Ca^{2+}$ stores. Ryanodine selectively depletes ryanodine-sensitive intracellular $Ca^{2+}$ stores. Following depletion of internal $Ca^{2+}$ stores with thapsigargin, bradykinin-evoked responses in the apical dendritic regions of VN neurons were diminished, but not totally suppressed, even after 20 trials (Figure 13A). After depletion of ryanodine-sensitive stores, dendritic responses to bradykinin stimulation was unchanged, however, in the somata region there was a significant reduction during the decay phase of the response (Figure 13B). Note that caffeine, which selectively binds to intracellular ryanodine receptors, was not able to evoke a $Ca^{2+}$ response in ryanodine-pretreated tissue. Similar results were obtained with ESS stimulation after thapsigargin and ryanodine $Ca^{2+}$ store depletion. These findings suggest a segregation of two different types of calcium stores and the presence of a thapsigargin-insensitive $Ca^{2+}$ transient in the most apical portions of these neurons.

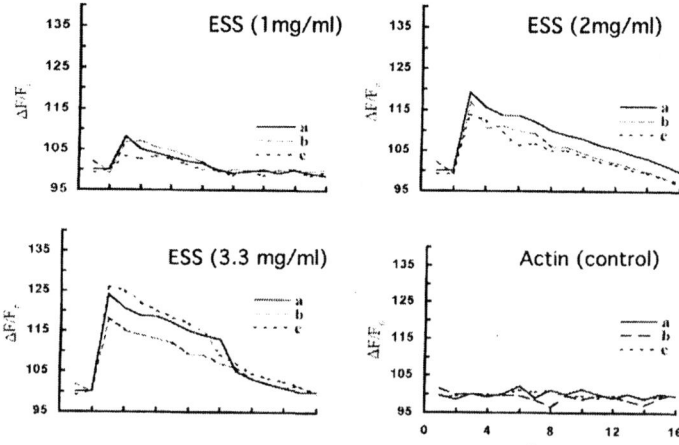

Figure 11. $Ca^{2+}$ transients evoked by ESS at three different concentrations and a control substance, actin, applied to a VN sensory epithelial slice. a, b and c represent values obtained from three different locations in the slice preparation. Time course of $[Ca^{2+}]_i$, over baseline levels ($\Delta F/F_0$) following application of ESS or actin. The first point of plots corresponds to basal cytosolic values. Subsequent values are graphed as responses observed on following frames (1 frame per second). From Cinelli et al. (2002) with permission.

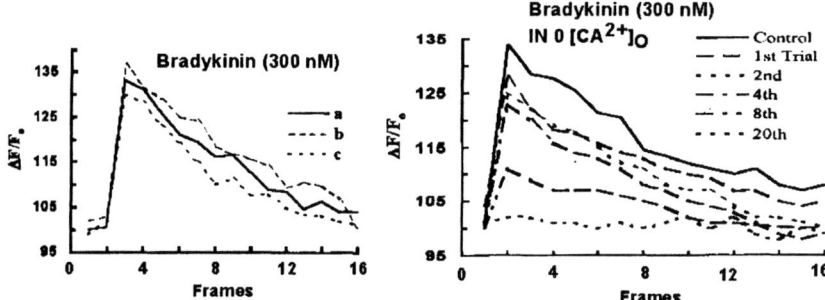

Figure 12. Left: response to bradykinin applied to VN sensory epithelial slice (See Figure 11 for details). Right: response to bradykinin applied to VN sensory epithelial slice in the absence of extracellular calcium. Note that on the first trial the early peak is diminished. On subsequest trials the response is further reduced until after 20 trials no response remains. From Cinelli et al. (2002) with permission.

Since $Ca^{2+}$ release from intracellular stores appears to play an important role during chemosensory transduction we also evaluated the mechanisms for recovery of baseline $Ca^{2+}$ concentrations in VN neurons. To determine the role of the $Na^+/Ca^{2+}$ exchanger in the response of VN neurons to ESS stimulation, the exchanger was blocked by replacing external $Na^+$ with lithium or choline (Figure 14). Both of these monovalent cations caused a significant increase in response magnitude during the late phases of the response to ESS, suggesting that the $Na^+/Ca^{2+}$ exchanger is critical for clearance of excess $Ca^{2+}$ from the cytosol.

## BK-Responses after $Ca^{2+}$ Store Depletion

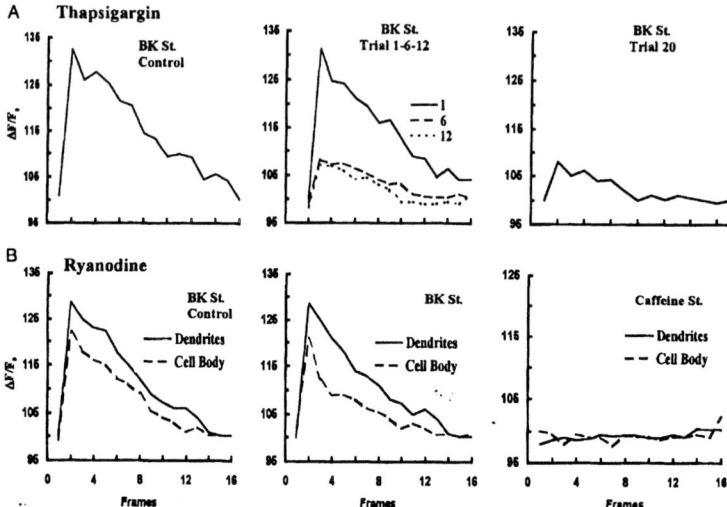

**Figure 13.** $Ca^{2+}$ signals recorded in VN neurons evoked by bradykinin (BK) following depletion of internal $Ca^{2+}$ stores by thapsigargin (A) and ryanodine (B). In thapsigargin treated tissue repeated stimulation with BK resulted in a reduction in the response recorded in the apical dendritic regions of VN neurons; however, a thapsigargin-resistant component remains even after 20 trials. In the ryanodine treated tissue, the response in the dendrites is unaltered compared to control levels; however, in the cell body region of the epithelium, there is a considerable reduction in the response. Caffeine stimulation is ineffective in ryanodine-treated tissue. From Cinelli et al. (2002) with permission.

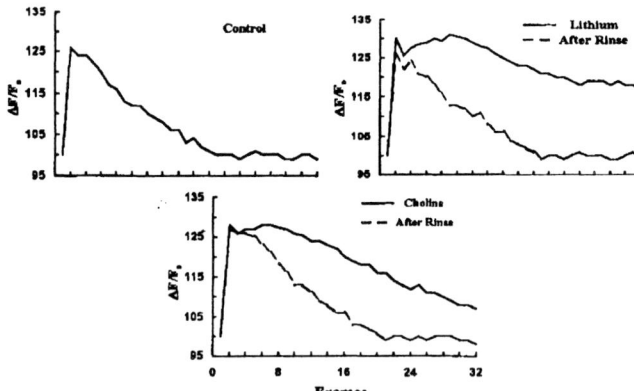

**Figure 14.** Demonstration of $Na^+/Ca^{2+}$ exchanger in the generation of ESS-elicited $Ca^{2+}$ transients. Top left: control recording; top right: $Na^+/Ca^{2+}$ exchanger blocked by replacing $Na^+$ with $Li^+$ in the bath. Bottom: $Na^+/Ca^{2+}$ exchanger blocked by replacing $Na^+$ with choline in the bath. When the $Na^+/Ca^{2+}$ exchanger is blocked by either method there is a significant increase in the magnitude of the response during its late phase. From Cinelli et al. (2002) with permission.

## 9. CHEMOATTRACTANT-ELICITED TRANSIENT PHOSPHORYLATION OF TWO MEMBRANE-BOUND PROTEINS

During ES20-elicited signal transduction, a number of proteins in the intact tissues of VN sensory epithelium were phosphorylated (Figure 15A), but the most prominent proteins phosphorylated were two membrane-bound proteins with molecular masses of 42 kDa and 44 kDa (Figure 15B; Liu et al., 1999). This ES20-elicited phosphorylation of p42/44 proteins was transient, reaching its highest level within 40 seconds and then decreasing thereafter (Figure 16). Phosphorylated p42/44 negatively modulated the hydrolysis of Gα-bound GTP. The phosphorylation of p42/44 was countervailingly regulated by membrane-bound enzymes: a protein kinase and a protein phosphatase. The activity of the kinase was inhibited by calcium ions (Figure 17), but was insensitive to specific inhibitors of $Ca^{2+}$-dependent protein kinase C (PKC) and could not be augmented by phorbol ester. ES20-induced phosphorylation of p42/44 proteins was mimicked by DAG. Thus, the finding that the protein kinase responsible for phosphorylation of p42/44 is activated by DAG, is not inhibited by $Ca^{2+}$-dependent PKC inhibitors, and is not activated by phorbol esters, suggests that it is probably a novel member of the PKC family. The identity of this protein kinase remains unknown.

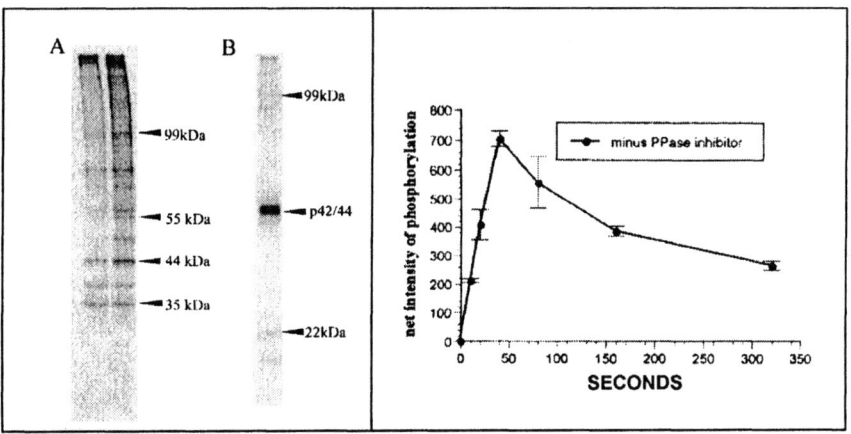

Figure 15 (left). (A) Autoradiograph of phosphorylated proteins in intact tissues of snake VN sensosry epithelium. Left lane control, right lane treated with 2.8 μM ES20. (B) Autoradiograph of phosphorylated proteins in the VN membrane fraction. Figure 16 (right) Time course of p42/44 phosphorylation in the absence of protein phosphatase inhibitors, NaF and $Na_3VO_4$. From Liu et al. (1999) with permission.

The activity of the membrane-bound protein phosphatase, which dephosphorylates the phosphorylated p42/44 proteins, is fully inhibited by fluoride ions, but is insensitive to $Na_3VO_4$ and okadaic acid (Liu et al., 1999), implying that this is a member of the protein phosphatase 2C family (PP2C). A full-length complementary 4119-bp DNA containing an open reading frame of 1146-bp that encodes a protein of 382 amino acids

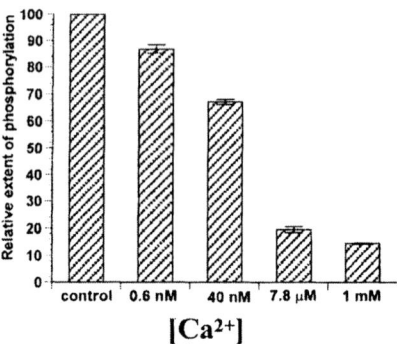

Figure 17. Effect of calcium ions on p42/44 phosphorylation in membrane fractions of VN sensory epithelium. From Liu et al. (1999) with permission.

with a molecular mass of 49123 Da was obtained from a VN cDNA library of garter snakes (Wang et al., 2002). It shows high amino acid identity to PP2C from several species.

## 10. THE BIOCHEMICAL ROLE OF CALCIUM IN PREY CHEMICAL SIGNAL TRANSDUCTION

Calcium ion is the most common second messenger in neurons. In snakes, $Ca^{2+}$ is critical for ligand-receptor binding and for the behavioral response to purified chemoattractant (Luo et al., 1994). As mentioned above, $AC_{VN}$ is $Ca^{2+}$-sensitive based on its homology to ACVI. The levels of intracellular cAMP in snake VN neurons is determined by the activity of $AC_{VN}$ and the activity of this cyclase is biphasically modulated by calcium ions (Wang et al., 1988) and calcium ions negatively modulate the ES20-induced phosphorylation of membrane-bound p42/44 proteins (Liu et al., 1999). Since the binding of chemoattractant to its receptors results in an increase in the levels of $IP_3$ (Luo et al., 1994), which, in turn, induces an increase in $[Ca^{2+}]_i$ transients[16], $Ca^{2+}$, among other functions, could modulate the activity of $AC_{VN}$[17] suggesting that $AC_{VN}$ probably also participates in ES20-induced signal transduction. However, because of the rapid increase of $Ca^{2+}$ transients during ES20-induced signal transduction, the changes in $Ca^{2+}$ concentration likely modulate the activity of $AC_{VN}$, i.e., at the initial increase of $[Ca^{2+}]_i$, the activity of $AC_{VN}$ might be increased. As $[Ca^{2+}]_i$ increases further, $Ca^{2+}$ probably acts as an inhibitor of $AC_{VN}$ lowering the $[cAMP]_i$, since $AC_{VN}$ shows a biphasic response toward changes in calcium concentration (Wang et al., 2002), i.e., at low calcium concentrations $AC_{VN}$ activity is optimal and as the calcium concentration is increased the activity of $AC_{VN}$ is inhibited. So far cAMP has not been demonstrated to exert any effects on either calcium signals or electrical signals generated by chemoattractants (Taniguchi et al., 2000; Cinelli et al., 2002). The exact roles of cAMP and $AC_{VN}$ remain to be established. Since $Ca^{2+}$ inhibited the ES20-induced phosphorylation in a concentration dependent manner, it is tempting to suggest that cAMP may play a role in the ES20-elicited signal transduction cascade following mobilization by $IP_3$. The fact that ES20-induced $[Ca^{2+}]_i$ transients correlate with the inhibitory activity of calcium on $AC_{VN}$ suggests that $AC_{VN}$ may be involve in the process of desensitization in ES20-elicited signal transduction.

## 11. SUMMARY

Our understanding of the chemoattractant-induced signal transduction in the VN system of garter snakes has been considerably advanced, but a number of questions remain to be answered. We summarize our findings as follows (Figure 18): the binding of chemoattractant to its G-protein-coupled receptors initiates the signal transduction cascade that involves a series of chemical reactions and voltage changes in VN chemoreceptor neurons. The series of chemical reactions consist of ligand-elicited activation of G proteins resulting in dissociation of the heterotrimeric G-proteins into $G\alpha$-GTP (active) and $G\beta\gamma$ subunits. The $G\alpha$-GTP subunit activates phospholipase C (PLC) which in turn hydrolyzes phosphatidylinositol bisphosphate ($PIP_2$) to generate two second messengers: inositol 1,4,5-triphosphate ($IP_3$) and diacylglycerol (DAG). Under

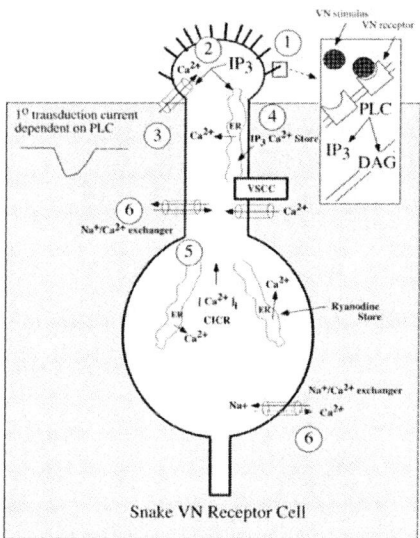

**Figure 18.** Current model of mechanisms postulated to participate in generation of calcium transients during stimulation of snake VN neurons with chemoattractants in ESS. Initially ESS activates receptors that trigger $IP_3$ production, mediating an increase in cytosolic $Ca^{2+}$ through simultaneous release of $Ca^{2+}$ from intracellular stores and influx through a PLC-dependent channel in the plasma membrane. There is additional $Ca^{2+}$ release from ryanodine-sensitive stores through a calcium-induced calcium release mechanism. Normal cytosolic levels of $Ca^{2+}$ are restored by activation of the Na+/ $Ca^{2+}$ exchanger.

whole cell patch-clamp, snake VN receptor neurons in slice preparations exhibit an inward, depolarizing current from intracellular dialysis of $IP_3$. $IP_3$ reacts with its receptor ($IP_3R$) on the endoplasmic reticulum (ER) causing a release of intracellularly stored calcium ions into the cytosol. The extent of chemoattractant-elicited $[Ca^{2+}]_i$ is augmented by $Ca^{2+}$-induced calcium-release (CICR) via the action of $Ca^{2+}$ on ryanodine receptors on the ER. $IP_3$ may also induce the opening of a $Ca^{2+}$ channel, such as the transient receptor potential channel (TRPC), to effect $Ca^{2+}$ influx from the extracellular compartment. The magnitude and duration of the chemoattractant-induced $Ca^{2+}$ transients are modulated by $Ca^{2+}$-ATPase and a $Na^+/Ca^{2+}$ exchanger mechanism. On the other hand, the second messenger, DAG, activates a novel membrane-bound protein kinase which catalyzes the phosphorylation of two membrane-bound proteins, p42/44. These proteins negatively regulate the hydrolysis of $G\alpha$-bound GTP when they are in the phosphorylated state. The phosphorylated p42/44 proteins are dephosphorylated by a membrane-bound protein, phosphatase 2C. In addition, $Ca^{2+}$ appears to negatively modulate the phosphorylation of p42/44 proteins, although the mechanism involved remains unknown. Chemoattractant-receptor binding also modulates the generation of the second messenger cAMP, which is known to activate protein kinase A (PKA), and the enzyme $AC_{VN}$ catalyzes cAMP formation, which is sensitive to $Ca^{2+}$ regulation. The roles of these latter components in the chemoattractant-elicited signal transduction pathways remain to be resolved, although they may be involved in desensitization. Electrophysiologically, the binding of chemoattractant to its receptors in VN epithelium of garter snakes evokes action potentials in the mitral cells of the accessory olfactory bulb (AOB), the first projection site. The relationship between stimulus-induced $Ca^{2+}$ transients and the generation of electrical signals in the AOB remains to be established, although we have evidence that chemoattractants induce changes in membrane potential, as revealed by using voltage sensitive dyes, which mimic the time course and shape of the stimulus-induced $Ca^{2+}$ transients (unpublished data). This change in membrane potential transients is ESS concentration-dependent.

## 12. ACKNOWLEDGEMENTS

This research was supported by grants DC00104, DC02531 and DC03735 from the National Institute of Deafness and Communication Disorders to MH and by grant IBN-9905700 from the National Science Foundation to DW.

## 13. REFERENCES

Begun, D., Kubie, J. L., O'Keefe, M. P., and Halpern, M., 1988, Conditioned discrimination of airborne odorants by garter snakes (*Thamnophis radix* and *T. sirtalis sirtalis*), *J. Comp. Psychol.* **102**:35-43.

Cinelli, A. R., Wang, D., Chen, P., Liu, W., and Halpern, M., 2002, Calcium transients in the garter snake vomeronasal organ, *J. Neurophys.* **87**:1449-1472.

Halpern, M., and Frumin, N., 1979, Roles of the vomeronasal and olfactory systems in prey attack and feeding in adult garter snakes, *Physiol. Behav.* **22**:1183-1189.

Halpern, M., and Kubie, J. L., 1980, Chemical access to the vomeronasal organs of garter snakes, *Physiol. Behav.* **24**:367-371.

Halpern, M., and Kubie, J. L., 1983, Snake tongue flicking behavior: Clues to vomeronasal system functions, in: *Chemical Signals III*, R. M. Silverstein and D. Müller-Schwarze, eds., Plenum Press, New York, pp. 45-72.

Halpern, M., Halpern, J., Erichsen, E., and Borghjid, S., 1997, The role of nasal chemical senses in garter snake response to airborne odor cues from prey, *J. Comp. Psychol.* **111**:251-260.

Jiang, X. C., Inouchi, J., Wang, D., and Halpern, M., 1990, Purification and characterization of a chemoattractant from earthworm electric shock-induced secretion, its receptor binding, and signal transduction through the vomeronasal system of garter snakes, *J. Biol. Chem.* **265**:8736-8744.

Kubie, J. L., and Halpern, M., 1978, Garter snake trailing behavior: Effects of varying prey extract concentration and mode of prey extract presentation, *J. Comp. Physiol. Psychol.* **92**:362-373.

Kubie, J. L., and Halpern, M., 1979, The chemical senses involved in garter snake prey trailing, *J. Comp. Physiol. Psychol.* **93**:648-667 (1979).

Liu, J., Chen, P., Wang, D., and Halpern, M., 1999, Signal transduction in the vomeronasal organ of garter snakes: ligand-receptor binding-mediated protein phosphorylation, *Biochem. Biophys. Acta* **1450**:320-330.

Liu, W., Wang, D., Chen, P., and Halpern, M., 1997, Cloning and expression of a gene encoding chemoattractive protein from earthworm secretion to garter snakes, *J. Biol. Chem.* **272**:27378-27381.

Liu, W., Wang, D., Liu, J., Chen, P., and Halpern, M., 1998, Cloning of a gene encoding adenylate cyclase from vomeronasal organ of garter snakes, *Arch. Biochem. Biophys.* **358**:204-210.

Luo, Y., Lu, S., Chen, P., Wang, D., and Halpern, M., 1994, Identification of chemoattractant receptors and G proteins in the vomeronasal system of garter snakes, *J. Biol. Chem.* **269**:16867-16877.

Martinez-Marcos, A., Lanuza, E., and Halpern, M., 2002, Neural substrates for processing chemosensory information in snakes, *Brain Res. Bull.* **57**:543-546.

Martínez-Marcos, A., Ubeda-Bañón, I., and Halpern, M., 2001, Neural substrates for tongue-flicking behavior in snakes, *J. Comp. Neurol.* **432**:75-87.

Taniguchi, M., Wang, D., and Halpern, M., 2000, Chemosensitive conductance and inositol 1,4,5 trisphosphate-induced conductance in snake vomeronasal receptor neurons, *Chem. Senses* **25**:67-76.

Wang, D. Chen, P., Li, C.-S., and Halpern, M., 1988, Chemosignal transduction in the vomeronasal organ of garter snakes: $Ca^{2+}$-dependent regulation of adenylate cyclase, *Arch. Biochem. Biophys.* **267**:459-466.

Wang, D., Chen, P., Liu, W., Li, C.-S., and Halpern, M., 1997, Chemosignal transuction in the vomeronasal organ of garter snakes: $Ca^{2+}$-dependent regulation of adenylate cyclase, *Arch. Biochem. Biophys.* **348**:96-106.

Wang, D., Liu, W., Liu, J., Chen, P., Quan, W., and Halpern, M., 2002, Molecular cloning and characterization of protein phosphatase 2C of vomeronasal sensory epithelium of garter snakes, *Arch. Biochem. Biophys.* **408**:184-191.

# MODE OF DELIVERY OF PREY-DERIVED CHEMOATTRACTANTS TO THE OLFACTORY AND VOMERONASAL EPITHELIA RESULTS IN DIFFERENTIAL FIRING OF MITRAL CELLS IN THE MAIN AND ACCESSORY OLFACTORY BULBS OF GARTER SNAKES

Cheng-Shu Li, John Kubie and Mimi Halpern[*]

## 1. INTRODUCTION

One of the central issues concerning differences between the main olfactory and vomeronasal systems is the mechanism of odor delivery. To some extent the gross morphological differences between these two systems suggest that they will differ in the nature of the odorants that optimally stimulate them. Whereas the olfactory epithelium of terrestrial vertebrates is always located in the nasal cavity with the nares allowing airborne odorants ready access to the epithelium, the relationship of the vomeronasal organ to the external environment is variable. In some animals the vomeronasal epithelium and main olfactory epithelium are continuous structures; in other animals they are totally separate. In the latter condition, as a general rule, the opening of the organ to the external environment is extraordinarily small suggesting that the animal must in some way actively enlarge it to receive stimulation or actively introduce chemicals into the organ (Hamlin, 1929). In addition, the lumen of the vomeronasal organ is quite narrow when compared to the cavernous spaces of the nasal chamber and probably is always filled with liquid. These differences in purely morphological characteristics suggest that the optimal stimulus, mode of stimulation, physical state of the stimulus and threshold quantity of stimulating substances may differ for these two systems.

The mode of stimulation of these two systems is different. It is well established that land dwelling vertebrates deliver gaseous odorants to the olfactory mucosa by inhaling or sniffing with the nose. The molecules that stimulate the main olfactory apparatus must

---

[*] SUNY Downstate Medical Center, Department of Anatomy, 450 Clarkson Avenue, Brooklyn, NY 11203.

be both volatile and adsorbent to the main olfactory mucosa. The mechanism of odorant transport to the vomeronasal system is poorly understood and is not consistent across species. It involves a vascular pump in hamsters (Meredith and O'Connell, 1979; Meredith et al., 1980), head bobbing, sniffing and licking in guinea pigs (Wysocki et al., 1980), nuzzling in opossums (Poran et al., 1993), tongue compression strokes in bulls (Jacobs et al., 1981), flehmen, a species-typical lip curl (Hart, 1983; Estes, 1972), in goats (Ladewig and Hart, 1980) and elephants (Rasmussen et al., 1982) and tongue flicking in snakes and lizards (Kahmann, 1932; Halpern and Kubie, 1980; Graves and Halpern, 1989). Thus, different delivery mechanisms may be an important distinction between the vomeronasal system and the main olfactory system and may segregate different types of molecules.

The two sensory epithelia may have differential sensitivites to different types of odorants. Several studies in mammals and snakes confirm that vomeronasally mediated substances are non-volatile, high molecular weight proteinaceous materials, high molecular weight non-volatiles associated with proteins, or volatiles bound to larger carrier molecules (Singer et al., 1986; Burghardt et al., 1988; Wang et al., 1988; Jiang et al., 1990; Jemiolo et al., 1986; Moss et al., 1997; Moss et al., 1998, Novotny et al., 1990, 1999a, 1999b). Since volatility is a *sine quo non* for olfactory stimulation, this represents an important distinction between the two systems. However, it should be noted that volatiles may also stimulate the vomeronasal system (Meredith, 1991; O'Connell and Meredith, 1984).

Olfactory and vomeronasal bipolar neurons may differ in their responsiveness to stimulating substances and these differences in response characteristics could provide important clues to their differential functions. In garter snakes a spectrum of chemical stimuli (standard odorants, amino acids and prey products, including purified proteins from earthworms, a favored prey of garter snakes) applied to the vomeronasal epithelium modify unit responses in the AOB (Inouchi et al., 1993; Wang et al., 1993; Taniguchi et al., 1998, 2000).

There have been few studies using electrophysiological methods to characterize and compare the properties of main olfactory and vomeronasal systems. Electroolfactogram and electrovomeronasogram recordings from garter snakes revealed that the olfactory epithelium was more sensitive to vapor of amyl acetate, butanol and earthworm wash than the vomeronasal epithelium (Inouchi et al., 1993). Reliable EVG responses were obtained only from amyl acetate. The responses in the vomeronasal epithelium were similar in shape to those recorded in the olfactory epithelium although about 10 times smaller. In contrast, saturated vapor of earthworm wash -- a potent snake chemoattractant, response to which is dependent on a functional vomeronasal system -- did not produce responses from the vomeronasal epithelium. This anomalous result may be related to the medium in which the prey odor was delivered, i.e. as a vapor rather than as a liquid. Is the mode of delivery of odorants critical to the segregation of chemicals stimulating the olfactory and vomeronasal systems? We address this issue here by describing experiments in which we compare the effects of liquid and airborne delivery of odorants to the main olfactory and vomeronasal epithelia on evoked responses in the main and accessory olfactory bulbs, respectively.

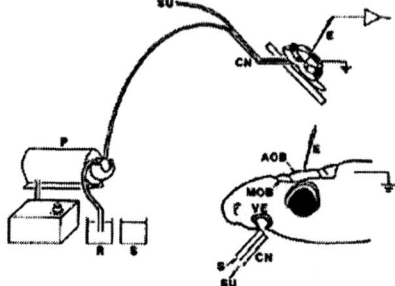

Figure 1. Experimental set up.

## 2. MATERIALS AND METHODS

Twenty-two cells from 22 adult garter snakes, *Thamnophis sirtalis*, were monitored for activation by prey stimuli. Prior to an experiment each snake was tested to ascertain that it would accept both earthworm and goldfish prey. Only snakes that ate both types of prey were used in this study.

Snakes were anaesthetized with urethane (2 g/kg) injected i.p. and secured in a Kopf sterotaxic instrument equipped with a specially designed snake restraint. The vomeronasal epithelium was exposed by removing the medial palate and mushroom body. The nasal cavity was exposed by removing part of the mucosa of the roof of the mouth. For recording single units in the MOB and AOB, the cranium was opened dorsally and the bulb exposed (Figure 1).

Mitral cells of both the MOB and AOB were identified by antidromic stimulation combined with histological examination (Figure 2 and 3). A stimulating electrode was placed into the nucleus sphericus or lateral olfactory tract for antidromic stimulation of mitral cells in the AOB or MOB, respectively. Single barrel micropipettes filled with a 2% (w/v) solution of Chicago Blue in 0.5 M sodium acetate were used for extracellular single unit recordings from the mitral cell layers of the AOB and MOB. Extracellular potentials were filtered to eliminate frequencies below 300 Hz, amplified, displayed on a dual beam storage oscilloscope, and recorded on magnetic tape. Offline, the output of a window discriminator was fed to a personal computer equipped with Brain Wave Software to create peristimulus-time histograms.

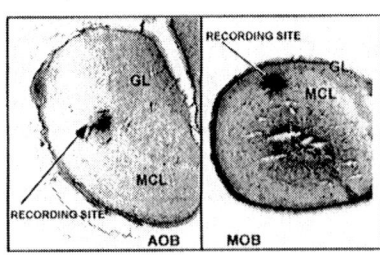

Figure 2. Recording sites in the AOB and MOB. GL=glomerular layer, MCL=mitral cell layer.

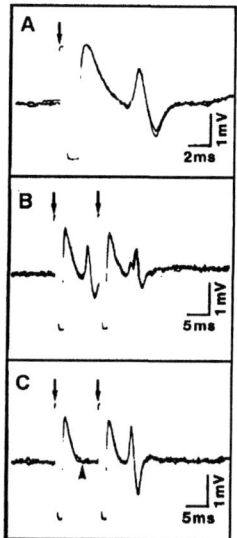

**Figure 3.** Action potentials from an AOB neuron in the mitral cell layer. Superimposed oscilloscope traces illustrate fulfillment of criterion for antidromic activation from the nucleus sphericus. Responses occur at a constant latency to the antidromic invasion (Arrow A), follow closely paired stimulation pulses (B), and are cancelled (arrowhead) by collision with spontaneously generate potentials (C).

The criteria for antidromic invasion were constant latency of mitral cell response to antidromic shock stimulus, ability to follow closely paired stimulus pulses at greater than 100 Hz, and collision between spontaneously generated spikes and spike-triggered stimulus-activated spikes (Figure 3). The immobilized snake was tilted 45° to permit simultaneous recording from the bulbs and delivery of odorants to the mucosae. Liquid odorants were delivered to the VNO or nasal cavity by a micropipette attached to capillary tubing at a rate of 2.4 ml/min. Ringer's solution flowed over the epithelia constantly (2.4 ml/min) and was interrupted momentarily for presentation of stimulus solutions. To avoid residual effects of stimuli remaining in the VNO or the nasal cavity, the fluid in the organs was aspirated and the organs were washed with Ringer's solution for more than 2 min following each stimulus delivery. For vapor delivery of chemoattractants to the VNO or the nasal cavity, an air delivery system was used that delivered a 2 s odorant pulse (2 ml/min). The odor pulse was an air stream of saturated odorant mixed with a dilution stream of filtered air. Stimulus concentrations were regulated using flow meters that permitted adjustment of the flow ratios between odor-saturated and filtered air streams.

Earthworm wash ($10^0$) was prepared at a concentration of 1.86 mg lyophilized earthworm powder to 100 ml snake Ringer's solution. This solution was diluted 1:10 to produce earthworm wash at a $10^{-1}$, and diluted 1:100 to produce earthworm wash at $10^{-2}$. Goldfish wash ($10^0$) was prepared at a concentration of 6 gms goldfish to 20 ml snake Ringer's solution. This solution was diluted as described above for earthworm wash to produce the logarithmic dilutions used in the experiments. Control stimuli consisted of Ringer's solution for liquid delivery and clean, humidified, air for vapor delivery.

Order of stimulus presentation was systematically varied. For some animals an ascending concentration series was used, for some a descending concentration series was used, for some goldfish-derived stimuli preceded earthworm-derived stimuli and for some vice-versa. Control stimuli were systematically interspersed in the delivery sequence such that they appeared at the beginning, middle or end of a series. When recording from the AOB logarithmic dilutions of liquid stimuli were used and only the highest concentration of air borne odors was tested. Conversely, when recording from the MOB logarithmic dilutions of airborne odors were used and only the highest concentration of liquid stimuli was tested.

At the completion of an experiment, a 10μA DC current was passed through the recording electrode. The animal was injected with an overdose of anesthetic and killed by perfusion through the heart with saline followed by fixative. Histological sections were made and stained with Neutral Red to identify the location of the recording electrode tip (Figure 2).

Spike waveforms were digitized with a DataWave data acquisition system (DataWave Technology, Boulder Colorado) at 10 kHz. The DataWave system collects 32 data points for each spike. Each waveform has 8 parameters extracted for spike sorting, with the most useful parameters being "peak" (the highest value of the waveform), "valley" (the lowest value following the peak) and "A-to-D value" (the A-to-D value at a user-selected time on the waveform). Most cells had effective spike discrimination using the two parameters of "peak" and "valley". The cluster plots for these two parameters were clear and few or no data points were ambiguous (Figure 4). For one cell, adding the A-to-D parameter was necessary for effective spike separation.

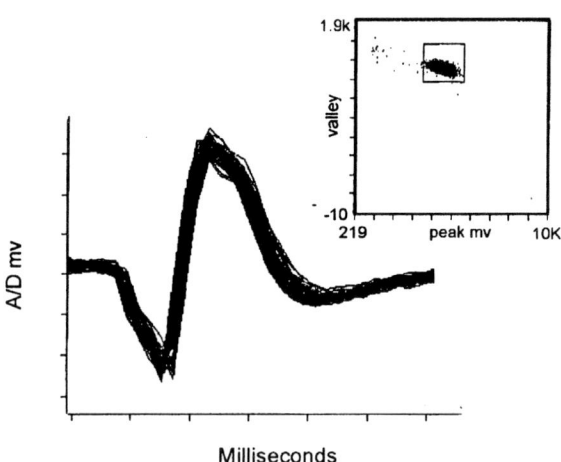

Figure 4. Cluster analysis (see text for description).

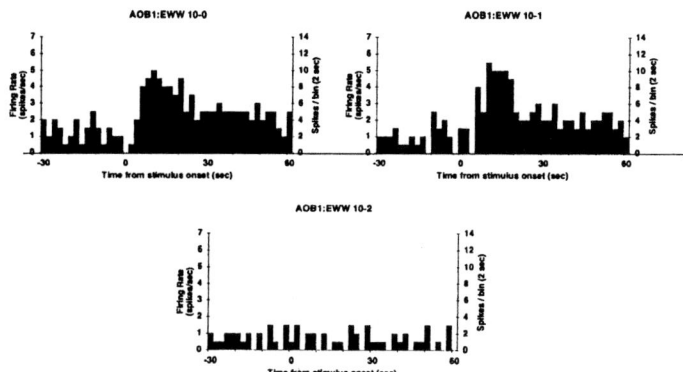

**Figure 5.** Spike frequency histograms of firing of AOB mitral cell to earthworm wash (EWW) at different concentrations.

A cell was determined to have responded to a stimulus if its mean firing rate during the minute following stimulus delivery equaled or exceeded two standard deviations above the mean firing rate during the 30 seconds preceding stimulus delivery. Concentration/response plots were constructed using the peak firing rate during the 30 seconds following stimulus delivery.

## 3. RESULTS

Twenty-two cells, one from each snake, were monitored for activation by prey stimuli. All cells responded to at least one stimulus. We report here the results on nine of these neurons. These cells were systematically tested with different concentrations of stimulating substances and with appropriate control stimuli. In all cases, air delivery of prey odors activated MOB mitral cells but not AOB mitral/tufted cells. Conversely, liquid delivery of prey odors activated AOB mitral/tufted cells, but not MOB mitral cells. Control substances, pure air and snake Ringer's solution, failed to activate any cells (Figures 5-10).

The responses of neurons to activating stimuli were concentration-dependent; firing frequency increased with increased concentration for both airborne odorants and liquid stimuli delivered to the MOB and AOB, respectively (Figure 11). The firing rates to the lowest concentrations of stimulating substances approached firing rates to control stimuli and to nonstimulating substances (liquid stimuli for the MOB and airborne stimuli for the AOB).

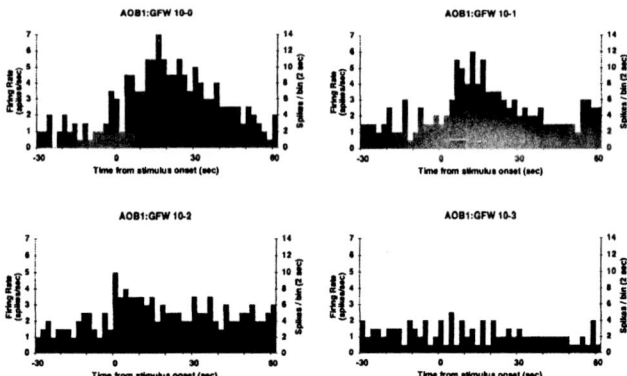

Figure 6. Spike frequency histograms of firing of AOB mitral cell to goldfish wash (GFW) at different concentrations.

Although the constraints of our odor delivery system precluded accurate latency measures, it is apparent from examination of Figures 5, 6, 8 and 9 that mitral cells in the MOB displayed faster onset and offset kinetics than some AOB mitral cells. MOB cells typically reached their maximum firing rate within the first two seconds after stimulus onset whereas some AOB cells took as long as 10-20 seconds to reach their maximum firing rate. Similarly MOB cells returned to baseline levels within approximately 4-6 seconds whereas some AOB cells remained above baseline firing levels as long as 30-50 seconds. Some of these differences can be ascribed to the time course of stimulus delivery, however, it is unlikely that the major portion of the differences is solely related to the physical conditions of the experiment since the duration of odor delivery for stimuli delivered both as air streams and liquids was at least 2 min and several AOB neurons showed activation during the first 2 seconds following initial odor delivery.

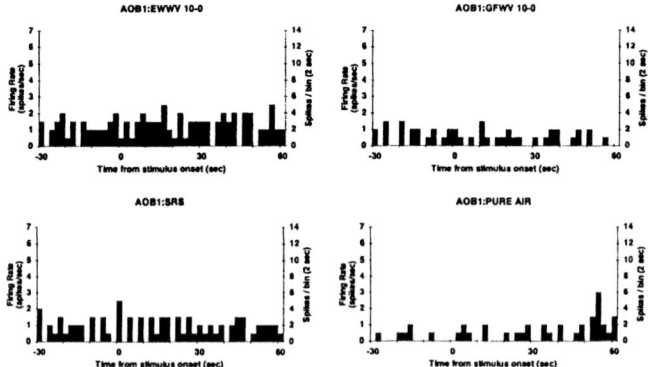

Figure 7. Spike frequency histograms of firing of AOB mitral cell to earthworm wash vapor (EWWV), goldfish wash vapor (GFWV), snake Ringer's solution (SRS), and pure air.

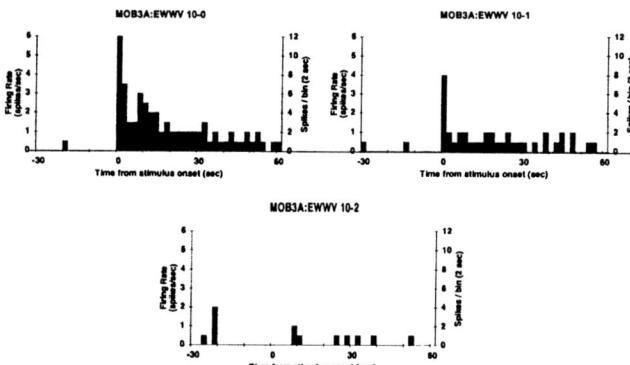

Figure 8. Spike frequency histograms of firing of MOB mitral cell to EWWV at different concentrations.

Since snakes used in these experiments ate both earthworms and goldfish, it is interesting to note that the responsiveness of the recorded cells frequently differed for the two prey sources. Cells in the AOBs of some snakes responded better to earthworm wash and other responded better to gold fish wash. Similarly, cells in the MOBs of some snakes responded more vigorously to earthworm wash vapors and some to goldfish wash vapors. Without threshold testing of intact animals and recordings from multiple cells in the olfactory bulbs of the test animals it is not possible to determine whether the apparent bias of some of these cells has a functional significance.

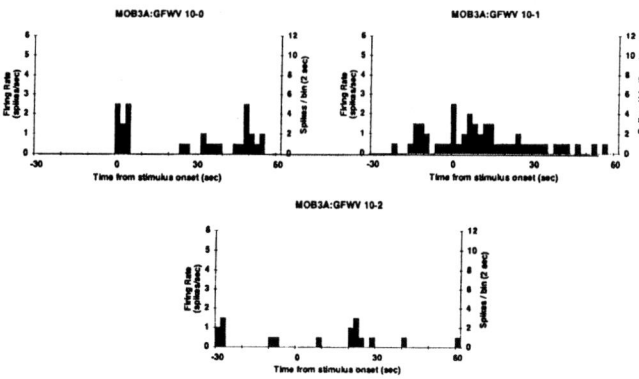

Figure 9. Spike frequency histograms of firing of MOB mitral cell to GFWV at different concentrations.

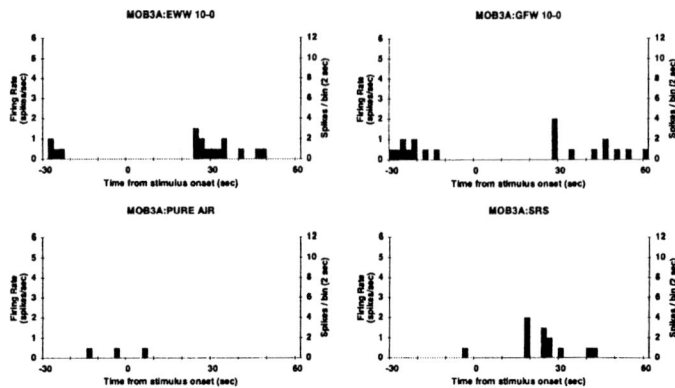

Figure 10. Spike frequency histograms of firing of MOB mitral cell to EWW, GFW, pure air and SRS.

## 4. DISCUSSION

We have demonstrated that irrigation of the vomeronasal organ with liquid stimuli derived from natural prey of garter snakes causes an increase in firing rate of AOB mitral cells. No similar increase could be evoked with airborne prey odors or the control stimuli. This activation is concentration-dependent, with increased concentrations of the liquid stimuli causing an increased firing rate and, in some neurons, an extended excitation period. Concentrations do not appear to have a systematic effect on latency to respond. Delivery to the olfactory epithelium of vapors derived from prey exacts resulted in a rapid increase in firing rate of MOB mitral cells. No similar increase could be evoked with liquid delivery of prey odors or control stimuli. These findings strongly support the hypothesis that the vomeronasal system of snakes is differentially sensitive to liquid delivery of biologically significant chemical stimuli and the main olfactory system is differentially sensitive to airborne delivery of these same stimuli.

Whereas volatility is a constraint for gaseous delivery of odorants to chemosensory surfaces, no such constraint exists for odorants delivered in a liquid state. It is, therefore, not surprising to find that large, non-volatile substances have access to the fluid-bathed VN sensory epithelium, and that some of these, usually pheromones, have been identified as vomeronasal stimulants.

Non-volatile substances labeled with tritium or fluorescent dyes have been found in the vomeronasal organs of salamander (Dawley and Bass, 1989), snakes (Halpern and Kubie, 1980), lizards (Graves and Halpern, 1989), guinea pigs (Wysocki et al., 1980), goats (Ladewig and Hart, 1980), opossums (Poran et al., 1993), mice, voles and rats (Wysocki et al., 1985) when the animals have made direct contact with the substances. These non-volatile molecules may be as large as 66,000 Daltons (Wysocki et al., 1985). In most cases an observable, species-specific behavior is used by the animal to deliver the stimulating substance to the VNO or its vicinity. These behaviors include nose tapping in salamanders (Dawley and Bass, 1989), tongue-flicking in snakes and lizards (Halpern and Kubie, 1980; Graves and Halpern, 1989), head bobbing in guinea pigs (Wysocki et al.,

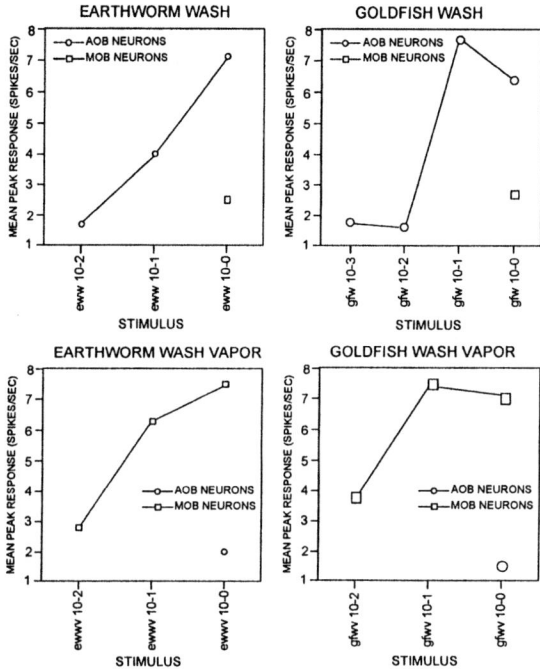

Figure 11. Concentration-response curves for responses of AOB and MOB mitral cells to EWW, GW, EWV and GFWV.

1980), flehmen in goats (Ladewig and Hart, 1980), and nuzzling in opossums (Poran et al., 1993).

There is, in addition, evidence that these large, non-volatile molecules are biologically significant and that their behavioral effect is mediated via the VN system. Female mouse urine contains a non-volatile pheromone to which male mice respond with ultrasonic vocalizations (Nyby et al., 1979; Nyby and Zakeski, 1980). Direct contact with the fresh urine is required for this response (Sipos et al., 1995). Hamster vaginal fluid contains compounds larger than 10,000 Daltons to which male hamsters respond, but only when direct contact is permitted (Singer et al., 1984). Similarly, when direct contact occurs, male guinea pigs investigate, with head bobbing, female urine, but do not respond significantly when they have access only to the volatiles emanating from the urine (Beauchamp et al., 1982). Garter snakes respond to prey-derived substances of 20,000 Daltons with increased tongue flicks and attack (Wang et al., 1988; Jiang et al., 1990) and require direct contact with the prey wash to make discriminated responses (Kubie and Halpern, 1979). All of these behavioral responses to large molecular weight, biologically significant compounds or substances require a functional vomeronasal system (Beauchamp et al., 1982, 1985; Kubie and Halpern, 1979; Clancy et al., 1984).

In sum, non-volatile, large molecular weight vomeronasal stimuli gain access to the vomeronasal organ via behavioral adaptations that deliver the odorants to the liquid-filled organ where they may contact the sensory epithelium and initiate the transduction process that will eventually result in a behavioral response. As we have shown in this report, for snakes liquid delivery of odorants is the preferred mechanism of VNO access and this may be true as well for other species. For example, in the aquatic turtle, *Geoclemys reevesii*, aquatic solutions reach the vomeronasal epithelium, but do not reach the olfactory epithelium (Shoji et al., 1994). Similarly, Prosimian mouse lemurs, *Microcebus murinus*, communicate via chemosignals in urine. Liquid urine produces activation of the main and accessory olfactory bulbs as measured by 2-deoxyglucose uptake, but urine in the volatile phase activates only the main olfactory bulb (Schilling et al., 1990).

Unfortunately, no other study to our knowledge has directly compared the effects of mechanism of odor delivery to the VN epithelium and main olfactory epithelium on the response of accessory and main olfactory bulb neurons. We can therefore make no direct comparisons between our data and that of others.

Numerous studies have reported evoked responses to chemical stimuli in the main and accessory olfactory bulbs of reptiles. Single unit recordings in the main olfactory bulb of gecko (*Gekko gecko*), four-toed tortoise (*Testudo horsifieldi*) and cayman (*Caiman crocodilus*), reveal excitation, suppression or complex responses to airborne delivery of a variety of odorants to olfactory epithelium (Shibuya et al., 1977). Summated activity recorded from both the accessory and main olfactory bulbs of the turtle *Geoclemys reevesii* increased in response to epithelial stimulation with standard chemicals such as *n*-Amyl acetate dissolved in Ringer's solution. Hatanaka and Hanada (1986, 1987) tested a number of chemical solutions and airborne odorants for their ability to induce a wave of excitation on the surface of the AOB when applied to the VN epithelium of the turtles *Pseudemys scripta*, *Geoclemys reevesii* and *Clemmys japonica*. Most liquid and a few airborne odors yielded responses. Single unit activity in the AOB of *G. reevesii* was modified by vapor or aqueous delivery of a large number of chemicals including salts, sugars and acids (Hatanaka and Shibuya, 1989).

To our knowledge, this is the first study that has used natural odors as stimuli to compare responsiveness of vomeronasal and olfactory neurons. Previous studies have almost exclusively employed organic, biologically insignificant stimuli.

## 5. ACKNOWLEDGEMENTS

This research was supported by grants DC02531 and DC03735 from the National Institute on Deafness and other Communication Disorders to M.H.

## 6. REFERENCES

Beauchamp, G. K., Martin, I. G., Wysocki, C. J., and Wellington, J. L., 1982, Chemoinvestigation and sexual behavior of male guinea pigs following vomeronasal organ removal, *Physiol. Behav.* 29:329-336.

Beauchamp, G. K., Wysocki, C. J., and Wellington, J. L., 1985, Extinction of response to urine odor as a consequence of vomeronasal organ removal in male guinea pigs, *Behav. Neurosci.* 99:950-955.

Burghardt, G. M., Goss, S. E., and Schell, F. M., 1988, Comparison of earthworm- and fish-derived chemicals eliciting prey attack by garter snakes (*Thamnophis*), *J. Chem. Ecol.* 14:855-881.

Clancy, A. N., Macrides, F., Singer, A. G., and Agosta, W. C., 1984, Male hamster copulatory responses to a high molecular weight fraction of vaginal discharge: effects of vomeronasal organ removal, *Physiol. Behav.* 33:653-660.

Dawley, E. M., and Bass, A. H., 1989, Chemical access to the vomeronasal organ of a plethodontid salamander, *J. Morph.* **200**:163-174.
Estes, R. D., 1972, The role of the vomeronasal organ in mammalian reproduction, *Mammalia* **36**:315-341.
Graves, B. M., and Halpern, M., 1989, Chemical access to the vomeronasal organs of the lizard, *Chalcides ocellatus, J. Exper. Zool.* **249**:150-157.
Halpern, M., and Kubie, J. L., 1980, Chemical access to the vomeronasal organs of garter snakes, *Physiol. Behav.* **24**:367-371.
Hamlin, H. E., 1929, Working mechanisms for the liquid and gaseous intake and output of Jacobson's organ, *Amer. J. Physiol.* **191**:201-205.
Hart, B. L., 1983, Flehmen behavior and vomeronasal organ function, in: *Chemical Signals in Vertebrates 3*, D. Muller-Schwarze and R. M. Silverstein,eds., Plenum Press, New York, pp. 87-103.
Hatanaka, T., and Hanada, T., 1986, Structure of the vomeronasal system and induced wave in the accessory olfactory bulb of red eared turtle, in: *Proceed. 20th Japanese Symp. Taste and Smell*, T. Shibuya and S. Saito, eds., JASTS, Gifu, Japan, pp. 183-186.
Hatanaka, T., and Hanada, T., 1987, Activity of the accessory olfactory bulb to chemical stimuli delivered to the vomeronasal mucosa in turtles, *Zool. Science (Tokyo)* **4**:964.
Hatanaka, T., and Shibuya, T., 1989, Odor response patterns of single units in the accessory olfactory bulb of the turtle, *Geoclemys reevesii, Comp. Biochem. Physiol.* **92A**:505-512.
Inouchi, J., Wang, D., Jiang, X. C., Kubie, J. L., and Halpern, M., 1993, Electrophysiological analysis of the nasal chemical senses in garter snakes, *Brain Behav. Evol.* **41**:171-182.
Jacobs, V. L., Sis, R. F., Chenoweth, P. J., Klemm, W. F., and Sherry, C. J., 1981, Structures of the bovine vomeronasal complex and its relationship to the palate: Tongue manipulation, *Acta Anat.* **110**:48-58.
Jemiolo, B., Harvey, S., and Novotny, M., 1986, Promotion of the Whitten effect in female mice by synthetic analogs of male urinary constituents, *Proc. Natl. Acad. Sci. USA* **83**:4576-4579.
Jiang, X. C., Inouchi, J., Wang, D., and Halpern, M., 1990, Purification and characterization of a chemoattractant from electric shock-induced secretion, and its receptor binding and signal transduction through vomeronasal system of garter snakes, *J. Biol. Chem.* **265**:8736-8744.
Kahmann, H., 1932, Sinnesphysiologische Studien an Reptilien: I. Experimentelle Untersuchungen über das Jakobsonische Organ der Eidechsen und Schlangen, *Zool. Jahr., Abt. für Allge. Zool. Physiol. der Tiere* **51**:173-238.
Kubie, J. L., and Halpern, M., 1979, The chemical senses involved in garter snake prey trailing. *J. Comp. Physiol. Psychol.* **93**:648-667.
Ladewig, J., and Hart, B. L., 1980, Flehmen and vomeronasal organ function in male goats, *Physiol. Behav.* **24**:1067-1071.
Meredith, M., 1991, Sensory processing in the main and accessory olfactory systems: comparisons and contrasts, *J. Steroid Biochem. Mol. Biol.* **39**:601-614.
Meredith, M., and O'Connell, R. J., 1979, Efferent control of stimulus access to the hamster vomeronasal organ, *J. Physiol.* **286**:301-316.
Meredith, M., Marques, D. M., O'Connell, R.J., and Stern, F. L., 1980, Vomeronasal pump; significance for male hamster sexual behavior, *Science* **207**:1224-1226.
Moss, R. L., Flynn, R. E., Shi, J., Shen, X. M., Dudley, C., Zhou, A., and Novotny, M., 1998, Electrophysiological and biochemical responses of mouse vomeronasal receptor cells to urine-derived compounds: possible mechanism of action, *Chem. Senses* **23**:483-489.
Moss, R. L., Flynn, R. E., Shen, X. M., Dudley, C., Shi, J., and Novotny, M., 1997, Urine-derived compound evokes membrane responses in mouse vomeronasal receptor neurons, *J. Neurophysiol.* **77**:2856-2862.
Novotny, M., Jemiolo, B., and Harvey, S., 1990, Chemistry of rodent pheromones: Molecular insights into chemical signalling in mammals, in: *Chemical Signals in Vertebrates 5*, D. W. Macdonald, D. Muller-Schwarze, and S. E. Natynczuk, eds., Oxford University Press, New York, pp. 1-22.
Novotny, N. V., Jemiolo, B., Wiesler, D., Ma, W., Harvey, S., Xu, F., Xie, T.-M., and Carmack, M., 1999, A unique urinary constituent, 6-hydroxy-6-methyl-3-heptanone, is a pheromone that accelerates puberty in female mice, *Chem. Biol.* **6**:377-383.
Novotny, M. V., Ma, W., Wiesler, D., and Zidek, L., 1999, Positive identification of the puberty-accelerating pheromone of the house mouse: the volatile ligands associating with the major urinary protein, *Proc. R. Soc. Lond. B* **266**:2017-2022.
Nyby, J., Wysocki, C. J., Whitney, G., Dizinno, G., and Schneider, J., 1979, Elicitation of male mouse (*Mus musculus*) ultrasonic vocalizations: I. Urinary cues. *J. Comp. Physiol. Psychol.* **93**:957-975.
Nyby, J., and Zakeski, D., 1980, Elicitation of male mouse ultrasounds: Bladder urine and aged urine from females, *Physiol. Behav.* **24**:737-740.
O'Connell, R. J., and Meredith, M., 1984, Effect of volatile and nonvolatile chemical signals on male sex behaviors mediated by the main and accessory olfactory systems, *Behav. Neurosci.* **98**:1083-1093.

Poran, N. S., Vandoros, A., and Halpern, M., 1993, Nuzzling in gray short-tailed opossum I: Delivery of odors to the vomeronasal organ, *Physiol. Behav.* **53**:959-967.

Rasmussen, L. E., Schmidt, M., Henneous, R., and Groves, D., 1982, Asian bull elephants: Flehmen-like responses to extractable components in female elephant estrous urine, *Science* **217**:159-162.

Schilling, A., Serviere, J., Gendrot, G., and Perret, M., 1990, Vomeronasal activation by urine in the primate *Microcebus murinus*: a 2 DG study, *Exper. Brain Res.* **81**:609-618.

Shibuya, T., Aihara, Y., and Tonosaki, K., 1977, Single cell responses to odors in the reptilian olfactory bulb, in: *Food Intake and Chemical Senses*, Y. Katsuki, ed., University of Tokyo Press, Tokyo, pp. 23-32.

Shoji, T., Abe, Y., Furihata, E., and Kurihara, K., 1994, High sensitivity of the turtle olfactory system to nonvolatile substances: comparison of response properties with those in gustatory systems. *Brain Res.* **666**:68-76.

Singer, A.G., Clancy, A. N., Macrides, F., and Agosta, W. C., 1984, Chemical studies of hamster vaginal discharge: Effects of endocrine ablation and protein digestion on behaviorally active macromolecular fractions, *Physiol. Behav.* **33**:639-643.

Singer, A. G., Macrides, F., Clancy, A. N., and Agosta, W. C., 1986, Purification and analysis of a proteinaceous aphrodisiac pheromone from hamster vaginal discharge, *J. Biol. Chem.* **261**:13323-13326.

Sipos, M. L., Wysocki, C. J., Nyby, J. G., Wysocki, L, and Nemura, T. A., 1995, An ephemeral pheromone of female house mice: perception via the main and accessory olfactory systems, *Physiol. Behav.* **58**:529-534.

Taniguchi, M., Wang, D., and Halpern, M., 1998, The characteristics of the electrovomeronasogram (EVG): Its loss following vomeronasal axotomy in garter snake, *Chem. Senses* **23**:653-659.

Taniguchi, M., Wang, D., and Halpern, M., 2000, Chemosensitive conductance and inositol 1,4,5 trisphosphate-induced conductance in snake vomeronasal receptor neurons, *Chem. Senses* **25**:67-76.

Wang, D., Chen, P., and Halpern, M., 1988, Isolation from earthworms of a proteinaceous chemoattractant to garter snakes. *Arch. Biophys. Biochem.* **267**:459-466.

Wang, D., Jiang, X. C., Chen, P., Inouchi, J., and Halpern, M., 1993, Chemical and immunological analysis of prey-derived vomeronasal stimulants, *Brain Behav. Evol.* **41**:246-254.

Wysocki, C. J., Bean, N. J., and Beauchamp, G. K., 1985, The mammalian vomeronasal system: Its role in learning and social behaviors, in: *Chemical Signals in Vertebrates IV, Ecology, Evolution and Comparative Biology*, D. Duvall, D. Muller-Schwarze, and R. M. Silverstein, eds., Plenum Press, New York, pp. 471-485.

Wysocki, C. J., Wellington, J. L., and Beauchamp, G. K., 1980, Access of urinary nonvolatiles to the mammalian vomeronasal organ, *Science* **207**:781-783.

# COMMUNICATION BY MOSAIC SIGNALS:
## Individual recognition and underlying neural mechanisms

Robert E. Johnston*

## 1. INTRODUCTION

Many aspects of social behavior depend on long-term relationships between individuals. From the individual's point of view, it is essential to remember neighbors, rivals, potential mates, etc., so as to promote one's own reproductive success. Naturalists and experimental scientists have been aware of this for a long time, and there have been numerous studies of individual differences in vocalizations, odors, visual appearance and other cues that could be used for such recognition (Colgan, 1983; Halpin, 1980). Similarly, there have been many investigations of the ability to discriminate between individuals of the same species (Colgan, 1983; Halpin, 1980; Halpin, 1986; Johnston et al., 1993; Johnston and Jernigan, 1994). Less common are attempts to investigate the complete repertoire of cues that provide individual information in a species, the types of knowledge that individuals have about one another, and how knowledge of other individuals influences responses to other signals, such as threats or signals used to attract mates. Although recognition of others is important for an individual's reproductive success and for social behavior and organization, we still know relatively little about the basic processes involved in individual recognition.

One reason for this is that the signals themselves are necessarily more complex than signals used for many other purposes. For example, chemical signals that attract a mate or influence the endocrine system may be single chemical compounds ("pheromones") or mixtures of a few (2-5) chemical compounds in relatively specific ratios ("pheromone blends" (Johnston, 2000; Linn and Roelofs, 1989; Sorensen and Stacy, 1999). In order to discriminate between individuals, however, information must be available that is characteristic of one individual but different from other individuals, while still retaining relevant species-typical features. Such information is invariably provided by complex,

---

* Robert E. Johnston, Cornell University, Ithaca, New York, 14853

multi-dimensional stimuli, such as the appearance of the face, the tonal quality of the voice or other sounds, or the quality of the odor of a chemical secretion or excretion (Bruce, 1991; Johnston, 2000; Smith et al., 2001).

Odors provide individually distinctive information and are used for discrimination between individuals in a wide range of species. Although many species use visual or auditory cues, odor cues may be the most widespread and most important source of individually specific information among mammals. The nature of individual differences in odors, however, has been investigated in only a few mammalian species. Studies of individual recognition in a marmoset, *Callithrix jaccus*, have shown that a secretion from one individual differs from that of another individual in the proportions of chemical compounds in the mixture and that this information is sufficient for animals and a mathematical algorithm to discriminate between these odors (Smith et al., 2001). This study also showed that differences between individuals were not restricted to a particular subset or class of chemical constituents within a secretion. Rather, the compounds that differed between one pair of marmosets were different from those that differed between other pairs. These same principles seem to apply to similar discriminations in other species. Most notably, kin or colony discrimination in insects depends on the proportions of different hydrocarbons on the surface of the body (Bonavita-Cougourdan et al., 1987; Gamboa et al., 1996; Howard, 1993; Smith and Breed, 1995).

I have coined the term "mosaic signal" to characterize this type of chemical signal because, like a real mosaic made of small tiles, each unit (or each tile) has no meaning by itself. Rather, the individual units provide information only because of their relationship to other units (Johnston, 2000). I coined this term because I think it is important to make clear the distinction between mosaic signals, "pheromones" and "pheromone blends". In addition to the difference in the chemical nature of the signal, there must also be differences in the neural mechanisms underlying the perception of pheromones and mosaic signals and in the mechanisms leading to responses to these signals. Most importantly, (1) mosaic signals must be analyzed by a complex pattern-recognition system rather than a few dedicated receptors, receptor cells and labeled-line central circuits, and (2) the significance of mosaic signals must be learned rather than, at the extreme, entirely genetically specified (Johnston, 2000; Johnston, 2003).

Studying the neural mechanisms underlying individual recognition by odors should be valuable as a counterpoint to studying the neural mechanisms underlying responses to classic pheromones. Presumably the mechanisms underlying responses to pheromones are relatively hard-wired and simple. The responses to mosaic signals, in contrast, will be more complex and probably involve different pathways and regions of the central nervous system. Since the majority of tasks carried out by the sense of smell in vertebrates probably involves discriminating between complex mixtures of compounds, e.g., discrimination and recognition of foods, the condition of foods (ripe vs. not ripe or spoiled), evaluation of potential mates (reproductive condition, hormonal status, health, social status, etc.), the investigation of the neural mechanisms underlying individual recognition may be a good model system for investigation of higher-order processes in the olfactory system.

## 2. SOURCES OF INDIVIDUALLY DISTINCTIVE ODORS

In order to understand individual recognition it is important to determine the sources of individually distinctive information that are used for this purpose. Some years ago we investigated whether individually specific odors were distributed all over the body or were restricted to specific locations on the body. On the one hand, one might expect that grooming would distribute secretions from scent glands or other sources of odors over the entire body. In addition, the secretions of sebaceous glands on every hair follicle of mammals could be a source of individually distinctive information, as could urine and feces, which animals may contact and get traces of these substances on the body (perhaps especially for laboratory animals kept in small cages). On the other hand, if identifying information is all over the body, why do individuals of almost all species concentrate on sniffing particular locations on the body when two animals meet?

To investigate the distribution of information about individual identity on the surface of the body, we tested golden hamsters in a habituation-dishabituation task. Subjects were exposed to freshly collected samples of the odor from one particular source of a donor animal and we measured the time that subjects investigated this odor. We found, as expected, that over repeated trials, investigation time decreased. Then, on a test trial, we exposed subjects to the same type of odor from a second individual. When such odors were taken from flank glands, ear glands (inside the pinna), vaginal secretions, urine or feces, subjects investigated the novel scent significantly longer than the familiar scent on the previous trial, indicating that they noticed the difference. In contrast, no such increase was found when scent from six other regions of the body were tested : saliva, feet, fur on the ventral surface, fur from between the shoulders, fur behind the ears, or fur from the flank region after the flank gland had been removed (Johnston et al., 1993). These results indicate that sources of individually specific information are located in specific places on the body and were not found all over the body. Similar experiments with a species of dwarf hamster, *Phodopus campbelli*, yielded similar results (Lai and Johnston, 1994). Interestingly, the ability to distinguish between the odors of different individuals does not appear to be highly constrained by species-typical features of the odors. Both golden hamsters and Campbell's hamster easily discriminated between odors of individuals from the other species (Johnston and Robinson, 1993).

These studies indicate that hamsters in at least two genera readily notice the difference between the odors of two individuals and remember these odor qualities over at least 15 minutes. Other experiments show that after 5 such habituation trials, golden hamsters remember the odor quality of the first donor and increase their investigation of a novel hamster's odor when tested 10 days later (Johnston, 1993). Since individuals of these two hamster species also distinguish between odors of individuals in another genus, this suggests that these discriminations are probably due to a generic ability to make distinctions between subtle differences in complex mixtures, rather than to perceptual mechanisms specialized for conspecific odors.

I am not aware of any other systematic attempt to investigate the complete repertoire of individually distinctive odors within a species, although more than one source of such cues has often been studied. For example, experiments with the dwarf mongoose suggest that the anal gland secretion is individually distinctive whereas the cheek gland secretion is not (Gorman, 1976; Rasa, 1973). Similarly, in both black-tailed dear and the North American beaver, one source of odor has been shown to be individually distinctive but other secretions are not (Müller-Schwarze, 1974; Müller-Schwarze, 1999). It is interesting to note that we have also tested some odors in golden hamsters for sex and reproductive state information, and we have found that there is an unpredictable pattern of types of information across odor sources. That is, some secretions that are not individually distinctive do contain information about the sex of the animal or about the reproductive state, whereas others do not. Likewise, some sources, such as saliva, contain information about sex but not individuals (Johnston, 2003).

## 3. COMPARISON OF METHODS USED TO ASSESS DISCRIMINATION BETWEEN ODORS OF INDIVIDUALS

I use the term discrimination to refer to the basic sensory ability to distinguish between two odors, in this case the odors of two individuals. These abilities can be directly assessed by a number of methods, but the most common are trained discriminations and habituation-dishabituation methods.

Habituation methods involve exposing a subject to repeated samples of odor from the same source (e.g., flank gland odor from one male hamster) on a series of trials and measuring the time spent investigating the odor. With repeated trials with the same odor, investigation time decreases. On the test trial, the subject is exposed to the flank gland of a different donor or to the flank gland secretion from the same male and a different male simultaneously). Investigation time generally increases to the novel stimulus, indicating that the animals do notice the difference between the two samples. Positive results from habituation studies allow one to conclude that animals can discriminate between two odors or other stimuli. Furthermore, they show that animals spontaneously do make such discriminations, that is, with no external reward or punishment, subjects behave differently in response to odor from the new individual. Negative results in such experiments (after habituation to one odor, no increase in investigation of a new individual's odor), however, are less readily interpreted. Such results could be due to a lack of ability to discriminate but they could also be due to the lack of motivation (i.e., the subjects don't care about a perceived difference). That is, negative results in habituation tests do not show that the animals can not discriminate, only that they did not discriminate. One can assess the likelihood that lack of motivation is the explanation by testing the subjects with additional stimuli that the animal has discriminated in the past or by manipulating the animal's motivational state (Todrank et al., 1998; Todrank et al., 1999a).

Conditioning experiments, in which reward (or punishment) is made contingent on the ability to choose one odor over another, are a more direct means of assessing the sensory capacities of the animal in question. For example, it has been shown that when the odors of two individuals are extremely similar due to similar genotype (Major Histocompatibility Complex (MHC) types), rats do not discriminate when tested in a habituation-dishabituation task, but they do discriminate when they are trained to do so (Brown et al., 1987). On the other hand, the results of studies using trained discriminations have the disadvantage that they may not reflect how animals spontaneously behave in nature. That is, positive results showing that animals do discriminate between two odors after training do not prove that animals actually make such discriminations in the wild. They may be capable of such discriminations, but if the discrimination has not been associated with some consequence, they may not make it.

Habituation methods have the advantage that they do reflect the natural, untrained behavior of animals. They can also reflect the effects of relevant experience. In habituation studies aimed at discrimination between the odors of closely related individuals (full siblings), we found that golden hamsters and Turkish hamsters (*Mesocricetus brandti*) did spontaneously discriminate between the odors two of their own brothers or between two foster siblings (full siblings of each other) that the subjects were raised with. However, they did not discriminate between either two brothers from another family that the subjects had not interacted with or two of their own full siblings that the subjects were not familiar with (Heth et al., 1999; Todrank et al., 1998). If the subjects had a few brief interactions with these brothers before the habituation tests were carried out, they did discriminate between these full siblings (Todrank et al., 1999b). These data are consistent with the interpretation that the odors of full siblings are very similar and that without experience with the donors the differences between their odors are not noticed. Once the male subjects had interacted with the scent donors, however, they learned these subtle differences (or they learned the significance of these subtle differences). The degree of similarity in odor quality of full siblings may in part be due to the highly inbred nature of domestic stocks of golden hamsters (all derived from one family (Gattermann, 2000; Murphy, 1985), but the same degree of inbreeding is not the case for domestic stocks of Turkish hamsters, that show similar effects (Heth et al., 1999).

## 4. THE NATURE OF THE REPRESENTATIONS OF INDIVIDUALS

The ability to discriminate between cues from different individuals and remember these cues indicates that the basic information is available to allow an animal to remember individuals, but it does not demonstrate this ability. That is, memory of a specific individual's odor, or song, or visual pattern (e.g., face) does not prove that an animal recognizes an individual as such; it just indicates that the animal remembers that sensory pattern. Indeed, many experiments that have been touted as demonstrating individual recognition merely show recognition of familiar sensory information compared to unfamiliar sensory information (Johnston and Jernigan, 1994). What, in addition, might be involved in recognition of individuals as individuals? At the very least, animals should have a multi-component representation of other individuals in memory; that is, the memory should consist of more than one individually distinctive attribute. Furthermore, the attributes of one individual should be linked in memory and be separate from similar, linked memories of other individuals. Having this type of memory could be thought of as having a concept of other individuals. Of course, the types of information that animals remember about other individuals may be much more complex than just a list of linked sensory features. For example, animals in many species no doubt remember specific interactions with other individuals or the social/emotional significance of these interactions; they may also remember the friends or supporters of the target individual, the relative ranks of individuals, etc. (Cheney and Seyfarth, 1990a; de Waal, 1982). Humans remember an immense amount about other individuals, even about individuals they have not interacted with and know only by hearsay or other indirect means (e.g., books, magazines, TV).

It is not easy to study the nature of the knowledge that animals have of one another. We developed one method that has proven useful in discovering what cues an individual remembers about another individual, based on techniques used to discover what monkeys know about their world (Cheney and Seyfarth, 1988; Cheney and Seyfarth, 1990b). This method is a variation on our standard habituation technique. First, we allowed male subjects to interact with two different females in a series of 4 encounters over 4 days. Then, as in earlier tests aimed at discovering the sources of odors that were individually distinctive, we habituated a male to one odor (e.g., vaginal secretion) from one of the females he was familiar with. On the test trial, we gave this male the flank scent of either the same female that had provided the vaginal secretion or the flank scent of the other familiar female and measured investigation of the odor (Johnston and Jernigan, 1994). These two odors have very different odor qualities to humans, and thus one would expect that, to hamsters, both flank scents would be quite different from the habituation odor and would be investigated more than the habituation odor on the last habituation trial. On the other hand, if, during habituation, the subject became habituated to the odor of the vaginal secretion of female A and to the female it represented, then the subject should show less investigation of the flank gland secretion from the "same" female (A) compared to that from the "different" female (B). This is what we found (Figure 1; Johnston and Jernigan, 1994). In this case the habituation odor, vaginal secretion, is investigated more than the flank gland odor because it is a source of sexually arousing stimuli, but the difference in the response to flank gland secretion from the same versus a new female is clear.

An alternative explanation of these results could be that there were common features in flank and vaginal secretions from the same individual and thus the results were due to a generalization from one odor to the other. This interpretation is unlikely. First, the two secretions are quite different in their chemical composition and odor quality. The vaginal

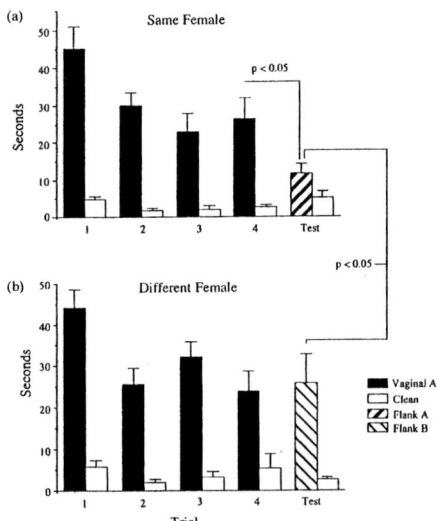

**Figure 1.** The mean (± SEM) time that males spent investigating the scents from familiar females (black bars) and the clean side of the plate (open bars); the scent was vaginal secretion during habituation trials (black bars) and flank gland secretion on the test trial (diagonally striped bars). In the test trials of (a), males investigated the flank scent from the same female that had provided the vaginal secretion, whereas in (b) they investigated the flank secretion of a different, familiar female.

secretion is a mucoid secretion that also contains sloughed cellular debris that has been subject to bacterial action (O'Connell, 1983; O'Connell and Meredith, 1984; O'Connell et al., 1978). This secretion has a rich, pungent odor to humans. In contrast, the secretion from the flank gland is a thick, waxy, sebaceous secretion, but also contains a variety of smaller molecules (W. Ma & M. Novotny; personal communication). It has little or no odor to humans. Second, if the observed cross-habituation effect was due to stimulus similarity, we should see the same cross-habituation effect in animals that did not interact with and get to know the stimulus donors. Control experiments of this type show that males that have not interacted with the donors do not show cross-habituation between vaginal and flank secretions (Johnston and Bullock, 2001). Thus, the most likely explanation of the cross-habituation effect is that it is due to habituation to the individual as well as to the specific odor during the 4 exposures to the odor (Johnston and Bullock, 2001; Johnston and Jernigan, 1994).

Subsequent experiments have further characterized this cross-habituation effect. Male hamsters show this effect when vaginal and ear gland secretions of familiar females are used, and females also show this effect when flank and ear gland secretions from familiar males are used (Johnston and Bullock, 2001). A curious exception is that cross-habituation was not obtained with male subjects between urine and either flank or vaginal secretions, despite the fact that urine is individually distinguishable. We suggest that the reason for this is that hamsters may not use urine as a means of communication because such use would waste water in this desert-dwelling species. Golden hamsters produce extremely thick, concentrated urine and they do not use urine to scent mark (Johnston and Bullock, 2001), suggesting that water conservation is a high priority.

We also investigated the nature of the experience that is necessary for hamsters to show this cross-habituation effect. We tested the hypothesis that for a solitary species it might be sufficient to explore another animal's home living area in order to form a multi-odor representation of that individual. Even if other individuals have left scent marks in such an area, the odors of the resident would predominate. After exposure to the home cages of two females on four days, males were tested for cross habituation between vaginal and flank secretions. There was, however, no difference in their response to the flank gland secretions from the two females, indicating no cross-habituation. In contrast, exposure of males to the scent-donor females across a wire-mesh barrier did result in cross-habituation, indicating that some limited interaction between individuals is necessary in order to form a multi-odor memory of that individual (Johnston and Bullock, 2001).

## 5. INDIVIDUAL RECOGNITION IN A SOCIAL CONTEXT

The cross-habituation experiments described above indicate that hamsters are not just reacting to familiar versus unfamiliar odors, but rather have integrated memories of individuals that incorporate at least several different cues. These experiments did not, however, address the issue of what significance another individual has for the subject, which, in the long run, is the most important aspect of such recognition in that it can influence an individual's reproductive success. The significance of other individuals must be studied in a realistic social context. We have developed several methods to do this. The basic plan for these experiments is to first allow subjects to interact with one or two individuals and later compare their reactions to the cues from familiar individuals that they had different types of experiences with as well as from unfamiliar stimulus animals (Lai and Johnston, 2002; Lai, 2003).

In one set of experiments of this type, adult male subjects fought with another adult male in a small arena (36 x 30 x 16 cm) with an open top. We did not enclose the males because we wanted to allow the loser of a fight to escape whenever he wanted. This design was based on observations in nature of *Mesocricetus raddei*, in which aggressive encounters were quite brief and terminated with the flight of one animal (Johnston, unpublished). Upon first meeting, males typically sniffed one another, engaged in a variety of sideways and upright postures, and got into a "rolling fight" (Floody and Pfaff, 1977; Johnston, 1976). After a brief, sometimes intense fight, one male would quickly flee and jump out of the arena (mean latency to flee, 49.0 sec; (Lai and Johnston, 2002). Males were returned to their home cages for 3 min and then placed into the arena together again; 3 min after this encounter, the two were paired a third time. In these encounters the original loser fled and jumped out of the arena immediately after encountering the familiar winner (mean latency to jump out less than 10 sec).

In the test phase (30 min to 7 days after fighting), the behavior of subjects was observed and recorded in a Y-maze. First, males were placed in a clean Y maze with no stimuli present and their behavior was recorded during a 3-min trial. At the end of this trial males were confined in the start box of the Y-maze and the fan that drew air through the maze was turned off for about 40 seconds while we placed a stimulus male into a compartment at the end of one arm of the Y. The fan was turned on and 20 sec later the subject was allowed to explore the Y-maze again for 3 min. We recorded time spent in various parts of the maze, latency to reach the ends of each arm of the maze, and overall activity (Lai and Johnston, 2002).

Characteristic results are shown in Figure 2. One day after the interaction phase, males that lost fights spent more time in the base of the Y-maze, near the start box, than in either of the arms of the Y and they spent very little time near the male that had beaten them in the series of three encounters. In contrast, males that did not fight or those that were tested with an unfamiliar winner spent less time in the base of the Y and much more time near the stimulus male. Males that lost also took much longer to approach the end of the Y containing the familiar winner than the clean arm of the maze. Similarly, the latency to approach a stimulus male was much greater for males tested with a familiar winner versus those tested with an unfamiliar winner. Similar results were found when we tested males 30 min, 1 day, 3 days or 7 days after the three brief encounters (Lai and Johnston, 2002).

These results indicate that males learn the characteristics of other males that they interact with and later adjust their reactions to these individuals depending on the type of experience they had with the stimulus animal. Although we can not prove that odors were the only cues being used by the subjects in these experiments, it is likely that recognition was based predominantly or entirely on odors. Although hamsters do produce ultrasonic vocalizations, they occur primarily in sexual contexts and there is no evidence for distinctive individual differences or even sex differences (Floody and Pfaff, 1977a.; Floody and Pfaff, 1977c; Floody et al., 1977b). Only minimal visual cues were available, and only when the subjects were immediately adjacent to the stimulus compartment. Thus we conclude that males in these experiments recognized familiar males by odor cues and responded appropriately, depending on the experience that they had with the stimulus animal. It is noteworthy that males that lost and were tested with an unfamiliar stimulus male showed some hesitancy to approach but were significantly less hesitant than when tested with a familiar winner (Lai and Johnston, 2002). Thus, the behavior of subjects toward familiar winners was not due to generalized anxiety or to fear of any male hamster, but rather their fear was specific to stimuli from the individual that had beaten them. These data contrast with those obtained with the conditioned defeat method, in which a generalized fear is shown after much longer aggressive encounters (e.g., three interactions of 15 min; (Huhmun et al., 1992; Jasnow and Huhman, 2001).

These experiments demonstrate that our method does elicit different responses to different stimulus animals, and thus does measure recognition of individuals with different significance to the subject. Further experiments support this conclusion. In these experiments we tested the reactions of males that had a series of aggressive encounters with one male, as described above, and a series of neutral encounters across a wire mesh with a second male. Males reactions depended on the stimulus male that they were tested with: males were attracted to the familiar neutral stimulus male but avoided and showed fearful reactions to the familiar winner (Lai, 2003). Thus, males showed unique responses to other males depending on their past history with them (Lai, 2003).

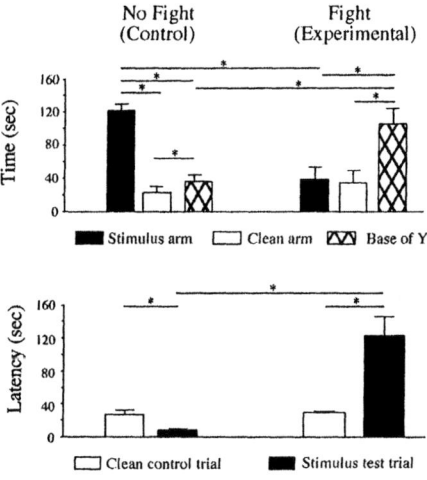

**Figure 2.** The behavior of male hamsters in a Y-maze: subjects had either fought and lost and were tested in a Y-maze or were exposed to an arena and were tested in a Y maze. Males that fought and lost (right side) spent most of their time in the start box (top right) and took a long time to approach the familiar winner (bottom right). Males that were just exposed to a clean arena spent most time near the unfamiliar winner (top left) and approached this male more quickly than the clean arm of the Y (bottom left). (* - $p < 0.05$; Lai and Johnston, 2002)

## 6. NEURAL MECHANISMS OF INDIVIDUAL DISCRIMINATION AND RECOGNITION

### 6.1. Main Olfactory System and Accessory Olfactory System

The most thoroughly described neural mechanisms related to learning about odors are those involved in the reactions of female mice to a familiar mate versus an unfamiliar male (Kaba et al., 1989; Keverne, 1998; Lloyd-Thomas and Keverne, 1982). If female mice are exposed within four days after mating to a male or the odors of a male that was not the mate, implantation of the fertilized eggs fails to take place in the majority of females (the Bruce or pregnancy block effect). In an elegant series of experiments over many years, this effect has been shown to be dependent on events taking place in the accessory olfactory bulb (Brennan and Keverne, 2000; Keverne, 1998). These findings are something of a surprise, however, because it is generally believed that the main olfactory system is the one that is primarily responsible for the perception of complex odor mixtures (Johnston, 2000; Johnston and Rasmussen, 1984; Matochik, 1988; Wilson and Stevenson, 2003).

Two lines of evidence may account for this apparent inconsistency. First, the vomeronasal organ in many species is often particularly important for reception of and response to large, relatively non-volatile molecules (Meredith, 1980; Wysocki and Meredith, 1987; Wysocki et al., 1980) and there is evidence that mouse urinary proteins (MUPS) are one source of individually specific information (Brennan and Keverne, 2000; Hurst et al., 2001). Second, the vomeronasal and main olfactory systems often both make contributions to the processing of a specific chemical signal (Johnston, 1998). For example, in investigations of the ability to discriminate between individual odors among hamsters we have found that the vomeronasal organ may contribute to such discrimination, but only for some sources of odor and only for discrimination by males (Johnston, 1998; Johnston and Peng, 2000). In these experiments we used habituation experiments and tested both males and females for their discrimination of several odors (flank glands, vaginal secretions, urine). Removal of the vomeronasal organ from female hamsters had no effect on their discrimination of odors of other individuals. In contrast, removal of the vomeronasal organ from males did compromise their performance in tests with flank gland secretion, but not with urine or vaginal secretion (Johnston and Peng, 2000). Thus, in hamsters the VNO may have different roles in males and females. In addition, in male hamsters the VNO appears to influence discrimination between individuals using some odors but not others.

### 6.2. Central Nervous System

For the majority of cases of individual discrimination and recognition, it is likely that the main olfactory system is the primary system involved and that higher-order olfactory processing areas are essential. Thus one would expect that projection areas of the main olfactory system should be involved in individual discrimination. Recent neurophysiological studies of the anterior piriform cortex indicate a role for this region in discrimination of learned odors, including odors of mixtures (Wilson, 2002; Wilson and Stevenson, 2003). In one set of lesion studies we showed that one area that may be important for discrimination of individual odors is the lateral entorhinal cortex and the surrounding para-hippocampal area (peri-rhinal cortex, temporal cortex, and subiculum).

Lesions of this area resulted in failure to discriminate between odors of two individuals in habituation tests, whereas such lesions did not affect behaviors that depended on discrimination between males and females (Petrulis et al., 2000). In contrast, lesions of the medial amygdala eliminated differences in responses to odors of males and females but did not influence the ability to discriminate between odors of two individuals (Petrulis and Johnston, 1999). These results suggest that information about male odors versus female odors may be processed in circuits that are partly separate from those dealing with individual differences.

In a recent set of experiments we have been trying to identify the circuits involved in individual recognition and, simultaneously, the appropriate responses to these individuals. We are using the behavioral methods described above for studying recognition of familiar opponents in aggressive interactions and imaging brain activity with the use of immediate early genes (IEGs) as markers for cell activity. In our initial experiment, we compared the activity in 20 brain areas using immunohistochemistry for the proteins produced by the activation of two IEGs, *c-fos and egr-1*. As in the experiments described above, males in the exposure phase had three brief aggressive encounters with a stimulus male. One day later they were tested for their responses to the odors of this male in a Y-maze. The behavior of these males was like that described above; males tested with a familiar winner avoided this stimulus male. One hour after the Y-maze test, males were sacrificed for histochemistry. The brain tissue of this experimental group was compared to those of two control groups: one that was exposed to a clean area during the exposure phase and a clean Y-maze during the test phase and one group in which the males remained in their home cages at all times.

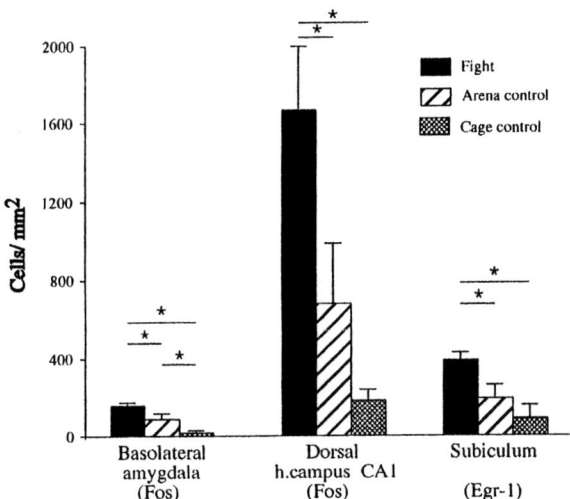

Figure 3. Mean (±SEM) density of cells stained for Fos or Egr-1 proteins in the basolateral amygdala, CA1 region of anterior dorsal hippocampus, and posterior dorsal subiculum in groups of males (n = 6 per group) that fought and lost and were later exposed to the odors of the male that won the fight (Fight group), males that were merely exposed to the arena and exposed to a clean Y maze (Arena control group), or males that remained in their home cages (Cage control group). (* - $p < 0.05$; Lai, 2003; Lai et al., 2004).

The most interesting findings relevant to individual recognition were the results from three areas of the brain in which the density of stained cells for the males that fought, lost and were exposed to the familiar winner was significantly greater than the other two groups (Figure 3). Two of these areas were the anterior CA1 region of dorsal hippocampus and more posterior parts of the dorsal subiculum (Lai, 2003; Lai et al., 2004). Given that the hippocampus is involved in many different kinds of memory processes, our interpretation is that the CA1 region of the hippocampus and the dorsal posterior subiculum are involved in retrieval of the memory of the familiar opponent. The other area that showed significantly different levels of cell staining was the basolateral nucleus of the amygdala (Figure 3). This area has been implicated in learned fear in conditioning studies (Davis, 1992; LeDoux, 1993; Maren, 2001), and thus the activation we observed is probably related to the emotional and motivational response to the familiar winner, although we cannot rule out the possibility that this area is involved in the memory for a familiar winner as well.

## 7. CONCLUSION

Although the mechanisms of individual recognition have not been thoroughly studied, they are central to understanding how the brain regulates social behavior and emotional reactions to others. Such reactions necessarily involve the integration of a variety of different processes. By studying recognition in a variety of different social contexts, however, it may be possible to untangle some of the complex webs of interacting brain areas. In particular, the study of individual recognition by odors is likely to be a valuable model for the investigation of other behaviors that are guided by similar complex odors.

## 8. REFERENCES

Bonavita-Cougourdan, A., Clement, J. L., and Lange, C., 1987, Nestmate recognition: the role of cuticular hydrocarbons in the ant *Camponotus vagus* Scop, *J. Entomol. Sci.* **22**:1-10.
Brennan, P. A., and Keverne, E. B., 2000, Neural mechanisms of olfactory recognition memory, in: *Brain, Perception, Memory*, J. Bolhuis, ed., Oxford University Press, Oxford, pp. 93-112.
Brown, R. E., Singh, P. B., and Roser, B., 1987, The major histocompatibility complex and the chemosensory recognition of individuality in rats, *Physiol. Behav.* **40**:65-73.
Bruce, V. ed. 1991, *Face Recognition*, Lawrence Erlbaum Associates, Hillsdale, New Jersey.
Cheney, D. L., and Seyfarth, R. M., 1988, Assessment of meaning and the detection of unreliable signals by vervet monkeys, *Anim. Behav.* **36**:477-486.
Cheney, D. L., and Seyfarth, R. M., 1990a, *How Monkeys See the World*. University of Chicago Press, Chicago.
Cheney, D. L., and Seyfarth, R. M., 1990b, The representation of social relations by monkeys, *Cognition* **37**:167-196.
Colgan, P., 1983, *Comparative Social Recognition*, John Wiley & Sons, New York.
Davis, M., 1992, The role of the amygdala in conditioned fear, in: *The Amygdala*, J. P. Aggleton, ed., Wiley-Liss, New York, pp. 255-306.
de Waal, F., 1982, *Chimpanzee Politics*, Harper and Row, New York.
Floody, O. R., and Pfaff, D. W., 1977, Aggressive behavior in female hamsters: the hormonal basis for fluctuations in female aggressiveness correlated with estrous state, *J. Comp. Physiol. Psych.* **91**:443-464.
Floody, O. R., and Pfaff, D. W., 1977a, Communication among hamsters by high frequency acoustic signals: I. Physical characteristics of hamster calls, *J. Comp. Physiol. Psychol.* **91**:794-806.
Floody, O. R., and Pfaff, D. W., 1977c, Communication among hamsters by high-frequency acoustic signals: III. Responses evoked by natural and synthetic ultrasounds, *J. Comp. Physiol. Psychol.* **91**:820-829.

Floody, O. R., Pfaff, D. W., and Lewis, C. D., 1977b, Communication among hamsters by high-frequency acoustic signals: II. Determinants of calling by females and males, *J. Comp. Physiol. Psychol.* **91**:807-819.

Gamboa, G. J., Grudzien, T. A., Espelie, K. E., and Bura, E. A., 1996, Kin recognition pheromones in social wasps: combining chemical and behavioural evidence, *Anim. Behav.* **51**:625-629.

Gattermann, R., 2000, 70 Jahre Goldhamster in menschlicher Obhut - wie gross sind die Unterschiede zu seinen wildlebenden Verwandten? *Tierlaboratorium* **23**:86-99.

Gorman, M. L., 1976, A mechanism for individual recognition by odour in Herpestes auropunctatus (Carnivora:Viverridae), *Anim. Behav.* **24**:141-145.

Halpin, Z. T., 1980, Individual odors and individual recognition: review and commentary, *Biol. Behav.* **5**:233-248.

Halpin, Z. T., 1986, Individual odors among mammals: origins and functions, *Adv. Study Behav.* **16**:39-70.

Heth, G., Todrank, J., and Johnston, R. E., 1999, Similarity in the qualities of individual odors among kin and species in Turkish (*Mesocricetus brandti*) and Golden (*Mesocricetus auratus*) hamsters, *J. Comp. Psychol.* **113**:321-326.

Howard, R. W., 1993, Cuticular hydrocarbons and chemical communication, in: *Insect Lipids: Chemistry, Biochemistry and Biology*, D. W. Stanley-Samuelson and D. R. Nelson, eds., University of Nebraska Press, Lincoln, pp. 179-226.

Huhmun, K. L., Moore, T. O., Mougey, E. H., and Meyerhoff, J. L., 1992, Hormonal responses to fighting in hamsters: separation of physical and psychological causes., *Physiol. Behav.* **51**:1083-1086.

Hurst, J. L., Payne, C. E., Nevinson, C. M., Marle, A. D., Humphries, R. E., Robertson, D. H. L., Cavaggioni, A., and Beynon, R. J., 2001, Individual recognition in mice mediated by major urinary proteins, *Nature* **414**:631-634.

Jasnow, A. M., and Huhman, K. L., 2001, Activation of $GABA_A$ receptors in the amygdala blocks the acquisition and expression of conditioned defeat in Syrian hamsters, *Brain Res.* **920**:142-150.

Johnston, R. E., 1976, The role of dark chest patches and upright postures in the agonistic behavior of male hamsters, *Mesocricetus auratus*, *Behav. Biol.* **17**:161-176.

Johnston, R. E., 1993, Memory for individual scent in hamsters (*Mesocricetus auratus*) as assessed by habituation methods, *J. Comp. Psychol.* **107**:201-207.

Johnston, R. E., 1998, Pheromones, the vomeronasal system, and communication, in: *Olfaction and Taste XII. An International Symposium. Ann. NY Acad. Sci.*, C. Murphy, ed., NY Academy of Sciences, New York, Vol. 855, pp. 333-348.

Johnston, R. E., 2000, Chemical communication and pheromones: the types of signals and the role of the vomeronasal system, in: *The Neurobiology of Taste & Smell, 2nd edition*, T. E. Finger, W. L. Silver, and D. Restrepo, eds., Wiley, New York, pp. 101-127.

Johnston, R. E., 2003, Chemical communication in rodents: from pheromones to individual recognition, *J. Mammal.* **84**(4):1141-1162.

Johnston, R. E., and Bullock, T. A., 2001, Individual recognition by use of odours in golden hamsters: the nature of individual representations, *Anim. Behav.* **61**:545-557.

Johnston, R. E., Derzie, A., Chiang, G., Jernigan, P., and Lee, H.-C., 1993, Individual scent signatures in golden hamsters: evidence for specialization of function, *Anim. Behav.* **45**:1061-1070.

Johnston, R. E., and Jernigan, P., 1994, Golden hamsters recognize individuals, not just individual scents, *Anim. Behav.* **48**:129-136.

Johnston, R. E., and Peng, M., 2000, The vomeronasal organ is involved in discrimination of individual odors by males but not by females in golden hamsters, *Physiol. Behav.* **70**:537-549.

Johnston, R. E., and Rasmussen, K., 1984, Individual recognition of female hamsters by males: Role of chemical cues and of the olfactory and vomeronasal systems, *Physiol. Behav.* **33**:95-104.

Johnston, R. E., and Robinson, T., 1993, Cross-species discrimination of individual odors by hamsters, *Ethology* **94**:317-325.

Kaba, H., Rosser, A., and Keverne, E. B., 1989, Neural basis of olfactory memory in the context of pregnancy block, *Neuroscience* **32**:657-662.

Keverne, E. B., 1998, Vomeronasal/accessory system and pheromonal recognition, *Chem. Senses* **23**:491-494.

Lai, S.-C., and Johnston, R. E., 1994, Individual odors in Djungarian hamsters, *Phodopus campbelli*, *Ethology* **96**:117-126.

Lai, W.-S., 2003, To fight or not to fight? Individual recognition and social memory in golden hamsters (*Mesocricetus auratus*). Ph.D., Cornell University.

Lai, W.-S., Chen, A., and Johnston, R. E., 2004, Functional neuroanatomy of recognition of a familiar opponent in golden hamsters, *Horm. Behav.* **44**:in press.

Lai, W.-S., and Johnston, R. E., 2002, Individual recognition after fighting by golden hamsters: A new method, *Physiol. Behav.* **76**:225-239.

LeDoux, J. E., 1993, Emotional memory systems in the brain, *Behav. Brain Res.* **58**:69-79.
Linn, C. E., and Roelofs, W. L., 1989, Response specificity of male moths to multicomponent pheromones, *Chem. Senses* **14**:421-437.
Lloyd-Thomas, A., and Keverne, E. B., 1982, Role of the brain and accessory olfactory system in the block to pregnancy in mice, *Neuroscience* **7**:907-913.
Maren, S., 2001, The neurobiology of Pavlovian fear conditioning, *Annu. Rev. Neurosci.* **24**:897-931.
Matochik, J. A., 1988, Role of the main olfactory system in recognition between individual spiny mice., *Physiol. Behav.* **42**:217-222.
Meredith, M., 1980, The vomeronasal organ and accessory olfactory system in the hamster, in: *Chemical Signals in Vertebrates and Aquatic Invertebrates*, D. Müller-Schwarze and R. Silverstein, eds., Plenum Press, New York, pp. 303-326.
Müller-Schwarze, D., 1974, Pheromones in black-tailed deer (*Odocoileus hemionus columbianus*), *Anim. Behav.* **19**:141-152.
Müller-Schwarze, D., 1999, Chemical signals in the beaver: One species, two secretions, many functions?, in: *Advances in Chemical Communication in Vertebrates*, R. E. Johnston, D. Müller-Schwarze, and P. W. Sorensen, eds., Kluwer Academic/Plenum Publishers, New York, pp. 281-288.
Murphy, M. R., 1985, History of the capture and domestication of the Syrian golden hamster (*Mesocricetus aruatus Waterhouse*), in: *The Hamster; Reproduction and Behavior*, H. I. Siegel, ed., Plenum Press, New York, pp. 3-20.
O'Connell, R. J., 1983, Estrous cycle modulation of the attraction to odors in female golden hamsters, *Behav. Neural Biol.* **37**:317-325.
O'Connell, R. J., and Meredith, M., 1984, Effects of volatile and non-volatile chemical signals on male sex behaviors mediated by the main and accessory olfactory systems, *Behav. Neurosci.* **98**:1083-1093.
O'Connell, R. J., Singer, A. G., Macrides, F., Pfaffmann, C., and Agosta, W. C., 1978, Responses of the male golden hamster to mixtures of odorants identified from vaginal discharge, *Behav. Biol.* **24**:244-255.
Petrulis, A., and Johnston, R. E., 1999, Lesions centered on the medial amygdala impair scent-marking and sex-odor recognition but spare discrimination of individual odors in female golden hamsters, *Behav. Neurosci.* **113**:345-357.
Petrulis, A., Peng, M., and Johnston, R. E., 2000, The role of the hippocampal system in social odor discrimination and scent marking in female golden hamsters (*Mesocricetus auratus*), *Behav. Neurosci.* **114**:184-195.
Rasa, O. A. E., 1973, Marking behaviour and its social significance in the African Dwarf Mongoose, *Helogale undulata rufula*, *Z. Tierpsych.* **32**:293-318.
Smith, B. H., and Breed, M. D., 1995, The chemical basis for nestmate recognition and mate discrimination in social insects, in: *Chemical Ecology of Insects 2*, R. T. Carde' and W. J. Bell, eds., Chapman & Hall, New York.
Smith, T. E., Tomlinson, A. J., Mlotkiewicz, J. A., and Abbott, D. H., 2001, Female marmoset monkeys (*Callithrix jaccus*) can be identified from the chemical composition of their scent marks, *Chem. Senses* **26**:449-458.
Sorensen, P. W., and Stacy, N. E., 1999, Evolution and specialization of fish hormonal pheromones, in: *Advances in Chemical Communication in Vertebrates*, R. E. Johnston and P. W. Sorensen, eds., Plenum Publishers, New York, pp. 15-47.
Todrank, J., Heth, G., and Johnston, R. E., 1998, Kin recognition in golden hamsters: evidence for kinship odors, *Anim. Behav.* **55**:377-386.
Todrank, J., Heth, G., and Johnston, R. E., 1999a, Kin and individual recognition: odor signals, social experience, and mechanisms of recognition, in: *Advances in Chemical Signals in Vertebrates*, R. E. Johnston, D. Müller-Schwarze, and P. W. Sorensen, eds., Plenum Press, New York, pp. 289-297.
Todrank, J., Heth, G., and Johnston, R. E., 1999b, Social interaction is necessary for discrimination between and memory for odours of close relatives in golden hamsters, *Ethology* **105**:771-782.
Wilson, D. A., 2002, Odor specificity of habituation in the rat anterior piriform cortex, *J. Neurophysiol.* **83**:139-145.
Wilson, D. A., and Stevenson, R. J., 2003, The fundamental role of memory in olfactory perception, *Trends Neurosci.* **26**:243-247.
Wysocki, C. J., and Meredith, M., 1987, The vomeronasal system, in: *Neurobiology of Taste and Smell*, T. E. Finger and W. L. Silver, eds., John Wiley & Sons, New York, pp. 125-150.
Wysocki, C. J., Wellington, J. L., and Beauchamp, G. K., 1980, Access of urinary nonvolatiles to the mammalian vomeronasal organ, *Science* **207**:781-783.

# SEXUAL DIMORPHISM IN THE ACCESSORY OLFACTORY BULB AND VOMERONASAL ORGAN OF THE GRAY SHORT-TAILED OPOSSUM, *MONODELPHIS DOMESTICA*

Jennifer H. Mansfield, Wei Quan, Changping Jia, and Mimi Halpern[*]

## 1. INTRODUCTION

Sex differences are present in the vomeronasal systems of many species, both in the vomeronasal organ (VNO) and in the CNS structures that form part of its central projection system. For example, males of the salamander species *Plethodon cinereus* have significantly larger VNOs than do females (Dawley, 1992; Dawley and Crowder, 1995). The bed nucleus of the stria terminalis (BNST) is larger in male than in female guinea pigs (Hines et al., 1985). Extensive studies in the rat by Segovia and Guillamón (Guillamón and Segovia 1993, 1996, 1997; Segovia and Guillamón, 1993, 1996) have revealed sex differences in many vomeronasal (VN) system structures, including the VNO (Segovia and Guillamón, 1982), the accessory olfactory bulb (AOB) (Roos et al., 1988; Valencia et al., 1992), the bed nucleus of the accessory olfactory tract (Collado et al., 1990), the BNST (del Abril et al., 1987; Guillamón et al., 1988), the posteromedial cortical nucleus of the amygdale (Vinader-Caerols et al., 1998), and non-vomeronasal structures such as the locus coeruleus (LC) (Guillamón et al., 1988). These sex differences appear during the early postnatal period and, as they can be reversed by gonadectomy of perinatal males and androgenization of perinatal females, appear to be induced by gonadal hormones (Roos et al., 1988; Segovia et al., 1984; Segovia et al., 1986).

The vomeronasal system is necessary in mediating a variety of reproductive behaviors--behaviors which are clearly sexually dimorphic. A male garter snake, for example, requires an intact vomeronasal system to court a female (Kubie et al., 1978). Female mice undergo several effects in response to pheromones of male conspecifics, including induction of estrus, estrus synchrony (in group-housed females), and pregnancy

---

[*] SUNY Downstate Medical Center, 450 Clarkson Avenue, Brooklyn, NY 11203.

block caused by strange male odors. All of these effects depend upon, and are mediated by, a functional vomeronasal system (Halpern, 1987).

The two components of the VN system examined in the present study are the vomeronasal organ and the accessory olfactory bulb. The VNO is a paired, chemoreceptive structure present at the base of the nasal septum in most terrestrial mammals, amphibians and reptiles. The VNO's bipolar receptor neurons detect pheromonal signals (Halpern, 1987; Farbman, 1992).

The AOB, which processes vomeronasal input (Halpern, 1987; Farbman, 1992), is located on the dorsocaudal surface of the main olfactory bulb (MOB) and, in addition to VN receptor axons, receives centrifugal afferents from nuclei of the vomeronasal system (Halpern, 1987; Farbman, 1992; Davis et all., 1978; DeOlmos et al., 1978; Raisman, 1972). The AOB is comprised of several distinct cell layers; its laminar pattern is similar to, but simpler than, that of the MOB (Halpern, 1987; Farbman, 1992). The deepest layer of the AOB is composed of light and dark granule cells. The lateral olfactory tract lies dorsal to the granule cell layer, and ventral to the mitral/tufted cell layer. Superficial to the mitral cells is an external plexiform layer, above which lies the glomerular layer. The most superficial lamina of the AOB consists of fiber bundles of the vomeronasal nerve, which terminate in the glomerular layer (Halpern, 1987; Farbman, 1992; McLean and Shipley, 1992).

The glomerular layer of the AOB can be divided into anterior and posterior (rostral and caudal) halves on the basis of their chemoarchitecture (Halpern et al., 1995): the anterior stains more darkly than the posterior when treated with the lectin VVA (*Vicia villosa* agglutinin) (Shapiro et al., 1995), NADPH-diaphorase (Shapiro and Halpern, 1998), and antibodies to olfactory marker protein (OMP) (Shnayder et al., 1993) and the G-protein $G_{i2\text{-alpha}}$ (Halpern et al., 1995). The posterior half is selectively stained by $G_{o\text{-alpha}}$ antibodies (Halpern et al., 1995; Jia and Halpern, 1996). A retrograde HRP labeling study has suggested, and another using immunohistochemical technique has confirmed, that these halves receive projections from distinct laminae of the VNO (Jia and Halpern, 1996; Shapiro et al., 1995).

Since the vomeronasal system is involved in pheromone recognition, a sexually dimorphic activity, and because prior studies have reported sex differences in the VNO and AOB of rats, we pursued this question in the opossum *Monodelphis domestica*. An additional interest was to determine if the two parts of the vomeronasal system are equally dimorphic.

This study was designed, therefore, to determine if there is sexual dimorphism in the volume of the AOB and VNO of the Brazilian gray short-tailed opossum. To accomplish this, we determined the volumes of the sensory epithelium of the VNO, and of the mitral/tufted layer as well as the anterior and posterior halves of the nerve fiber/glomerular layer of the AOB.

## 2. MATERIALS AND METHODS

Six adult male-female pairs of litter mates from six different mothers of the opossum species *Monodelphis domestica* were used for this study. At five months of age, the body weight and length of each animal was recorded immediately before it was deeply anesthetized and transcardially perfused with saline followed by Bouin's fixative.

The vomeronasal organs and brains were removed and post-fixed in Bouin's fixative overnight.

The brain was weighed, and its length with and without the olfactory bulb was measured prior to post-fixing. After post-fixing, each brain was cut parasagittally into right and left halves, paraffin-embedded, cut into 10 micron parasagittal sections, and mounted on slides. All sections lost during the cutting procedure were recorded. A series was Nissl stained with 1% cresyl violet to determine where the AOB began and ended in each animal, and to visualize the granule cell layer.

Every sixth section containing the AOB was retained for olfactory marker protein immunocytochemistry. For the OMP-staining procedure, slides were deparaffinized with xylene, a series of ethanol baths, and a .3% $H_2O_2$ in methanol bath to quench endogenous peroxidases in the tissue (Polack and Van Noorder, 1997). The sections were then rehydrated in phosphate-buffered saline (pH 7.4). Standard ABC immunocytochemical methods were employed using a Vectastain kit from Vector Laboratories, Burlingame, CA. Hydrogen peroxide and 3,3'-diaminobenzidine (DAB) were used to visualize the staining. The primary goat OMP antibody was diluted to 1:10,000. A parallel series was stained following the same procedure, but using a primary antibody against the G protein Go□. Sections were coverslipped and examined under a light microscope. A small number of sections were lost during the staining procedure; these were replaced by adjacent sections.

Because OMP is differentially expressed in the anterior and posterior halves of the AOB NF/G layer (Shapiro and Halpern, 1998), OMP antibody staining caused the two halves to be visually distinguishable (Figure 1A). The area of the anterior and posterior halves was calculated for each section containing AOB using the NIH imaging program Scion Image 1.57 on a Macintosh computer. The area of each section was summed and integrated to find the anterior, posterior, and total volumes for both the left and right sides of the brain.

The M/TC layer was measured in the same manner, using the $G_{o\alpha}$-stained slides to discern it (Figure 2). The granule layer, visible in Nissl stained sections, could not be reliably measured because it merges with the granule cell layer of the MOB.

Each vomeronasal organ was embedded whole in paraffin, cut into 10 micron coronal sections, and mounted on slides. Every thirtieth section was Nissl stained with 1% cresyl violet (Figure 1B), and the area of the vomeronasal sensory epithelium determined using the same computer program described above. Damaged sections were replaced by adjacent sections. VNO epithelial areas for each section were integrated to determine the volume of the vomeronasal sensory epithelium for each animal.

## 3. RESULTS

The mean total volume of the nerve fiber and glomerular layer (NFL/GL) is larger in male opossums than in females ($p=0.03$, one tailed; Figure 3). The anterior NFL/GL of males is larger in males ($p=0.008$, one tailed; Figure 4), however the posterior NFL/GL does not differ in volume between the sexes (Figure 5). The mitral/tufted cell layer (M/TCL) is significantly larger in males than in females ($p=0.01$, one tailed; Figure 6).

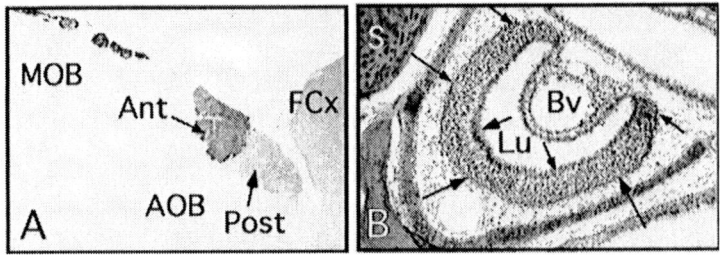

Figure 1. A. Sagittal section of the anterior telencephalon stained with antibody to olfactory marker protein. Ant=anterior AOB, AOB=accessory olfactory bulb, FCx=frontal cortex, MOB=main olfactory bulb, Post=posterior AOB. Rostral is to the left. B. Vomeronasal organ stained by the Nissl method. Bv=blood vessel, Lu=lumen, S=septum; arrows point to sensory epithelium.

The volume of the VNO sensory epithelium is significantly larger in male than in female opossums (p=0.0016, one tailed; Figure 7).

Since male opossums are considerably larger than female opossums (Mean=101.52 vs 59.73 respectively, t=9.339, p=0,0001 one-tailed), we normalized the volume measurements for the AOB and VNO by dividing these volumes by the weight of each animal or the weight of the brain to determine if the volume differences remained. As indicated in Figures 8 and 9, when such normalization occurs using body weight, the NG of females is larger per gram body weight than that of males (p=0.0012, two-tailed test). The same is true of the anterior NGL, posterior NGL and mitral/tufted cell layer of the AOB (P=0.0023, 0.0015, 0.0003 respectively, two-tailed tests), but not of the VNO. The brains of males weighed more than females (Mean=0.630 vs 0.569 gms respectively, t=2.764, p=0.0198, one-tailed). When brain weight is used to normalize the volume measurements, no male/female brain volume differences are significant, but the male VNOs are significantly larger per gram brain weight than female VNOs (p=0.0137 two-tailed test). As noted above, the posterior NGL did not differ significantly between males and females. Using the posterior NGL to normalize the data for the NGL layer, anterior NGL, and VNO were significantly larger in males than in females (p=0.031, 0.031 and 0.0002 respectively), but the M/T layer did not differ significantly in males and females.

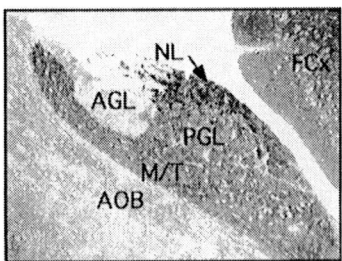

Figure 2. Sagittal section through the accessory olfactory bulb of an opossum stained with antibody to $G_{\alpha\alpha}$. AOB=accessory olfactory bulb, AGL=anterior glomerular layer, PGL=posterior glomerular layer, NL=nerve layer, M/T=mitral/tufted cell layer, FCx=frontal cortex.

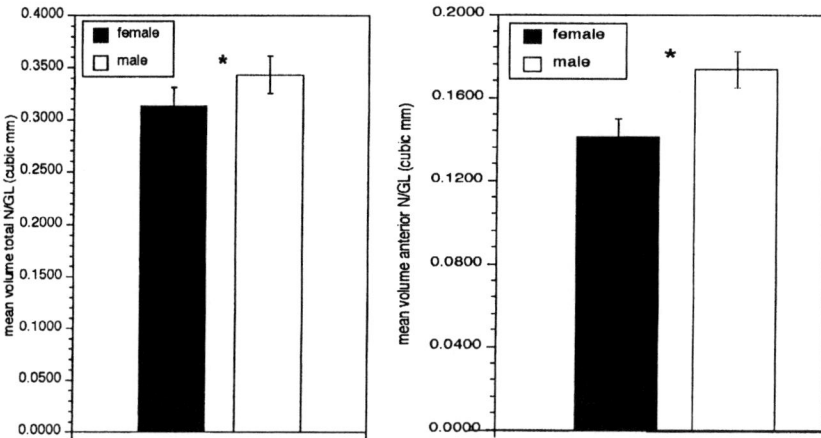

**Figure 3 (left).** Mean volume ± SEM of nerve/glomerular layer of male and female opossums. * indicates that the differences are significant (p=.03). **Figure 4 (right).** Mean volume ± SEM of anterior nerve/glomerular layer of male and female opossums. * indicates that the differences are significant (p=.008).

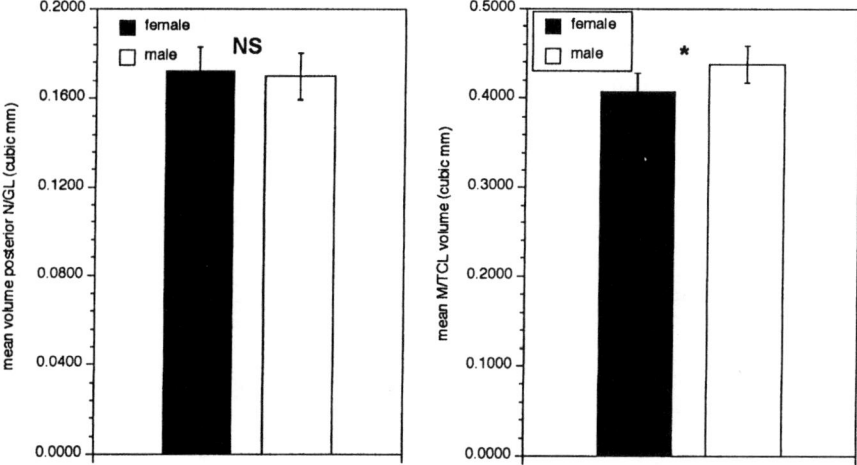

**Figure 5 (left).** Mean volume ± SEM of posterior nerve/glomerular layer of male and female opossums. NS indicates that the differences are not significant. **Figure 6 (right).** Mean volume ± SEM of mitral/tufted cell layer of male and female opossums. * indicates that the differences are significant (p=.01).

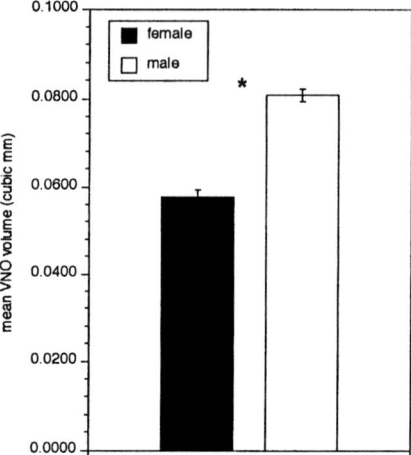

**Figure 7.** Mean volume ± SEM of sensory epithelium of the vomeronasal organ of male and female opossums. * indicates that differences are significant (p=.0016).

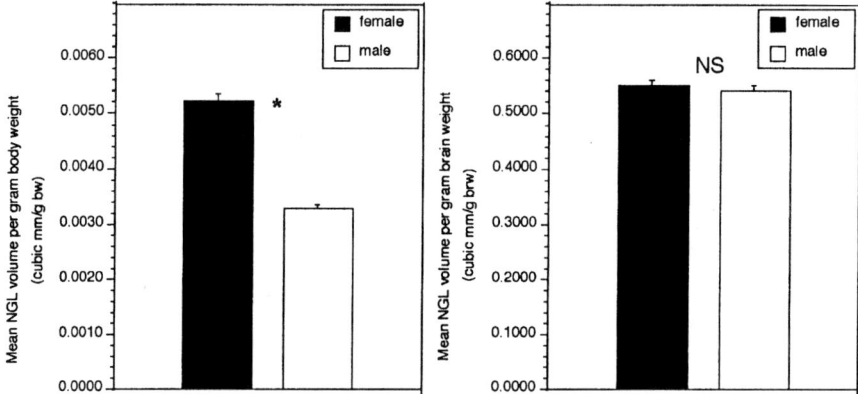

**Figure 8 (left).** Mean volume ± SEM of nerve/glomerular layer of male and female opossums normalized for body weight. * indicates that the differences are significant p = 0.0012. **Figure 9 (right).** Mean volume ± SEM of nerve/glomerular layer of male and female opossums normalized for brain weight. NS indicates that the differences are not significant.

## 4. CONCLUSIONS

Sexual dimorphism was found in the volumes of the M/TC and NF/G layers of the accessory olfactory bulb and in the vomeronasal organ of the opossum, males being larger than females in all cases. These results support similar findings in the rat by Segovia and Guillamón (Guillamón and Segovia 1993, 1996, 1997; Segovia and Guillamón, 1993, 1996). In addition, we found that in the NFL/GL of the AOB, dimorphism is apparent in the anterior but not in the posterior half. Since these halves are defined on the basis of their differing chemoarchitecture, the present findings provide further evidence for a functional difference between the two parts of the AOB.

Although we did not attempt to determine the cause of the observed sex differences, previous studies suggest that they are due to differences in circulating gonadal hormones. Determining this could be a subject for future study, as could be determination of the stage in development when these sex differences arise.

Finally, given the presence of dimorphism in the anterior half of the AOB NFL/GL and its absence in the posterior half, it would be useful to determine whether the same situation exists in the VNO receptor cell populations that topographically project to the two halves.

It should be noted, however, that when the VNO and AOB volumes were normalized to body weight, the female structures were larger per gram body weight than the males and when the volumes were normalized to brain weight no differences between males and females were found. The one structure whose volume did not differ between males and females was the posterior NGL. Using this measure, which is probably the preferred measure to use, to normalize the data, all structures except the MTC layer continued to be sexually dimorphic with males having larger vomeronasal system structures than females. This finding suggests that between sex differences should be normalized to some structure that does not differ between males and females.

## 5. ACKNOWLEDGEMENTS

This research was supported by a grant to M.H. from the National Institute of Deafness and Communication Disorders, DC02745.

## 6. REFERENCES

Collado, P., Guillamón, A., Valencia, A., and Segovia, S., 1990, Sexual dimorphism in the bed nucleus of the accessory olfactory tract in the rat, *Dev. Brain Res.* **56**:263-268.

Davis, B. J., Macrides, F., Youngs, W. M., Schneider, S. P., and Rosene, L. L., 1978, Efferents and centrifugal afferents of the main and accessory olfactory bulbs in the hamster. *Brain Res. Bull.* **3**:59-72.

Dawley, E. M., 1992, Sexual dimorphism in a chemosensory system: The role of the vomeronasal organ in salamander reproductive behavior, *Copeia* **1992**:113-120.

Dawley, E. M., and Crowder J., 1995, Sexual and seasonal differences in the vomeronasal epithelium of the red-backed salamander (*Plethodon cinereus*), *J. Comp. Neurol.* **359**:382-390.

del Abril, A., Segovia, S., and Guillamón A., 1987, The bed nucleus of the stria terminalis in the rat: regional sex differences controlled by gonadal steroids early after birth, *Dev. Brain Res.* **32**:295-300.

DeOlmos, J., Hardy, H., and Heimer, L., 1978, The afferent connections of the main and the accessory olfactory bulb formations in the rat: An experimental HRP-study, *J. Comp. Neurol.* **181**:213-244.

Farbman, A. I., 1992, *Cell Biology of Olfaction*, Cambridge University Press, Cambridge.

Guillamón, A., and Segovia, S., 1993, Sexual dimorphism in the accessory olfactory system, in: *The Development of Sex Differences and Similarities in Behavior*, NATO ASI Series, Vol. 73, M. Haug, R. E. Whalen, C. Aron, and K. L. Olsen, eds., Kluwer Academic, Dordrecht/Norwell, Massachusetts, pp. 363-376.
Guillamón, A., and Segovia, S., 1996, Sexual dimorphism in the CNS and the role of steroids, in: *CNS Neurotransmitters and Neuromodulators: Neuroactive Steroids*, T.W. Stone, ed., CRC Press, Boca Raton, Florida, pp. 127-152.
Guillamón, A., and Segovia, S., 1997, Sex differences in the vomeronasal system, *Brain Res. Bull.* **44**:377-382.
Guillamón, A., Segovia, S., and del Abril, A., 1988, Early effects of gonadal steroids on the neuron number in the medial posterior region and the lateral division of the bed nucleus of the stria terminalis in the rat, *Dev. Brain Res.* **44**:281-290.
Guillamón, A., de Blas, M. R., and Segovia S., 1988, Effects of sex steroids on the development of the locus coeruleus in the rat, *Dev. Brain Res.* **40**:306-310.
Halpern, M., 1987, The organization and function of the vomeronasal system, *Ann. Rev. Neurosci.* **10**:325-362.
Halpern, M., Shapiro, L. S., and Jia, C-P., 1995, Differential localization of G proteins in the opossum vomeronasal system, *Brain Res.* **677**:157-161.
Hines, M., Davis, F. C., Coquelin, A., Goy, R. W., and Gorski, R. A., 1985, Sexually dimorphic regions in the medial preoptic area and the bed nucleus of the stria terminalis of the guinea pig brain: A description and an investigation of their relationship to gonadal steroids in adulthood, *J. Neurosci.* **5**:41-47.
Jia, C., and Halpern, M., 1996, Subclasses of vomeronasal receptor neurons: Differential expression of G proteins (Giα2 and Goα) and segregated projections to the accessory olfactory bulb, *Brain Res.* **719**:117-128.
Kubie, J. L., Vagvolgyi, A., and Halpern, M., 1978, The roles of the vomeronasal and olfactory systems in the courtship behavior of male garter snakes, *J. Comp. Physiol. Psychol.* **92**:627-641.
McLean, J. H., and Shipley, M. T., 1992, Neuroanatomical Substrates of Olfaction, in: *Science of Olfaction*, M. J. Serby and K.L. Chobar, eds., Springer-Verlag, New York, pp. 126-171.
Polack, J. P., and Van Noorden, S., 1997, *An introduction to immunocytochemistry: Current techniques and problems. Royal Microscopical Society Miscroscopy Handbooks*, Oxford University Press, Oxford.
Raisman, G., 1972, An experimental study of the projection of the amygdala to the accessory olfactory bulb and its relationship to the concept of the dual olfactory system, *Exp. Brain Res.* **14**:395-408.
Roos, J., Roos, M., Schaeffer, C., and Aron, C., 1988, Sexual differences in the development of accessory olfactory bulb in the rat, *J. Comp. Neurol.* **270**:121-131.
Segovia, S., and Guillamón, A., 1982, Effects of sex steroids on the development of the vomeronasal organ in the rat, *Dev. Brain Res.* **5**:209-212.
Segovia, S., and Guillamón, A., 1993, Sexual dimorphism in the vomeronasal pathway and sex differences in reproductive behaviors, *Brain Res. Rev.* **18**:51-74.
Segovia, S., and Guillamón, A., 1996, Searching for sex differences in the vomeronasal pathway, *Horm. Behav.* **30**:618-626.
Segovia, S., Paniagua, R., Nistal, M., and Guillamón, A., 1984, Effects of postpuberal gonadectomy on the neurosensorial epithelium of the vomeronasal organ in the rat, *Dev. Brain Res.* **14**:289-291.
Segovia, S., Valencia, A., Cales Jose, M., and Guillamón, A., 1986, Effects of sex steroids on the development of two granule cell populations in the rat accessory olfactory bulb, *Dev. Brain Res.* **30**:283-286.
Shapiro, L. S., and Halpern, M., 1998, Development of NADPH-diaphorase expression in chemosensory systems of the opossum, *Monodelphis domestica. Dev. Brain Res.* **111**:51-63.
Shapiro, L. S., Ee, P-L., and Halpern, M., 1995, Lectin histochemical identification of carbohydrate moieties in the opossum chemosensory systems during development, with special emphasis on VVA-identified subdivisions in the accessory olfactory bulb, *J. Morphol.* **224**:331-349.
Shapiro, L. S., Li, C-S., Jia, C-P., and Halpern, M., 1995, Histochemical, immunocytochemical and tract tracing studies investigating the heterogeneity of the primary vomeronasal pathway in the Brazilian gray short-tailed opossum, *Monodelphis domestica, Chem. Senses* **20**:777-778.
Shnayder, L., Schwanzel-Fukuda, M., and Halpern, M., 1993, Differential OMP expression in opossum accessory olfactory bulb, *NeuroReport* **5**:193-196.
Valencia, A., Collado, P., Coles Jose, M., Segovia, S., and Laso Carmen P., 1992, Postnatal administration of dihydrotestosterone to the male abolishes sexual dimorphism in the accessory olfactory bulb: a volumetric study, *Dev. Brain Res.* **68**:132-135.
Vinader-Caerols, C., Collado, P., Segovia, S., and Guillamón, A., 1998, Sex differences in the posteromedial cortical nucleus of the amygdala in the rat, *NeuroReport* **9**:2653-2656.

# THE NEUROBIOLOGY OF ODOR-BASED SEXUAL PREFERENCE
## The case of the Golden hamster

Aras Petrulis*

## 1. INTRODUCTION

It is well-known that chemical cues produced by conspecifics can serve to attract opposite-sex members of the same species and thereby facilitate mate choice and reproduction (Brown and Macdonald, 1985). What is less known is how these decisions and preferences are instantiated in the nervous system and what mechanisms and factors modulate and produce adaptive behavioral responses to sex-specific odors. This short review will attempt to summarize what is known about the neural mechanisms underlying sex odor preferences in Golden hamsters (*Mesocricetus auratus*) and what outstanding issues remain to be addressed by future research. The Golden hamster is a particularly well-studied species with regard to its chemosensory-guided behavior and the neurobiology underlying it (Johnston, 1990; Wood and Swann, 2000).

Sexual odor preferences in hamsters can be inferred primarily by observation of chemosensory investigation of conspecifics and their isolated odors and also by considering differential scent marking responses toward male and female odors. Not surprisingly, both male and female hamsters spend much more time investigating the ano-genital, head and flank areas of opposite-sex awake or anesthetized individuals as well as isolated odors from these areas (Johnston, 1979; Landauer et al., 1978). In addition, both male and female hamsters often show more flank marking toward same-sex odors, supporting the proposed competitive signaling function of this behavior (Johnston, 1975, 1977). In contrast, female hamsters show considerably more vaginal marking, a proceptive behavior that functions in attracting males for mating, in response to male odors than to female odors (Johnston, 1977; Petrulis and Johnston, 1997). These sexual dimorphisms in preference behavior require gonadal steroid action during early development as well as circulating levels of sex-appropriate steroids (Gregory et al.,

---

* Aras Petrulis, Department of Psychology, Georgia State University, Atlanta, GA, 30303.

1975; Gregory and Pritchard, 1983; Powers and Bergondy, 1983); this mirrors the well-developed connections between chemosensory and hormone-sensitive brain regions in hamsters (Wood and Swann, 2000).

## 2. VOMERONASAL VERSUS OLFACTORY SYSTEM CONTROL

### 2.1. Anatomy

Social odors cues are detected and processed by at least two distinct chemosensory systems: the vomeronasal or accessory olfactory system (VNS) and the main olfactory system (MOS). The anatomy and function of the VNS and MOS have been reviewed extensively elsewhere (Cleland and Linster, 2003; Halpern and Martinez-Marcos, 2003; Jonhston, 1998). In brief, the vomeronasal and olfactory systems are separate at the periphery: the vomeronasal sensory neurons are located in an enclosed tubular structure (vomeronasal organ; VNO) with limited access to air-borne odorants whereas the main olfactory sensory neurons are located on nasal turbinates that are readily exposed to volatile odorants. These two separate sets of sensory neurons project to separate regions of the olfactory bulb: the accessory olfactory bulb (AOB) for VNS afferents and the main olfactory bulb (MOB) for projections from the olfactory sensory epithelium. Cells in the AOB in turn project to a restricted set of nuclei including the anterior (MeA) and posterior (MeP) regions of the medial amygdala, the bed nucleus of the stria terminalis (BNST) and the postermedial cortical nucleus (PMCo). In addition to reciprocal connections between these structures, the AOB targets project to neurons in the medial preoptic area (MPOA) and hypothalamus, including ventromedial nucleus (VMN). In contrast, the MOB sends extensive projections to many regions including the piriform cortex (Pir), anterior (ACo) and posterolateral (PLCo) cortical nuclei and the lateral entorhinal cortex (LEnt). Some MOB target structures innervate other cortical structures such as the orbitofrontal cortex (OFC) and the hippocampus (H). Convergence exists between the two systems at the level of the amygdala: (1) the cortical amygdala structures (ACo, PLCo, PMCo) are all interconnected with each other and MeA, and (2) deep layers of Pir project to MeA (Coolen and Wood, 1998; Gomez and Newman, 1992).

### 2.2. Male Preference: Chemoinvestigation and Scent Marking

The vast majority of studies on male hamster chemosensory investigation have focused on ano-genital investigation (AGI) of the female as well as responses to vaginal secretion, the primary source of ano-genital odors that is both attractive and sexually-stimulating to male, but not female, hamsters (Johnston, 1974, 1975; Petrulis and Johnston, 1995). Male hamsters with ablations of both olfactory bulbs (OBX), or with cuts to OB efferent nerve bundles will not engage in AGI, copulatory behavior, or show preferences for investigating conspecific females or their vaginal secretions (Devor, 1973; Murphy, 1980), indicating that chemosensory stimulation is critical for male attraction to females and that other sensory systems cannot compensate for this loss of MOS and/or VNS activity.

Removing the VNO or cutting the vomeronasal nerves (VN) does not reliably impair AGI or attraction to vaginal secretion. For example, although VN cuts reduce investigation of anesthetized males with applied vaginal secretion (Powers et al., 1979),

VNO removal does not effect AGI toward anesthetized female hamsters (O'Connell and Meredith, 1984) or reduce investigation of vaginal secretion (Steel and Keverne, 1985) nor do VN cuts influence the preference for investigating conspecific females or their vaginal secretions (Murphy, 1980). It does appear, however, that the chemoinvestigatory behavior of sexually inexperienced males is more vulnerable to VNS damage than if the animals had copulated prior to surgery (Meredith, 1986). Reducing MOS function, via nasal ZnSO4 irrigation, decreases investigation of isolated vaginal secretion (Powers et al., 1979) but may or may not reduce AGI of anesthetized female surrogates (Powers et al., 1979; O'Connell and Meredith, 1984). Taken together, the data suggest that the MOS may be preferentially involved in attraction over a distance as well as initiating AGI, and that the VNS may be involved in maintaining persistent chemoinvestigation, especially in sexually-naïve male hamsters.

Although much is known about the neurobiological, experiential and hormonal control of flank marking (Albers et al., 2002), we know very little about the sensory processing that leads to differential scent marking to male versus female odors. What is known suggests that flank marking depends on the functional integrity of the olfactory bulbs (Murphy and Schneider, 1970) and marking toward male odors is primarily regulated by the MOS as ZnSO4 olfactory lesions greatly reduce flank marking but VNO removal has no deleterious effects on marking (Johnston and Mueller, 1990).

## 2.3. Female Preference: Chemoinvestigation and Scent Marking

Female hamsters with transections of the lateral olfactory tract (LOT; efferents and afferents of MOB and AOB) at the level of the anterior olfactory nucleus (AON) are not anosmic, but do show very little interest in male and female odors (Petrulis et al., 1999). This result indicates that chemosensory structures caudal to the cut, such as Pir, MeA and LEnt, are critical for female hamster sex odor preferences and interest in conspecific odors. Removing the VNO has no effect on a female's preferential investigation of male volatile odorants from a male versus a female's odor but did reduce the overall amount of investigation of male flank scent when direct contact was allowed (Petrulis et al., 1999). It appears that, similar to male hamsters, preferential attraction by females toward opposite sex odors at a distance is mediated by the MOS but that the VNS may play a role in chemoinvestigation in circumstances where direct contact with odorants is allowed.

Female flank marking, like that of males, is virtually eliminated by either bilateral OBX or by transection of LOT (Petrulis et al., 1999; Kairys et al., 1980). The MOS mediates this effect as peripheral MOS damage, but not VNO removal, reduces flank marking in areas containing male and female odors (Johnston, 1992). However, VNO removal does cause more subtle deficits; removing the VNO eliminates the differential marking response to male and female odors (Petrulis et al., 1999). This suggests that maintaining high levels of flank marking requires MOS input, but also that the VNS may be necessary for transmitting information about sexually-differentiated odors to neural structures that generate flank marking output.

Vaginal marking in response to male odors is also greatly reduced by OBX (Kairys et al., 1980) and to a lesser degree by peripheral MOS damage (Johnston, 1992). However, LOT cuts do not cause a drastic reduction in overall vaginal marking. Instead, females with LOT transections do not show evidence of discriminatory vaginal marking: females mark at similar rates to male and female odors (Petrulis et al., 1999). This pattern is replicated by VNO removal (Petrulis et al., 1999) although overall levels of vaginal marking are unimpaired after VNO removal (Johnston, 1992).

Females can use both types of marking to target appropriate audiences (female competitors, male suitors) and this kind of sexual discrimination requires both MOS input and VNS input. Although the VNS appears critical for discriminating between male and female odors using scent marking responses, MOS input may be required for maximal activation of the VNO pumping mechanism that brings chemical cues to the VNO (Meredith, 1998). Therefore, interrupting MOS processing may indirectly reduce appropriate chemical stimulation of the VNO and thereby prevent VNS input to areas that regulate scent marking responses.

## 3. CENTRAL OLFACTORY SYSTEM STRUCTURES

### 3.1. Male Preference: Chemoinvestigation and Scent Marking

LOT transections caudal to the olfactory tubercle (OT) eliminate or reduce AGI by male hamsters indicating that OT, anterior Pir and AON are not sufficient to support chemoinvestigation of conspecifics (Devor, 1973). Consequently, chemosensory input to the corticomedial amygdala, posterior Pir and/or LEnt must be critical for sex odor preferences. Although direct evidence is not available for assessing the role of LEnt and posterior Pir in attraction to sex odors, lesioning Pir projections to OFC decreases AGI of females and eliminates the normal preference for investigating female odors over those from males (Sapolsky and Eichenbaum, 1980). However, using the immediate-early gene Fos as a "marker" for neuronal activation, no sex differences or increases above baseline were detected in response to vaginal secretion in MOS structures (Fernandez-Fewell and Meredith, 1994; Fiber and Swann, 1996). In general, the data suggest that Pir connections with OFC mediate sex preferences in sexually-experienced male hamsters, perhaps as part of a more general network that assigns and alters valence toward desired objects and conspecifics (Petrulis and Eichenbaum, 2003).

Although flank marking is greatly reduced or eliminated by OBX or LOT cuts (Murphy and Schneider, 1970; Macrides et al., 1976), very little additional information is available as to which MOS structures might be critical for odor-elicited marking. Nevertheless, regions of MPOA that regulate flank marking receive projections from ACo, PLCo and MeA (Albers et al., 2002), suggesting that MOB and AOB afferents to these areas are responsible for odor-elicited flank marking.

### 3.2. Female Preference: Chemoinvestigation and Scent Marking

Females with LOT cuts are not attracted to conspecific odors and do not show a preference for male over female odor (Petrulis et al., 1999) whereas females with VNO removal still approach and investigate male odors preferentially. A reasonable conclusion is that sex preference in females requires MOB input to posterior Pir, corticomedial amygdala and /or LEnt. Although lesions to posterior Pir have not been reported, females with lesions of OFC, unlike males, show normal opposite-sex preferences (Petrulis et al., 1998) suggesting a sex difference in neural control of odor preference. Similarly, damage to LEnt does not alter a female's preference for male odors nor does subcortical disconnection of H (Petrulis et al., 2000). Preference for male odors, however, may be mediated by olfactory input to the corticomedial amygdala as lesions of MeA/MeP eliminate differential investigation of male odors (Petrulis and Johnston, 1999).

Since flank marking, like chemoinvestigation, is regulated to a great degree by MOS processing, we might expect that structures like posterior Pir, MeA and LEnt would regulate flank marking. However, severing Pir connections to OFC does not influence flank marking behavior (Petrulis et al., 1998). Differential flank marking toward female odors is impaired after LEnt lesions, but not by subcortical disconnection of H (Petrulis et al., 2000). However, these LEnt lesions also damaged PMCo which does not allow us to separate the role of MOB input to LEnt from that of AOB input to PMCo. What is clear is that overall levels of flank marking are critically dependent on olfactory input to MeA as lesions to this area virtually eliminate flank marking behavior (Petrulis and Johnston, 1999).

MOS structures such as OFC or LEnt are not critical for preferential vaginal marking toward male odors, as removal of these structures does not impair vaginal marking (Petrulis et al., 1998, 2000). In contrast, vaginal marking is reduced by MeA/MeP lesions, but interestingly animals with these lesions still vaginal mark more toward male odors than to female odors (Petrulis and Johnston, 1999). So, VNS and MOS input to MeA might regulate overall levels of vaginal marking but not the bias in marking toward male odors. Other structures such as posterior Pir, ACo, PLCo or AOB inputs to BNST may regulate this sex bias in vaginal marking.

## 4. VOMERONASAL SYSTEM AND BASAL DIENCEPHALIC STRUCTURES

### 4.1. Male Preference and Scent Marking

Several lines of evidence point to central VNS structures (and MOS indirectly) as being critical for both generating appropriate levels of AGI toward females and for interest in vaginal secretion. First, lesions of MeA, BNST, but not MPOA, greatly impair AGI and/or vaginal secretion investigation (Lehman, et al., 1980; Powers et al., 1987). Second, Fos activation increased in the posterodorsal (pm) BNST, MeP and the magnocellular nucleus of the medial preoptic nucleus (MPNmag) following investigation of vaginal secretion (Fiber et al., 1993). Males and, to a lesser degree, females show increased Fos expression in BNSTpm and MeP; this dimorphism is partially dependent on circulating testosterone (T) levels (Fiber and Swann, 1996). However, MPNmag only shows odor-induced Fos in male hamsters and this difference is not dependent on sex differences in circulating T (Fiber and Swann, 1996). Other experiments have demonstrated that VNO-lesioned animals that do not mate but still chemoinvestigate show increased levels of Fos in BNSTpm (Fernandez-Fewell and Meredith, 1994); this suggests that BNSTpm may be critical for appropriately-directed investigation in male hamsters. Lastly, intracerebral replacement of T or estradiol in MeP, BNST and MPOA can reverse castration-induced declines in female-directed AGI and this requires integration with chemosensory information in these structures (Wood and Swann, 2000). Overall, it appears that MOS and VNS information may converge in the MeA and direct sexually-appropriate investigation through its connections with BNST. The MPOA may be less critical for orientation and approach toward sexual stimuli and may instead be more involved in connecting chemosensory processing to effector systems that generate stereotyped social behavior (e.g. copulation).

Connections from chemosensory regions do connect to neurochemically-defined circuits in the MPOA and hypothalamus that mediate flank marking (Albers et al., 2002) but there is no available evidence yet to show how chemical cues alter these neurons, let alone how this might lead to differential marking toward male odors.

### 4.2. Female Preference and Scent Marking

Very little information is available on the role of VNS or diencephalic control of a female's attraction to male odors. As indicated above, damage to MeA/MeP eliminates sex-odor preference; however, this deficit is due to elimination of MOS, not VNS, input to MeA (Petrulis et al., 1999; Petrulis and Johnston, 1999). Lesions to VMN do reduce an estrus female's approach toward caged males (Floody, 2002), indicating that MeA/MeP connections with VMN (Gomez and Newman, 1992) may mediate odor-based sexual attraction in female hamsters. Fos expression is prominent in MeA and MPOA of females following mounting by male hamsters but BNSTpm neurons only show Fos expression with intromissive stimulation (Ramos and DeBold, 2000). Although this suggests both commonalities (activation of MeA and MPOA) and differences (medial BNST activation) between male and female responses to opposite-sex odors, there is no evidence that this is due to odor exposure alone. Interestingly, vaginal secretion from other females can increase Fos expression in MeP and BNSTpm but not to levels observed in males unless females are treated with T (Fiber and Swann, 1996). This result does suggest that social odors may broadly activate MeP and BNSTpm in both sexes but the precise nature and function of this activation is unknown.

It appears that the odor cues that generate flank marking are mediated by both AOB and MOB input to the MeA (Petrulis and Johnston, 1999). Similarly, vaginal marking also appears to require AOB and MOB input to the MeA (Petrulis et al., 1999; Johnston, 1992; Petrulis and Johnston, 1999). As MeA/MeP has substantial input to MPOA and hypothalamic areas that are known to be responsible for the control of both vaginal and flank marking (Albers et al., 2002; Malsbury et al., 1977; Takahashi and Lisk, 1985), it is likely that odor modulation of scent marking is mediated by medial amygdala efferents to these specific diencephalic targets.

## 5. CONCLUSIONS AND FUTURE DIRECTIONS

Several conclusions and directions for further research are warranted by the existing data. First, it is clear that the neural mechanisms controlling approach and investigation toward opposite-sex odors are separate from those regulating more stereotyped responses toward the same odors and that this separation occurs relatively early in the (sensory) processing stream. For example, VNO removal eliminates vaginal marking preferences toward male odors but does not effect attraction to male odors from a distance (Petrulis et al., 1999) and MeA/MeP lesions in females eliminate attraction to male odors, but these females still vaginal mark more to male odors than to female odors (Petrulis and Johnston, 1999).

Second, the investigative and scent marking preferences most likely require the functional interplay of both MOS and VNS for the full, adaptive expression of these preferences. The nature of this interplay may depend heavily on the context in which odors are perceived (volatile vs. non-volatile; AGI vs. odor alone) as well as on the sociosexual experience of the investigating animal (Meredith, 1998).

Third, sex differences in odor preferences are apparent behaviorally and can be manipulated hormonally, but little is currently known about the neurobiology underlying these differences. We know much about the neurobiology of odor preferences by males, but we do not have comparable information for females. For example, does the BNST play a role in female odor preference as it does for males?

Lastly, odor preferences, like all vertebrate behaviors, are generated by a rich interplay between functionally different neuronal cell groups. Assessment of this kind of functional interconnectivity on a behaviorally relevant time-scale is not possible using currently used techniques (i.e. Fos expression) and so we do not know how the coding properties of neurons in important areas, such as the BNST, are altered by neuronal changes in other areas such as MeA/MeP. Even at a more basic level, we have little insight into the information-bearing properties or operating principles of neurons that might guide their responses to male, female or other odorants. We are currently addressing these issues with single-unit and ensemble electrophysiological recordings in behaving male and female hamsters that are responding to controlled presentations of same- and opposite-sex odors.

## 5. REFERENCES

Albers, H. E., Huhman, K. L., and Meisel, R. L., 2002, Hormonal basis of social conflict and communication, in: *Hormones, Brain and Behavior*, D. W. Pfaff, A. Arnold, A. Etgen, S. Fahrbach, and R. Rubin, eds., Academic Press, New York, pp. 393-433.
Brown, R. E., and Macdonald, D. W., 1985, *Social Odors in Mammals*, Clarendon Press, Oxford.
Cleland, T. A., and Linster, C., 2003, Central olfactory processing in: *Handbook of Olfaction and Gustation, Second Edition*, R. L. Doty, ed., Marcel Dekker, New York, pp. 165-180.
Coolen, L. M., and Wood, R. I., 1998, Bidirectional connections of the medial amygdaloid nucleus in the Syrian hamster brain: simultaneous anterograde and retrograde tract tracing, *J. Comp. Neurol.* **399**:189-209.
Devor, M., 1973, Components of mating dissociated by lateral olfactory tract transection in male hamsters, *Brain Res.* **64**:437-441.
Fernandez-Fewell, G. D., and Meredith, M., 1994, C-fos expression in vomeronasal pathways of mated or pheromone-stimulated male golden hamsters: Contributions from vomeronasal sensory input and expression related to mating performance, *J. Neurosci.* **14**:3643-3654.
Fiber, J. M., and Swann, J. M., 1996, Testosterone differentially influences sex-specific pheromone-stimulated Fos expression in limbic regions of Syrian hamsters, *Horm. Behav.* **30**:455-473.
Fiber, J. M., Adames, P., and Swann, J. M., 1993, Pheromones induce c-fos in limbic areas regulating male hamster mating behavior, *Neuroreport* **4**:871-874.
Floody, O. R., 2002, Time course of VMN lesion effects on lordosis and proceptive behavior in female hamsters, *Horm. Behav.* **41**:366-376.
Gomez, D. M., and Newman, S. W., 1992, Differential projections of the anterior and posterior regions of the medial amygdaloid nucleus in the Syrian hamster, *J. Comp. Neurol.* **317**:195-218.
Gregory, E., and Pritchard, W. S., 1983, The effects of neonatal androgenization of female hamsters on adult preference for female hamster vaginal discharge, *Physiol. Behav.* **31**:861-864.
Gregory, E., Engel, K., and Pfaff, D., 1975, Male hamster preference for odors of female hamster vaginal discharges: studies of experiential and hormonal determinants, *J. Comp. Physiol. Psychol.* **89**:442-446.
Halpern, M. and Mártinez-Marcos, A., 2003, Structure and function of the vomeronasal system: an update. *Prog. Neurobiol.* **70**:245-318.
Johnston, R. E., 1974, Sexual attraction function of golden hamster vaginal secretion, *Behav. Biol.* **12**:111-117.
Johnston, R. E., 1975, Scent marking by male golden hamsters (*Mesocricetus auratus*) I. Effects of odors and social encounters, *Z. Tierpsychol.* **37**:75-98.
Johnston, R. E., 1975, Sexual excitation function of hamster vaginal scretion, *Anim. Learn. Behav.* **3**:161-166.

Johnston, R. E., 1977, The causation of two scent-marking behaviour patterns in female hamsters (*Mesocricetus auratus*), *Anim. Behav.* 25:317-327.
Johnston, R. E., 1979, Olfactory preferences, scent marking, and "proceptivity" in female hamsters, *Horm. Behav.* 13:21-39.
Johnston, R. E., 1990, Chemical communication in golden hamsters: From behavior to molecules and neural mechanisms, in: *Contemporary Issues in Comparative Psychology*, D. A. Dewsbury, ed., Sinauer, Sunderland, Massachusetts, pp. 381-412.
Johnston, R. E., 1992, Vomeronasal and/or olfactory mediation of ultrasonic calling and scent marking by female golden hamsters, *Physiol. Behav.* 51:437-448.
Johnston, R. E., 1998, Pheromones, the vomeronasal system, and communication: From hormonal responses to individual recognition, *Ann. N. Y. Acad. Sci.* 855:333-348.
Johnston, R. E., and Mueller, U. G., 1990, Olfactory but not vomeronasal mediation of scent marking by male golden hamsters, *Physiol. Behav.* 48:701-706.
Kairys, D. J., Magalhaes, H., and Floody, O. R., 1980, Olfactory bulbectomy depresses ultrasound production and scent marking by female hamsters, *Physiol. Behav.* 25:143-146.
Landauer, M. R., Banks, E. M., and Carter, C. S., 1978, Sexual and olfactory preferences of naive and experienced male hamsters, *Anim. Behav.* 26:611-621.
Lehman, M. N., Winans, S. S., and Powers, J. B., 1980, Medial nucleus of the amygdala mediates chemosensory control of male hamster sexual behavior, *Science* 210:557-560.
Macrides, F., Firl, A. C., Schneider, S. P., Bartke, A., and Stein, D. G., 1976, Effects of one-stage or serial transections of the lateral olfactory tracts on behavior and plasma testosterone levels in male hamsters, *Brain Res.* 109:97-109.
Malsbury, C. W., Kow, L.-M., and Pfaff, D. W., 1977, Effects of medial hypothalamic lesions on the lordosis responses and other behaviors in female golden hamsters, *Physiol. Behav.* 19:223-237.
Meredith, M., 1986, Vomeronasal organ removal before sexual experience impairs male hamster mating behavior, *Physiol. Behav.* 36:737-743.
Meredith, M., 1998, Vomeronasal, olfactory, hormonal convergence in the brain. Cooperation or coincidence? *Ann. N. Y. Acad. Sci.* 855:349-361.
Murphy, M. R., 1980, Sexual preferences of male hamsters: Importance of preweaning and adult experience, vaginal secretion, and olfactory or vomeronasal sensation, *Behav. Neural Biol.* 30:323-340.
Murphy, M. R., and Schneider, G. E., 1970, Olfactory bulb removal eliminates mating behavior in the male golden hamster, *Science* 167:302-304.
O'Connell, R. J., and Meredith, M., 1984, Effects of volatile and nonvolatile chemical signals on male sex behaviors mediated by the main and accessory olfactory systems, *Behav. Neurosci.* 98:1083-1093.
Petrulis, A., and Johnston, R. E., 1995, A re-evaluation of dimethyl disulfide as a sex attractant in golden hamsters, *Physiol. Behav.* 57:779-784.
Petrulis, A., and Johnston, R. E., 1997, Causes of scent marking in female golden hamsters (*Mesocricetus auratus*): specific signals or classes of information? *J. Comp. Psychol.* 111:25-36.
Petrulis, A., and Johnston, R. E., 1999, Lesions centered on the medial amygdala impair scent-marking and sex-odor recognition but spare discrimination of individual odors in female golden hamsters, *Behav. Neurosci.* 113:345-357.
Petrulis, A., and Eichenbaum, H., 2003, Olfactory memory, in: *Handbook of Olfaction and Gustation*, R. L. Doty, ed., Marcel Dekker, New York, pp. 409-438.
Petrulis, A., Peng, M., and Johnston, R. E., 1999, Lateral olfactory tract transections impair discrimination of individual odors, sex odor preferences and scent-marking in female golden hamsters (*Mesocricetus auratus*), in: *Advances in Chemical Communication in Vertebrates*, R. E. Johnston, D. Muller-Schwarze, and P. Sorensen, eds., Plenum Press, New York, pp. 549-562.
Petrulis, A., Peng, M., and Johnston, R. E., 2000, The role of the hippocampal system in social odor discrimination and scent-marking in female golden hamsters (*Mesocricetus auratus*), *Behav. Neurosci.* 114:184-195.
Petrulis, A., Peng, M., and Johnston, R. E., 1999, Effects of vomeronasal organ removal on individual odor discrimination, sex-odor preference, and scent marking by female hamsters, *Physiol. Behav.* 66:73-83.
Petrulis, A., DeSouza, I., Schiller, M., and Johnston, R. E., 1998, Role of frontal cortex in social odor discrimination and scent-marking in female golden hamsters (*Mesocricetus auratus*), *Behav. Neurosci.* 112:199-212.
Powers, J. B., and Bergondy, M. L., 1983, Androgenic regulation of chemoinvestigatory behaviors in male and female hamsters, *Horm. Behav.* 17:28-44.
Powers, J. B., Fields, R. B., and Winans, S. S., 1979, Olfactory and vomeronasal system participation in male hamsters' attraction to female vaginal secretions, *Physiol. Behav.* 22:77-84.

Powers, J. B., Newman, S. W., and Bergondy, M. L., 1987, MPOA and BNST lesions in male Syrian hamsters: Differential effects on copulatory and chemoinvestigatory behaviors, *Behav. Brain Res.* **23**:181-195.

Ramos, S. M., and DeBold, J. F., 2000, Fos expression in female hamsters after various stimuli associated with mating, *Physiol. Behav.* **70**:557-566.

Sapolsky, R. M., and Eichenbaum, H., 1980, Thalamocortical mechanisms in odor-guided behavior. II. Effects of lesions of the mediodorsal thalamic nucleus and frontal cortex on odor preferences and sexual behavior in the hamster, *Brain Behav. Evol.* **17**:276-290.

Steel, E., and Keverne, E. B., 1985, Effect of female odour on male hamsters mediated by the vomeronasal organ, *Physiol. Behav.* **35**:195-200.

Takahashi, L. K., and Lisk, R. D., 1985, Estrogen action in the anterior and ventromedial hypothalamus and modulation of heterosexual behavior in female golden hamsters, *Physiol. Behav.* **34**:233-239.

Wood, R. I., and Swann, J. M., 2000, Neuronal integration of chemosensory and hormonal signals in the control of male sexual behavior, in: *Reproduction in Context: Social and Environmental Influences on Reproduction*, K. Wallen and J. E. Schneider, eds., MIT Press, Cambridge, Massachusetts, pp. 423-444.

# RETENTION OF OLFACTORY MEMORIES BY NEWBORN INFANTS

Richard H. Porter and John J. Rieser[*]

## 1. INTRODUCTION

Healthy newborn human infants are endowed with a highly sensitive sense of smell. Moreover, there are documented accounts of olfactory learning during the early postpartum period. In the present chapter, we present a brief overview of the relevant research literature and suggest tentatively that olfactory learning may be facilitated by neurochemical activities associated with labor and delivery, and memory traces of odors learned shortly after birth may be retained more efficiently than early postnatal memories involving other sensory modalities (i.e., vision and audition).

## 2. EARLY LEARNING/MEMORY RETENTION

### 2.1. Olfaction

In the 1960s, Engen and Lipsitt (1965; Engen et al., 1963) conducted a pioneering series of experiments in which 1-3 day old neonates were exposed repeatedly to particular olfactory stimuli. A clear decrement in responses was observed over trials with the same odorant (inter-trial interval ~1 min.). Such diminished responsiveness was believed to reflect habituation rather than sensory adaptation since presentation of a novel odor resulted in a recovery of responses. Differential responses to the training odor versus a novel scent imply that the infants had become familiar with the former stimulus and remembered it over the brief inter-trial interval. Evidence of short-term habituation to

---

[*] Richard H. Porter, UMR 6073 INRA-CNRS-Université de Tours, 37380 Nouzilly, France. John J. Rieser, Dept. of Psychology & Human Development, George Peabody College of Vanderbilt University

odors has likewise been reported for a sample of pre-term babies tested at 32-36 weeks gestational age (Sarnat, 1978).

More recent investigations provide preliminary insights into the durability of olfactory memories established within the first hours of human infants' postnatal existence. In one study using a classical conditioning paradigm, infants were exposed to an odor cue (citrus) associated with gentle stroking, on the day of birth (Sullivan et al., 1991). The following day, the babies that underwent this conditioning treatment turned their head preferentially in the direction of the citrus odor, whereas control infants showed no overt signs of attraction to that scent. Thus, the learned association between the conditioned odor stimulus and the stroking reinforcement was retained over a period of approximately 24 hrs. An alternative method to assess olfactory memory involves mere exposure to an odorant followed by subsequent test trials with that stimulus after it had been removed for a specific time interval. Within the first 2 days after birth, babies were exposed to an artificial scent that remained in their nursery bassinet for ~22 hrs. (Davis and Porter, 1991; see also Balogh and Porter, 1986). As soon as the training session ended, the odor was eliminated and the babies had no further contact with that scent until they were tested on day 16-18 postpartum. At that time, the exposure odor and a novel odor were presented simultaneously to the infants, suspended along opposite sides of their face. Overall, the neonates spent significantly more time oriented towards the exposure odorant than to the novel control odor, indicating that they had retained a memory trace of the familiar olfactory stimulus over a two-week interval. It was subsequently demonstrated that mere exposure to an odor for a considerably shorter period is sufficient for the establishment of early olfactory learning and memory (Varendi et al., 2002). Beginning several minutes after birth, babies received 30-min. of continuous exposure to a training odor. Despite the brevity of the immediate post-partum exposure session, infants in this treatment condition - but not control infants who had no prior odor training - responded discriminatively to the exposure odor when tested later at a median age of 80 hr. (range = 41 - 125 hrs.).

## 2.2. Vision

In contrast with the above summarized studies of early olfactory learning, we have found no reports of enduring memories resulting from visual learning occurring during the first few hours after birth. Rather, the relevant empirical literature includes descriptions of visual learning by somewhat older infants with assessments of short-term memory only. For example, "as early as postnatal day 3", infants displayed preferential fixation to a familiar visual stimulus over a retention interval lasting 2 min. (Pascalis and de Schonen, 1994; see also Simion et al., 2002). The human face is a particularly salient visual stimuls for human neonates (Fantz, 1963; Turati and Simion, 2002); Bushnell et al. (1989) demonstrated that 2 day-old infants are capable of recognizing their mother's face. The babies in their study (mean age = 49 hrs.) spent more time looking at the face of their mother than at a stranger after being separated from their mother for at least 5-min. (maximum separation interval = 30 min.). Infants in an additional experiment were habituated to a face and their recognition of that stimulus was later assessed (Pascalis et al., 1998). Evidence of recognition memory was found following a 24-hr. retention interval, but the subject infants were 3-month old.

The lack of data concerning visual learning and memory immediately after birth is likely to reflect methodological problems that arise when attempting to assess these issues. Moreover, visual discrimination tasks may be relatively difficult shortly after birth because acuity, accomodation and contrast sensitivity are poorly developed at that age (Gorski et al., 1987; Haith, 1986; Turati and Simion, 2002).

## 2.3. Audition

On turning to the research literature on neonatal auditory learning, there once again appear to be no discussions of learning and the establishment of (relatively) long-term memories during the initial postnatal hours. This is particularly curious since human fetuses appear to be capable of learning and remembering specific auditory (DeCasper and Spence, 1986; Hepper, 1991) and olfactory (Hepper, 1995; Schaal et al., 2000) stimuli to which thay are exposed during the final weeks of gestation. Auditory memories may nevertheless develop within several days after birth, but there is no clear evidence that they are retained beyond 24-hrs. Thus, infants that were exposed to a specific speech sound when 2-days old displayed behavioral evidence (directional head turning) that they retained a memory of that same sound 24 hrs. later (Swain et al., 1993). More recently, a group of sleeping infants whose ages ranged from 1-7 days was taught to discriminate between speech sounds during an evening learning session. When tested the next morning, these babies still responded discriminatively to the training sounds (Cheour et al., 2002). An additional experiment investigated preferences for familiar vs. novel nursery rhymes using a non-nutritive sucking procedure (Spence, 1996). Three days after the final familiarization trial, "infants as young as 1 month of age" recognized a nursery rhyme to which they had been repeatedly exposed. In striking contrast, as mentioned above, olfactory memories were retained over a comparable 3-day retention interval even when the learning sessions occurred within the first hour after the babies were born (Varendi et al., 2002).

## 3. ROLE OF THE BIRTH PROCESS IN EARLY OLFACTORY LEARNING?

During the fetal period, major physiological functions such as thermogenesis, gastrointestinal activity and breathing are inhibited, and sleeping appears to be the most common behavioral state (Lagercrantz, 1996). However, this changes abruptly with birth; immediately following vaginal delivery newborn infants are typically awake, alert and responsive to stimulation (Lagercrantz, 1996; Lagercrantz and Slotkin, 1986). The same heightened state of postnatal arousal is not observed in infants who are delivered by cesarean section before the onset of labor. Rather, they tend to be relatively inactive, cry less frequently, and spend more time sleeping. Neuroendocrine processes that are triggered by the rhyhmic uterine contractions that characterize labor, and the mechanical pressure associated with passage through the birth canal, are a major source of the differences observed between vaginally-delivered and c-section neonates. Uterine labor contractions are believed to induce brief periods of fetal hypoxia and head compression that result, in turn, in the release of catecholamines (primarily norepinephrine, NE) from the adrenal medulla and paraganglia (Lagercrantz, 1996; Lagercrantz and Slotkin, 1986; Marchini, Lagercrantz, Winberg and Uvnas-Moberg, 1988). Beginning during the initial stages of labor, there is an enormous surge in fetal peripheral NE, some of which crosses the blood-brain barrier (Lagercrantz, 1996). Plasma levels of NE reach a peak at parturition and remain elevated until ~2 hrs. after vaginal delivery is completed.

Labor contractions may concomitantly activate the locus coeruleus, a brain nucleus that functions as a general arousal center (Jacobs, 1990; Svensson, 1987). The dense population of noradrenergic cells in the locus coeruleus project diffusely throughout the CNS (Shipley et al., 1985; Svensson, 1987). The marked increase in the turnover of NE observed in the brain of rat pups shortly after birth most likely reflects activitation of that structure (Lagercrantz et al., 1992; Foote, Berridge, Adams and Pineda, 1991; Jacobs, 1990), and may contribute further to the highly alert state of newborn infants and their sensitivity to sensory input (Lagercrantz, 1996). As seen below, NE and locus coeruleus acivity have both been implicated in neonatal olfactory learning (reviewed by Gervais, Holley and Keverne, 1988; Nelson and Panksepp, 1998).

### 3.1. Experimental Evidence: Animal Models

Sullivan and her colleagues (1989) demonstrated that training sessions involving presentation of an artificial odorant paired with an NE receptor agonist (isoproterenol) on days 1-18 postpartum was sufficient for rat pups to develop a conditioned preference for the training odor. In this context, exposure to the conditioned odorant alone did not result in olfactory learning. Moreover, infusing an NE receptor antagonist (propranolol) directly into the olfactory bulbs of 5-day old pups blocked the acquisition of such conditioned odor preferences (Sullivan et al., 1992).

In the rat locus coeruleus, all neurons are noradrenergic, and a minimum of 40% of those cells project to the olfactory bulbs (Shipley et al., 1985). To assess the role of the locus coeruleus in early olfactory learning, one-week old rat pups were exposed to an odor paired with a pharmacological manipulation that presumably stimulates that structure (Sullivan et al., 2000). The following day, pups in this treatment condition oriented preferentially to the training odor in a simultaneous 2-choice test situation. In contrast, there was no evidence of olfactory learning in a group of pups that underwent similar pharmacological treatments and training procedures after being injected with an NE antagonist. It was therefore concluded that odor exposure associated with locus coeruleus activiation and the release of NE is sufficient for the development of a learned olfactory preference.

### 3.2. Experimental Evidence: Human Neonates

The role of NE and the locus coeruleus in early olfactory learning by rat pups, along with the massive surge of catecholamines and (putative) locus coeruleus activation triggered by uterine labor contractions in human infants, provide a basis for questioning whether pre-delivery labor might have a positive effect on olfactory learning by newborns of our own species. To evaluate this hypothesis, Varendi et al. (2002) compared the performance of a group of babies who were delivered by cesarean section after the onset of maternal uterine contractions (mean duration of labor preceding c-section = 6.5 hrs.) and that of a sample of neonates born of mothers who experienced no labor contractions before undergoing elective c-section. Beginning within several minutes after birth, each baby was exposed to a particular odorant during a single training session lasting 30-min. The odor stimulus was immediately removed at the end of the exposure session. One to five days later, all the babies were tested for their responses to the exposure odor versus a novel scent in a simultaneous 2-choice situation.

The results of this olfactory choice test differed between the two conditions; the babies who experienced pre-delivery labor displayed a significant orientation preference for the exposure odor, whereas the neonates born without preceding labor oriented indiscriminately to the exposure odor and the novel scent. Thus, behavioral evidence of olfactory learning and memory was only evident for the babies in the labor condition. Analysis of blood samples collected from the umbilical artery immediately after birth revealed a significant positive correlation between NE levels and duration of pre-delivery labor. In addition, the blood levels of NE were reliably greater for the babies who oriented preferentially to the exposure odor than for the babies who did not show an exposure-odor preference. These data indicate that a brief period of mere exposure to an odorant immediately after birth is sufficient for the development of a learned preference for that stimulus - provided that the fetus had been subjected to uterine labor contractions. Moreover, they suggest that the observed positive effect of labor may be mediated, at least partially, by heightened NE levels and locus coeruleus activity, as previously demonstrated in analogous experiments with rat pups.

## 4. CONCLUSIONS

The acquisition of enduring olfactory memories (i.e., retained over a period of 24-hrs. or longer) has been reported for human infants at an earlier postnatal age than the establishment of comparable memory traces for visual or auditory stimuli. At the present time, the interpretation of these differences in the neonatal-learning research literature across stimulus modalities remains somewhat ambiguous. On one hand, olfactory learning and memory encoding may indeed be more robust, or develop more readily, than learning of auditory or visual cues during the first hours after birth. Alternatively, early auditory, visual and olfactory learning may be equally efficient, even though this conclusion does not necessarily follow from the limited relevant research literature that is currently available. Simply put, this question has not been adequately studied.

None the less, infants are clearly capable of olfactory learning and memory formation beginning within the first hour after birth, and those memories may be retained over an interval of at least several days. Neurophysiological processes that occur naturally during parturition prepare the newborn infant to become familiarized rapidly with salient odors in the postnatal environment. Neonates may be particularly sensitive to stimulus input during a brief period immediately after birth as a result of elevated NE levels and brain arousal at that time.

A final question concerns the adaptive significance of early olfactory learning/memory; in what way does this precocial capacity contribute to the baby's fitness? Perhaps the most apparent benefit of neonatal olfactory learning involves its role in the development of mother-infant recognition and mutual attachment. As early as days 2-3 postpartum, infants respond discriminatively to their own mother's individual odor signature (see Porter and Schaal, 2003 for a recent review). Converging data indicate that such olfactory recognition is mediated by rapid familiarization with the mother's distinctive phenotype. Thus, two-week old bottle-fed infants displayed no overt signs that they recognized their mother's odor, whereas breast-fed infants of the same age oriented preferentially to those cues (Cernoch and Porter, 1985). Furthermore, breast-fed infants did not respond discriminatively to odor samples from their own father versus a comparison adult male. This pattern of results can be explained by the unique interactions between nursing mothers and their breast-fed neonates. Effective breastfeeding, which may be initiated within the first hour after delivery (Righard and Alade, 1990; Varendi et al., 1994; Widstrom et al., 1987), assures that the sucking infant has recurring periods of exposure to the mother's bare flesh and scent. During feeding bouts, the mother's odor signature will also be associated with various reinforcing stimuli (warmth, food intake, physical contact) that should further facilitate learning of that cue. In comparison, bottle-fed infants will typically have less opportunity to become acquainted with the individual odor of their mother. A similar lack of sufficient contact with the father could explain 2-week-old (breast-fed) infants' failure to discriminate that parent's odor. Additional evidence of learned recognition of maternal odors is provided by an experiment in which nursing mothers applied perfume to their breasts (Schleidt and Genzel, 1990). Newborn infants who sucked from perfumed breasts for one week developed a preference for that familiar odorant.

Individual recognition is a necessary preliminary step in the development of specific, enduring social attachments. The formation of a unique bond with another individual implies that that person can be distinguised from others. Because it develops shortly after birth, olfactory recognition of the mother may be an important factor in the initial phases of the attachment process (Porter and Winberg, 1999). Newborns' recognition of their mother may also have a positive impact on reciprocal maternal attachment to offspring. Infants can demonstrate an ability to recognize their mother by preferentially smiling and gazing at her rather than other individuals. Mothers commonly report that such overt signs of recognition have a positive effect on their attitude toward their infant (Robson and Moss, 1970). In this way, infants can begin to exert some control over the primary caregiver and thereby enhance the likelihood of continuing parental investment. The sense of smell is uniquely suited for early recognition of the mother since her salient odor phenotype is continuously accessible near the skin surface regardless of whether she is sleeping or awake, production of the recognizable odor does not require her active participation (cf. vocal recognition), and detection of that cue does not vary according to the ambient level of illumination (cf. visual recognition).

## 5. REFERENCES

Balogh, R. D., and Porter, R. H., 1986, Olfactory preferences resulting from mere exposure in human neonates, *Infant Behav. Develop.* 9:395-401.
Bushnell, I. W. R., Sai, F., and Mullin, J. T., 1989, Neonatal recognition of the mother's face, *Brit. J. Develop. Psychol.* 7:3-15.
Cernoch, J. M., and Porter, R. H., 1985, Recognition of maternal axillary odors by infants, *Child Dev.* 56:1593-1598.
Cheour, M., Martynova, O., Naatanen, R., Erkkola, R., Sillanpaa, M., Kero, P., Raz, A., Kaipio, M.-L., Hiltunen, J., Aaltonen, O., Savela, J.,and Hamalainen, H., 2002, Speech sounds learned by sleeping newborns, *Nature* 415:599-600.
Davis, L. B., and Porter, R. H., 1991, Persistent effects of early odor exposure on human neonates, *Chem. Senses* 16:169-174.
DeCasper, A. J., and Spence, M. J., 1986, Prenatal maternal speech influences newborns' perception of speech sounds, *Infant Behav. Develop.* 9:133-150.
Engen, T., and Lipsitt, L. P., 1965, Decrement and recovery of responses to olfactory stimuli in the human neonate, *J. Comp. Physiol. Psychol.* 59:312-316.
Engen, T., Lipsitt, L. P., and Kaye, H., 1963, Olfactory responses and adaptation in the human neonate, *J. Comp. Physiol. Psychol.* 56:73-77.
Fantz, R., 1963, Pattern vision in newborn infants, *Science* 140:296-297.

Foote, S. L., Berridge, C. W., Adams, L. M., and Pineda, J. A., 1991, Electrophysiological evidence for the involvement of the locus coeruleus in alerting, orienting, and attending, *Prog. Brain Res.* **88**:521-532.

Gervais, R., Holley, A., and Keverne, B., 1988, The importance of central noradrenergic influences on the olfactory bulb in the processing of learned olfactory cues, *Chem. Senses* **13**:3-12.

Gorski, P. A., Lewkowicz, D. J., and Huntington, L., 1987, Advances in neonatal and infant behavioral assessment: Toward a comprehensive evaluation of early patterns of development, *J. Develop. Behav. Pediatr.* **8**:39-50.

Haith, M. M., 1986, Sensory and perceptual processes in early infancy, *J. Pediatr. Suppl.*, **109**, 158-171.

Hepper, P. G., 1991, An examination of fetal learning before and after birth, *Irish J. Psychol.* **12**:95-107.

Hepper, P. G., 1995, Human fetal "olfactory" learning, *Int. J. Prenatal Perinat. Psychol. Med.* **7**:147-151.

Jacobs, B., 1990, Locus coeruleus neuronal activity in behaving animals, in: *The Pharmacology of Noradrenaline in the Central Nervous System*, D. J. Heal and C. A. Marsden, eds., Oxford University Press, Oxford, pp. 248-264.

Lagercrantz, H., 1996, Stress, arousal, and gene activation at birth, *News Physiol. Sci.* **11**:214-218.

Lagercrantz, H., and Slotkin, T. A., 1986, The "stress" of being born, *Scient. Amer.* **254**:92-102.

Marchini, G., Lagercrantz, H., Winberg, J., and Uvnas-Moberg, K., 1988, Fetal and maternal plasma levels of gastrin, somatoatatin and oxytocin after vaginal delivery and elective cesarean section, *Early Human Develop.* **18**:73-78.

Nelson, E. E., and Panksepp, J., 1998, Brain substrates of infant-mother attachment: Contributions of opioids, oxytocin, and norepinephrine, *Neurosci. Biobehav. Rev.* **22**:437-452.

Pascalis, O., de Haan, M., Nelson, C. A., and de Schonen, S., 1998, Long-term recognition memory for faces assessed by visual paired comparison in 3- and 6-month-old infants, *J. Exp. Psychol. Learn. Mem. Cog.* **24**:249-260.

Pascalis, O. and de Schonen, S., 1994, Recognition in 3-4-day-old human neonates, *Neuroreport* **5**:1721-1724.

Porter, R. H., and Schaal, B., 2003, Olfaction and the development of social behavior in neonatal mammals, in: *Handbook of Olfaction and Gustation, Second Edition*, R. L. Doty, ed., Marcel Dekker, Monticello, New York, pp. 309-327.

Porter, R. H., and Winberg, J., 1999, Unique salience of maternal breast odors for newborn infants, *Neurosci. Biobehav. Rev.* **23**:439-449.

Righard, L., and Alade, M. O., 1990, Effect of delivery room routines on success of first breastfeeding, *Lancet* **336**:1105-1107.

Robson, K. S., and Moss, H. A., 1970, Patterns and determinants of maternal attachment, *J. Pediat.* **77**:976-985.

Sarnat, H. B., 1978, Olfactory reflexes in the newborn infant, *J. Pediatr.* **92**:624-626.

Schaal, B., Marlier, L., and Soussignan, R., 2000, Human foetuses learn odours from their pregnant mother's diet, *Chem. Senses* **25**:729-737.

Schleidt, M., and Genzel, C., 1990, The significance of mother's perfume for infants in the first weeks of their life, *Ethol. Sociobiol.* **11**:145-154.

Shipley, M. T., Halloran, F. J., and de la Torre, J., 1985, Surprisingly rich projection from locus coeruleus to the olfactory bulb in the rat, *Brain Res.* **329**:294-299.

Simion, F., Farroni, T., Cassia, V. M., Turati, C., and Barba, B. D., 2002, Newborns' local processing in schematic facelike configuration, *Brit. J. Develop. Psychol.* **20**:465-478.

Spence, M. J., 1996, Young infants' long-term auditory memory: evidence for changes in preference as a function of delay. *Develop. Psychobiol.* **29**:685-695.

Sullivan, R. M., Stackenwalt, G., Nasr, F., Lemon, C., and Wilson, D. A., 2000, Association of an odor with activation of olfactory bulb noradrenergic B-receptors or locus coeruleus stimulation is sufficient to produce learned approach responses to that odor in neonatal rats, *Behav. Neurosci.* **114**:957-962.

Sullivan, R. M., Taborsky-Barba, S., Mendoza, R., Itano, A., Leon, M., Cotman, C. W., Payne, T. F., and Lott, I., 1991, Olfactory classical conditioning in neonates, *Pediatrics* **87**:511-518.

Sullivan, R. M., Wilson, D. A., and Leon, M., 1989, Norepinephrine and learning-induced plasticity in infant rat olfactory system, *J. Neurosci.* **9**:3998-4006.

Sullivan, R. M., Zyzak, D. R., Skierkowski, P., and Wilson, D. A., 1992, The role of olfactory bulb norepinephrine in early olfactory learning, *Develop. Brain Res.* **70**:279-282.

Svensson, T. H., 1987, Peripheral, autonomic regulation of locus coeruleus noradrenergic neurons in brain: putative implications for psychiatry and psychopharmacology, *Psychopharm.* **92**:1-7.

Swain, I. U., Zelazo, P. R., and Clifton, R. K., 1993, Newborn infants' memory for speech sounds retained over 24 hours, *Develop. Psychol.* **29**:312-323.

Turati, C., and Simion, F., 2002, Newborns' recognition of changing and unchanging aspects of schematic faces, *J. Exp. Child Psychol.* **83**:239-261.

Varendi, H., Porter, R. H., and Winberg, J., 1994, Does the newborn baby find the nipple by smell? *Lancet* **344**:989-990.

Varendi, H., Porter, R. H., and Winberg, J., 2002, The effect of labor on olfactory exposure learning within the first postnatal hour, *Behav. Neurosci.* **116**:206-211.

Widstrom, A. M., Ransjo-Arvidson, A. B., Christensson, K., Matthiesen, A. S., Winberg, J., and Uvnas-Moberg, K., 1987, Gastric suction in healthy newborn infants, *Acta Paediat. Scand.* **76**:566-572.

# HUMAN SWEATY SMELL DOES NOT AFFECT WOMEN'S MENSTRUAL CYCLE

Lixing Sun, Wendy A. Williams, and Corinna Avalos[*]

## 1. INTRODUCTION

Pheromonal effects of human body secretions have been closely examined in the study of human chemical communication since the 1970s (e.g., McClintock, 1971; Russell et al., 1980; Preti et al., 1987; Weller et al., 1995; Stern and McClintock, 1998). Men's body odor has been shown to regularize women's menstrual cycle (e.g., Cutler et al., 1986). However, attempts to identify the active constituents in complex mixtures of body odors are sporadic. The available publications are mainly focused on steroids (e.g., Cowley and Brooksbank, 1991; Maiworm and Langthaler, 1992; Grammer, 1993) and the results are often ambiguous. Therefore, much has yet to be done to identify active compounds in human body secretions and their effects on women's menstrual cycle.

Carboxylic (or Fatty) Acids are responsible for the sweaty odor of men's axillary secretion, with the major component being $(E)$-3-methyl-2-hexenoic acid (3M2H) (Zeng et al. 1991, 1992). Despite being copious in men's axillary secretion, its effect on women's menstrual cycle is unknown. In this study, we tested the possible regulatory effect of 3M2H on women's menstrual cycle by observing changes in the length of the menstruation and its variation over a period of three cycles.

## 2. METHODS

We chemically synthesized 3M2H using the Wittig methodology (see Pierce, Jr. et al., 1996 for the protocol). Heterosexual female volunteers (aged: 18-40), who did not use oral contraceptives, served as the subjects. Participants were asked to fill out a form to

---

[*] Lixing Sun, Department of Biological Sciences, Central Washington University. Wendy A. Williams, Department of Psychology, Central Washington University. Corinna Avalos, Medical School, Howard University.

provide relevant information about their daily habits and interactions with both male and female friends and relatives. Then, they were tested for the threshold of detecting 3M2H, which was serially diluted. With a double-blind procedure, the subjects were then randomly assigned to either the control or the experimental group. The experimental group received 10 ppm 3M2H - isopropyl alcohol solution, whereas the control, pure isopropyl alcohol.

Both groups were instructed to apply a drop of the given solution onto the upper lip daily before going to bed for 3 cycles, starting from the first day of the first menses. At the end of the fourth menses, we collected their reports about their periods. We encouraged them to mark down any special occasions including failure in application or use of medication that may affect the results. We also encouraged volunteers to report honestly by assuring them that there was absolutely no negative consequence if they failed to complete the session.

With such stringent criteria, we could only include 16 subjects who had completed the entire session without interruption or without using any medication in our analysis. We analyzed the lengths (independent t-test) and variations (F-test) of the menses, inter-menses, and menstrual cycle. A repeated measures ANOVA was used to test for the cumulative effect of the treatment over the entire period of the three consecutive cycles. All tests were two-tailed, and the level of significance was set at $P < 0.05$.

## 3. RESULTS

Figure 1 diagrams the effect of 3M2H on menses duration reported by our subjects. Despite some interesting and compelling trends, we failed to find any significant difference in the lengths of the menses, inter-menses and menstrual cycle. We also did not find any significant difference in the variation of the menses, inter-menses and menstrual cycle. Using a repeated-measures ANOVA to analyze the cumulative effect of the treatment, we found the only significant difference was in the third menses ($F_{1,7} = 14.80$, $p = 0.006$). This difference, however, was not shown in the fourth menses. Therefore, based on our study, we have no conclusive evidence suggesting that 3M2H serves as a pheromonal compound that may regularize women's menstrual cycles.

## 4. DISCUSSION

The results of the present study failed to positively identify 3M2H as the primary pheromonal compound responsible for the regularization of women's menstrual cycle. Thus, the sweaty smell frequently encountered in a hot, humid, and crowded environment may not cause any significant effect on women's reproductive cycle. However, the interesting and compelling (albeit not significant) trends seen in Figure 1 warrant brief mention. First, the overall average length of each of the four reported menses was longer for the experimental group relative to controls. Second, the period between each menses (inter-menses) was slightly shorter across all three cycles for the experimental subjects. And finally, the overall duration of the entire cycle was shorter for experimental subjects across all three cycles. It is important to note that despite the consistent average trends,

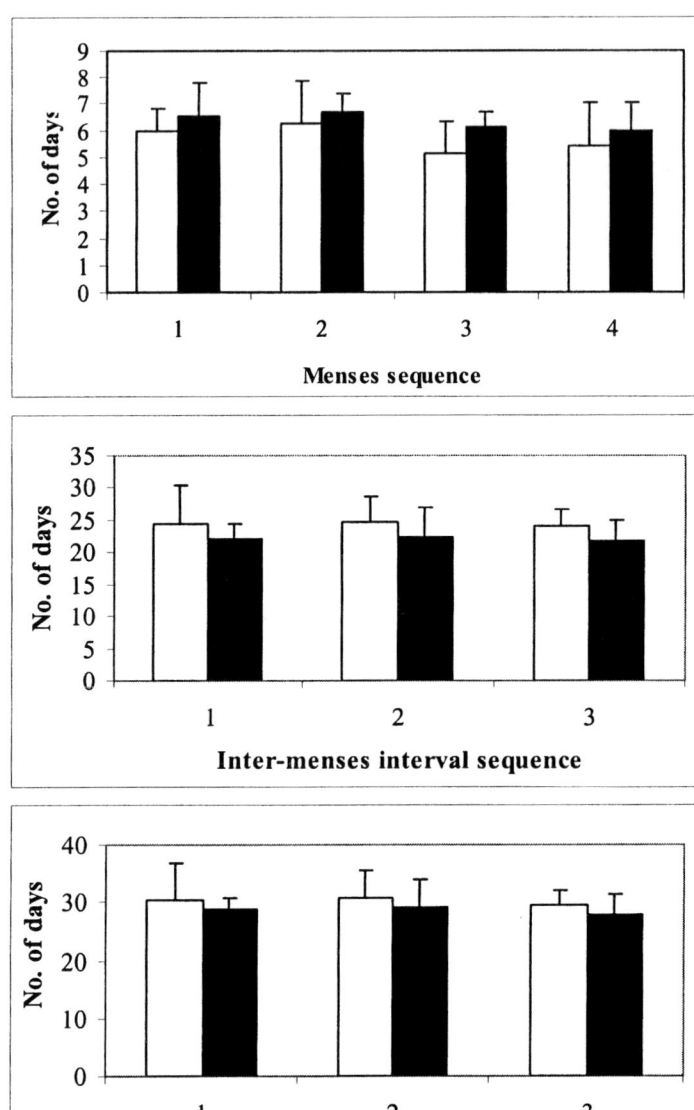

**Figure 1.** The differences between the experimental group and the control group for the menses (top), inter-menses (center), and the entire cycle (bottom). Labels: white, control; black, experimental. Bars are standard deviations.

variances tended to be large and prevented the mean differences from reaching statistical significance. As a result, we have to interpret the results with care because there were several methodological caveats to our study.

First, the chemical synthesis and purification process may have failed to completely eliminate the Z- isomers (see Pierce et al., 1995, 1996). Thus, the effect may be confounded by the mixture of two isomers, rather than a single compound. In the worst case, if the physiological effects on women's menstrual cycle of these two compounds are antagonistic, the overall effects may be cancelled out. This may be one of the reasons for our failure in detecting the treatment effect.

A second and more plausible explanation for the failure to reach statistical significance may have been due to the small sample size. This is a likely cause since all three general trends (menses duration, inter-menses duration and cycle duration) were generally consistent across all three cycles. Many volunteers failed to complete the project or were excluded in our analyses due to occasional failure to apply the solutions or use of medication. This resulted in an increase in the likelihood of a Type II error and reduced power of the test. As such, future replications must ensure a larger sample of subjects who complete the entire project.

A third concern, the concentration of 3M2H used in our study may have caused potential problems. The threshold of human perception of axillary odors varies greatly (Baydar et al., 1992). In our test for the threshold, some subjects reported that they could faintly detect the 3M2H. Thus, we believe that 10 ppm is about the average detection threshold for women exposed to 3M2H. A concentration lower than 10 ppm may become ineffective, and a higher concentration may cause bias because the subjects would know they were in the treatment group. This was the reason for us to decide to use 10 ppm. However, our estimate of 10 ppm may have been too low. Further investigation is needed to clarify whether there is a relationship between the concentration and effect of 3M2H.

Finally, several other factors, such as age, smoking habit, sexual activity, reliability of the reporting, may have also affect the results. Unfortunately, most of these variables remain beyond our control despite efforts to minimize them. However, a larger sample size will allow us to examine these elusive variables more closely in the future. Therefore, the synthesis process, sample size and 3M2H concentration detection thresholds are all factors which can be addressed in future replications.

In conclusion, the question of whether 3M2H represents the primary pheromonal compound responsible for the regularization of women's menstrual cycles remains an interesting and important one. The present investigation represents a first attempt to isolate the effects of 3M2H on various menstrual variables, and stands as a valuable pilot study which may provide direction for future investigations.

## 5. ACKNOWLEDGEMENTS

We thank all of the volunteers who participated in our research. We are also grateful to Justin Mallonee who collected some data in the early stage of our study. The research was funded by a grant from Central Washington University. The use of human subject was approved by Central Washington Human Subject Review Committee.

## 6. REFERENCES

Baydar, A. E., Petrzilka, M., and Schott, M.-P., 1992, Perception of characteristic axillary odors, *Perfum. Flav.* **17**:2-9.

Cowley, J. J., and Brooksbank, B. W. L., 1991, Human exposure to putative pheromones and changes in aspects of social behaviour, *J. Steroid. Biochem. Molec. Biol.* **39**:647-659.

Cutler, W. B., Preti, G., Krieger, A., Huggins, G. R., Garcia, C. R., and Lawley, H. J., 1986, Human axillary secretions influence women's menstrual cycles: the role of donor extract from men, *Horm. Behav.* **20**:463-73.

Grammer, K., 1993, 5-α-androst-16en-3α-on: A male pheromone? A brief report, *Ethol. Sociobiol.* **14**:201-208.

Maiworm, R. E., and Langthaler, W. U., 1992, Influence of androstenol and androsterone on the evaluation of men of varying attractiveness levels, in: *Chemical Signals in Vertebrates VI*, R. L. Doty and D. Müller-Schwarze, eds., Plenum Press, New York, pp. 575-579.

McClintock, M., 1971, Menstrual synchrony and suppression, *Nature* (Lond.) **229**:244-245.

Pierce, Jr., J. D., Blank, D. H., Aronov, E. V., Guo, Z., Preti, G., and Wysocki, C. J., 1996, Cross-adaptation of sweaty-smelling 3-methyl-2-hexenoic acid by its ethyl esters is determined by structural similarity, *J. Soc. Cosmet. Chem.* **47**:363-75.

Pierce, Jr., J. D., Zeng, X.-N., Aronov, E. V., Preti, G., and Wysocki, C. J., 1995, Cross-adaptation of sweaty-smelling 3-methyl-2-hexenoic acid by a structurally-similar, pleasant-smelling odorant, *Chem. Senses* **20**:401-411.

Preti, G., Cutler, W. B., Christensen, C. M., Lawley, H., Huggins, G. R., and Garcia, C.-R., 1987, Human axillary extracts: Analysis of compounds form samples which influence menstrual timing, *J. Chem. Ecol.* **13**:717-731.

Russell, M. J., Switz, G. M., and Thompson, K., 1980, Olfactory influences on the human menstrual cycle, *Pharmacol. Biochem. Behav.* **13**:737-738.

Stern, K., and McClintock, M., 1998, Regulation of ovulation by human pheromones, *Nature* (Lond.) **392**:177-179.

Weller, L., Weller, A., and Avinir, O., 1995, Menstrual synchrony: only in roommates who are close friends? *Physiol. Behav.* **58**:883-895.

Zeng, X., Leyden, J. J., Lawley, H. J., Sawano, K., Nohara, I., and Preti, G., 1991, Analysis of characteristic odors from human male axillae, *J. Chem. Ecol.* **17**:1469-92.

Zeng, X., Leyden, J. J., Brand, J. G., Spielman, A. I., McGinley, K. J., and Preti, G., 1992, An investigation of human apocrine gland secretion for axillary odor precursors, *J. Chem. Ecol.* **18**:1039-1055.

# LOCAL PREDATION RISK ASSESSMENT BASED ON LOW CONCENTRATION CHEMICAL ALARM CUES IN PREY FISHES: EVIDENCE FOR THREAT-SENSITIVITY

Grant E. Brown[*]

## 1. INTRODUCTION

Individuals that can reliably detect and avoid potential predators should gain significant fitness benefits (Lima and Dill, 1990). Early detection and avoidance of potential predation threats should, therefore, be strongly selected among prey species. Predator avoidance, however, comes with a significant cost associated with the loss of time available for other fitness related activities such as foraging, mating and/or territorial defence (Lima and Dill, 1990; Lima and Bednekoff, 1999). Thus, antipredator behaviour can be conceptualized as a series of 'threat-sensitive' trade-offs between the benefits associated with predator avoidance and those of other fitness related activities. Individuals that are able to reliably assess local predation risk should be at a selective advantage, as they would be capable of maximizing these threat-sensitive trade-offs through context appropriate behavioural responses (Brown, 2003).

One of the underlying assumptions of the threat-sensitivity model is that individuals should have sufficient flexibility to exhibit graded responses to a perceived predation threat (Helfman, 1989). Individuals should exhibit stronger responses to a higher perceived predation risk, as this would maximize potential survival benefits. Conversely, if the perceived predation risk is low, individuals should exhibit a less intense antipredator response, as this would allow them to continue gaining benefits from other activities. By adjusting the intensity of antipredator responses, individuals would be able to continue, for example, foraging while still gaining some antipredator benefits.

Assessment of local predation risk based on damage-released chemical alarm cues is widespread among freshwater fishes (Chivers and Smith, 1998; Smith, 1999; Wisenden, 2000). Such alarm cues are typically released following mechanical damage to prey, as

---
[*] Department of Biology, Concordia University, 7141 Sherbrooke W., Montréal, Quebéc, H4B 1R6 Canada

2000). Such alarm cues are typically released following mechanical damage to prey, as would occur during a predation attempt. When detected by nearby conspecifics and some sympatric heterospecifics, these alarm cues can elicit dramatic, short-term increases in species typical antipredator behaviours (Chivers and Smith, 1998). Recent studies by Mirza and Chivers (2001, 2002, 2003) and Chivers et al. (2002) have shown that individuals responding to chemical alarm cues gain significant survival benefits associated with an increase in antipredator behaviour.

In a recent review, Smith (1999) classified the response of individuals to conspecific and/or heterospecific chemical alarm cues as being either 'overt' or 'covert'. Overt responses are immediate, short-term changes in behaviour leading to increased predator avoidance. Such overt responses include increased shoal cohesion, area avoidance, dashing and/or freezing and reduced foraging and/or mating (Chivers and Smith, 1998; Smith, 1999). Covert responses are longer-term effects, which may not be readily observable. Such covert responses include induced morphologies and life history strategies and acquired recognition of novel predators (Chivers and Smith, 1998; Smith, 1999; Brown, 2003).

My goal in this paper is to briefly describe the role of chemical alarm cues in local risk assessment, focusing on threat-sensitive trade-offs. Specifically, I will address the following questions: (1) do prey fish show graded responses in 'overt' antipredator responses with decreasing stimulus concentration, (2) are prey fish able to detect chemical alarm cues below their minimum overt behavioural response threshold, and (3) do prey fish exhibit threat-sensitive changes in behaviour in response to alarm cues at concentrations below the minimum overt response threshold?

## 1.1 Overt Responses and Stimulus Concentration

According to the threat-sensitivity hypothesis, as the relative concentration of chemical alarm cues detected by an individual decreases, the intensity of the overt antipredator response should also decrease. However, recent studies suggest that there is no such graded response. Brown et al. (2001a) exposed shoals of fathead minnows (*Pimephales promelas*) to hypoxanthine-3-*N*-oxide (H3NO), the putative Ostariophysan alarm 'pheromone' (Pfeiffer et al., 1985; Brown et al., 2000, 2001b, 2003) ranging in concentration from 6.7 to 0.1 nM. Minnows exhibited consistent antipredator behaviour responses when exposed to H3NO at concentrations of 0.4 nM and above. At concentrations below this point, there was no measurable change in overt antipredator behaviour (Brown et al., 2001a). Similar results have been shown for juvenile rainbow trout (*Oncorhynchus mykiss*; Mirza and Chivers, 2003). Thus, when examining overt antipredator responses (Smith, 1999), there appears to be a minimum behavioural response threshold, below which, prey individuals do not exhibit an alarm response. In addition, these data suggest that if chemical alarm cues above this threshold are detected, individuals respond in an 'all-or-nothing' fashion (i.e. not graded; Figure 1).

Recent evidence has shown that individuals are able to detect chemical alarm cues well below the overt behavioural response threshold. Fathead minnows were able to acquire the recognition of the chemical cues of a novel predator (yellow perch, *Perca flavenscens*) if the predator odour was paired with H3NO at concentrations as low as 0.1 nM (25% of the previously demonstrated minimum overt response threshold) (Brown et al., 2001c). Mirza and Chivers (2003) likewise found that juvenile rainbow trout did not exhibit any overt antipredator response (i.e. not different from a distilled water control) when a stock solution of trout skin extract was diluted below 1:250 (in distilled water). However, when exposed to conspecific skin extracts at dilutions as low as 1:500, individual trout still gained significant survival benefits when exposed to live predators (northern pike, *Esox lucius*). These data demonstrate that chemical alarm cues are detectable well below the point at which they elicit overt antipredator responses (Figure 1).

### 1.2 Are Overt Antipredator Responses Threat-Sensitive?

To date, there is no compelling evidence suggesting that the overt antipredator responses of prey fishes to decreasing concentrations of chemical alarm cues are graded, as predicted by models of threat-sensitivity. There are, however, several lines of evidence, which indirectly support the existence of a graded response to low concentration chemical alarm cues. First, several species of prey fishes are known to exhibit flexible responses to conspecific chemical alarm cues, depending upon hunger level. Iowa darters (*Etheostoma exile*; Smith, 1981), fathead minnows (Brown and Smith, 1996) and finescale dace (*Phoxinus neogaeus*; Brown and Cowan, 2000) fail to exhibit species typical antipredator responses when food deprived for periods of 24 to 48 hours. In all cases, food deprived individuals did not exhibit any evidence of a graded response pattern, rather they responded in an 'all-or-nothing' fashion. Thus individuals under high energy constraints continued to forage under increased perceived predation risk. However, when allowed to forage freely and exposed to the same predation threat, they responded with an overt antipredator response (Smith, 1981; Brown and Smith, 1996; Brown and Cowan, 2000).

Secondly, at least one prey species (fathead minnows) has been shown to exhibit population specific minimum overt response thresholds. Minnows tested by Brown et al. (2001a) originated from a low predation risk environment. Conspecifics from a high predation environment exhibit a minimum response threshold approximately one order of

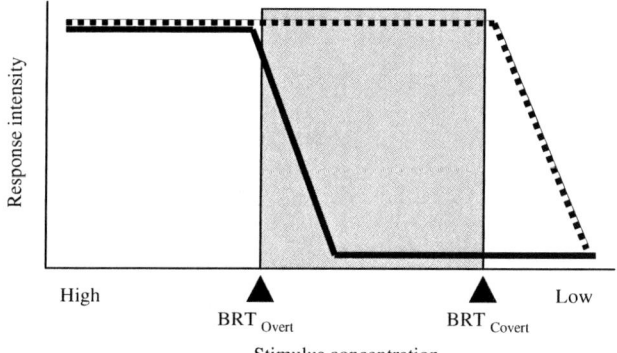

Figure 1. A simplified graphical representation of the behavioural response intensity of individual prey as a function of relative chemical alarm cue concentration. Bold line denotes overt antipredator response curve, dashed line denotes covert response curve. Shaded area between overt response threshold (BRT$_{overt}$) and covert response threshold (BRT$_{covert}$) represents concentration range in which we would expect to see threat-sensitive behavioural responses.

magnitude lower (Brown et al., 2001c). Similarly, population differences in the form and intensity of antipredator behaviour due to local predation risks have been widely documented (Magurran, 1993).

Finally, the difference between the overt and covert response thresholds (Figure 1) suggests that there exists a range of concentrations over which individual prey can detect an alarm cue, but do not exhibit an overt antipredator response. Recent studies have begun to examine the role of chemical alarm cues over this low concentration range.

## 2. EVIDENCE FOR THREAT-SENSITIVE BEHAVIOURS IN RESPONSE TO LOW CONCENTRATIONS OF CHEMICAL ALARM CUES

There remains the possibility that sub-threshold concentrations of chemical alarm cues may provide information leading to immediate changes in behaviour, even in the absence of an overt antipredator response. These threat-sensitive changes in behaviour may include: 1) increased vigilance towards secondary predator cues such as visual information, 2) an increase in risk-aversive foraging tactics and 3) context dependent behavioural shifts in response to conspecific chemical alarm cues.

### 2.1 Increased Vigilance Towards Secondary Predator Cues

One prediction of the threat-sensitivity model is that individuals detecting low predation threats should increase vigilance towards secondary predator cues (Treves, 2000). Several prey fishes engage in visual alarm signaling upon detecting a conspecific chemical alarm cue. Starry gobies (*Asterropteryx semipunctatus*), for example, engage in 'head-bobbing' behaviour when they detect a chemical alarm cue. Glowlight tetras (*Hemigrammus erythrozonus*) likewise engage in 'fin-flicking' behaviour as part of their species-typical overt antipredator response (Brown et al., 1999). The function of these conspicuous visual displays is two-fold. Initially, they act as a predator deterrent, reducing the likelihood of an attack (Brown et al. 1999). Secondly, such visual displays elicit increased antipredator behaviour in nearby conspecifics, providing increased survival for both the visual signal sender and receiver (Smith, 1992; Chivers et al., 1995; Brown et al., 1999).

Recently, Brown et al. (unpublished data[13]) directly tested this hypothesis by exposing shoals of glowlight tetras to a distilled water control or chemical alarm cues at concentrations above or below the overt antipredator response threshold. Test shoals were then exposed to the sight of a solitary tetra (in an adjacent tank) exhibiting an obvious antipredator response (a visual alarm cue). When exposed to the chemical cues only, tetras exposed to the sub-threshold concentration of chemical alarm cues did not exhibit any antipredator response (i.e. were not different from the distilled water controls). However, when subsequently exposed to the sight of a visually displaying conspecific, tetras initially exposed to the subthreshold concentration of alarm cue significantly increased their antipredator behaviour, similar to those initially exposed to the suprathreshold concentration. Those initially exposed to the distilled water control also increased their antipredator behaviour in response to the visual display, but at a

---

[13] Brown, G.E., Poirier, J.-F. and Adrian, J.C., Jr. in review. Assessment of local predation risk: the role of sub-threshold concentrations of chemical alarm cues. Behav. Ecol.

significantly lower intensity. These data demonstrate that low concentrations of chemical alarm cues may provide information to prey, leading to increased reliance on secondary predator cues; a threat-sensitive modification to the individual's behaviour patterns.

## 2.2 Threat-Sensitive Foraging Tactics

Under increased predation risk, individuals often engage in threat-sensitive foraging tactics (Lima and Dill, 1990). Such tactics allow individuals to continue foraging, often at a lower rate, under an elevated risk of predation. Body position during foraging bouts may directly influence an individual's risk of predation (Krause and Godin, 1996). For many prey fishes, foraging in a 'head-down' position can be an energetically efficient foraging tactic, however, it is risky due to a reduced ability to detect and respond to potential predators (Krause and Godin, 1996). Conversely, utilizing a head-up posture may be considered threat-sensitive, as individuals are at a lower risk of predation. This increased antipredator benefit, however, comes at a cost of foraging efficiency.

Recent evidence suggests that sub-threshold concentrations of chemical alarm cues elicit a switch to such a threat-sensitive foraging tactic in juvenile convict cichlids (*Archocentrus nigrofasciatus*). Foam et al. (unpublished data[14]) exposed pairs of juvenile cichlids to conspecific chemical alarm cues above and below the minimum overt antipredator response threshold (and a distilled water control). Cichlids were then allowed to forage on a horizontal and vertical food patches. When exposed to supra-threshold alarm cues, cichlids exhibited significantly increased antipredator behaviour (reduced foraging and aggression) relative to either sub-threshold alarm cue or distilled water treatments. Cichlids exposed to the distilled water control directed a significantly higher proportion of foraging attempts at the horizontal food patch (an energy maximizing tactic), while those exposed to either the super- or sub-threshold alarm cue treatments directed a significantly higher proportion of foraging attempts towards the vertical food patch (a threat-sensitive tactic). In addition, Foam et al. found no difference between the proportions of foraging attempts towards the vertical patch. These data demonstrate that even though cichlids exposed to the sub-threshold alarm cue treatment did not show an 'overt' antipredator response, they did adopt a threat-aversive tactic. Such a tactic would allow them to continue foraging under an elevated risk of predation risk.

## 2.3 Concentration Dependant Shifts in Behavioural Response

Juvenile centrarchids, such as largemouth bass (*Micropterus salmoides*) and green sunfish (*Lepomis cyanellus*) undergo an ontogenetic shift in their response to heterospecific chemical alarm cues (Brown et al., 2001d; Brown et al., 2002; Golub and Brown, 2003). As inshore juveniles, many centrarchids are members of the cyprinid prey guild (Brown et al., 2001d) and hence benefit from increasing antipredator behaviour upon detecting heterospecific alarm cues. However, once they have made the shift to piscivory, typically within the first one to two years of growth (Olson, 1996; Mittleback and Persson, 1998) juvenile bass and sunfish may benefit from increased foraging responses to these cues (Brown et al., 2001d; Golub and Brown, 2003). The point at

---

[14] Foam, P.E., Brown, G.E. and Harvey, M.C. in review. Heads up: juvenile convict cichlids switch to risk-aversive foraging tactics based on chemical cues. Ethology.

which they shift from antipredator to foraging responses is dependent upon a threat-sensitive trade-off (Brown et al., 2002; Golub and Brown, 2003). Golub and Brown (2003) demonstrated that green sunfish do not show a similar shift in response to conspecific alarm cues, since conspecific cues likely convey a greater risk.

Marcus and Brown (in press) have shown that alarm cue concentration may significantly influence the behavioural decision (threat-sensitive trade-off) associated with switching from antipredator to foraging responses in juvenile centrarchids. They exposed pumpkinseed sunfish (*Lepomis gibbosus*) above and below the size at which they shift to piscivory (approximately 110 mm; Mittlebach and Persson, 1998; Golub and Brown, 2003) to conspecific chemical alarm cues at varying concentrations. As the relative concentration of the alarm cue was reduced, small sunfish exhibited consistent antipredator responses until a minimum overt antipredator response threshold was reached. Below this concentration, small sunfish displayed no change in behaviour. Piscivorous sub-adult sunfish likewise exhibited significant antipredator responses when exposed to the highest concentrations. However, as the relative concentration was decreased, piscivorous individuals began to exhibit significant foraging responses. These data demonstrate that relative concentration, especially low concentrations, strongly influences the threat-sensitive trade-offs between antipredator and foraging benefits associated with the detection of chemical alarm cues by juvenile centrarchids.

## 3. SUMMARY

Previously, the behavioural response to chemical alarm cues by prey fishes has been characterized as either overt or covert (Smith, 1999). Above some minimum concentration, prey fish typically respond with an overt species-typical antipredator response. Though this response threshold appears to be flexible (i.e. Brown et al., 2001c), there appears to be little variability in the overt response patterns of individuals as a function of stimulus concentration. Below this overt antipredator response threshold, covert responses such as learned predator recognition are common (Brown, 2003). Recent studies, however, suggest the presence of a third category of threat-sensitive behaviour. Individuals detecting an alarm cue at low concentrations (i.e. below the point at which we might observe an overt antipredator response) appear to adjust their behaviour patterns in a risk sensitive fashion. Evidence suggests that such threat-sensitive tactics would allow individuals to gain some benefit associated with increased detection and avoidance of potential predators while still engaging in other fitness related activities such as foraging, territorial defense and/or mating.

The relative concentration of alarm cue detected by individuals thus appears to provide a great deal of information regarding local predation risk. When the detected concentration of alarm cues is sufficiently high, individual prey may benefit from engaging in a full antipredator response, ceasing foraging, territorial defense and mating, thus maximizing potential predator avoidance benefits. As the relative concentration of alarm cues decreases (below $BRT_{overt}$; Figure 1), individual prey may engage in a series of threat-sensitive responses. This would allow them to maximize the otherwise costly trade-offs between antipredator behaviour and other fitness related activities. As a result, the use of relative concentration as an information source should be selected within prey species.

## 4. ACKNOWLEDGEMENTS

James Grant, Douglas Chivers, Reehan Mirza, Patricia Foam, Justin Golub, Mark Harvey, Isabelle Désormeaux and Antoine Leduc provided helpful comments on earlier versions of this manuscript. Financial support was provided by NSERC of Canada and Concordia University.

## 5. REFERENCES

Brown, G. E., 2003, Learning about danger: chemical alarm cues and local risk assessment in prey fishes, Fish Fisheries 4:227-234.
Brown, G. E., and Cowan, J., 2000, Foraging trade-offs and predator inspection in an Ostariophysan fish: switching from chemical to visual cues, Behaviour 137:181-196.
Brown, G. E., and Smith, R. J. F., 1996, Foraging trade-offs in fathead minnows (Pimephales promelas): acquired predator recognition in the absence of an alarm response, Ethology 102:776-785.
Brown, G. E., Adrian, J. C., Jr., and Shih, M. L., 2001a, Behavioural responses of fathead minnows (Pimephales promelas) to hypoxanthine-3-N-oxide at varying concentrations, J. Fish Biol. 58:1465-1470.
Brown, G. E., Adrian, J. C. Jr., Patton, T., and Chivers, D. P., 2001c, Fathead minnows learn to recognize predator odour when exposed to concentrations of artificial alarm pheromones below their behavioural response thresholds, Can. J. Zool. 79:2239-2245.
Brown, G. E., Adrian, J. C. Jr., Smyth, E., Leet, H., and Brennan, S., 2000, Ostariophysan alarm pheromones: laboratory and field tests of the functional significance of nitrogen oxides, J. Chem. Ecol. 26:139-154.
Brown, G. E., Adrian, J. C., Jr., Naderi, N. T., Harvey, M. C., and Kelly, J. M., 2003, Nitrogen oxides elicit antipredator responses in juvenile channel catfish, but not in convict cichlids or rainbow trout: conservation of the ostariophysan alarm pheromone, J. Chem. Ecol. 29:1781-1796.
Brown, G. E., Adrian, J. C. Jr., Kaufman, I. H., Erickson, J. L., and Gershaneck, D, 2001b, Responses to nitrogen-oxides by Characiforme fishes suggest evolutionary conservation in Ostariophysan alarm pheromones, in: Chemical Signals in Vertebrates, Vol. 9, A. Marchlewska-Koj, J.J. Lepri, and D. Müller-Schwarze, eds., Plenum Press, New York, pp. 305-312.
Brown, G.E., Godin, J. G. J., and Pederson, J., 1999, Fin flicking: a visual antipredator signal in a characin fish (Hemigrammus erythrozonus), Anim. Behav. 59:469-475.
Brown, G. E., Leblanc, V. J., and Porter, L. E., 2001d, Ontogenetic changes in the response of largemouth bass (Micropterus salmoides, Centrarchidae, Perciformes) to heterospecific alarm pheromones, Ethology 107:401-414.
Brown, G. E., Gershaneck, D. L., Plata, D. L., and Golub, J. L., 2002, Ontogenetic changes in response to heterospecific alarm cues by juvenile largemouth bass are phenotypically plastic., Behaviour 139:913-927.
Chivers, D. P., and Smith, R. J. F., 1998, Chemical alarm signaling in aquatic predator-prey systems: a review and prospectus, Écoscience 5:338-352.
Chivers, D. P., Brown, G. E., and Smith, R. J. F., 1995, Familiarity and shoal cohesion in fathead minnows (Pimephales promelas): implications for antipredator behaviour, Can. J. Zool. 73:995-960.
Chivers, D. P., Mirza, R. S., and Johnston, J. G., 2002, Learned recognition of heterospecific alarm cues enhances survival during encounters with predators, Behaviour 139: 929-938.
Golub, J. L., and Brown, G. E., 2003, Are all signals the same?: Ontogenetic change in the response to conspecific and heterospecific alarm signals by juvenile green sunfish (Lepomis cyanellus), Behav. Ecol. Sociobiol. 54:113-118.
Helfman, G. S., 1989, Threat-sensitive predator avoidance in damselfish-trumpetfish interaction, Behav. Ecol. Sociobiol. 24:47-58.
Krause, J., and Godin, J. G. J., 1996, Influence of prey foraging posture on flight behavior and predation risk: predators take advantage of unwary prey, Behav. Ecol. 7:264-271.
Lima, S. L., and Bednekoff, P. A., 1999, Temporal variation in danger drives antipredator behavior: the predation risk allocation hypothesis, Am. Nat. 153:649-659.
Lima, S. L., and Dill, L. M., 1990, Behavioral decisions made under the risk of predation: a review and prospectus. Can. J. Zool. 68:619-640.
Magurran, A. E., 1993, Individual differences and alternative behaviours, in: Behaviour of Teleost Fishes, 2nd ed., T. J. Pitcher, ed., Chapman and Hall, New York, pp. 441-477.

Marcus, J. P., and Brown, G. E., in press, Response of pumpkinseed sunfish to conspecific chemical alarm cues: an interaction between ontogeny and stimulus concentration, *Can. J. Zool.*

Mittelbach, G. G., and Persson, L., 1998, The ontogeny of piscivory and its ecological consequences, *Can. J. Fish. Aquat. Sci.* **55**:1454-1465.

Mirza, R. S., and Chivers, D. P., 2001, Do chemical alarm signals enhance survival of aquatic vertebrates: an analysis of the current research paradigm, in: *Chemical Signals in Vertebrates, Vol. 9.*, A. Marchlewska-Koj, J.J. Lepri, and D. Müller-Schwarze, eds., Plenum Press, New York, pp. 19-26.

Mirza, R. S., and Chivers, D. P., 2002, Brook char (*Salvelinus fontinalis*) can differentiate chemical alarm cues produced by different age/size classes of conspecifics, *J. Chem. Ecol.* **28**:555-564.

Mirza, R. S., and Chivers, D. P., 2003, Response of juvenile rainbow trout to varying concentrations of chemical alarm cues: response thresholds and survival during encounters with predators, *Can. J. Zool.* **81**:88-95.

Olson, M. H., 1996, Ontogenetic niche shifts in largemouth bass: variability and consequences for first-year growth, *Ecology* **77**: 179-190.

Pfeiffer, W., Riegelbauer, G., Meier, G., and Scheibler, B., 1985, Effect of hypoxanthine-3-$N$-oxide and hypoxanthine-1-$N$-oxide on central nervous excitation of the black tetra, *Gymnocorymbus ternetzi* (Characidae, Ostariophysi, Pisces) indicated by dorsal light response, *J. Chem. Ecol.* **11**:507-523.

Smith, R. J. F., 1981, Effect of food deprivation on the reaction of Iowa darters (*Etheostoma exile*) to skin extracts, *Can. J. Zool.* **59**:558-560.

Smith, R.J.F., 1989, The response of *Asterropteryx semipunctatus* and *Gnatholepis anjerensis* (Pisces, Gobbidae) to chemical stimuli from injured conspecifcs, and alarm response in gobies, *Ethology* **81**:279-290.

Smith, R. J. F., 1992., Alarm signals in fishes, *Rev. Fish Biol. Fish.* **2**:33-63.

Smith, R. J. F., 1999, What good is 'smelly stuff' in the skin? Cross-function and cross-taxa effects in fish 'alarm substances', in: *Advances in Chemical Signals in Vertebrates*, R.E. Johnston, D. Müller-Schwarze, and P.W. Sorensen, eds., Kluwer Academic, New York, pp. 475-488.

Treves, A., 2000, Theory and method in studies of vigilance and aggregation, *Anim. Behav.* **60**:711-722.

Wisenden, B. D, 2000, Olfactory assessment of predation risk, *Phil. Trans. R. Soc.* **355**:1205-1208.

# LEARNED RECOGNITION OF HETEROSPECIFIC ALARM CUES BY PREY FISHES: A CASE STUDY OF MINNOWS AND STICKLEBACK

M. S. Pollock[1], D. P. Chivers[1], R. C. Kusch[1], R. J. Tremaine[1], R. G. Friesen[1], X. Zhao[1], and G. E. Brown[2]

## 1. INTRODUCTION

Numerous aquatic organisms use the odour of damaged prey to assess predation risk (Chivers and Smith, 1998). Cues are released during a predation event when an individual is either damaged or consumed. When confronted with this indication of impending predation, nearby neighbours engage in anti-predator behaviour (Pollock et al., 2003). In fact, the response to damage released alarm cues has been found to be nearly ubiquitous in most groups of freshwater fishes tested (Chivers and Smith, 1998). One of the most common groups used in predator/prey studies, the Ostariophysans, comprise 75% of all North American freshwater fishes and contain specialized epidermal club cells which are thought to contain the alarm substance (Brown et al., 2000).

Fishes have been documented to respond with anti-predator behaviour to cues of both conspecifics as well as ecologically similar heterospecifics with which they co-occur (Chivers and Smith, 1998). The response to heterospecific alarm cues is innate in closely related species, and learned in distantly related species (Smith, 1982; Pollock et al., 2003). Prey 'warned' by conspecific and heterospecific cues may have increased survival over prey that are not 'warned' (Chivers et al., 2002). As well as increasing survival, heterospecific cues are also known to affect the timing of reproduction and the reproductive output of some species (Pollock et al., unpublished data).

With the importance of heterospecific alarm cues well established, studies have focused on mechanisms of learned recognition. Such studies have documented that individual learning can occur through the diet of a known predator (Mirza and Chivers, in press). Learning can also occur through the association of a novel heterospecific cue

---

[1] University of Saskatchewan, Saskatoon, Saskatchewan, Canada, S7N 5E2.
[2] Concordia University, Montréal, Quebéc, Canada H4B 1R6.

with the odour of a known conspecific alarm cue (Mirza and Chivers, 2001), and may occur through cultural transmission from one individual to another. At the population level, ecological factors known to affect learning include the ratio between two prey species (Pollock and Chivers, in press) and the habitat characteristics in which the prey occur (Pollock and Chivers, 2003).

In this paper we review learned recognition of heterospecific alarm cues by prey fishes. We do this by providing a case study of the fathead minnow (*Pimephales promelas*)/brook stickleback (*Culaea inconstans*) alarm systems. Fathead minnows and brook stickleback commonly occur together in a diversity of water bodies. They share a similar suite of predators and consequently cross-species responses to alarm cues should be highly advantageous.

## 2. DEMONSTRATING THE ABILITY TO LEARN

Pollock et al (2003) were the first to document the ability of minnows to learn to recognize heterospecific cues as an indication of predation. In their study a naturally occurring population of fathead minnows allopatric with brook stickleback did not respond to the skin extract of stickleback with an antipredator response. Stickleback fish were then introduced into the pond with the minnows and the two species were left to co-exist for a period of five years. Following the period of co-existence, minnows were tested in the laboratory for a response to stickleback alarm cues. Not only did minnows now respond to stickleback cues, but they did so with the same intensity as they did to their own conspecific cues. A field experiment was also able to document a significant avoidance of stickleback skin extract in the wild.

To be sure that individual learning and not rapid selection on the population accounted for the documented response, Pollock et al. (2003) collected minnow eggs from the pond and reared them in the laboratory. Once the fry had reached several months of age they were tested for a response to stickleback cues as well as to their own conspecific cues. Results indicated that while the fish responded to their own cues they failed to respond to stickleback cues. Taken together, these studies indicate that minnows are able to acquire the ability to respond to stickleback alarm cues, and that the response occurs through learning and is not an innate response.

Previous to this study several researchers had inferred that fishes learned to respond to sympatric heterospecific alarm cues because larger more experienced fishes tended to respond to sympatric heterospecific cues with greater intensity than did smaller, less experienced fish (e.g., Chives and Smith, 1994).

## 3. HOW INDIVIDUALS LEARN

Learning through association has been a common mode of training in many behavioural experiments. All studies outlined below involve the pairing of a known stimulus with an unknown stimulus, followed by testing to document the occurrence of learning.

## 3.1 Learning Through the Diet of a Predator

Mirza and Chivers (in press) found that fathead minnows had the ability to learn to recognize a novel heterospecific odour as an indication of predation if the cue was present in the diet of a known predator. In their study, minnows were exposed to chemical stimuli collected from a tank containing a known predator (northern pike, *Esox lucius*) fed one of two unknown prey, stickleback or swordtails (*Xiphophorous helleri*). In subsequent behavioural tests, the minnows were exposed to either swordtail skin extract or stickleback skin extract. Minnows exposed to the odour of pike fed stickleback responded to stickleback skin extract with an anti-predator response but did not respond to swordtail skin extract. Similarly, minnows exposed to pike fed swordtail cues responded to swordtail skin extract with an antipredator response but did not respond to stickleback skin extract. This study demonstrated that minnows had the ability to learn to recognize a novel cue in the diet of a known predator, whether that cue is from a species that commonly co-occurs with minnows (the stickleback) or an allopatric tropical species with which it has never co-occurred.

## 3.2 The Importance of a Mixed Diet

A second study by Mirza and Chivers (2001) investigated whether or not minnows could learn to recognize a novel heterospecific cue (stickleback cue) when the cue was associated with a conspecific alarm cue in the diet of an unknown predator. Mirza and Chivers (2001) conditioned fathead minnows with chemical stimuli from a predatory yellow perch (*Perca flavescens*) fed a mixed diet of either minnows and brook stickleback, or swordtails and stickleback. Minnows were then exposed to chemical alarm cues of injured stickleback alone. Those minnows previously conditioned with perch fed a diet of minnows and stickleback increased their use of shelter. They also "froze" significantly more often than minnows conditioned with perch fed a diet of swordtails and stickleback or those exposed to distilled water. This study demonstrates another mechanism by which fishes can learn to recognize a novel heterospecific alarm cue as an indication of predation.

## 3.3 The Benefits of Learned Responses

Chivers et al. (2002) tested whether or not learned responses could result in a survival benefit, besides confirming that fish could learn unknown heterospecific cues through the diet or a predator. In a two-part study, fathead minnows were exposed to chemical stimuli collected from rainbow trout (*Oncorhynchus mykiss*) fed a mixed diet of either minnows and brook stickleback, or swordtail and stickleback. To test if the minnows had acquired recognition of stickleback alarm cues, Chivers et al. (2003) exposed the fish to stickleback alarm cues and introduced an unknown predator, yellow perch or northern pike. Both perch and pike took longer to initiate an attack on minnows that were previously exposed to trout fed minnows and stickleback than those previously exposed to trout fed swordtails and stickleback. These results show again that fishes are able to learn novel cues through association with known cues in a predator's diet. Furthermore, it shows that anti-predator responses to these newly learned cues could result in a survival benefit.

## 4. CULTURAL TRANSMISSION

Mathis et al. (1996) showed that minnows can learn to recognize novel odours via cultural transmission. Minnows naïve to northern pike gave fright responses to chemical stimuli from a pike when paired with a pike-experienced minnow but did not when paired with a pike-naïve minnow. The conditioned minnows not only retained the fright response to pike odour when tested alone, but were also able to transmit their fright responses to other pike naïve minnows in subsequent trials. In a final test, Mathis et al. (1996) determined that the same process worked across species barriers. Stickleback naïve to pike responded with an anti-predator response only when paired with pike experienced minnows. These same stickleback were then able to transfer this fright response back to naïve minnows. This study, while not demonstrating the learning of a heterospecific alarm cue, indicates the potential for such a cue to be culturally transmitted between species. Future studies should test whether fishes can acquire recognition of heterospecific alarm cues using the same mechanism.

## 5. ECOLOGICAL FACTORS THAT AFFECT LEARNING

We have a reasonable understanding of mechanisms of learned recognition of novel heterospecific odours. However, much less attention has been given to understanding how ecological factors may influence a population's ability to learn.

### 5.1 The Effect of Density on Learning

Pollock and Chivers (in press) hypothesized that the ability of minnows to learn to recognize stickleback alarm cues should increase with increasing stickleback density because there would be more opportunity for minnows to associate the heterospecific alarm cue with a predation threat. To test this hypothesis they stocked stickleback-naïve minnows into large 18,000 l outdoor pools with no stickleback, low numbers of stickleback, or high numbers of stickleback. All pools contained a predator (northern pike) known to the minnows. Following a 14-day conditioning period, minnows were taken into the laboratory and tested for a response to skin extract of stickleback, minnow, or an unknown heterospecific (swordtail). Minnows from pools with large numbers of stickleback learned to respond to stickleback alarm cues, and responded with the same intensity as they did to their own conspecific cues, while minnows from pools with low numbers of stickleback, or no stickleback, did not. These results show that density can have an impact on a population's ability to learn to use the alarm cues of heterospecifics.

### 5.2 The Effect of Habitat on Learning

Following from the studies on density, Pollock and Chivers (2003) hypothesized that the type of habitat occupied by two species may likewise play a role in the ability to learn heterospecific alarm cues. Brook stickleback were introduced into one of two types of outdoor pools containing stickleback-naïve fathead minnows and one predatory pike. Fishes were stocked at a ratio known to facilitate learning in the previous study. The two pool types were designed to represent complex habitat, consisting of various types of cover and shelter, and open habitat with no cover present. Fishes were first conditioned in the pools for 8 days and subsequently tested in the laboratory to determine if they had learned to recognize stickleback cues. Results of the laboratory indicated that minnows from the open habitat had learned to recognize stickleback cues as an indication of predation, while minnows from the complex habitat had not. Pollock and Chivers (2003) proposed that one possible explanation for their result is the interaction between visual and chemical cues that occurred in the open habitat but not the complex habitat. This study, as well as the previous study (Pollock and Chivers, in press), indicate that ecological factors play a significant role in the ability of co-occurring species to learn to use heterospecific alarm cues.

## 6. FACTORS AFFECTING THE USE OF HETEROSPECIFIC CUES

One area of research that has received little attention is the understanding of factors that affect a fish's 'decision' to respond to a known heterospecific cue. To address these short comings, Pollock et al. (unpub. data) and Friesen et al. (unpub. data) are conducting long term field experiments testing the responses of both minnows and sticklebacks from various populations to several skin extracts, including minnow, stickleback, swordtail (unknown control), morpholine (unknown non-biological odour) and distilled water (blank control). Preliminary results indicate that the responses to both conspecific and heterospecific cues in the field are context dependent. Results from the first year of study indicate that factors such as body condition, age, gender, breeding condition and time of year all play a role in a fish's "decision" to respond to a heterospecific, or conspecific, cue. Further studies are attempting to put these factors into a predictable context that can explain the fish's avoidance behaviour.

## 7. AREAS NEEDING FURTHER STUDY

### 7.1 How Individuals Respond

Research is needed to determine if prey fish can culturally transfer the knowledge of a heterospecific alarm cue to naïve individuals. While a similar study was conducted by Mathis et al. (1996) using predator odours, the ecological pressure to learn a heterospecific cue may not be as high as it may be to learn a predator odour. Therefore, cultural transmission may not be an effective method of learning.

A second area in need of study is the interaction between visual and chemical stimuli and the learning of chemical cues. Field (Wisenden et al., in press) and laboratory studies (Abrahams and Hartman, 2000) have indicated that a fish's behaviour is affected when given the opportunity to use visual and chemical cues together versus being presented with chemical cues alone. In response to the results outlined both in the literature and in section 5.2, Pollock and Chivers are currently exploring the interaction between learning and the two sensory modalities as a potential explanation for why learning does not occur as readily when populations occur in complex habitat.

## 7.2 The Benefits of Learned Responses

While Mirza and Chivers (2002) have established that learning heterospecific cues through the diet of a predator is possible and can result in increased survival, a correlation between mode of learning and survival benefit may be enlightening. Several modes of learning have been examined and it may be that these differing modes of learning offer differing benefits in terms of memory retention or survival advantage.

## 7.3 How Ecological Factors Affect Learning

This area of research is ripe for study; several ecological factors may influence learning. Areas that need further study include understanding the interaction between multiple factors such as density and habitat type. Other factors that could act alone or in conjunction to influence learning include the type or rate of predation, type of water body, time of year, species specific learning rates, etc. A fish preyed upon by a bird may release less alarm cue then an individual preyed upon by another fish or invertebrate. Therefore, populations experiencing predominantly bird predation may have faster or slower rate of learning then a population exposed to either fish or invertebrate predators.

## 7.4 Factors Affecting the 'Decision' of Fishes to Respond to Heterospecific Cues

This area, above all others, is still in its infancy. Length of fish (Chivers and Smith, 1994) and hunger level (Brown and Smith, 1996) have been documented as factors that affect a fish's response to conspecific cues and in some cases heterospecific cues. However, thorough long term studies, such as those of Pollock et al. (unpub. data) and Friesen et al. (unpub. data), on the multitude of individual factors that go into a fish's "decision" to respond, are needed before further work incorporating interactions or manipulating individuals can be done.

## 8. CONCLUSIONS

The response of prey fishes to heterospecific alarm cues has received much attention in the past decade. While much of the research has been conducted in the laboratory recent work has begun to emphasize field studies (Chivers and Smith, 1998). The fact that the responses can either be innate for closely related species (Smith, 1982; Mirza and Chivers, 2001) or learned in distantly related species (Pollock et al., 2003) has also been established. Researchers are now concentrating on three main areas of study: 1) mechanisms of learning; 2) ecological factors affecting a population's ability to learn; and 3) factors that affect the 'decision' of prey to respond to the cues. Perhaps the most puzzling enigma will be deciphering and putting into context those factors an individual takes into consideration when formulating a response to a known heterospecific cue.

## 9. REFERENCES

Brown, G. E., and Smith, R. J. F., 1996, Foraging trade-offs in fathead minnows (*Pimephales promelas*, Osteichthyes, Cpyrinidae): Acquired predator recognition in the absence of an alarm response, *Ethology* **102**:776-785.
Brown, G. E., Adrian, J. C., Jr., Smyth, E., Leet, H., and Brennan, S., 2000, Ostariophysan alarm pheromones: laboratory and field tests of the functional significance of nitrogen oxides, *J. Chem. Ecol.* **26**:139-154.
Chivers, D. P., Mirza, R. S., and Johnston, J. G., 2002, Learned recognition of heterospecific alarm cues enhances survival during encounters with predators, *Behaviour* **139**:929-938.
Chivers, D. P., and Smith, R. J. F., 1994, Intra- and interspecific avoidance of areas marked with skin extract from brook sticklebacks in a natural habitat, *J. Chem. Ecol.* **20**:1517-1524.
Chivers, D. P., and Smith, R. J. F., 1998, Chemical alarm signaling in aquatic predator-prey systems: A review and prospectus, *Ècoscience* **5**:338-352.
Hartman, E. J., and Abrahams M. V., 2000, Sensory compensation and the detection of predators: the interaction between chemical and visual information, *Proc. Royal Soc. Lond. B.* **267**:571-575.
Mathis, A., Chivers, D. P. and Smith, R. J. F., 1996, Cultural transmission of predator recognition in fishes: intraspecific and interspecific learning, *Anim. Behav.* **51**:185-201.
Mirza, R. S., and Chivers, D. P., 2001, Learned recognition of heterospecific alarm signals: the importance of a mixed predator diet, *Ethology* **107**:1007-1018.
Mirza, R. S., and Chivers, D. P., 2001, Are chemical alarm cues conserved within the salmonid fishes? *J. Chem. Ecol.* **27**: 1641-1655.
Mirza, R. S., and Chivers, D. P., *in press*, Fathead minnows learn to recognize heterospecific alarm cues they detect in the diet of a known predator, *Behaviour*.
Pollock, M. S., and Chivers, D. P., *in press*, The effects of density on the learned recognition of heterospecific alarm cues, *Ethology*.
Pollock, M. S., and Chivers, D. P., 2003, Does habitat complexity influence the ability of fathead minnows to learn heterospecific chemical alarm cues? *Can. J. Zool.* **81**:923-927.
Pollock, M. S., Chivers, D. P., Mirza, R. S. and Wisenden, B. D., 2003, Fathead minnows, *Pimephales promelas*, learn to recognize chemical alarm cues of introduced brook stickleback, *Culaea inconstans*, *Env. Biol. Fish.* **66**:313-319.
Smith, R. J. F., 1982, Reaction of *Percina nigrofasciata, Ammocrypta beani*, and *Etheostoma swaini* (Percidae, Pisces) to conspecific and intergeneric skin extract, *Can. J. Zool.* **60**:1067-1072.
Wisenden, B. D., Pollock, M. S., Tremaine, R. J., Webb, J. M., Wismer, M. E., and Chivers, D. P., *in press*, Synergistic interactions between chemical alarm cues and the presence of conspecific and heterospecific fish shoals, *Behav. Ecol. Sociobiol.*

# THE RESPONSE OF PREY FISHES TO CHEMICAL ALARM CUES: WHAT RECENT FIELD EXPERIMENTS REVEAL ABOUT THE OLD TESTING PARADIGM

Robyn J. Tremaine, Michael S. Pollock, Robert G. Friesen, Robin C. Kusch, and Douglas P. Chivers[1]

## 1. INTRODUCTION

Upon detection or capture by predators, numerous species of aquatic vertebrates release chemical alarm cues into the environment (Chivers and Smith, 1998). The cues act to 'warn' conspecifics and heterospecifics of danger. Individuals that are 'warned' survive longer in the presence of predators than individuals that are not warned (Mathis and Smith, 1992; Mirza and Chivers, 2001). Long-term exposure to chemical alarm cues causes prey to exhibit an adaptive change in body morphology (Stabell and Lwin, 1997). Moreover, exposure to alarm cues has been shown to mediate learning of unknown predators (Chivers and Smith, 1994 a, b). To date, most studies of alarm cues have been carried out in relatively simplistic chemical environments in the laboratory. The animals are often fed to satiation and tested in clean water, free of pollution and chemical signals from competitors, mates etc. Moreover, they are often tested in small monospecific groups and only after they have been held in the absence of predation for a considerable period of time. It is imperative that such laboratory experiments undergo field verification in the complex chemical environment of the animal's natural habitat.

Mathis and Smith (1992) pioneered an excellent trap experiment technique for studying the responses of freshwater littoral fishes to chemical alarm cues. Mathis and Smith's experiment, and most of those that followed, used Gee's Improved Minnow Traps, roughly cylindrical wire enclosures (43 cm length x 22 cm diameter) with a funnel located at each end leading into the trap. Cube-shaped artificial cellulose sponges saturated with the experimental treatment were threaded onto stainless steel wire, and

---

[1] University of Saskatchewan, Saskatoon, Saskatchewan, Canada, S7N 5E2

then attached near the trap entrance. The traps were then placed into the water for a set period and the number of fish captured in different chemical treatments was quantified.

In this paper we provide a complete review of the past and current trap experiment paradigm and suggest areas in need of future study and clarification.

## 2. FIELD EXPERIMENTS USING DISTILLED WATER AS A CONTROL

The original field study by Mathis and Smith (1992) consisted of traps marked with either fathead minnow (*Pimephales promelas*) skin extract or a control of glass-distilled water. The minnows exhibited significant avoidance of traps marked with the skin extract. The response was very dramatic with only 27 fish (less than 4% of the total) captured in 16 experimental traps versus 822 fish captured in 16 control traps. They also noted a nonsignificant tendency for a higher number of larger, more experienced minnows in control traps suggesting a possible role of experience or developmental factors in determining their response to alarm substance. While this pioneering experiment increased the knowledge of the behavior of fishes in the field, the one major limitation was comparing the fish's response to the smell of something (i.e. the odour of a damaged conspecific) to the smell of nothing (i.e. distilled water). What this really tells us is that fish can smell something over nothing and will respond accordingly.

In the past decade the protocol in most trap experiments has involved the use of distilled water as the sole control treatment. For example, Chivers and Smith (1994c) demonstrated that brook stickleback (*Culaea inconstans*) avoided conspecific skin extract over distilled water. The stickleback captured in traps marked with skin extract were significantly smaller than those individuals captured in traps marked with water, implicating experience or physiological development as critical factors in the development of anti-predator behavior.

Brook stickleback and fathead minnows occupy the same habitat and are vulnerable to common predators. Individuals that detect alarm cues of co-habiting species may benefit by gaining early warning of danger. A field study by Mathis and Smith (1993) demonstrated that skin extract from fathead minnows is effective at inducing avoidance responses by stickleback. Brook stickleback exploit the alarm system of minnows and thus reduce their own risk of predation (Mathis and Smith, 1993; Wisenden et al., 1994). In a trap experiment by Wisenden et al. (1994), the duration of area avoidance by brook stickleback of areas marked with fathead minnow alarm substance was measured. They found that stickleback continued to avoid locations associated with predation risk after the source of the cue was removed, and only after 2-4 hours did the fish resume use of these risky areas. In a follow-up experiment, Wisenden et al. (1995) determined that fishes naïve to the association of an area with alarm cue were the first to migrate into the risky area. Fishes present at the time of cue release did not return for 7 to 8 hours after the cue was removed. Perhaps the chief beneficiaries of chemical alarm cues are only those individuals present at the time of cue release.

Trap experiments have also been used to determine the chemical composition of Ostariophysan alarm cues. In a study by Brown et al. (2000), traps labeled with fathead minnow extract, hypoxanthine-3-N-oxide or pyridine-N-oxide caught significantly fewer fishes (including dace and minnow) than those labeled with distilled water. These field results validate the outcome of laboratory studies in which significant increases in anti-predator behavior in both fathead minnows and finescale dace were observed when they were exposed to conspecific skin extract, hypoxanthine-3-N-oxide, and pyridine-N-oxide. Both the field and laboratory data strongly suggest that it is the nitrogen oxide functional group that elicits behavioral responses.

In summary, most published trap experiments have shown that brook stickleback avoid fathead minnow or stickleback skin extract over distilled water (Wisenden et al., 1994; Chivers and Smith, 1994c; Wisenden et al., 1995; Mathis and Smith, 1992, 1993). Similarly, fathead minnows avoid stickleback and minnow cues over distilled water (Wisenden et al., 1995; Mathis and Smith, 1992; Brown et al., 2000). While novel and informative, these studies have the drawback of using distilled water as the only control.

## 3. FIELD EXPERIMENTS USING UNKNOWN HETEROSPECIFIC CONTROL

Most laboratory studies examining the responses of fishes to chemical alarm cues have shown that the anti-predator responses exhibited to conspecific cues are specific to conspecifics and are not generalized responses to any injured fish cue or any novel odor (Chivers and Smith, 1998). Consequently, most of the field tests have used distilled water as the control. However, more recently, several studies now include an unfamiliar heterospecific skin extract (usually swordtail, *Xiphophorus helleri*) or novel odors, such as morpholine, as additional controls. In some cases the results are somewhat surprising.

One such study (Chivers et al., 1995) tested the response of two populations of minnows to traps labelled with either darter skin extract (*Etheostoma exile*) or swordtail skin extract (control). One population of fathead minnows was sympatric with darters, while the other was allopatric. In the darter-sympatric population, significantly fewer and smaller minnows were captured in traps marked with darter skin extract than traps marked with the control. In the darter-allopatric populations there was no difference in the number or size of minnows captured in experimental versus control traps. These results indicate that fathead minnows recognize and avoid areas where darter alarm cues are detected and that this may be a learned response.

A more recent study by Mirza and Chivers (2001) found that in one of two experiments brook charr (*Salvelinus fontinalis*) avoided conspecific alarm cues significantly more than an unfamiliar heterospecific cue (swordtail) under natural conditions. As well as adding an unfamiliar heterospecific control, this was the first study to document a behavioral response of a salmonid to damage-released alarm cues in the wild.

Recent studies (Friesen et al. unpublished data) include both the unfamiliar eterospecific skin extract to control for a response to generalized injured fish cue and morpholine to control for response to an unfamiliar odor (Tables 1 and 2). The purpose of this design is to further discriminate the fish's response, from a response to a novel damaged fish versus a novel non-biological odor.

## 4. CONTRADICTORY RESULTS

Populations of fathead minnows and stickleback are known to respond to alarm cues in both the laboratory setting, as well as in the field (Pollock et al., 2003; Chivers and Smith, 1995). In some cases, however, they display no preferential avoidance between conspecific or heterospecific cues in the field (Pollock et al., unpublished data), or respond contrary to prediction (Tables 1 and 2).

One such example is a series of experiments by Pollock et al. (unpublished) in which minnows more often than not showed no preferential avoidance between their own cues and the cues of an unknown heterospecific. In one experiment, contrary to prediction, minnows avoided swordtail cues over minnow cues, and avoided swordtail cues over stickleback cues (Table 1). Similarly, sticklebacks avoid their own cues over a familiar heterospecific, although they more often than not showed no differential avoidance. Again, contrary to the predictions of one experiment, sticklebacks avoided fathead minnow and swordtail cue over their own.

There may be several explanations for these incongruous results. The complex natural habitat of these species includes competing chemical cues from mates and competitors and the test stimulus must be of a high enough concentration to override this chemical background. It may also be that in a natural setting, the response, or lack of response of a fish to an alarm cue, may be influenced by various factors including reproductive state, age, the presence of other fish, as well as physical and biological properties of the pond. Several experiments have set out to investigate the interactions between these factors affecting decision making.

**Table 1.** Summary of fathead minnow (*Pimephales promelas*) response in trap experiments (SB=stickleback; FHM=fathead minnow; SWT=swordtail; DW=distilled water).

|  | SB cue | SWT cue | DW | Morpholine |
|---|---|---|---|---|
| Minnows avoid FHM cue over: | 2/8 studies | 0/7[1] studies | 4/6 studies | 1/2 studies |
| Minnows avoid SB cue over: | N/A | 1/9[2] studies | 1/4 studies | 0/2 studies |

[1] In one experiment, minnows avoided swordtail cue over fathead minnow cue, contrary to prediction.
[2] In one experiment, minnows avoided swordtail cue over stickleback cue, contrary to prediction.

**Table 2.** Summary of stickleback (*Culaea inconstans*) responses in trap experiments (SB=stickleback; FHM=fathead minnow; SWT=swordtail; DW=distilled water).

|  | FHM cue | SWT cue | DW | Morpholine |
|---|---|---|---|---|
| Stickleback avoid SB cue over: | 1/9[1] studies | 0/8[2] studies | 2/4 studies | 0/2 studies |
| Stickleback avoid FHM cue over: |  | 0/6 studies | 3/3 studies | 0/2 studies |

[1] In one experiment, stickleback avoided fathead minnow cue over stickleback cue, contrary to prediction.
[2] In one experiment, stickleback avoided swordtail cue over stickleback cue, contrary to prediction.

## 5. VALIDATION OF ECOLOGICAL FACTORS

### 5.1. Response to Chemical Cues in the Presence of Shoals

Shoal membership reduces a predator's ability to visually isolate individual prey, and allows fellow prey to act as shields against direct attack (Wisenden et al., in press). Therefore, the presence of conspecific shoals, or shoals of ecologically or morphologically similar heterospecifics are attractive to prey species. A trap experiment conducted by Wisenden et al. (in press) investigated the interaction between the presence of fish shoals (held within a transparent jar) and chemical alarm cues. The presence of chemical alarm cues alone caused prey to avoid traps, while the presence of a fish shoal alone attracted fish to traps. When chemical alarm cues were combined with the presence of a shoal, fish tended to increase shoal cohesion and enter the traps, despite the fact that the traps were the source of the alarm cue. An understanding of the interaction between visual and chemical cues is critical in the evaluation of anti-predator behavior of natural populations of fishes.

### 5.2. Response to Chemical Cues under Varying Light Intensities

Both chemical and visual information is available when a prey species assesses local predation risk. However, contradictory models exist regarding the relative importance of visual versus chemical cues during risk assessment. Laboratory experiments conducted by Hartman and Abrahams (2000) demonstrated that fathead minnows reacted to alarm cues under low light conditions with significantly more dashes than they did in clear water. They concluded that vision was the primary source of information used in predator assessment (visual compensation model), as the minnows would respond to alarm cues with greater intensity when visibility was reduced.

In an attempt to verify the visual compensation model in the field, a day/night trap experiment was conducted by Kusch et al. (unpublished) on an established community of fathead minnows and brook stickleback. The Kusch et al. study found no interaction between light intensity and treatment. The light level did not affect the avoidance pattern of either species to the various concentrations of fathead minnow alarm cue.

Fishes exist in a complex context-dependent system that is based on multiple sensory inputs that could lead to a number of variable, yet appropriate, anti-predator responses.

## 6. CONCLUSIONS

Field studies are necessary to validate anti-predator responses of prey fishes in the chemically complex natural environment. While several of the early trap experiments demonstrated the effectiveness of alarm cues in the field (i.e., Mathis and Smith, 1992), the more complicated recent studies often demonstrate no preferential avoidance between familiar and unfamiliar cues or, alternatively, results are contradictory to predictions. Trap experiments have therefore brought into question previous knowledge based primarily on laboratory studies (i.e., visual compensation model). Anti-predator defense strategies of fishes are clearly context-dependent. To decipher and explore the complexities of this dependent response further, field studies investigating the various biotic and abiotic factors affecting a fish's response are required.

## 7. ACKNOWLEDGMENTS

This research was funded by grants to D.P. Chivers from NSERC and the University of Saskatchewan.

## 8. REFERENCES

Brown, G. E., Adrian, J. C. Jr., Smyth, E., Leet, H., and Brennan, S., 2000, Ostariophysan alarm pheromones: laboratory and field tests of the functional significance of nitrogen oxides, *J. Chem. Ecol.* **26**:139-154.

Chivers, D. P., and Smith, R. J. F., 1994a, Fathead minnows, *Pimephales promelas*, acquire predator recognition when alarm substance is associated with the sight of unfamiliar fish, *Anim. Behav.* **48**:597-605.

Chivers, D. P., and Smith, R. J. F., 1994b, The role of experience and chemical alarm signaling in predator recognition by fathead minnows, *Pimephales promelas, J. Fish Biol.* **44**:273-285.

Chivers, D. P., and Smith, R. J. F., 1994c, Intra- and interspecific avoidance of areas marked with skin extract from brook sticklebacks (*Culaea inconstans*) in a natural habitat, *J. Chem. Ecol.* **20**:1517-1524.

Chivers, D. P., and Smith, R. J. F., 1998, Chemical alarm signaling in aquatic predator-prey systems: A review and prospectus, *Écoscience* **5**:338-352.

Chivers, D. P., Wisenden, B. D., and Smith, R. J. F., 1995, The role of experience in the response of fathead minnows (*Pimephales promelas*) to skin extract of Iowa darters (*Etheostoma exile*), *Behaviour* **132**:665-674.

Hartman, E. J., and Abrahams, M. V., 2000, Sensory compensation and the detection of predators: the interaction between chemical and visual information, *Proc. Roy. Soc. Lon.* **267**:571-575.

Mathis, A., and Smith, R. J. F., 1992, Avoidance of areas marked with a chemical alarm substance by fathead minnows (*Pimephales promelas*) in a natural habitat, *Can. J. Zool.* **70**:1473-1476.

Mathis, A. and Smith, R. J. F., 1993, Intraspecific and cross-superorder responses to chemical alarm signals by brook stickleback, *Ecology* **74**:2395-2404.

Mirza, R. S., and Chivers, D. P., 2001, Chemical alarm signals enhance survival of brook charr (*Salvelinus fontinalis*) during encounters with predatory chain pickerel (*Esox niger*), *Ethology* **107**:989-1005.

Pollock, M. S., Chivers, D. P., Mirza, R. S., and Wisenden, B. D., 2003, Fathead minnows, *Pimephales promelas*, learn to recognize chemical alarm cues of introduced brook stickleback, *Culaea inconstans*, *Env. Biol. Fish.* **66**:313-319.

Stabell, O. B., and Lwin, M. S., 1997, Predator-induced phenotypic changes in crucian carp are caused by chemical signals from conspecifics, *Env. Biol. Fishes* **49**:145-149.

Wisenden, B. D., Chivers, D. P., Brown, G. E., and Smith, R. J. F., 1995, The role of experience in risk assessment: Avoidance of areas chemically labelled with fathead minnow alarm pheromone by conspecifics and heterospecifics, *Écoscience* **2**:116-122.

Wisenden, B. D., Chivers, D. P., and Smith, R. J. F., 1994, Risk-sensitive habitat use by brook stickleback (*Culaea inconstans*) in areas associated with minnow alarm pheromone, *J. Chem. Ecol.* **20**: 2975-2983.

Wisenden, B. D., Pollock, M. S., Tremaine, R. J., Webb, J. M., Wismer, M F., and Chivers, D. P., *in press*, Synergistic interactions between chemical alarm cues and the presence of conspecific and heterospecific fish shoals, *Behav. Ecol. Sociobiol.*

# RESPONSE OF JUVENILE GOLDFISH *(CARASSIUS AURATUS)* TO CHEMICAL ALARM CUES: RELATIONSHIP BETWEEN RESPONSE INTENSITY, RESPONSE DURATION, AND THE LEVEL OF PREDATION RISK

Xiaoxia Zhao and Douglas P. Chivers[*]

## 1. INTRODUCTION

How prey animals balance their activities against antipredator demands has been a major issue in modern behavior ecology (Sih, 1984; 1992; Berejikian et al., 1999). Prey animals in nature experience a broad range of predation risk. Failure of prey animals to respond to predators results in a high risk of being attacked or captured. In contrast, prey that overrespond to a threat may waste time and energy that would otherwise be directed towards other activities (Sih, 1992; Lima and Dill, 1990; Chivers et al., 2001). Helfman (1989) proposed the threat-sensitive predator avoidance hypothesis to reflect this dynamic balance; the intensity of antipredator response of prey should reflect the level of predation threat (Helfman, 1989; Sih, 1992; Chivers et al., 2001; Chivers and Mirza, 2001).

Recent studies indicate that chemical cues provide a wealth of information for aquatic prey in assessing predation risk (Chivers and Smith, 1998; Kats and Dill, 1998; Chivers and Mirza, 2001). Aquatic media are well suited for chemical signals because a large number of compounds can dissolve in water, providing a large number of potential chemical signals for detection (Hara, 1994). Regarding chemical cues, the concentration of alarm cue that prey animals detect may be used to distinguish the degree of predation threat. The ability to exhibit threat-sensitive avoidance has important implications. By being able to differentiate between different cue concentrations, prey animals will not waste time and energy responding to predators that do not pose an imminent threat.

---

[*] Department of Biology, University of Saskatchewan, 112 Science Place, Saskatoon, Saskatchewan, S7N 5E2, Canada.

In fishes, damage-released alarm cues are best characterized in the Superorder Ostariophysi (Smith, 1992). Goldfish, *Carassius auratus*, like other Ostariophysans, possess chemical alarm cues, which are released from epidermal club cells on injury and serve to warn nearby individuals of potential threat. When conspecifics detect the alarm cue, they respond with anti-predator behavior, including shoaling, reduced foraging, freezing or dashing (e.g., Wisenden and Smith, 1997; Chivers and Smith, 1998; Mirza and Chivers, 2003a, b). In this study we exposed juvenile goldfish to varying concentrations of conspecific alarm cue and quantified the behavioural response. We predicted that fish would respond with a greater intensity and duration to higher concentration chemical alarm cues than to lower concentration cues. The intensity and duration of the response to different chemical alarm cue concentrations has not been reported. Such studies are critical in understanding the ecological significance of responses to chemical alarm cues.

## 2. METHODS

We obtained juvenile goldfish from a commercial supplier in May, 2003. The fish were kept in 622-L artificial-stream tanks at about 14°C for 10 days prior to the experiment. Fresh dechlorinated water was introduced into the holding tank at a rate of 1 L/min. Fish were fed daily on commercial fish pellets.

Fish alarm cue was prepared from 8 small goldfish (4.23 ± 0.18 cm standard length (mean ± SE)). Fish were humanely killed with a single blow to the head in accordance with guidelines set by the Canadian Council on Animal Care. A fillet of skin was removed from both sides of each fish and placed in 100 mL of chilled glass-distilled water. A total of 31.85 cm$^2$ of skin was collected which was then homogenized, filtered through filter floss to remove the larger particles, and then diluted in distilled water to make a final volume of 620 mL. This represented our base solution (1:1). We then diluted the appropriate volumes of the base solution with distilled water to create the following dilutions: 1:10, 1:100, 1:1000, and 1:5000. Stimulus solutions were pipetted into sample bags in 30-mL aliquots and frozen at -20°C until use. Distilled water was pipetted in 30-mL aliquots and frozen as well. During each behavioural trial (see below) we introduced 10 mL of either goldfish stimulus or distilled water.

We conducted a total of 90 trials, fifteen replicate trials in each of six treatments, five alarm cues (1:1, 1:10, 1:100, 1:1000, 1:5000) and one distilled water control treatment. Each trial consisted of three randomly chosen fish, thus a total of 270 juvenile goldfish were used. Each group of fish was used only once, and the order of the treatments was randomized.

The experimental set-up and procedure were similar to previous studies in our laboratory (e.g., Mirza and Chivers, 2003a, b). The trials were performed in 74-L aquaria (60 x 30 x 40 cm$^3$). Each tank had two horizontal, 10-cm-long lines on the front of the tank which divided the tank into 3 vertical areas. Each of the test tanks contained a single airstone mounted in the centre of the end wall. A piece of airline was attached to the airstone.

Fish were allowed to acclimate in the test chambers for 48 h prior to trials. Each trial was 18 min in length and consisted of an 8-min pre-stimulus period and an 8-min post-stimulus period, with a 2-min stimulus-introduction period between the pre- and post-stimulus periods. Dye trials indicated that it took approximately 40 seconds for the stimulus to disperse throughout the test tank. Prior to the prestimulus period, we removed and discarded 60 ml of water through the stimulus injection tube. We then removed and retained an additional 60 mL of water. After the prestimulus observation period, we injected either 10 mL of goldfish stimulus or distilled water and flushed it through with the 60 mL of the previously retained water. The tanks were drained and thoroughly rinsed after each use.

During both the pre- and post-stimulus periods we recorded shoaling index and vertical area use. Shoaling index was scored from 1 (no fish within one body length of another) to 3 (all 3 individuals within one body length of each other) and was recorded every 15 second. An area use score was also recorded every 15 seconds. The score was the sum of each fish's score (1= the lower third of the water column, 2 = the middle third of the water column, and 3 = the upper third of the water column). A score of 3 would indicate that all fish were in the bottom third of the tank and a score of 9 would indicate that all fish were in the top third of the tank. The occurrence of freezing (fish drop to the bottom and remain motionless for at least 30 seconds) and dashing (a rapid burst of apparently disoriented swimming) (Chivers and Smith, 1998) behaviour was also recorded in each trial.

To assess the effect of variation in alarm cue concentration on the intensity of alarm responses, we calculated the average shoaling index and vertical area scores for each group of fish for both the 8 minute pre-stimulus and the 8 minute post-stimulus periods. Changes between the pre- and post-stimulus periods for each response variable were calculated as post-stimulus minus pre-stimulus. These differences were compared using a one way ANOVA followed by Tukey post-hoc tests (Zar, 1999). The occurrences of freezing and dashing were compared using a Chi-square test for independence (Zar, 1999).

In order to assess the effect of variation in alarm cue concentration on the duration of alarm responses, we calculated the average shoaling index and vertical area use score for each group of fish during the pre-stimulus period to determine a baseline level of response. Following exposure to the stimulus we calculated the shoaling index and vertical area scores for each group of fish for 1 minute intervals of the 8 minute post-stimulus period. We used repeated measures ANOVA to determine the effects of treatment on the response of the fish and whether there was a treatment by time interaction.

## 3. RESULTS

### 3.1 Intensity of Alarm Response

Results from the one way ANOVA followed by Tukey post-hoc tests show that exposure to higher concentrations of alarm cue (1:1, 1:10, 1:100) caused a significant increase in shoaling and area use compared with lower concentrations of alarm cue (1:1000 and 1:5000) and distilled water (Figures 1 and 2). There were no significant differences between any two adjacent concentrations in both shoaling and area use behaviours (Figures 1 and 2).

Goldfish showed a greater amount of freezing and dashing as the concentration of alarm cues increased. Goldfish exhibited at least one period of freezing in all 15 trials when exposed to an alarm-cue concentration of 1:1 compared with 12 out of 15 trials to a concentration of 1:10, 9 out of 15 trials to a concentration of 1:100, and 5 out of 15 trials to a concentration of 1:1000. There were no occurrences of freezing when fish were exposed to conspecific alarm cue concentrations of 1:5000 or distilled water ($x^2 = 48.75$, df = 5, $p$ <0.001). Fish exhibited dashing in all 15 trials as well when exposed to alarm-cue concentration of 1:1 compared with 13 out of 15 trials to a concentration of 1:10 and 7 out of 15 times to a concentration of 1:100. There was significantly less dashing when fish were exposed to conspecific alarm cue concentrations of 1:1000 or 1:5000 or distilled water (4, 2, 1 respectively) ($x^2 = 45.54$, df = 5, $p$ <0.001).

### 3.2 Duration of Alarm Response

Results from repeated measures ANOVA show there was a significant treatment effect and that there was a treatment by time interaction in terms of both shoaling index (F = 154.90, $p$ < 0.001, Figure 3) and area use (F = 233.05, $p$ <0.001, Figure 4). At higher concentrations the responses were longer and stronger than those at lower concentrations.

## 4. DISCUSSION

In our experiment, we were able to determine the behavioural response intensity of juvenile goldfish exposed to different concentrations of conspecific alarm cues. Goldfish exposed to higher concentrations of alarm cues significantly increased shoaling and dashing more than goldfish exposed to lower concentrations of alarm cue, while those exposed to distilled water and the lowest concentration of alarm cues showed no response. These results indicate that responses to different concentrations of chemical cues are not generalized, but instead are graded to reflect the degree of threat.

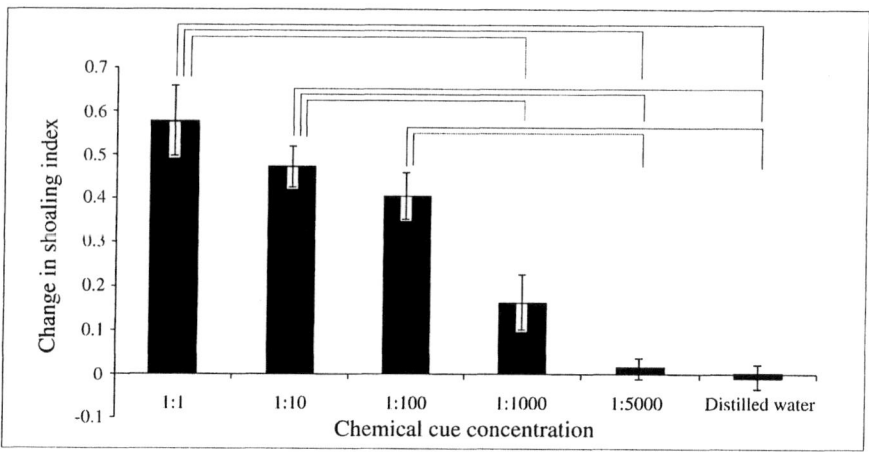

**Figure 1.** Mean (± SE) changes in the shoaling index of goldfish exposed to different concentrations of alarm cues. Lines over bars denote significance at $p < 0.05$ (see text for details).

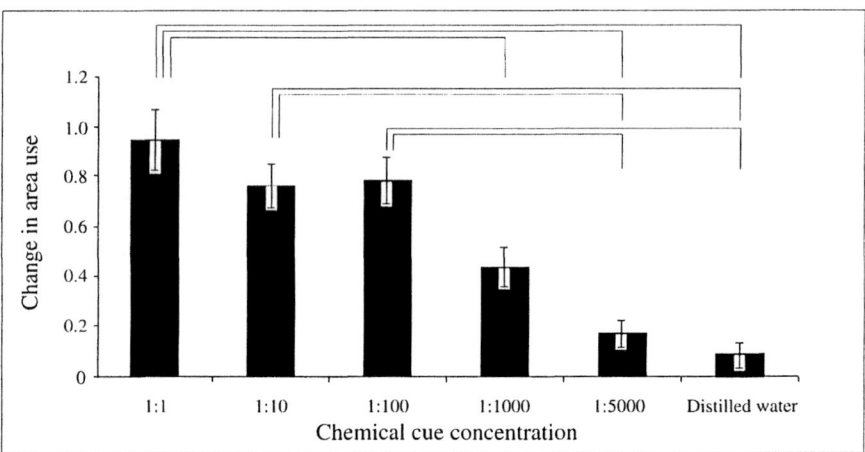

**Figure 2.** Mean (± SE) changes in the area use of goldfish exposed to different concentrations of alarm cues. Lines over bars denote significance at $p < 0.05$ (see text for details).

Our data emphasize the importance of including a temporal component when measuring behavioural responses to alarm cues. Conclusions based on a minute by minute analysis clearly demonstrate that fish exposed to higher concentrations of alarm cue will exhibit a longer duration response than those exposed to lower concentrations of alarm cues. This provides another line of evidence that fishes exhibit an anti-predator response that matches the threat. Had we used the average response of the fish during the

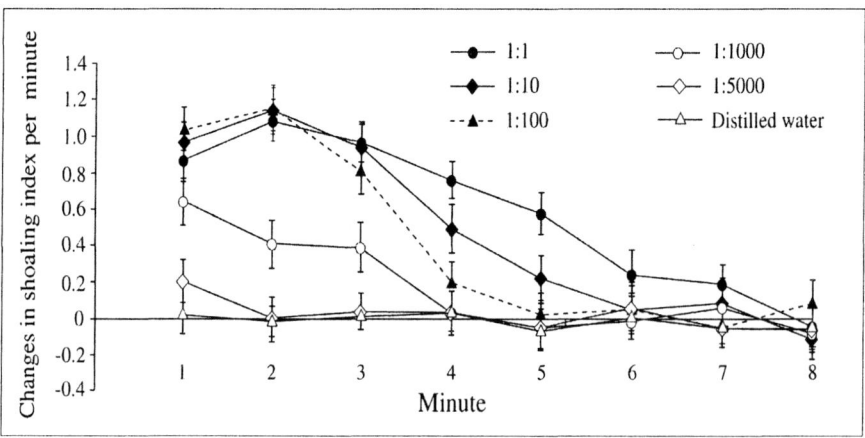

**Figure 3.** Changes in shoaling-index score per minute (mean ± SE) for goldfish exposed to chemical alarm cue of different concentrations and distilled water.

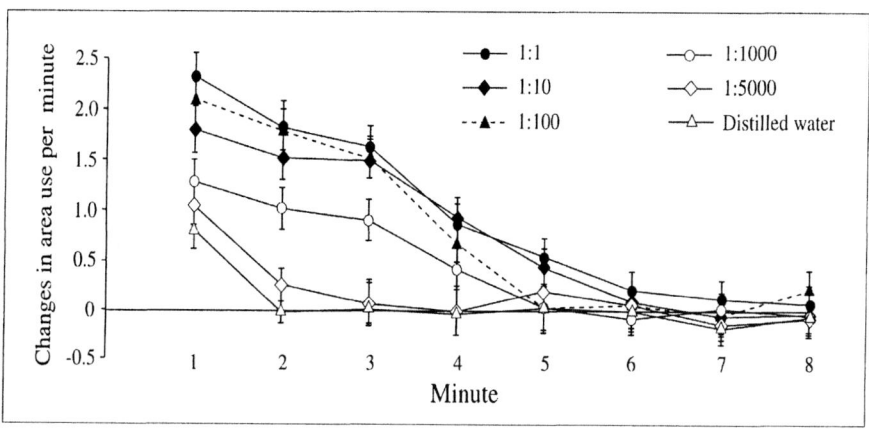

**Figure 4.** Changes in area use score per minute (mean ± SE) for goldfish exposed to chemical alarm cue of different concentrations and distilled water.

8 minute post-stimulus periods, we would have missed this important temporal component, as have most previous studies of alarm cues (review Chivers and Smith, 1998). Reporting the average response would also miss the fact that fish returned to normal baseline activity within minutes of being exposed to the threat.

Combining information on the amount of alarm cue released by prey during a predation event with information on the active space of an alarm cue may indicate the distance at which animals can detect a nearby predation threat (Mirza and Chivers, 2003a). No studies have been conducted to determine the amount of alarm cue released during a predation event, but some authors have attempted to determine the active space of alarm cues. Lawrence and Smith (1989) calculated the active space of 1 $cm^2$ of fathead minnow, *Pimephales promelas*, skin as being greater than 58, 823 L. Likewise, Mirza and Chivers (2003a) found that juvenile rainbow trout, *Oncorhynchus mykiss*, showed a behavioral response threshold at 1 $cm^2$ skin in 134, 255 L of water. Our results suggest that 1 $cm^2$ of goldfish skin creates an active space of 122, 637 L. Mirza and Chivers (2003a) argue that these values are overestimates because they are calculated based on the final concentration that would be achieved if the stimulus was completely dispersed through the test tank. We also need to be cautious when drawing conclusions about the response thresholds of prey to alarm cues because the prey's physiological response threshold and its behavioural response threshold may be different. Both Brown et al. (2001) and Mirza and Chivers (2003a) provide evidence that prey fishes have the ability to use chemical alarm cues to assess their level of risk in the absence of an observable antipredator response. They found that prey exposed to concentrations of alarm cue below their behavioral-response threshold were still using the information to assess predation risk (Mirza and Chivers, 2003a). Notwithstanding the cautions outlined above, our data suggest that prey animals have the ability to detect predation-risk cues at a long distance.

Our data suggest that chemical alarm cues at different concentrations may provide important information for prey to assess and mediate their risk of predation. This general predation risk information is valuable even if information regarding the actual predator is lacking. It may be common for a prey animal to be alerted to the presence of a potential risk before the predator is detected. In this way prey animals are able to make an accurate assessment of their level of predation risk and to adjust the intensity of their antipredator response to reflect this risk. By responding to different intensity signals, the prey may obtain the advantage of alertness and lower susceptibility, thereby reducing their risk of mortality.

In nature prey animals might face sustained periods of risk when predators are abundant and in close proximity; in contrast, they might experience low risk when predators are sparse and wide ranging (Sih and McCarthy, 2002). Also, depending on the characteristics of the water body, it is possible that alarm cues could be diluted to lower concentrations as well as mixed with other odours from the environment, which may decrease the intensity of prey's antipredator response. Therefore, additional studies aimed at understanding the importance of graded responses to different concentrations of chemical cue in nature are needed.

## 5. REFERENCES

Berejikian, B. A., Smith, R. J. F., Tezak, E. P., Schroder, S. L., and Knudsen, C. M., 1999, Chemical alarm signals and complex hatchery rearing habitats affect antipredator behavior and survival of Chinook salmon (*Oncorhynchus tshawytscha*) juveniles, *Can. J. Fish. Aquat. Sci.* **56**:830-838.

Brown, G. E., and Smith, R. J. F., 1998, Acquired predator recognition in juvenile rainbow trout (*Oncorhynchus mykiss*): conditioning hatchery- reared fish to recognize chemical cues of a predator, *Can. J. Fish. Aquat. Sci.* **55**:611-617.

Brown, G. E., Adrian, J. C. Jr., Patton, T., and Chivers, D. P., 2001, Fathead minnows learn to recognize perch when exposed to concentrations of alarm pheromone below their behavioural response threshold, *Can. J. Zool.* **79**:2239-2245.

Chivers, D. P., and Mirza, R. S., 2001, Predator diet cues and the assessment of predation risk by aquatic vertebrates: a review and prospectus, in: *Chemical Signals in Vertebrates IX*, A. Marchlewska-Koj, J. J. Lepri, and D. Müller-Schwarze, eds., Plenum Press, New York, pp. 277-284.

Chivers, D. P., and Smith, R. J. F., 1998, Chemical alarm signalling in aquatic predator-prey systems: a review and prospectus, *Écoscience* **5**:338-352.

Chivers, D. P., Mirza, R. S., Bryer, P. J., and Kiesecker, J. M., 2001, Threat-sensitive predator avoidance by slimy sculpins: understanding the role of visual versus chemical information, *Can. J. Zool.* **79**:867-873.

Hara, T. J., 1994, The diversity of chemical stimulation in fish olfaction and gestation, *Rev. Fish Biol. Fish.* **4**:1-35.

Helfman, G. S., 1989, Threat-sensitive predator avoidance in damselfish-trumpetfish interactions, *Behav. Ecol. Sociobiol.* **24**:47-58.

Kats, L. B., and Dill, L. M., 1998, The scent of death: Chemosensory assessment of predation risk by prey animals, *Ecoscience* **5**:361-394.

Laurila, A., 2000, Behavioral responses to predator chemical cues and local variation in antipredator performance in *Rana temporaria* tadpoles, *Oikos* **88**:159-168.

Lawrence, B. J., and Smith, R. J. F., 1989, Behavioral response of solitary fathead minnows, *Pimephales promelas*, to alarm substance, *J. Chem. Ecol.* **15**:209-219.

Lima, S. L., and Dill, L. M., 1990, Behavioural decisions made under the risk of predation: a review and prospectus, *Can. J. Zool.* **68**:619-640.

Mirza, R. S., and Chivers, D. P., 2003a, Response of juvenile rainbow trout to varying concentrations of chemical alarm cue: response threshold and survival during encounters with predators, *Can. J. Zool.* **81**:88-95.

Mirza, R. S., and Chivers, D. P., 2003b, Predator diet cues and the assessment of predation risk by juvenile brook charr: do diet cues enhance survival? *Can. J. Zool.* **81**:126-132.

Mirza, R. S., and Chivers, D. P., 2000, Predator-training enhances survival of brook trout: evidence from laboratory and field enclosure studies, *Can. J. Zool.* **78**:2198-2208.

Mirza, R. S., and Chivers, D. P., 2002, Brook charr (*Salvelinus fontinalis*) can differentiate chemical alarm cues produced by different size classes of conspecifics, *J. Chem. Ecol.* **28**:555-564.

Pettersson, L. B., and Brönmark, C., 1999, Energetic consequences of an inducible morphological defence in crucian carp, *Oecologia* **121**:12-18.

Sih, A., 1984, The behavioral response race between predator and prey, *Am. Nat.* **123**:143-150.

Sih, A., 1992, Prey uncertainty and the balancing of antipredator and feeding needs, *Am. Nat.* **139**:1052-1069.

Sih, A., and McCarthy, T. M., 2002, Prey responses to pulses of risk and safety: testing the risk allocation hypothesis, *Anim. Behav.* **63**:437-443.

Smith, R. J. F., 1992, Alarm signals in fishes, *Rev. Fish Biol. Fish.* **2**:33-63.

Wisenden, B. D., and Smith, R. J. F., 1997, The effect of physical condition and shoalmate familiarity on proliferation of alarm substance cells in the epidermis of fathead minnows, *J. Fish. Biol.* **50**:799-808.

Zar, J. H., 1999, Biostatistical analysis, 4$^{th}$ ed., Prentice-Hall, Inc., Upper Saddle River, New Jersey.

# THE EFFECTS OF PREDATION ON PHENOTYPIC AND LIFE HISTORY VARIATION IN AN AQUATIC VERTEBRATE

Robin C. Kusch, Reehan S. Mirza, Michael S. Pollock, Robyn J. Tremaine, and Douglas P. Chivers[*]

## 1. INTRODUCTION

Predation has the capacity to influence prey directly through modulating the density and size structure of the population (Brönmark et al., 1995) and/or indirectly by altering the population by causing changes in growth, survival and fecundity rates (Fraser and Gilliam, 1992). As a result, prey species have evolved a number of defense mechanisms to minimize their risk of being detected, caught and consumed by predators (Sih, 1987; Lima and Dill, 1990; Chivers and Smith, 1998; Kats and Dill, 1998). These defenses include adaptations in behaviour, morphology and life history traits (Chivers and Smith, 1998; Chivers and Mirza, 2001).

Several studies have shown that prey have the ability to alter their phenotypes in response to predation threat (Brönmark and Miner, 1992; Trussel, 1996; Reimer and Tedengre, 1996; Weber and Declerck, 1997; Van-Buskirk and Schmidt, 2000; Kappes and Sinsch, 2002). Such phenotypic plasticity has important ecological consequences as the changes result in the modification of species interactions. For example, when exposed to caged predators (*Aeschna*, dragonfly larvae) two species of larval newts (*Triturus alpestris* and *T. helveticus*) developed darker tail fin pigmentation, larger heads, larger tails and spent more time hiding in the leaf litter in comparison with newts in predator-free ponds. The individuals with the predator-induced phenotype survived significantly longer during survival trials when exposed to free dragonfly larvae (Van-Buskirk and Schmidt, 2000). In another study, Brönmark and Miner (1992) found that crucian carp (*Carassius carassius*) in the presence of piscivorous pike (*Esox lucius*) developed deeper bodies than those not exposed to this gape limited predator. The increase in size resulted in a refuge from predation.

---

[*] University of Saskatchewan, Saskatoon, Saskatchewan, Canada, S7N 5E2.

In addition to changing their phenotype during a specific life stage prey may also have the ability to alter the amount of time they remain in each life cycle stage. The ecological theory of life history trait plasticity is that the timing of the transition between one life stage and the next will vary with the costs and benefits associated with each stage (Werner, 1996). If the mortality to growth rate ratio is lower in the succeeding stage early transition will be exhibited and if it is higher then delayed transition should occur (Chivers et al., 2001). For example, when raised in the presence of chemical alarm cues from injured conspecifics, western toad (*Bufo boreas*) tadpoles metamorphose earlier than those raised in the absence of such predation cues (Chivers et al., 1999). Studies have shown that several species of amphibians possess the ability to facultatively adjust hatching time (Sih and Moore, 1993; Moore et al., 1996; Chivers et al., 1999; Warkentin, 2000; Warkentin et al., 2001; Chivers et al., 2001; Kiesecker et al., 2002; Altwegg, 2002) and recently one study illustrated the ability in whitefish, *Coregonus* sp. (Wedekind, 2002).

Anti-predator adaptations are often mediated or induced by chemical cues (Kats and Dill, 1998), especially in aquatic systems where visual cues are limited (Smith, 1992). Chemical cues function well in this medium as a large number of compounds can dissolve in water allowing for the production of a great number of possible signals (Hara, 1994). Research, in the past decade, has indicated that the assessment of these chemical cues is highly sophisticated (reviews Chivers and Mirza, 2001). Logically, the ability to accurately assess the risk of predation would be beneficial as each anti-predator defense has an innate cost to the user and the effectiveness of each response option is dependent on the context of the encounter and the specific predator.

The objective of the following laboratory experiments was to examine the influence of predation cue exposure on a prey species' phenotype and life history. Specifically, what is the effect of predation cue exposure on the incubation period and fry phenotype of fathead minnows (*Pimephales promelas*). The early life stages are often the most vulnerable to predation. Egg predation is common, therefore, there should exist strong selection pressure for the development of defense mechanisms specifically for this vulnerable stage. Little research has been conducted on the possible defense mechanisms employed by embryos themselves and how these adaptations are environmentally triggered. In one study, Chivers et al. (2001) show that amphibian embryos altered their hatching time in response to chemical cues of leeches and also to cues of injured eggs. In our study we simulated predation by exposing the embryos to injured egg cues.

## 2. MATERIALS AND METHODS

Fathead minnows in breeding condition were collected from Briarwood Lake in Saskatoon, Saskatchewan. The first collection of minnows occurred in the middle of June, 2002. Fish were artificially spawned in the laboratory and eggs were separated using a weak solution of tannic acid (200 mg/L) and left to water-harden in treated tap water for two hours. Eggs were then placed in individual 500 ml cups filled with 250 ml of treated tap water. The cups were randomly assigned a treatment and randomly place throughout the room. There were three treatment conditions: (1) a predation cue of injured egg, (2) a control of injured shrimp (previously frozen brine shrimp (*Artemia sp*)), and (3) a control of distilled water. Eggs were treated twice a day and monitored every four hours except for a six hour period of dark between 12:00 am and 6:00 am. All treatments were an equivalent volume of 5 ml to control for disturbance. The room temperature was controlled at 25°C and the light cycle was 18:6 L:D. The injured egg cue was prepared fresh for each use by homogenizing 450 eggs in 1125 ml of distilled water. Each treatment was equivalent to exposure to two injured eggs. The shrimp cue was kept as consistent as possible to the egg cue by homogenizing an equivalent mass of shrimp per unit of water. The 450 eggs weighed 0.236 g. The experiment ended after all eggs had either died or hatched. Hatchlings were removed using a pipette, euthanized using a lethal dose of tricaine methane sulfonate (MS222), placed into a vile of buffered formalin and stored for morphological analysis.

The second experiment was designed to examine the effect of stimulus concentration on the hatching and phenotypic response. The setup and protocol were the same as experiment one, except only 125 ml of water was added to the treatment containers; changing the volume allowed us to produce high injured-egg concentrations later in the season when fewer eggs were available to make stimulus. Minnows for the second experiment were collected in early July, 2002. The four treatment groups were (1) a predation cue twice the concentration of first experiment, (2) a predation cue half the concentration of experiment one, (3) a predation cue equivalent to that in experiment one and (4) a control of distilled water.

Minnows used in the third experiment were collected at the end of July, 2002. Treatment containers were filled with 125 ml of water and there were three treatment conditions: (1) a predation cue of injured egg equivalent to that in experiment one, (2) a control of injured shrimp and (3) a control of distilled water.

During April, 2003 the length of the fry from each experiment was measured and one-way ANOVA analysis was conducted on the three sets of data followed by Tukey tests. Kolmogorov-Smirnov tests were conducted on the hatch time data to determine if the distributions differed between the treatments. The P-value was adjusted to 0.03 using the bonferroni method to account for the multiple comparisons being made. Chi-square tests were conducted to determine a hatch time bias for the period of dark.

## 3. RESULTS

### 3.1 Hatching Time

Despite the fact that temperature and photoperiod were constant, the mean hatching time among the three experiments were substantially variable. In experiment one and three the hatch times were long in relation to experiment two (median = 132.2 hrs in experiment one, 125.0 hrs for experiment two and 138.0 for experiment three). The great majority of fish hatched during the period of dark (experiment one: 89.7%, experiment two: 91.5% and experiment three: 95.9%). Chi-square tests conducted on all three sets of data revealed that significantly more eggs hatched during the period of dark ($P<0.001$) and that there was no difference among treatments in the percent of fish that hatching at night. Furthermore, Kolmogorov-Smirnov tests revealed that there were no significant differences in hatching time among the treatments for any of the three experiments.

A retrospective power analysis conducted on the comparisons between distilled water and injured egg for all three experiments revealed that we can be 95% confident that the null hypothesis is true for a significant effect size of 5% or greater (i.e., all three experiments were powerful enough to detect a change of 5% or greater).

## 3.2 Hatchling Phenotype

ANOVA analysis on the phenotype data showed that there was a significant difference in length in experiment one (P<0.001) and experiment three (P<0.001), but not in experiment two (0.731; Figure 1). The experiments that showed significant phenotypic differences among treatments correspond to the two experiments with the longer hatch times. Tukey tests conducted on the data from experiment one revealed that fry treated with injured egg cues during the egg stage were significantly shorter than those reared in the injured shrimp (P=0.001) treatment and tended to be shorter than those in the distilled water treatment. Similarly, in experiment three there was a significant difference in length among the fry from the three treatments (P=0.01). Tukey tests revealed that the fry treated with injured egg cues during the egg stage were significantly shorter than those reared in the distilled water (P=0.014) or injured shrimp (P=0.047) treatments.

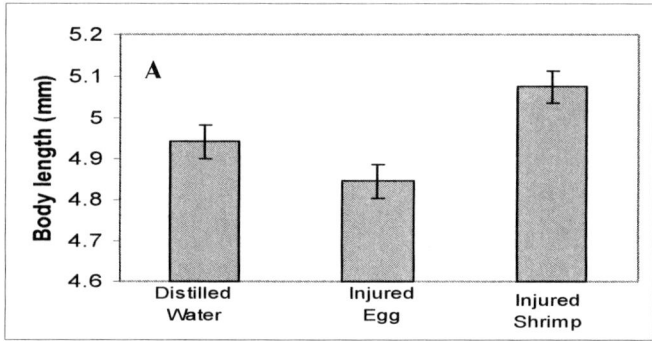

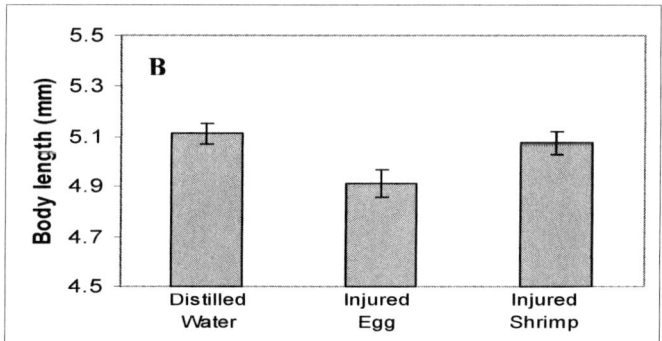

**Figure 1.** Mean body length of fathead minnow hatchlings for experiment one (A) and experiment three (B), influenced by cue exposure during the embryo stage.

## 4. DISCUSSION

Predation has long been recognized as a strong selection force influencing behaviour, morphology and life history traits of prey species. A particularly vulnerable stage, one would assume, is that of embryos as they are a major prey item for many invertebrate and vertebrate predators. Empirical studies of predator-induced changes in hatching time are still rare in most taxa, but it has been well established that amphibians can hatch earlier in response to high predation threat levels (Warkentin 1995, 2000; Warkentin et al., 2001). Chivers et al. (2001) suggest that both Cascades frogs and Pacific treefrogs show considerable plasticity in hatch time that could be induced solely based on chemosensory assessment of predation risk. Likewise, Sih and Moore (1993) showed that flatworm chemical cues alone induced eggs to delay their hatching. We found no evidence of predator-induced alterations in hatching time in fathead minnows. We are 95% confident that a 5% or greater change in hatch time could have been detected during our investigation if induced by the levels of threat simulated. One must consider that the pressure to hatch during the dark could be a factor in their inability to alter hatching time. We are confident that the concentration of chemical cue used was high enough to be perceived as a threat, since phenotypic variation was caused. Therefore, in order to ensure that predation-induced hatching does not occur in fathead minnows additional experiments with predators should be conducted and steps should be taken to record hatching times during the daily period of dark.

Facultative adjustment of hatching time may not be possible in some species or under certain circumstances. Changes in hatching time are not the only defense available to embryos. Laurila et al. (2001) found no consistent evidence that predator presence altered the hatching time of the common frog (*Rana temporaria*), but that hatchlings raised in the presence of a predator had relatively shorter bodies and deeper tail fins than those reared in a predator absent environment. Such morphological alterations are believed to improve the anti-predator capacities of the prey, for example escape propulsion. We also found evidence of phenotype alteration due to predation threat in experiments one and three. In experiment two there were no phenotypic differences due to treatment, which may be due to a constraint imposed by its overall shorter hatch time. In experiment one hatchlings exposed to an injured egg predation cues during the egg stage were significantly shorter than hatchlings exposed to injured shrimp and tended to be shorter than those in the distilled water treatment. Likewise, in experiment three hatchlings exposed to injured egg predation cues during the egg stage were significantly shorter than hatchlings exposed to injured shrimp or distilled water. It appears injured egg cues were interpreted as a predation threat and responded to by the slowing of development. In other words, they remained in the egg stage for the same amount of time as the other treatments, but hatched less developed. Phenotypic changes are highly predator specific and function to decrease the prey's risk of being detected, caught and/or consumed. If the threat is temporally variable, the shorter body length may translate into a size refuge from predators that prefer larger prey items. Alternatively, the driving force behind the shorter body length could be that smaller prey animals are not as easily detected by predators. Perhaps the alteration made to the developmental rate was not brought about to change the phenotype of the hatchling, but instead was a byproduct of the prey decreasing their metabolic rate to avoid releasing chemicals that attract predators. It seems logical that an egg batch with a higher metabolic rate would produce more chemical cue than a batch with a lower metabolic rate and have a higher chance of being detected. Reduction in metabolic rate would result, indirectly, in less developed hatchlings. This possibility remains to be investigated.

## 5. ACKNOWLEDGMENTS

This study was funded by grants to D.P. Chivers from NSERC and the University of Saskatchewan.

## 6. REFERENCES

Altwegg, R., 2002, Predator-induced life-history plasticity under time constraints in pool frogs, *Ecology* 83:2542-2551.
Brönkmark, C., and Miner, J.G., 1992, Predator-induced phenotypical change in body morphology in crucian carp, *Science* 258:1348-1350.
Brönkmark, C., Paskowksi, C. A., Tonn, W. M., and Hargeby, A., 1995, Predation as a determinant of size structure in populations of crucian carp (*Carassius carassius*) and tench (*Tinca tinca*), *Ecol. Freshw. Fish.* 4:85-92.
Chivers, D. P., and Smith, R. J. F., 1998, Chemical alarm signaling in aquatic predator-prey systems: a review and prospectus, *Écoscience* 5:338-352.
Chivers, D. P., Kiesecker, J. M., Marco, A., Wildy, E. L., and Blaustein, A. R., 1999, Shifts in life history as a response to predation in western toads (*Bufo boreas*), *J. Chem. Ecol.* 25:2455-2463.
Chivers, D. P., Kiesecker, J. M., Adolfo, M., DeVito, J., Anderson, M. T., and Blaustein, A. R., 2001, Predator-induced life history changes in amphibians: egg predation induces hatching, *Oikos* 92:135-142.
Chivers, D. P., and Mirza, R. S., 2001, Predator diet cues and the assessment of predation risk by aquatic vertebrates: a review and prospectus, in: *Chemical Signals in Vertebrates 9*, A. Marchlewsk-Koj, J. Lepri, and D. Müller-Schwarze, eds., Plenum Press, New York, pp. 277-284.
Fraser, D. F., and Gilliam, J. F., 1992, Non-lethal impacts of predation invasion: facultative suppression of growth and reproduction, *Ecology* 73:959-970.
Hara, T. J., 1994, The diversity of chemical stimulation in fish olfaction and gustation, *Rev. Fish Biol. Fish.* 4:1-35.
Johansson, F., Stoks, R., Rowe, L., and De-Block, M., 2001, Life history plasticity in damselfly: Effects of combined time and biotic constraints, *Ecology* 82:1857-1869.
Kappes, H., and Sinsch, U., 2002, Temperature and predator-induced photypic plasticity in *Bosmina cornuta* and *B. Pellucida* (Crustacea: Cladocera), *Fresh. Biol.* 47:1944-1955.
Kats, L. B., and Dill, L. M., 1998, The scent of death: Chemosensory assessment of predation risk by prey animals, *Écoscience* 5:361-394.
Kiesecker, J. M., Chivers, D. P., Anderson, M. T., and Blaustein, A. R., 2002, Effect of predator diet on life history shifts of red-legged frogs, *Rana aurora, J. Chem. Ecol.* 28:1007-1015.
Laurila, A., Crochet, P., and Merilä, J., 2001, Predator-induced effects on hatchling morphology in the common frog (*Rana temporaria*), *Can. J. Zool.* 79:926-930.
Lima, S., and Dill, L. M., 1990, Behavioral decisions made under the risk of predation: a review and prospectus, *Can. J. Zool.* 68:619-640.
Mathis, A., and Smith, R. J. F., 1993, Chemical labeling of northern pike (*Esox lucius*) by the alarm pheromone of fathead minnows (*Pimephales promelas*), *J. Chem. Ecol.* 19:1967-1979.
Moore, R. D., Newton, B., and Sih, A., 1996, Delayed hatching as a response of streamside salamander eggs to chemical cues from predatory sunfish, *Oikos* 77:331-335.
Reimer, O., and Tedengren, M., 1996, Phenotypical improvement of morphological defenses in the mussel *Mytilus edulis* induced by exposure to the predator *Asterias rubens, Oikos* 75:383-390.
Sih, A., and Moore, R. D., 1993, Delayed hatching of salamander eggs in response to enhanced larval predation risk, *Am. Nat.* 142:947-960.
Sih A., 1987, Predator and prey lifestyles: an evolutionary and ecological overview, in: *Predation: Direct and Indirect Impacts on Aquatic Communities*, W. C. Kerfoot and A. Sih, eds., University Press of New England, Hanover, New Hampshire, pp. 203-224.
Smith, R. J. F., 1992, Alarm signals in fishes, *Rev. Fish Biol. Fish.* 2:33-63.

Trussel, G. C., 1996, Phenotypic plasticity in an intertidal snail: The role of a common crab predator, *Evolution* **50**:448-454.

Van-Buskirk, J., and Schmidt, B. R., 2000, Predator-induced phenotypic plasticity in larval newts: Trade-offs, selection, and variation in nature, *Ecology* **81**:3009-3028.

Warkentin, K. M., 1995, Adaptive plasticity in hatching age: A response to predation risk trade-offs, *Ecology* **92**:3507-3510.

Warkentin, K. M., 2000, Wasp predation and wasp-induced hatching in red-eyed treefrog eggs, *Anim. Behav.* **60**:503-510.

Warkentin, K. M., Currie, C. R., and Rehner, S. A., 2001, Egg-killing fungus induces early hatching of red-eyed tree frog eggs, *Ecology* **82**:2860-2869.

Weber, A., and Declerck, S., 1997, Phenotypic plasticity in Daphnia life history traits in response to predator kairomones: Genetic variability and evolutionary potential, *Hydrobiologia* **360**:89-99.

Wedekind, C., 2002, Induced hatching to avoid infectious egg disease in whitefish, *Curr. Biol.* **12**:69-71.

Werner, E. E., 1996, Amphibian metamorphosis: growth rate, predation risk and the optimal size at transformation, *Am. Nat.* **128**:319-341.

# NOCTURNAL SHIFT IN THE ANTIPREDATOR RESPONSE TO PREDATOR-DIET CUES IN LABORATORY AND FIELD TRIALS

Aaron M. Sullivan, Dale M. Madison, and John C. Maerz[*]

## 1. INTRODUCTION

Prey species may react to predator chemical traces in the environment with a variety of antipredator behaviors (Weldon, 1990; Chivers and Smith, 1998; Kats and Dill, 1998). Such responses to predator chemical cues might reduce predation risk, but also may result in lost foraging or mating opportunities (Lima, 1998a,b), so it is not surprising that some species adjust their responses based on the degree of perceived predation threat. These modifications may be based on chemical information gathered directly from the predator or its recent prey (Madison et al., 1999a; Chivers and Mirza, 2001). The mosaic of chemical products released at a predation site, and dispersed by the predator, could allow nearby prey to assess predation threat and fine-tune their responses.

Among the chemical cues released from predators during and after a predation event, those associated with predator diet have emerged as an important factor in predator assessment by both aquatic (Chivers and Mirza, 2001) and terrestrial prey species (Madison et al., 1999a,b). However, in most circumstances it is unclear whether the active components of the cue are from the prey, the predator, or some combination of prey and predator cues (e.g., Madison et al., 2002). As one example of this complexity, the interaction between the terrestrial red-backed salamander, *Plethodon cinereus*, and its garter snake predator, *Thamnophis sirtalis*, shows that during the day salamander avoidance of snake cues occurs independent of snake diet, but late at night avoidance is apparently restricted to cues from snakes feeding on red-backed salamanders (Madison et al. 1999a,b). One limitation of these studies is that few diet treatments were tested, leaving the possibility that alternative diets might also elicit late night avoidance.

---

[*] A. M. Sullivan and D. M. Madison, Department of Biological Sciences, State University of New York at Binghamton, Binghamton, New York, 13902 USA. J. C. Maerz, Cornell University, Ithaca, NY 14853 USA.

We hypothesize that the late night discrimination by red-backed salamanders of predator diet cues is most likely the result of either the "phylogenetic relatedness" or "ecological relatedness" between the test subjects and the prey of the snake. Phylogenetic relatedness is the degree of genetic similarity between species, who share a similar chemical composition (Smith, 1992; Brown et al., 2000). According to this concept, one would predict that a target prey species would avoid chemical cues from predators feeding on a congeneric prey because of shared prey chemistry and defense mechanisms, despite possible allopatry between the test species and the prey species. On the other hand, the "ecological relatedness" hypothesis for diet discrimination may be defined as the degree of habitat and trophic overlap among sympatric prey species that likely share the same predators (Mathis and Smith, 1993; Chivers et al., 1995). For example, two species sharing similar refuges that are members of the same prey guild would have a high degree of ecological relatedness, despite not being closely related phylogenetically.

Our goals in this study were: 1) to test the efficacy of the phylogenic relatedness hypothesis by determining whether the chemical component from salamanders that "labels" garter snakes as an increased risk is conserved within the salamander family Plethodontidae, 2) to examine the ability of red-backed salamanders to detect and discriminate among predator chemical cues under laboratory and field conditions, and 3) to detect and accurately describe the temporal shift in responsiveness of red-backed salamanders to predator diet cues during early and late night trials. To achieve these goals, we tested red-backed salamanders in the laboratory and field to chemical traces from garter snakes fed red-backed salamanders and two-lined salamanders, *Eurycea bislineata*.

## 2. METHODS

### 2.1. Collection and Maintenance of Test Animals

For laboratory experiments, we collected red-backed salamanders and two lined salamanders from the Binghamton University Nature Preserve, and the Binghamton University Natural Area (Broome Co., New York), respectively. All salamanders were maintained in individual petri dishes (15 cm diameter × 1.5 cm high) with moistened paper towels at 17°C and a 14L:18D photoperiod. We collected five adult garter snakes as predator stimulus donors and housed them in 38-L aquaria with a water dish, heating element, and crumpled paper towels for shelter. We kept snakes in a separate room at 22°C and a 14L:18D photoperiod. Snakes were fed salamanders of the appropriate species each week for three weeks prior to collecting treatment rinses.

### 2.2. Collection of Predator Chemical Cues

A distilled water rinse of each garter snake was used as the predator stimulus. Immediately prior to collecting a rinse, we fed source garter snakes three adult female and three adult male salamanders of the appropriate species, and then transferred the snake to a 4 L beaker that had been cleaned with Alcojet detergent and rinsed in 2% nitric acid and distilled water. After transfer, we covered the beaker with cheesecloth and placed the beaker on a heating block for 72 h. We removed the snake and rinsed the beaker with 200 ml of distilled water. We filtered the rinse through a 0.45-μm filter to remove solid materials, froze samples in liquid nitrogen, and stored them in a freezer. The entire collection process was repeated to provide enough snake stimuli for laboratory and field experiments.

### 2.3. Experiment 1: Laboratory Study of Predator Cue Avoidance

We examined the laboratory responses of red-backed salamanders to two predator diet treatments at two different time periods to determine whether the chemical "label" that identifies individual garter snakes as an elevated risk is conserved within the Plethodontidae, and to examine a possible temporal shift in red-backed salamander antipredator behavior. The "early" trials were conducted between 2100 and 2230 hours, and the "late" trials were conducted between 2330 and 0100 hours. In each of these trial periods, test salamanders were given the choice between *T. sirtalis* fed *P. cinereus* ($TS_{Pc}$) or *E. bislineata* ($TS_{Eb}$) versus distilled water. To assess the relative strength of each snake diet treatment, test salamanders also were given the choice between $TS_{Pc}$ and $TS_{Eb}$.

We used the behavioral assay described in Sullivan et al. (2003) to examine salamander responses to predation-related chemical cues. In brief, we placed two filter paper semicircles on opposite sides of 15 cm petri dishes while maintaining a 3 mm gap between each semicircle. We added 1.5 ml of treatment solution ($TS_{Pc}$, $TS_{Eb}$, or distilled water) to each semicircle using a 10 ml tuberculin syringe, and randomly distributed treatment dishes on an 8 × 7 grid on the floor of our experimental room. We transferred each salamander from its home dish to the assigned experimental dish with a cotton swab, and placed a 15 mm collar of brown paper around the dish to visually isolate each salamander within test dishes during the set-up period. After all salamanders were distributed (~15 min), the lights were turned off and the trial was recorded for 60 min in complete darkness with a video camera sensitive to infrared light.

We recorded the side occupied by each salamander every three minutes for one hour. Since red-backed salamanders tend to be relatively inactive and their behavioral response is bimodally distributed (Madison et al., 1999a; Sullivan et al., 2003), we used binomial and Chi-square goodness-of-fit tests to analyze salamander avoidance responses. We considered salamanders located on the treatment substrate ≤ 6 out of 21 observations as avoiding the treatment because this ratio is below an alpha value of 0.05 using a binomial test (Siegel & Castellan, 1988), and those located on the treatment side ≥ 15 out of 21 observations as being attracted to the treatment. Individuals on the treatment substrate for > 6 and < 15 observations were considered indifferent to the treatment, and were excluded from the analyses. We used Chi-square tests to determine whether the number of salamanders avoiding the treatment differed significantly from random expectation.

### 2.4. Experiment 2: Field Study of Predator Cue Avoidance

We conducted a field experiment to examine whether a temporal shift in antipredator responses to predators based on the predator's diet would be observed in a more natural context. We performed the "early" experiment (conducted between 2030 and 2330 hours) and "late" experiment (conducted between 2330 and 0230 hours) on a single night in July 2003 following a day of rain so that leaf litter was wet and air was saturated with moisture. Plethodontid activity is generally restricted to moist conditions within a few days of rain (Grover, 1998; Feder, 1983), and the movement of salamanders among shelters at this site is positively correlated with precipitation (Maerz and Madison, 2000). Therefore, salamander movement into and out of shelters was not restricted by low moisture during the trials.

We exposed salamanders found under rocks to $TS_{Pc}$, $TS_{Eb}$, or distilled water during early and late experiments. If a salamander was present beneath the rock, we removed it by hand and recorded distinguishing characteristics including sex (males were identified by the enlargement of the mental gland and swollen cloacal vent), distinctive patterns in stripe, and tail autotomy. While salamander characteristics were being recorded, the rock was numbered with flagging and the soil beneath the rock was uniformly sprayed with 2 ml of snake rinse or water. We returned the rock to its original position, and then released the salamander under the rock edge. We re-examined the cover rocks for the presence of salamanders beginning two hours after initiating the experiment. We examined the rocks in the same sequence, but were "blind" to treatment during our search. For the early trial, we exposed thirty salamanders to each of the three treatments, and for the late trial, we exposed twenty salamanders to each treatment. We recorded whether the original or new salamander was under each test rock.

We examined the effects and possible interactions between snake diet treatment and salamander sex using a binomial regression with a log-log canonical link in Statistica's (StatSoft, Inc., 2001) Generalized Linear Model (GZLM). We tested the full factorial model, which examined the effects of treatment ($TS_{Pc}$, $TS_{Eb}$, distilled water) and sex of the test salamander on salamander responses. We tested for significant effects using the Wald statistic (analogous to least-squares estimates).

## 3. RESULTS

### 3.1. Results of Laboratory Experiments

During "early" night trials, red-backed salamanders significantly avoided both $TS_{Pc}$ ($\chi^2 = 8.0$, $P<0.01$) and $TS_{Eb}$ ($\chi^2=6.12$, $P=0.01$). When the two snake diet treatments were paired against one another to test for their relative effect, we found no significant avoidance of $TS_{Pc}$ or $TS_{Eb}$ ($\chi^2=0.14$, $P=0.71$) (Table 1). During "late" night trials, salamanders still significantly avoided $TS_{Pc}$ ($\chi^2=12.5$, $P<0.01$), but not $TS_{Eb}$ ($\chi^2=0.36$, $P=0.54$). When the two snake treatments were paired against one another, $TS_{Pc}$ was significantly avoided relative to $TS_{Eb}$ ($\chi^2=4.5$, $P=0.03$) (Table 2).

Table 1. The number of salamanders that avoided, were attracted to, or showed no choice to chemical treatments during early night trials.

| Treatment: | | No. of salamanders choosing: | | | | |
|---|---|---|---|---|---|---|
| A | B | A | B | No choice | $\chi^2$ | P |
| $TS_{Pc}$ Vs $dH_2O$ | | 7 | 22 | 3 | 7.76 | 0.005 |
| $TS_{Eb}$ Vs $dH_2O$ | | 8 | 22 | 2 | 6.53 | 0.011 |
| $TS_{Pc}$ Vs $TS_{Eb}$ | | 13 | 15 | 4 | 0.14 | 0.705 |

**Table 2.** The number of salamanders that avoided, were attracted to, or showed no choice to chemical treatments during late night trials.

| Treatment: A | B | No. of salamanders choosing: A | B | No choice | $\chi^2$ | P |
|---|---|---|---|---|---|---|
| $TS_{Pc}$ Vs | $dH_2O$ | 5 | 19 | 8 | 8.17 | 0.004 |
| $TS_{Eb}$ Vs | $dH_2O$ | 11 | 14 | 7 | 0.36 | 0.589 |
| $TS_{Pc}$ Vs | $TS_{Eb}$ | 8 | 19 | 5 | 4.48 | 0.034 |

## 3.2. Results of Field Experiments

During the "early" field trials, the binomial regression on salamander response by treatment ($TS_{Pc}$, $TS_{Eb}$, or distilled water) and salamander sex showed a significant treatment effect (Table 3). Red-backed salamanders exposed to $TS_{Pc}$ were more likely to remain in refuges than individuals exposed to $TS_{Eb}$ or distilled water (Table 4). During the "late" field trials, the binomial regression on salamander response by treatment ($TS_{Pc}$, $TS_{Eb}$, or distilled water) and salamander sex showed no significant treatment, sex, or interaction effect (Table 3). During the "late" trial, red-backed salamanders tended to leave refuges regardless of treatment (Table 4).

**Table 3.** The results of the binomial regression of salamander sex and diet treatment on salamander response in early and late night field trials.

| Effect | Early night | | | Late night | | |
|---|---|---|---|---|---|---|
| | Df | Wald | P | df | Wald | P |
| Sex | 1 | 0.43 | 0.514 | 1 | 3.09 | 0.079 |
| Treatment | 2 | 6.02 | 0.049 | 2 | 0.34 | 0.819 |
| Treatment × sex | 2 | 5.56 | 0.062 | 2 | 1.09 | 0.581 |

**Table 4.** The number of salamanders that remained under rock refuges or left refuges when exposed to $H_2O$, $TS_{Eb}$, or $TS_{Pc}$ in early and late night field trials.

| Treatment | Early night | | Late night | |
|---|---|---|---|---|
| | No. remaining | No. leaving | No. remaining | No. leaving |
| $H_2O$ | 11 | 19 | 5 | 15 |
| $TS_{Eb}$ | 7 | 23 | 7 | 13 |
| $TS_{Pc}$ | 16 | 14 | 6 | 14 |

## 4. DISCUSSION

These results generally corroborate previous studies of red-backed salamanders, demonstrating that this species can distinguish among individual predators based on diet-related cues, but that the response is contingent on the diel cycle of the test salamanders (Madison et al., 1999a, b). In the current study, test salamanders discriminated among garter snakes fed phylogenetically related (confamilial and conspecific) prey in both laboratory and field settings. During the day and early evening, the risk imposed by garter snakes, regardless of their recent diet, dominates the chemosensory behavior of red-backed salamanders. Since garter snakes forage on *P. cinereus* and other salamander species (Hamilton, 1951; Carpenter, 1952; Moreno, 1989), and since these snakes can extend their diurnal activity into warm evenings during summer (Reichenbach and Dalrymple, 1986), the avoidance by red-backed salamanders of all snake chemical traces seems like an adaptive response to predator cues.

The results of our laboratory experiments suggest that the discriminatory abilities of red-backed salamanders are quite refined. During the early night trials, red-backed salamanders avoided chemical cues from garter snakes fed red-backed and two-lined salamanders when paired against distilled water, but did not avoid either when paired against each other. These results show that at the time of study, the two predator stimuli were probably considered equally repulsive by test salamanders, suggesting that the chemical cue "labeling" the predator as an increased risk is conserved within the Plethodontidae. However, during the late-night laboratory trials, red-backed salamanders avoided chemical cues from garter snakes fed red-backed salamanders, but not from snakes fed two-lined salamanders. In addition, when given the choice between the two predator stimuli, salamanders chose to occupy the substrate labeled with $TS_{Eb}$, suggesting that late at night, $TS_{Pc}$ is considered more aversive. These results imply either refined discrimination of predator diet below the family level, and/or greater familiarity with conspecifics by red-backed salamanders.

The lack of response to $TS_{Eb}$ late at night negates the phylogenetic relatedness hypothesis of chemical cue homology between red-backed and two-lined salamanders. This result is surprising since two-lined salamanders are preyed on by garter snakes (Ducey and Brodie, 1983; Dowdey and Brodie, 1989), are similar in size to adult red-backed salamanders (Petranka, 1998), belong to the Plethodontidae, and are seasonally syntopic with red-backed salamanders (Stewart, 1956; MacCollogh and Bider, 1975). However, two-lined salamanders are not syntopic to red-backed salamanders where our test *P. cinereus* were collected. Thus, the individual red-backed salamanders tested may not have been familiar with *E. bislineata*, nor been exposed to chemical cues from garter snakes foraging on two-lined salamanders. It is possible that test salamanders did not recognize or interpret $TS_{Eb}$ as a risk equivalent to $TS_{Pc}$ late at night.

The results of our field trials, expressed as a reduction in activity and a tendency to remain beneath rock refuges when exposed to aversive chemical cues (Sullivan at al., 2002), confirm the diet discrimination seen in the laboratory. However, the cues that elicited this discrimination in the field were unexpected. In the field, salamanders did not respond to $TS_{Eb}$ in the early night trial (despite laboratory avoidance), and did not respond to either $TS_{Pc}$ or $TS_{Eb}$ late at night. We hypothesize that cues from invertebrate prey were abundant in the field, but not in the laboratory, so salamanders may have disregarded all except the most threatening predator stimuli. The activity of the invertebrate prey consumed by red-backed salamander tends to be highest in the leaf litter at night (Jaeger, 1978; Holomuzki, 1980), so salamanders that remain under refuges when exposed to chemical cues may miss out on important foraging opportunities and incur a non-lethal cost of predation risk (Lima, 1998a,b). In addition, the threat of snake predation declines after sunset, especially late at night, when garter snakes tend to be relatively inactive (Reichenbach and Dalrymple, 1986; Madison et al., 1999).

In conclusion, our data show the ability of red-backed salamanders to discriminate among chemical cues from predators according to predator diet in both laboratory and field settings. This cue discrimination shows nocturnal variation that may be related to the fluctuation in predation risk due to the activity period of the predator, or the conflicting demands of foraging and antipredator responses. Furthermore, our findings seem to reject the hypothesis that phylogenetic relatedness among prey species contributes to the diet cue discrimination (Chivers and Mirza, 2001) by salamanders in our system, at least at the level of Plethodontidae. Rather, our data suggest that the discriminatory ability of red-backed salamanders may be related to the ecological relatedness among prey species as demonstrated by Chivers et al. (1997) who show that being ecologically familiar with the chemical traces of recent prey in a snake's diet, rather than chemically recognizing a phylogenetic relationship with these prey, may elevate the perceived risk of predation in salamanders in an ecological community. Further studies examining the mechanisms by which diet cue discrimination occurs should yield valuable insight into studies of chemically-mediated predator assessment.

## 5. ACKNOWLEDGMENTS

We thank B. Choudury, E. Schips, Stacy Magliaro, D. Horvath, M. Machura, and M. Villanella for laboratory and field assistance. Funding for this study was provided by an NSF grant (IBN 9974591) to DMM, an NSF DDIG (IBN 0206606) to AMS and DMM, and a Binghamton University Departmental Summer Research Award to AMS. All experiments comply with the current laws of the United States of America and the Binghamton University Animal Care and Use Committee (IUCAC 406-98 and 467-00), and the New York Department of Environmental Conservation (permit LCP00-471).

## 6. REFERENCES

Brown, G. E., Adrian, J. C., Smyth, E., Leet, H., and Brennan, S., 2000, Ostariophysan alarm pheromones: laboratory and field tests of the functional significance of nitrogen oxides, *J. Chem. Ecol.* **26**:139-154.

Carpenter, C. C., 1952, Comparative ecology of the common garter snake (*Thamnophis s. sirtalis*), the ribbon snake (*Thamnophis s. sauritus*), and Butler's garter snake (*Thamnophi butleri*) in mixed populations, *Ecol. Monogr.* **22**:235-258.

Chivers, D. P., Kiesecker, J.M., Anderson, M. T., Wildy, E. L., and Blaustein, A.R., 1997, Chemical alarm signalling in terrestrial salamanders: intra- and interspecific responses, *Ethology* **103**:599-613.

Chivers, D. P., and Mirza, R. S., 2001, Predator diet cues and the assessment of predation risk by aquatic vertebrates: a review and prospectus, in: *Chemical Signals in Vertebrates IX*, A. Marchlewska-Koj, J. Lepri, and D. Müller-Schwarze, eds., Klewer Academic/Plenum Publishers, New York, pp. 277-284.

Chivers, D. P., and Smith, R. J. F., 1998, Chemical alarm signalling in aquatic predator-prey systems: a review and prospectus, *Écoscience* **5**:338-352.

Chivers, D. P., Wisenden, B. D., and Smith, R. J. F., 1995, The role of experience in the response of fathead minnows (*Pimephales promelas*) to skin extract of Iowa darters (*Etheostoma exile*), *Behaviour* **132**:665-674.

Dowdey, T. G., and Brodie, E. D., Jr., 1989, Antipredator strategies of salamanders: individual and geographical variation in responses of *Eurycea bislineata* to snakes, *Anim. Behav.* **38**:707-711.

Ducey, P., and Brodie, E. D., Jr., 1983, Salamanders respond selectively to contacts with snakes: survival advantage of alternative antipredator strategies, *Copeia* **1983**:1036-1041.

Feder, M. E., 1983, Integrating the ecology and physiology of plethodontid salamanders, *Herpetologica* **39**:291-310.

Grover, M. C., 1998, Influence of cover and moisture on abundances of the terrestrial salamanders *Plethodon cinereus* and *Plethodon glutinosus*, *J. Herpetol.* **32**:489-497.

Hamilton, W. J., Jr., 1951, The food and feeding behavior of the garter snake in New York State. *Amer. Midl. Natur.* **46**:385-390.

Holomuzki, J. R., 1980, Synchronous foraging and dietary overlap of three species of plethodontid salamanders, *Herpetologica* **36**:109-115.

Jaeger, R., 1978, Plant climbing by salamanders: periodic availability of plant-dwelling prey, *Copeia* **1978**:686-691.

Kats, L. B., and Dill, L. M., 1998, The scent of death: chemosensory assessment of predation by prey animals, *Écoscience* **5**:361-394.

Lima, S. L., 1998a, Nonlethal effects of the ecology of predator-prey interactions, *BioScience* **48**:25-34.

Lima, S. L., 1998b, Stress and decision-making under the risk of predation: recent developments from behavioral, reproductive, and ecological perspectives, in: *Advances in the Study of Behavior*, A. P. Møller, M. Milinski, and P. J. B. Slater, eds., Academic Press, New York, pp 1-50.

MacCullogh, R. D., and Bider, J. R., 1975, Phenology, migrations, circadian rhythm and the effect of precipitation of the activity of *Eurycea b. bislineata* in Quebec, *Herpetologica* **31**:433-439.

Madison, D. M., Maerz, J. C., and McDarby, J. D., 1999a, Optimization of predator by salamanders using chemical cues: diet and diel effects, *Ethology* **105**:1073-1086.

Madison, D. M., Maerz J. C., and McDarby, J. D., 1999b, Chemosensory avoidance of snake odors by salamanders: freeze and flight contingencies, in: *Chemical Signals in Vertebrates VIII*, R. E. Johnston, D. Müller-Schwarze, and P. Sorensen, eds., Kluwer Academic/ Plenum Press, New York, pp. 508-516.

Madison, D. M., Sullivan, A. S., Maerz, J. C., McDarby, J. H., and Rohr, J.R., 2002, A complex, cross-taxon, chemical releaser of anti-predator behavior in amphibians, *J. Chem. Ecol.* **28**:2271-2282.

Maerz, J. C., and Madison, D. M., 2000, Environmental variation and territorial behavior in a terrestrial salamander, in: *The Biology of Plethodontid Salamanders*, R. C. Bruce, R. G. Jaeger, and L. D. Houck, eds., Plenum Publishing Corp., New York, pp 395-406.

Mathis, A., and Smith, R. J. F., 1993, Fathead minnows, *Pimephales promelas*, learn to recognize northern pike, *Esox lucius*, as predators on the basis of chemical stimuli from minnows in the pike's diet, *Anim. Behav.* **46**:645-656.

Moreno, G., 1989, Behavioral and physiological differentiation between the color morphs of the salamander, *Plethodon cinereus*, *J. Herpetol.* **23**:335-341.

Petranka, J., 1998, *Salamanders of the United States and Canada*, Smithsonian Institution Press, Washington D.C.

Reichenbach, N. G., and Dalrymple, G. H., 1986, Energy use, life histories, and the evaluation of potential competition in two species of garter snakes, *J. Herpetol.* **20**:133-153.

Siegel, S., and Castellan, N. J., Jr., 1988, *Nonparametric Statistics for the Behavioral Sciences*. McGraw-Hill, New York, N.Y.

Smith, R. J. F., 1992, Alarm signals in fishes, *Rev. Fish Biol. Fisher.* **2**:33-63.

Stewart, M., 1956, *Certain Aspects of the Natural History and Development of the Northern Two-lined Salamander, Eurycea bislineata bislineata (Green), in the Ithaca, New York Region*. Ph.D. thesis, Cornell University, Ithaca, N.Y.

Sullivan, A. M., Maerz, J. C., and Madison D. M., 2002, Anti-predator response of red-backed salamanders (*Plethodon cinereus*) to chemical cues from garter snakes (*Thamnophis sirtalis*): laboratory and field experiments. *Behav. Ecol. Sociobiol.* **51**:227-233.

Sullivan, A. M., Rohr, J. R., and Madison, D. M., 2003, Behavioural responses by red-backed salamanders to conspecific and heterospecific cues, *Behaviour* **140**:553-564.

Weldon, P. J. 1990, Responses of vertebrates to chemicals from predators, in: *Chemical Signals in Vertebrates V*, D. W. Macdonald, D. Müller-Schwarze, and R. M. Silverstein, eds., Plenum Press, New York, pp. 500-521.

# LONG-TERM PERSISTENCE OF A SALAMANDER ANTI-PREDATOR CUE

Michael P. Machura and Dale M. Madison[*]

## 1. INTRODUCTION

Since predators are ubiquitous in many environments, organisms that are capable of detecting and assessing predation risk should have a survival advantage over individuals that are oblivious to such threats (Lima and Dill, 1990). Terrestrial prey may assess this risk by detecting chemical cues from predators or prey conspecifics [see reviews by Kats and Dill (1998)]. Once a cue is released into the environment, a number of factors including temperature, humidity and wind velocity may contribute to its persistence over time (Alberts, 1992). A cue's endurance is an ecologically significant parameter, since prey organisms could loose foraging or mating opportunities if an anti-predator response persisted long after the threat had abated (Alberts, 1992; Sih, 1997; Maerz et al., 2001).

Red-backed salamanders (*Plethodon cinereus*) respond to distilled water rinses of garter snakes (*Thamnophis sirtalis*) that have recently attacked other red-backed salamanders (Madison et al., 2002). These anti-predator behaviors consist of avoidance (Madison et al., 1999a; Mcdarby et al., 1999) and altered activity (Madison et al., 1999a), and have been documented in the laboratory and in the field (Sullivan et al., 2002). The responses persist for up to 36 hours in laboratory trials, but no longer than 3 hours in the field (Sullivan et al., 2002). Whether the attenuated response in the field is due to a diminishing of the cue or to field-based shifts in salamander behavior remains unclear.

We hypothesized that microorganisms might contribute to the depletion of chemical signals in natural systems, and therefore be an important component in the snake-salamander predator-prey interaction. The microbial diversity of terrestrial environments can be overwhelmingly complex. Bacteria are associated with animal's integuments (Austin, 2000) and fecal material (Chiodini and Sundberg, 1981; Torsvik et al., 1996), and are copiously abundant in many soil types (Waksman, 1952). Almost every naturally occurring organic compound is vulnerable to microbial attack (Leadbetter, 2002; Schink, 2002; Schreiber et al., 2002). Many chemical secretions and pheromones are composed

---

[*] Department of Biological Sciences, State University of New York at Binghamton, Binghamton, New York, 13902, USA.

of glycoproteins (Feldhoff et al., 1999) and lipids (Oldak, 1976; Schell and Weldon, 1985; Graves et al., 1991), which could serve as potential energy sources for microorganisms (Atlas, 1997) and result in premature degradation of the chemical cues.

We examined the effects of ageing on garter snake rinse under sterile and non-sterile conditions and hypothesized that microorganisms, through biodegradation, would shorten the life of the cue. Because previous aging experiments involved microfiltration, which would effectively sterilize the cue (e.g. Sullivan et al., 2002), we predicted that not filtering the cue would shorten the duration of its effectiveness in the laboratory.

## 2. METHODS

### 2.1 Animal Collection and Maintenance

All animals were collected during the spring of 2003 from the Binghamton University Nature Preserve, Broome Co., New York. Salamanders were housed individually in 15-cm diameter perti dishes containing moistened gauze and maintained at 15°C on a 15L/9D photoperiod. Salamanders used in snake feedings were kept no longer than one month, and those used in behavioral bioassays were collected and tested within one week and then released. The snake was maintained in a 38-l glass aquarium and provided with a water dish, crumpled paper towels, and access to direct sunlight at ambient temperature (approximately 25°C). Although the same snake was used for all experiments, Madison et al. (1999b) demonstrated that salamanders respond similarly to snake cues irrespective of the individual snake producing the cue.

### 2.2 Experimental Treatments

Two experiments were conducted to examine the possible effects of ageing and microbial degradation. The cue was collected according to Madison et al. (1999a, b). In brief, we fed the snake two male and two female salamanders weekly for three weeks. After the third feeding, the snake was transferred to a 4-l glass beaker for 72-hours. It was then gently removed and the beaker was rinsed with 200 ml distilled water. The 200 ml rinse was filtered through Whatman #1 filter paper to remove fecal matter and other particulates, and then divided into two, 100 ml aliquots.

*2.2.1. Experiment 1*

One of the aliquots was sterilized by filtration through a 0.45 μm pore size membrane filter (Millipore HNWP04700) and classified as filtered ($F$), while the remaining rinse was not filtered ($NF$). The F- and NF-aliquots were then subdivided into three equal portions, which were each aged for either 0, 12, or 48 hours, resulting in six treatments: $F_0$, $F_{12}$, $F_{48}$, $NF_0$, $NF_{12}$, and $NF_{48}$. The rinses were then rapidly frozen in liquid nitrogen and stored at -20°C. Previous research has shown that salamanders consistently avoid snake rinse that has been frozen using this method, which suggests that rapid freezing does not chemically alter the sample (unpublished data).

### 2.2.2. Experiment 2

The second experiment was similar to the first, but differed in two details. Specifically, we used a 0.2 μm pore size filter (Millipore GNWP04700) to ensure removal of the smallest microorganisms, and we eliminated the intermediate 12-hour ageing period, resulting in only four treatments: $F_0$, $F_{48}$, $NF_0$, $NF_{48}$.

## 2.3 Behavioral Bioassays

The first experiment was conducted on four consecutive evenings during the week of 21 May 2003, and the second experiment was conducted during the evenings of 1 and 2 July 2003. We used a previously established behavioral bioassay (Sullivan et al., 2002) to test salamander responses to the treatments. Each trial consisted of an array of 54 petri dishes lined with two semi-circular pieces of Whatman filter paper. One side of the dish was moistened with 1.5 ml of the treatment, while the opposite side of the dish was moistened with 1.5 ml of distilled water. Dishes were prepared in this way for each of the treatments ($N$=36/treatment). All treatments were equally represented and randomly positioned in a 7 x 8 field of test dishes for each trial.

*Experiment 2* consisted of two additional conditions ($N$=36/treatment): one in which both sides of the dish were moistened with 1.5 ml of distilled water, and another in which both sides of the dish were moistened with 1.5 ml of a mixture of equal parts $F_0$, $F_{48}$, and $NF_0$. These two additional treatments were included to examine activity (see below) in response to completely low (water) or high (mixture) risk substrates. Ideally, the high-risk substrate would not have been a mixture, but limited rinse volumes prevented this.

A salamander was placed in the center of each dish using a cotton swab, and paper collars were placed around each dish to visually isolate the animal from adjacent dishes during the setting up of the experiment. The trial was recorded in total darkness with an infrared video camera for one hour beginning at 2300 hrs.

## 2.4 Analysis

All statistical analyses were performed using Statistica software (Statsoft, 2001). Substrate preferences were analyzed according to Madison et al. (1999a, b). We recorded the side of the dish occupied by each animal at three-minute intervals, resulting in a total of 21 position observations. If the animal straddled the midline, the side with more than half of the individual was considered the "occupied" side. We chose this criterion (as opposed to snout position) because red-backed salamanders tend to freeze when they detect a threatening substrate (Madison et al., 1999a), and it is conceivable that such a response could result in the animal freezing with its head on the treatment side. If location of the sensory organs were used in such a scenario, the animal might incorrectly be scored as "preferring" the treatment side.

An overall preference (water or treatment) for each individual was then assigned as the side of the dish that was occupied for at least 11 of the 21 observations. Since red-backed salamanders are relatively inactive, their responses to choice experiments tend to be bi-modally distributed (Madison et al., 1999a; Sullivan et al., 2003), precluding the use of parametric analysis. A chi-square goodness-of-fit test was therefore used to determine whether the total number of salamanders avoiding the treatment differed significantly from random expectation.

Activity was measured by counting the number of times each salamander crossed from one substrate to another (i.e. from water to treatment) during the first 15 minutes of the trial. We used this time interval, as opposed to the full 60 minutes, because salamander activity tends to diminish after 20 minutes in the test dish, and any activity differences between control and experimental treatments are most evident during this initial period (Madison et al., 1999b). A *post hoc* analysis revealed that the activity data was not normally distributed (Shapiro-Wilk test, $P<0.05$), so a Kruskal-Wallis test was used to compare activity between treatments.

## 3. RESULTS

In the first experiment, red-backed salamanders avoided all treatments except for the non-filtered snake rinse that had been aged 48-hours (Figure 1). Activity levels did not differ significantly between treatments [(Kruskal-Wallis: $H_{5,\ 211}=4.95$; $P=0.422$), Figure 2]. Sample sizes for some treatments (see figures) were less than the initial number of animals tested due to minor difficulties encountered while videotaping the trials.

In the second experiment, salamanders significantly avoided all treatments when paired against water, and no side preference was observed when the substrate was paired against itself (Figure 3). Because the substrate preference scores for the $NF_{48}$ treatment in both experiments were statistically similar (Fisher's exact test; $P=0.119$), the data were pooled ($N=72$) and a new $NF_{48}$ chi-square value was calculated ($\chi^2=12.500$; $P<0.001$), which showed highly significant avoidance of the $NF_{48}$ treatment.

Activity (Figure 4) was significantly different between treatments (Kruskal-Wallis: $H_{5,\ 216}=19.85$; $P=0.001$). A 2-tailed multiple comparisons test revealed that the activity response to the *W/W* treatment was significantly less than the response to $NF_0$ ($P=0.013$) and *Mx/Mx* ($P<0.001$). There were no significant differences between any of the aged treatments $F_0$, $F_{48}$, $NF_0$, $NF_{48}$.

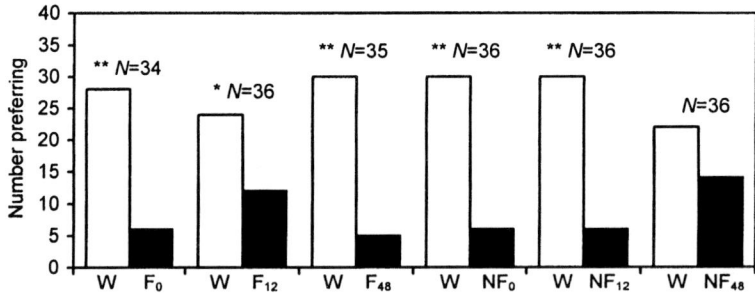

Figure 1. Number of salamanders in Experiment 1 that preferred substrates of distilled water (*W*), snake rinse that had been filtered and aged for 0- ($F_0$), 12- ($F_{12}$), or 48- ($F_{48}$) hours, or snake rinse that had not been filtered and aged for 0- ($NF_0$), 12- ($NF_{12}$), or 48- ($NF_{48}$) hours. Significance based on chi-square (*=$P<.05$ and **=$P<.001$) and sample size for each treatment is shown.

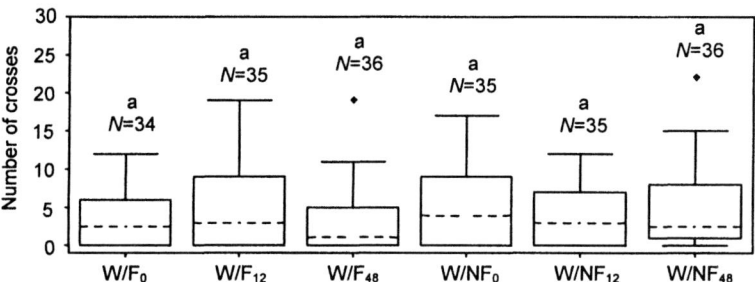

**Figure 2.** Experiment 1 median activity (dashed lines) measured as the number of crosses from one substrate to another. Also shown are the interquartile range (boxes), range (whiskers), and outliers (diamonds). One extreme outlier (35 crosses) in the $W/F_{48}$ treatment was excluded from the figure for clarity, but was not excluded from the analysis. Letters indicate significant differences between each treatment based on a Kruskal-Wallis test.

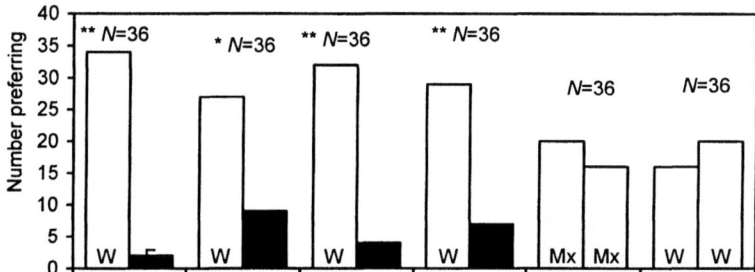

**Figure 3.** Number of salamanders in Experiment 2 that preferred substrates of distilled water ($W$), snake rinse that had been filtered and aged for 0- ($F_0$) or 48- ($F_{48}$) hours, snake rinse that had not been filtered and aged for 0- ($NF_0$) or 48- ($NF_{48}$) hours, or a mixture of $F_0$ + $NF_0$ + $F_{48}$ ($Mx$). Significance based on chi-square (*=$P<.05$ and **=$P<.001$) and sample size for each treatment is shown.

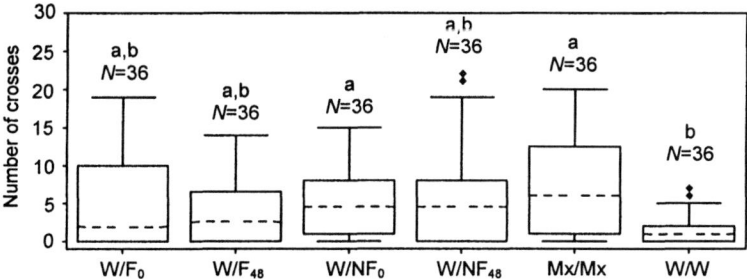

**Figure 4.** Activity levels shown as in Figure 2. Treatments are identical to Figure 3. Letters indicate significant differences for each treatment based on a Kruskal-Wallis and two-tailed multiple comparisons test.

## 4. DISCUSSION

This research did not demonstrate any conclusive evidence for microbial degradation of the garter snake chemical cue. While the results of the first experiment suggest that microorganisms might compromise the cue based on the decreased avoidance of the $NF_{48}$ treatment, the strong avoidance of $NF_{48}$ in the second experiment contradicted this. Furthermore, when the $NF_{48}$ data from both experiments were pooled, the treatment was significantly avoided.

When activity was examined, the only differences appeared in the second experiment, where animals exposed to $W/NF_0$ and $Mx/Mx$ were significantly more active than those on $W/W$. If activity is indicative of the level of perceived threat, then the heightened activity observed on the $Mx/Mx$ substrate may be a natural "escape" response. All other treatments included a "safe" water substrate on which the animal could freeze. Freezing on the "safe" side could explain the relatively uniform levels of activity between those treatments. The heightened activity on $W/NF_0$ suggests a slightly elevated perceived risk towards that treatment, but this increase was not observed in *Experiment 1*. The fact that activity did not differ significantly between the other ageing treatments suggests that the animals did not distinguish between the fresh, aged, filtered or not filtered cues.

Based on the strong avoidance and relatively uniform activity in response to all treatments, we conclude that the chemical cue in a garter snake rinse continues to elicit an avoidance response in red-backed salamanders for up to 48-hours in the laboratory, which is in agreement with previous studies by Sullivan et al., (2002). While our results do not support the microbial degradation hypothesis, further research is necessary before we can conclude that microbes are inconsequential to the cue. Other sources of microorganisms, such as the forest floor and soil that harbor a rich and diverse microbial community (Waksman, 1952), should be examined. Also, the hypotonic environment of the distilled water used to collect the snake rinse could have potentially caused cell lysis, destroying any bacteria in the rinse.

Previous research has examined the temporal domain of chemical signaling (Turner, 1996; Venzon et al., 2000; Barnes et al., 2002) and some have considered the role that microbes might play in the degradation of chemical cues (Polkinghorne et al., 2001; Turner and Montgomery, 2003). Much is known about the ability of soil bacteria to metabolize phenolic acids used by plants in allelopathic interactions (Blum, 1998; Souto et al., 2000). Still, our understanding of cue degradation in terrestrial animal systems is lacking. While controlled laboratory experiments are ideal for an initial investigation into the dynamics of cue ageing, field studies are necessary for a more accurate understanding of cue persistence in a natural behavioral context with the ambient microfauna.

## 5. ACKNOWLEDGEMENTS

We thank Katie Rovente, Becky Urban, Jeff Bohner and Elizabeth Brinck for their laboratory and field assistance. We also thank Aaron Sullivan for valuable comments on the manuscript. Funding for this study was provided by an NSF grant (IBN-0229523) to DMM. All experiments comply with the current laws of the United States of America, the Binghamton University Animal Care and Use Committee (IUCAC 513-02), and the New York Department of Environmental Conservation (LCP02-460).

## 6. REFERENCES

Alberts, A. C., 1992, Constraints on the design of chemical communication systems in terrestrial vertebrates, *Am. Nat.* **139**:S62-S89.
Atlas, R. M., 1997, *Principles of Microbiology*, Wm. C. Brown Publishers, Dubuque.
Austin, R. M., Jr., 2000, Cutaneous microbial flora and antibiosis in *Plethodon ventralis*, in: *The Biology of Plethodontid Salamanders*, R. C. Bruce, R. G. Jaeger, and L. D. Houck, eds., Plenum Publishers, New York, pp. 451-462.
Barnes, M. C., Persons, M. H., and Rypstra, A. L., 2002, The effect of predator chemical cue age on antipredator behavior in the wolf spider *Pardosa milvina* (Araneae: Lycosidae), *J. Insect Behav.* **15**:269-281.
Blum, U., 1998, Effects of microbial utilization of phenolic acids and their phenolic acid breakdown products on allelopathic interactions, *J. Chem. Ecol.* **24**:685-708.
Chiodini, R. J., and Sundberg, J. P., 1981, Salmonellosis in reptiles: a review, *Amer. J. Epidem.* **113**:494-499.
Feldhoff, R. C., Rollmann, S. M., and Houck, L. D., 1999, Chemical analysis of courtship pheromones in a Plethodontid salamander, in: *Advances in Chemical Signals in Vertebrates*, R. E. Johnston, D. Muller-Schwarze, and P. W. Sorensen, eds., Plenum Publishers, New York, pp. 117-125.
Graves, B. M., Halpern, M., and Friesen, J. L., 1991, Snake aggregation pheromones: source and chemosensory mediation in western ribbon snakes (*Thamnophis proximus*), *J. Comp. Psychol.* **105**:140-144.
Kats, L. B., and Dill, L. M., 1998, The scent of death: chemosensory assessment of predation risk by prey animals, *Ecosci.* **5**:361-394.
Leadbetter, E. R., 2002, Prokaryotic diversity: form, ecophysiology, and habitat, in: *Manual of Environmental Microbiology*, C. J. Hurst, R. L. Crawford, M. J. Mcinerney, G. R. Knudsen, and L. D. Stetzanbach, eds., ASM Press, Washington, D.C., pp. 19-31.
Lima, S. L., and Dill, L. M., 1990, Behavioral decisions made under the risk of predation: a review and prospectus, *Can. J. Zool.* **68**:619-640.
Madison, D. M., Maerz, J. C., and Mcdarby, J. H., 1999a, Chemosensory response of salamanders to snake odors, in: *Advances in Chemical Signals in Vertebrates*, R. E. Johnston, D. Muller-Schwarze, and P. W. Sorensen, eds., Plenum, New York, pp. 505-516.
Madison, D. M., Maerz, J. C., and Mcdarby, J. H., 1999b, Optimization of predator avoidance by salamanders using chemical cues: diet and diel effects, *Ethology* **105**:1073-1086.
Madison, D. M., Sullivan, A. M., Maerz, J. C., Mcdarby, J. H., and Rohr, J. R., 2002, A complex, cross-taxon, chemical releaser of antipredator behavior in amphibians, *J. Chem. Ecol.* **28**:2271-2282.
Maerz, J. C., Panebianco, N. L., and Madison, D. M., 2001, Effects of predator chemical cues and behavioral biorhythms on foraging activity of terrestrial salamanders, *J. Chem. Ecol.* **27**:1333-1344.
Mcdarby, J. H., Madison, D. M., and Maerz, J. C., 1999, Chemosensory avoidance of predators by red-backed salamanders, *Plethodon cinereus*, in: *Advances in Chemical Signals in Vertebrates*, R. E. Johnston, D. Muller-Schwarze, and P. W. Sorensen, eds., Plenum Publishers, New York, pp. 489-495.
Oldak, P. D., 1976, Comparison of the scent gland secretion lipids of twenty-five snakes: implications for biochemical systematics, *Copeia* 320-326.
Polkinghorne, C. N., Olson, J. M., Gallaher, D. G., and Sorensen, P. W., 2001, Larval sea lamprey release two unique bile acids to the water at a rate sufficient to produce detectable riverine pheromone plumes, *Fish Physiol. Biochem.* **24**:15-30.
Schell, F. M., and Weldon, P. J., 1985, $^{13}$C-NMR analysis of snake skin lipids, *Agric. Biol. Chem.* **49**:3597-3600.
Schink, B., 2002, Synergistic interactions in the microbial world, *Antonie van Leeuwenhoek* **81**:257-261.
Schreiber, J. V., Frackenpohl, J., Moser, F., Fleischmann, T., Kohler, H.-P. E., and Seebach, D., 2002, On the biodegradation of β-peptides, *Chembiochem* **3**:424-432.
Sih, A., 1997, To hide or not to hide? Refuge use in a fluctuating environment, *Trends Ecol. Evol.* **12**:375-376.
Souto, X. C., Chiapusio, G., and Pellissier, F., 2000, Relationships between phenolics and soil microorganisms in spruce forests: significance for natural regeneration, *J. Chem. Ecol.* **26**:2025-2034.
Sullivan, A. M., Maerz, J. C., and Madison, D. M., 2002, Anti-predator response of red-backed salamanders (*Plethodon cinereus*) to chemical cues from garter snakes (*Thamnophis sirtalis*): laboratory and field experiments, *Behav. Ecol. Sociobiol.* **51**:227-233.
Sullivan, A. M., Madison, D. M., and Rohr, J. R., 2003, Behavioural responses by red-backed salamanders to conspecific and heterospecific cues, *Behaviour* **140**:553-564.
Torsvik, V., Sorheim, R., and Goksoyr, J., 1996, Total bacterial diversity in soil and sediment communities: review, *J. Indust. Micro. Biotech.* **17**:170-178.
Turner, A. M., 1996, Freshwater snails alter habitat use in response to predation, *Anim. Behav.* **51**:747-756.

Turner, A. M., and Montgomery, S. L., 2003, Spatial and temporal scales of predator avoidance: experiments with fish and snails, *Ecology* **84**:616-622.
Venzon, M., Janssen, A., Pallini, A., and Sabelis, M. W., 2000, Diet of a polyphagous arthropod predator affects refuge seeking of its thrips prey, *Anim. Behav.* **60**:369-375.
Waksman, S. A., 1952, *Soil Microbiology*, John Wiley & Sons, New York.

# DECLINE IN AVOIDANCE OF PREDATOR CHEMICAL CUES: HABITUATION OR BIORHYTHM SHIFT?

Dale M. Madison, John C. Maerz, and Aaron M. Sullivan[*]

## 1. INTRODUCTION

Studies of the behavior of captive animals are widespread in the literature. Such practices may occur because animals are not seasonally available, do not occur near the home institution of the researcher, or because it is much easier to control for multiple variables in the laboratory. In addition, because of animal rarity or inaccessibility, repeat testing procedures are often used on small numbers of animals, potentially causing pseudoreplication problems (Ramirez et al., 2000). It is reasonable to expect that the laboratory environment at some point will alter the natural responsiveness of animals to particular stimuli, or result in behavior that may be an artifact of stress, housing, diet, and the testing apparatus (Hennig and Dunlap, 1978; Jarvi, 1990). Through differential mortality, captivity can also create laboratory populations of atypical animals most tolerant of captive conditions (Navas and Gomes, 2001). Even when captivity effects are detected, there are few opportunities or rewards for investigators to report shortcomings in their own methodology or in correcting long-standing methods widely in use. Field validation of results is one way of detecting possible laboratory artifact (e.g., Sullivan et al., 2002). However, field studies may not be possible, may not be directly comparable to laboratory studies, and may introduce other experimental effects (Rohr and Madison, 2001; Rohr et al., 2002). An alternative approach to detecting captivity effects is to record data on changes in behavior of captive animals through time (Rohr et al., 2003).

In seven years of studies, we have documented a highly predictable avoidance response in the red-backed salamander, *Plethodon cinereus*, to body rinses from garter snakes, *Thamnophis sirtalis*, that have been feeding on *P. cinereus* (standard $TS_{PC}$ stimulus; Madison et al., 1999a,b; McDarby et al., 1999; Maerz et al., 2001; Madison et al., 2002; Sullivan et al., 2002, 2003). While we always get avoidance of $TS_{PC}$, no such avoidance occurs in response to body rinses from garter snakes feeding on earthworms,

---

[*] Dale M. Madison and Aaron M. Sullivan, Department of Biological Sciences, State University of New York at Binghamton, Binghamton, NY, 13902 USA. John C. Maerz, Cornell University, Ithaca, NY 14853 USA

goldfish, or other salamander species when these tests are conducted late at night (Madison et al., 1999a,b; Sullivan et al., 2003), so avoidance is not simply to nitrogenous wastes. The response to $TS_{PC}$ occurs regardless of the individual adult garter snake used to collect the rinse (Madison et al., 1999b; Madison et al., 2002), and we have validated this response under field conditions (Sullivan et al., 2002). Even in occasional testing situations where salamanders were held captive for up to a month and were used in repeat-testing experimental designs, *P. cinereus* continued to selectively avoid $TS_{PC}$. Because of these observations, because *P. cinereus* are routinely held captive for 3 to 9 months during all seasons prior to behavioral studies involving chemical cues (e.g., Jaeger et al., 1982; Mathis, 1990; Mathis and Simons, 1994; Simons et al., 1994; Gabor and Jaeger, 1995; Jaeger et al., 1995; Mathis and Lancaster, 1998), and because the studies above have generally been field validated (e.g., Mathis, 1989), we expected captive *P. cinereus* to avoid $TS_{PC}$ throughout winter in our studies beginning Fall 1999. We did not expect that repeat testing at intervals greater than a week apart would affect the avoidance response, because habituation to predator stimuli rarely, if ever, occurs in animals (Blanchard et al., 1998). Instead of recording avoidance to $TS_{PC}$ throughout winter, we recorded a progressive weakening and loss of avoidance. This paper documents our findings in a long series of experiments before, during, and after the winter in question, and attempts to explain the cause of the winter decline.

## 2. MATERIALS AND METHODS

### 2.1. Collection and Maintenance of Animals

For our behavioral studies, adult *P. cinereus* (snout/vent length > 40 mm) were collected from the Binghamton University Nature Preserve (Broome Co., New York): 132 for multiple experiments throughout the summer and fall of 1999, 130 in late October and early November for our winter studies, and 48 animals in April 2000 to see if newly captured animals would respond to the $TS_{PC}$ used throughout the winter. Two adult garter snakes (*T. sirtalis*; snout/vent length > 36 cm) were collected from the Nature Preserve during the summer of 1999 to use in preparing $TS_{PC}$ for all the trials.

Salamanders and snakes were kept in separate rooms at 15 °C (salamanders) and 25 °C (snakes) on a natural photoperiod (14L:10D). Salamanders were housed individually in 15-cm diameter petri dishes lined with moistened paper towels. Long-term captive salamanders were fed 5-10, 5 mm-long cricket nymphs once per week, and the toweling was changed once every 3 wks. Snakes were housed individually in 38-L glass aquaria, provided with a heating block, given crumpled paper towels for cover, and maintained on a diet of earthworms, *Lumbricus* sp., until 3 weeks before the collection of the $TS_{PC}$ rinse used in our trials. After all tests were completed, we released all *P. cinereus* and *T. sirtalis* at their original capture locations.

### 2.2. Collection of $TS_{PC}$

The methods for preparing $TS_{PC}$ and testing salamander avoidance are similar to the laboratory techniques reported by Sullivan et al. (2002). Briefly, we obtained $TS_{PC}$ by collecting a 200 ml rinse of an adult garter snake that had been maintained on a diet of 4 *P. cinereus* (2 male, 2 female) per week for two weeks prior to sample collection. Immediately after the snake ate the last salamander, we transferred the snake to a 4-L beaker, covered the beaker with 8 layers of cheesecloth secured by rubber bands, and then placed a heating pad (26°C) under half the beaker. After 72 h, we removed the snake from the container. We then added 200 ml of distilled water to the beaker, swirled it for 10 min, and transferred it to a polypropylene container. We filtered the rinse through HPLC-grade nylon filters (0.45μm) to remove any solid materials, and then divided the filtered solution into smaller aliquots for rapid freezing in liquid nitrogen and freezer storage. Previous research has shown that salamanders still avoid $TS_{PC}$ that has been frozen using this method, and that individual differences in *T. sirtalis* secretions or excretions do not affect salamander avoidance of $TS_{PC}$ (Madison et al., 2002; Sullivan et al., 2002). All tests were conducted with samples that had just been removed from the freezer and thawed, so aging of thawed samples was not a factor in our studies.

## 2.3. Experimental Protocol

For testing $TS_{PC}$, we placed two filter paper semi-circles with a 3-mm gap between them into 15-cm petri dishes, inoculated the semi-circles with 1.5 ml of $TS_{PC}$ or distilled water, and then randomly placed these test dishes (position and orientation) on a 6 x 8 grid on a foam board on the floor of the test room. Salamanders were moved to the test room one hour prior to the experiments, which were conducted between 1030 and 1315 h. We transferred each salamander from its home dish to the test dish with a cotton swab, and then placed a 15-mm collar of brown paper around the covered dish to visually isolate the salamanders from each other during the transfer process. After all salamanders were distributed (~10 min), the lights were turned off and the trial was recorded in complete darkness using an infrared video camera (SONY TRV66) suspended 3 m above the dishes. From videotapes, we scored which semicircle the salamander occupied at 3-min intervals for one hour beginning a time 0, which gave 21 positions per salamander. When salamanders straddled the 3 mm gap, two criteria were used to judge side position. First, if the salamander was moving, the side into which the salamander was moving was scored. If the salamander was not moving, the side occupied by more than half the body was scored.

For those *P. cinereus* tested two or more times, salamanders were selected at random from the stock population no sooner than one week (including feeding and a change of toweling) after previous testing. Between 24 and 80 animals were given a choice of the $TS_{PC}$ vs. distilled water substrates during each trial day. We used a chi-square goodness-of-fit test (Sheskin, 2000) to compare the number of salamanders that spent more than 50% of their time on the treatment vs. the water side of the test dish. Over ninety percent of the salamanders spent at least 70 % of their time on one side or the other, so choices were very clear for most animals (see Sullivan et al., 2003). We could not use parametric statistics (e.g. *t*-tests, regression analyses on winter trends) to analyze the avoidance trials because *P. cinereus* most often showed scores that were near the response extremes, 0 or 21, and hence were bimodal. A 2 x 2 chi-square contingency test corrected for continuity was used to look for differences between trials (Siegel and Castellan, 1988).

To measure activity among salamanders, the number of times an animal crossed from one side of the petri dish to the other in 60 min was recorded. The activity data were not bimodal, and a *t*-test for two independent samples was used to test for differences in activity between samples, following the assumptions and guidelines of Sheskin (2000).

## 3. RESULTS

Except for one animal that died from unknown causes in January, all captive salamanders survived and consumed some of the crickets offered each week. The few animals that didn't feed weekly were not used in the behavioral trials. In total, 305 of the 310 original animals fed regularly and were used in the experiments.

The responses of 24 "captive" *P. cinereus*, previously tested 4 to 5 times with $TS_{PC}$ and held captive for 5 months from June 18 to November 9, 1999, were compared to the responses of 24 "fresh" salamanders captured on November 1, 1999 and tested for the first time on Nov 9, 1999. Both groups avoided the same $TS_{PC}$ rinse ("captive", $\chi^2$ = 4.17, $P < 0.05$; "fresh", $\chi^2 = 8.17$, $P < 0.01$), and there was no significant difference between the responses of the captive and fresh animals ($\chi^2 = 0.11$, $P > 0.70$) (Figure 1). There was also no difference in the activity levels of the two groups (captive mean ± SD = 3.17 ± 2.35; fresh mean = 3.25 ± 2.40; $t = 0.121$, $P > 0.80$, df = 46). We therefore conducted trials on salamanders captured in late October and early November through the winter, expecting continued avoidance of $TS_{PC}$.

Trials were conducted on 11 experimental days between November 1999 and April 2000, and the results for these trials showed a steady decline in the percent of the salamanders tested that avoided $TS_{PC}$ (Figure 1). Whereas significant avoidance was recorded in each of the first 5 trails up to January 26, trials after that failed to showed significant avoidance behavior. In no case were the data for adjacent testing days significantly different from each other, although there was a significant difference in preference scores of captive salamanders between the first and last trial days, November 9 and April 7 ($\chi^2 = 5.15$, $P < 0.05$). The activity levels for these two groups of salamanders also differed (Nov mean ± SD = 3.17 ± 2.35; Apr mean = 0.81 ± 1.38; $t = 5.33$, $P < 0.001$, df = 70).

To determine whether loss of $TS_{PC}$ potency or salamander non-responsiveness was responsible for the decline in behavioral avoidance during late winter, the responses of 48 "captive" *P. cinereus* were compared to those from 48 newly captured ("fresh") salamanders to the same $TS_{PC}$ rinse on 10 April 2000. The captive animals had been used in multiple $TS_{PC}$ tests throughout winter (Mean ± SD = 5.77 ± 1.19, Range 3 – 8). Unlike the captive salamanders tested in November after 5 months in summer/fall captivity, these over-winter captives in early April failed to avoid $TS_{PC}$ ($\chi^2 = 0.00$, $P = 1.00$; Figure 1), but the newly-captured salamanders tested with the same rinse avoided $TS_{PC}$ ($\chi^2 = 12.0$, $P < 0.001$), and the difference in responses between the two groups was significant ($\chi^2 = 5.38$, $P < 0.05$). The activity levels also differed between the two groups of salamanders (captive mean ± SD = 0.81 ± 1.381, fresh mean = 3.96 ± 3.07; $t = 6.48$, $P < 0.001$, df = 94).

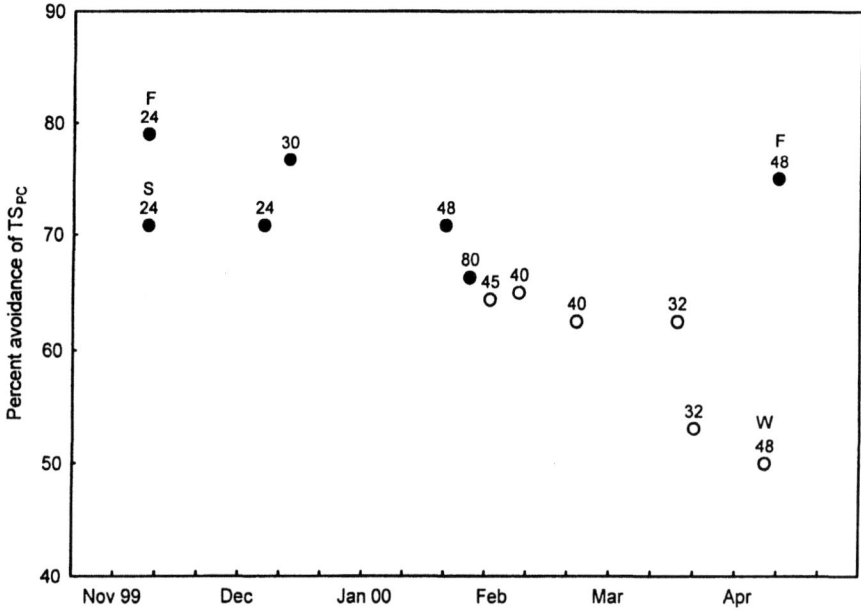

Figure 1. The percent of salamanders tested that avoided (had fewer recorded locations on) the $TS_{PC}$ side of the test dish during trials from November 1999 to April 2000. The number of salamanders tested on each date is indicated above each circle. Unlabelled circles represent results from salamanders that had previously been exposed to TSPC, with the number of exposures increasing with time through the winter. Salamanders that had just been captured and were tested for the first time are designated as fresh (F) salamanders. These fresh salamanders were tested with the same test samples at similar times as captive salamanders repeatedly exposed to TSPC for 5 months during summer (S) or winter (W). Significant statistical outcomes ($P < 0.05$) from random expectation (at 50 %) are indicated by solid circles; non-significant results, by open circles.

## 4. DISCUSSION

There was a clear, progressive loss of responsiveness to snake predator cues ($TS_{PC}$) in *P. cinereus* held over winter. It is not clear what caused the decline in responsiveness. The animals actively fed and appeared in good health throughout the winter. One explanation is habituation to the predator stimulus, since habituation is a widespread response to repetitive stimulation in the lower vertebrates (Goodman and Weinberger, 1973). However, we do not believe that habituation, at least by itself, was the cause of the decline for several reasons. First, the frequency of exposure to $TS_{PC}$ and the maintenance conditions of the summer and winter captives were comparable, and the summer animals did not show a decline in avoidance and activity. Second, the interval (1 week +) between exposures to $TS_{PC}$ is relatively long to induce habituation. Habituation typically occurs following continuous or frequent stimulation over shorter periods from several minutes or hours to a few days at most (Goodman and Weinberger, 1973). Third, prey species generally do not rapidly habituate, or even habituate at all, to predator stumuli, even following chronic exposure over many days (e.g., Sullivan et al., 1985; Magnhagen and Vestergaard, 1991; Epple et al., 1993; Holomuzki and Hatchett, 1994; Blanchard et al., 1998), and some prey may even become sensitized to predator stimuli following repeat exposure (Ducey et al., 1991). Finally, it seems unlikely that a salamander would habituate to predator cues in the wild when these are encountered less often than once per week.

We also considered whether the decline in responsiveness was a natural decrease in sensitivity to cues from a predator that is not a threat during winter, at least at our geographic location. Biorhythms have been reported for amphibians (e.g., Madison et al. 1999a,b; Maerz et al., 2001), and it is not unreasonable to expect that sensitivity to snake predators might also be linked to a circannual rhythm (e.g., Pancak and Taylor, 1983). Indeed, Hileman and Brodie (1994) showed that plethodontid responses to predators vary seasonally with changes in encounter rates with predators. Our original maintenance plan was to hold salamanders on a late summer photoperiod of 14L:10D and a temperature of 15 °C to keep them under environmental conditions where snakes would still be a natural threat. Inadvertently, we may have slowed but not arrested the seasonal onsets of winter (to early February) and spring (until after April 10) of the salamander's circannual rhythm, and hence prolonged predator avoidance until February and delayed seasonal resumption of predator avoidance until after early April. Some sensitivity of circannual rhythms to local external conditions has been reported in amphibians (Harri and Koskela, 1977), and circannual shifts in amphibian hormones are known to occur despite unchanging photoperiods (Pancak and Taylor, 1983). On a strict circannual cycle, both winter-captive and fresh-captured salamander groups should have responded similarly in April, but they didn't, and this is preliminary evidence that the two groups may have had circannual rhythms out of phase with each other. Testing two groups of captive salamanders throughout winter, one on a seasonally adjusted photoperiodic cycle, and a second on a constant late summer photoperiodic cycle, would be one way to test the delayed circannual rhythm hypothesis.

Whatever the reason is for the observed reduction in avoidance behavior in *P. cinereus* during winter, we hope our experience alerts other investigators to the possibility of seasonal or captivity and repeat-testing effects on the natural response patterns of amphibians, and that one way to detect possible effects is to record the same behavioral response through time with and without previous exposure to test stimuli. Further research is necessary concerning the influence of circannual rhythms on amphibian behavior in winter.

## 5. ACKNOWLEDGEMENTS

We thank Victor Lamoureux and Baishakhi Choudhury for assisting with salamander collections, experiments and data analysis. This study was licensed through the New York State Department of Environmental Conservation (License No. LCP99-500) and approved by the Institutional Animal Care and Use Committee of the State University of New York at Binghamton (Protocol No. 406-98 and 467-00). This study was funded by a grant from the National Science Foundation to DMM (IBN 99-74591) and a Faculty Grant from the Office of Research and Sponsored Funds of the State University of New York at Binghamton.

## 6. REFERENCES

Blanchard, R. J., Nikulina, J. N., Sakai, R. R., McKittrick, C., McEwen, B., and Blanchard, D. C., 1998, Behavioral and endocrine change following chronic predatory stress, *Physiol. Behav.* **63**:561-569.
Ducey, P. K., Anthony, C. D., and Brodie, E. D., Jr., 1991, Thresholds and escalation of antipredator responses in the Chinese salamander *Cynops cyanurus*: inter- and intra- individual variation, *Behav. Process.* **23**:181-191.
Epple, G., Mason, J. R., Nolte, D. L., and Campbell, D. L., 1993, Effects of predator odors on feeding in the mountain beaver (*Aplodontia rufa*), *J. Mamm.* **74**:715-722.
Gabor, C. R., and Jaeger, R. G., 1995, Resource quality affects the agonistic behaviour of territorial salamanders, *Anim. Behav.* **49**:71-79.
Goodman, D. A., and Weinberger, N. M., 1973, Habituation in "lower" vertebrates: Amphibia as vertebrate model systems, in: *Habituation, Volume 1*, H.V.S. Peeke, and M.J. Herz, eds., Academic Press, New York, pp. 85-140.
Harri, M. N. E., and Koskela, P., 1977, Terms of spawning in southern and northern Finnish populations of the common frog, *Rana temporaria*, under laboratory conditions, *Aquilo Ser Zool.* **17**:49-51.
Hennig, C. W., and Dunlap, W. P., 1978, Tonic immobility in *Anolis carolinensis*: Effects of time and conditions of captivity, *Behav. Neural Biol.* **23**:75-86.
Hileman, K. S., and Brodie, E. D., Jr., 1994, Survival strategies of the salamander *Desmognathus ochrophaeus*: interaction of predator avoidance and anti-predator mechanisms, *Anim. Behav.* **47**:1-6.
Holomuzki, J. R., and Hatchett, L., 1994, Predator avoidance costs and habituation to fish chemicals by a stream isopod, *Freshwater Biol.* **32**:585-592.
Jaeger, R. G., Kalvarsky, D., and Shimizu, N., 1982, Territorial behaviour of the red-backed salamander: Expulsion of intruders, *Anim. Behav.* **30**:490-496.
Jaeger, R. G., Schwarz, J., and Wise, S., 1995, Territorial male salamanders have foraging tactics attractive to gravid females, *Anim. Behav.* **49**:633-639.
Jarvi, T., 1990, Cumulative acute physiological stress in Atlantic salmon smolts: the effect of osmotic imbalance and the presence of predators, *Aquaculture* **89**:337-350.
Madison, D. M., Maerz, J. C., and McDarby, J. H., 1999a, Chemosensory responses of salamanders to snake odors: Flight, Freeze and Dissociation, in: *Advances in Chemical Communication in Vertebrates*, R. Johnston, D. Muller-Schwarze, and P. Sorensen, eds., Plenum Publishers, New York, pp. 505-516.
Madison, D. M., Maerz, J. C., and McDarby, J. H., 1999b, Optimization of predator avoidance by salamanders using chemical cues: diet and diel effects, *Ethology* **105**:1073-1086.
Madison, D. M., Sullivan, A. M., Maerz, J. C., and Rohr, J. R., 2002, A complex, cross-taxon chemical releaser of antipredator behavior in amphibians, *J. Chem. Ecol.* **28**:2271-2282.
Maerz, J. C., Panebianco, N. L., and Madison, D. M., 2001, Effects of predator chemical cues and behavioral biorhythms on foraging activity of terrestrial salamanders, *J. Chem. Ecol.* **27**:1333-1344.
Magnhagen, C., and Vestergaard, K., 1991, Risk taking in relation to reproductive investments and future reproductive opportunities in field experiments on nest-guarding common gobies *Pomatoschistus microps*, *Behav. Ecol.* **2**:351-359.
Mathis, A., 1989, Do seasonal spatial distributions in a terrestrial salamander reflect reproductive behavior or territoriality? *Copeia* **1989**:788-791.
Mathis, A., 1990, Territorial salamanders assess sexual and competitive information using chemical signals, *Anim. Behav.* **40**:953-962.
Mathis, A., and Lancaster, D., 1998, Response of terrestrial salamanders to chemical stimuli from distressed conspecifics, *Amphibia-Reptilia* **19**:330-335.
Mathis, A., and Simons, R. R., 1994, Size-dependent responses of resident male red-backed salamanders to chemical stimuli from conspecifics, *Herpetologica* **50**:335-344.
McDarby, J. H., Madison, D. M., and Maerz, J. C., 1999, Chemosensory avoidance of predators by red-backed salamanders, *Plethodon cinereus*, in: *Advances in Chemical Communication in Vertebrates*, R. Johnston, D. Muller-Schwarze, and P. Sorensen, eds., Plenum Publishers, New York, pp. 489-495.
Navas, C. A., and Gomes, F. R., 2001, Time in captivity as a confounding variable in herpetological research: An example from the metabolic physiology of treefrogs (*Scinax*), *Herpetol. Rev.* **32**:228-230.
Pancak, M. K., and Taylor, D. H., 1983, Seasonal and daily plasma corticosterone rhythms in American toads, *Bufo americanus*, *Gen. Comp. Endocrin.* **50**:490-497.
Ramirez, C. C., Fuentes-Contreras, E., Rodriguez, L. C., and Niemeyer, H. M., 2000, Pseudoreplication and its frequency in olfactometric laboratory studies, *J. Chem. Ecol.* **26**:1423-1431.
Rohr, J. R., and Madison, D. M., 2001, A chemically-mediated trade-off between predation risk and mate search in newts, *Anim. Behav.* **62**:863-869.

Rohr, J. R., Madison, D. M., and Sullivan, A. M., 2002, Sex differences and seasonal trade-offs in response to injured and non-injured conspecifics in red-spotted newts, *Notophthalmus viridescens*, *Behav. Ecol. Sociobiol.* **52**:385-393.

Rohr, J. R., Madison, D. M., and Sullivan, A. M., 2003, On temporal variation and conflicting selection pressures: A test of theory using newts, *Ecology* **84**:1816-1826.

Sheskin, D. J., 2000, *Handbook of Parametric and Nonparametric Statistical Procedures*, Chapman & Hall, Boca Raton.

Siegel, S., and Castellan, N. J., Jr., 1988, *Nonparametric Statistics for the Behavioral Sciences*, McGraw-Hill, New York.

Simons, R. R., Felgenhauer, B. E., and Jaeger, R. G., 1994, Salamander scent marks: site of production and their role in territorial defence, *Anim. Behav.* **48**:97-103.

Sullivan, A. M., Maerz, J. C., and Madison, D. M., 2002, Anti-predator response of red-backed salamanders (*Plethodon cinereus*) to chemical cues from snakes: laboratory and field experiments, *Behav. Ecol. Sociobiol.* **51**:227-233.

Sullivan, A. S., Madison, D. M., and Rohr, J. R., 2003, Behavioural responses by red-backed salamanders to conspecific and heterospecific cues, *Behaviour* **140**:553-564.

Sullivan, T. P., Nirdstrom, L. O., and Sullivan, D. S., 1985, Odors as repellents to reduce feeding damage by herbivores 1. Snowshoe hares *Lepus americanus*, *J. Chem. Ecol.* **11**:903-920.

# CHEMICALLY MEDIATED LIFE-HISTORY SHIFTS IN EMBRYONIC AMPHIBIANS

Reehan S. Mirza and Joseph M. Kiesecker[*]

## 1. ABSTRACT

The embryonic stage is often times overlooked as a simple developmental stage in the lifecycle of animals. However, recent work has shown that embryos have the ability to perceive risk in their local environments and respond by altering life-history traits. Shifts in life-history traits may be influenced by mechanical or chemical cues associated with predation. In this paper we review a number of studies that examine shifts in the hatching characteristics of embryonic amphibians in response to chemical cues. The embryonic responses tend to vary between different species and include: earlier hatching, altered morphological characteristics and delayed hatching. In most cases the timing of hatching is shaped by the type of predator present. Thus, selection shapes embryos to respond to cues that represent danger which are ecologically relevant. Ultimately this may lead to innate recognition of predators as well as recognizing cues from heterospecific embryos. The amphibian embryonic stage represents a complex and sophisticated set of physiological and behavioral responses which may alter the course of the remaining life cycle. Further work is needed to elucidate these effects.

## 2. INTRODUCTION

Amphibians employ complex life cycles consisting of both aquatic and terrestrial stages (Duellman and Trueb, 1994). Many amphibians typically breed in temporary ponds where the larvae grow during the spring and summer and then undergo metamorphosis to attain their terrestrial forms and vacate the water. These dramatic morphological, physiological and behavioral changes are accompanied by dramatic

---

[*] Reehan S. Mirza, Department of Biology, Concordia University, Montreal, QC, Canada H4B 1R6. Joseph M. Kiesecker, Department of Biology, The Pennsylvania State University, University Park, PA, 16802, USA.

ecological changes, including shifts in the nature of predation pressure. In order to survive the shift to terrestrial environments, amphibian larvae must be able to assess predation risk and respond accordingly. As with other organisms, larval amphibians possess plasticity in their developmental traits and can accelerate or delay shifts from one life-stage to the next depending on the type and level of predation risk. For example, common frog *Rana temporaria* and common toad *Bufo bufo* both metamorphose earlier in response to cues from dragonfly larvae fed tadpoles (Laurila et al., 1998). Likewise, western toad tadpoles *B. boreas* and red-legged frog tadpoles *R. aurora* metamorphose sooner when exposed to cues from predators fed tadpoles (Chivers et al., 1999; Kiesecker et al., 2002). Thus we do see that amphibians possess the plasticity to alter their developmental and growth rates in an effort to avoid predators.

The majority of studies that examine shifts in life-history traits in amphibians focus on the shift from the tadpole to the juvenile stage. One stage that is often overlooked is the embryonic stage. The embryonic stage is viewed as a simple developmental stage, however, there is evidence that shows that amphibian embryos can detect and respond to cues from predators in their environments. The aim of this paper is to provide a review of studies that have examined life-history shifts in amphibian embryos in response to chemical cues from predators. From the results of these studies, we will develop generalized hatching patterns that may be used to predict these shifts in other amphibian species. We will also discuss the complexity and specificity of the chemical cues involved and lastly we will examine those results which do not fit into our hypothesized patterns and address why there is variability in life-history responses.

## 3. LITERATURE SURVEY OF LIFE-HISTORY SHIFTS IN EMBRYONIC AMPHIBIANS

Examination of life-history shifts in embryonic amphibians has been a recent development. We have identified 18 studies between 1995 and 2003 that have tested the responses of larval amphibians to cues that represent potential danger. Of these 18 studies, 3 studies have tested mechanical cues from predators (Warkentin, 1995, 2000), 2 studies examine responses to pathogens (Warkentin et al., 2001; Kiesecker et al., unpublished data) and the remaining 13 studies examining chemical cues from predators (see Table 1). Chemosensory assessment of predation risk is widespread in aquatic systems (Chivers and Smith, 1998; Kats and Dill, 1998). Studies completed primarily in the last decade suggest that the level of this assessment is probably much more sophisticated than was previously thought. This review will show that this sophistication is also present in the embryonic stage.

## 4. HATCHING PATTERNS

Based on the results of the studies shown in Table 1, we see two generalized hatching patterns emerging. Embryos tend to hatch earlier when exposed to egg predators and tend to delay hatching when exposed to predators that prey on newly hatched larvae. For example, Chivers et al. (2001) exposed embryos from Pacific tree frogs (*Hyla regilla*) and Cascades frogs (*R. cascadae*) to chemical cues from leeches, non-predatory earthworms and from injured conspecific eggs. Treefrogs and Cascades frogs hatched earlier in response to chemicals from leeches (Glossiphonidae and Erpobdellidae), but not to cues from earthworms (*Lumbricus* spp.) or injured conspecific eggs. In response to injured-egg cues only treefrogs hatched earlier and both Cascades frogs and treefrogs hatched at a less developed stage. Similarly, Mirza et al. (unpublished data) found that wood frog embryos (*R. sylvatica*) hatch in less time when exposed to chemical cues from injured eggs. Wood frogs hatched out in 6 days compared to 10 days in the controls.

In response to larval predators we tend to see a delay in hatching time. Sih and Moore (1993) exposed streamside salamander embryos (*Ambystoma barbouri*) to chemical cues from flatworms (*Phagocotus gracilis*) or non-predatory isopods (*Lirceus fontinalis*). Embryos tended delay hatching in response to flatworm cues, but not to isopod cues. Moreover, hatchlings tended to be larger and more developed than control hatchlings.

Both of these patterns make intuitive sense. Several studies have shown that prey will avoid areas where predators are present or adopt tactics that will make them less conspicuous to the predator (see reviews Chivers and Smith, 1998; Kats and Dill, 1998). If a predator is consuming eggs, the best response would be to escape from the egg mass as soon as possible, although there is a cost associated with leaving early. Embryos tend to be smaller and have larger yolk sacs which could impair swimming ability (Chivers et al., 2001). Conversely, if a larval predator is present, it may be less costly to delay hatching until the predator is no longer present, or upon hatching, larvae have a body form that is better suited to escape predation altogether. Larger hatchlings may possess enhanced swimming capabilities that assist in escaping predators (Sih and Moore, 1993).

Not all studies show these general patterns, several studies do not report a change in hatching time, but report changes in morphology. Laurila et al. (2001, 2002) found that predator cues had no effect on hatching time in common frog (*Rana temporaria*) embryos, but embryos tended to have shorter tails with deeper tail fins and shorter deeper bodies. These morphological changes could also enhance survivability in the presence of a predator by increasing the likelihood of escape or outgrowing the gape of gape-limited predators. One study reports no evidence of life-history shifts. Anderson and Petranka (2003) found no change in either hatching time or morphology for wood frogs and spotted salamander (*A. maculatum*) embryos exposed to cues from predatory dragonfly larvae.

## 5. COMPLEXITY AND SOPHISTICATION OF CHEMICAL CUES

Over the last 10 years we have found that responses to chemical cues in aquatic systems are more sophisticated than we had previously thought. Prey animals that are able to distinguish between predatory and non-predatory cues would be better able to maximize time and energy usage (reviewed in Chivers and Mirza, 2001). Amphibian systems compared to other aquatic systems differ in one major aspect, the ability to recognize predators. In other systems prey must learn to recognize the identity of predators. Amphibians exhibit innate recognition of predators most likely due to different selection pressures (Kats et al., 1988; Kiesecker and Blaustein, 1997; Gallie et al., 2001). We find that there are three classes of chemical cues that elicit shifts in life-

Table 1. Summary of experiments that have tested predator-mediated life-history shifts in embryonic amphibians.

| Prey | Response |
| --- | --- |
| Pacific tree frog (Hyla regilla) | embryos hatch earlier in response to chemical cues from predatory leeches and injured eggs (Chivers et al., 2001) |
| Cascades frog (Rana cascadae) | embryos hatch earlier in response to predatory leeches (Chivers et al., 2001) |
| moor frog, (Rana arvalis) | Early hatching of embryos in response to chemical cues from leech, dragonfly nymph and fish predators (Laurila et al., 2002) |
| spotted salamander (Ambystoma maculatum) | Earlier hatching in response to caddisfly and frog tadpoles consuming eggs (Mirza et al., unpublished data) |
| | Delayed hatching in response to chemical cues from predatory caddisflies (Smyer et al., unpublished data) |
| | No change in hatching time in response to chemical cues from dragonfly fed eggs (Anderson and Petranka, 2003) |
| wood frog (Rana sylvatica) | Embryos exposed to cues from injured wood frog eggs and spotted salamander eggs hatch in a shorter period of time (Mirza et al., unpublished data) |
| | No change in hatching time in response to larval dragonfly predators (Anderson and Petranka, 2003) |
| streamside salamander (A. barbouri) | Delayed hatching in response to chemical cues from predatory flatworms and fish predators (Sih and Moore, 1993; Moore et al, 1996; Storfer, 1999) |
| small-mouthed salamander (A. texanum) | Delayed hatching in response to chemical cues from predatory flatworms and fish predators (Sih and Moore, 1993; Moore et al, 1996; Storfer, 1999) |
| green frog (Rana clamitans) | Delayed hatching in response to chemicals from leeches consuming eggs (Schalk et al., 2002) |
| common frog (Rana temporaria) | Delayed hatching in response to fish predators (Laurila et al., 2002) |
| | No change in hatching time, but morphological differences in response to cues from predatory diving beetles, leeches and dragonfly larvae (Laurila et al., 2001; 2002) |

history traits in amphibians: injury-released alarm cues, predator odors and diet cues (cues released from a predator fed a specific prey item) (see reviews Chivers and Smith, 1998; Kats and Dill, 1998; Chivers and Mirza, 2001). Responses to these classes of chemical cues have been documented in larval and juvenile amphibians; we will now relate these cues to responses in embryos.

By examining Table 1, we can see that life-history shifts have been documented in response to all three classes of chemical alarm cues. The shifts in life-history are not due to generalized chemical cues; the cues are very specific. Controls of injured eggs from other species, non-predatory animal odors and predators on different diets have verified this. The majority of studies show responses to predators fed a diet of frog eggs. Few studies have tested the response to injured egg cues or for innate recognition of predators.

Mirza et al. (unpublished data) have conducted two separate studies that test for the specificity of chemical cues used to elicit life-history shifts. In the first study, we tested for innate predator recognition in spotted salamander embryos. Embryos were exposed to chemical cues from caddisflies (*Ptilostomis postica*) fed salamander eggs, caddisflies fed bloodworms (*Tubifex* spp.), wood frog tadpoles fed salamander eggs, wood frog tadpoles fed alfalfa pellets or a blank control. Caddisflies are known predators of spotted salamander eggs (Rowe et al., 1994) while wood frog tadpoles are known to consume salamander eggs when resources are limited (Petranka et al., 1998). Salamander embryos hatched earlier in response to caddisflies on both diets and tadpoles fed eggs than tadpoles fed pellets and the blank control. Moreover, more embryos hatched sooner to caddisflies fed eggs than caddisflies fed worms or tadpole fed eggs. These results show that embryonic salamanders innately respond to caddisflies and that this response is more intense if caddisflies are consuming salamander eggs. Embryos did respond to tadpoles fed eggs but not tadpoles fed pellets showing that the diet cue is important in eliciting a life-history shift.

In a second study, Mirza et al. (unpublished data) examined shifts in life-history traits in wood frog embryos exposed to cues from injured eggs testing whether the chemical cues were phylogenetically conserved or ecologically relevant. Recent work has shown that fishes respond innately to injured cues from heterospecifics that are closely related and can learn to respond to heterospecific cues from members of the same prey guild i.e. those animals which share the same habitat and suite of predators (Mirza and Chivers, 2001). This relationship has not been examined in amphibians. Embryos were exposed to cues from injured woodfrog eggs, injured southern leopard frog eggs, *R. utriculata* (closely related phylogenetically), injured spotted salamander eggs (ecologically related) or distilled water. Spotted salamanders lay their eggs in the same waterbodies in close proximity to wood frog egg masses. Wood frog embryos hatched out in less time in response to injured woodfrog egg cues and injured salamander egg cues compared to injured leopard frog eggs and distilled water (6 days vs. 10 days). Therefore, wood frog embryos respond to cues that are ecologically relevant.

## 6. DISCUSSION

The results of the reviewed studies show timing to hatching is an important switch point that shows adaptive plasticity in response to environmental factors that cause egg hatching or mortality (Sih and Moore, 1993). Under the risk of threat animals can shift resources to alter their time of hatching to enhance survival. Although the majority of examples have been demonstrated in larval and juvenile animals, we have shown that embryonic amphibians are also capable of these shifts. Moreover, the shifts are in response to specific cues and not generalized responses to odors in their environments. Enhanced survival at the embryonic stage may translate to greater survival and fitness benefits during the latter part of the life-cycle at both the individual and population levels.

### 6.1 Variability in Life-history Responses

We have attempted to place the documented studies into two categories dependent upon whether the predator consumes eggs or newly hatched larvae. Although for the most part this can be done, we still have to account for studies that do not fit neatly into our patterns. Moreover, we tend to see variability in the responses of the same species to different types of predators or different species to similar types of predator. Responses may be species specific. In one study, Chivers et al. (2001) show that Cascades frog and treefrog tadpoles hatch earlier when exposed to predatory leeches which are believed to be an egg predator. In another study, Schalk et al. (2002) find that green frog tadpoles (*R. clamitans*) delay hatching in response to cues from predatory leeches that are also believed to be egg predators.

Another example of contrasting responses is seen in spotted salamander embryos in response to larval predators. Smyer et al. (unpublished data) found that salamander embryos delayed hatching when exposed to cues from predatory caddisflies (Limniphilidae), but Anderson and Petranka (2003) found no influence on hatching time in spotted salamander embryos exposed to cues from predatory dragonflies (*Aeschna* spp). This result may be due to population differences. Smyer et al. conducted their study in the northeastern US while Anderson and Petranka conducted theirs in the southeastern US.

What if embryos respond the same to both egg and larval predators? Laurila et al. (2002) found that moor frog embryos (*R. arvalis*) hatched earlier in response to leech, dragonfly and fish predators. Dragonfly larvae and fishes tend to be larval predators and should cause a delay in hatching according to our predators, yet this is not what we find.

One possible explanation for the "between studies" variation is that the results that we are seeing represent the current level of predation risk in that particular population. For example, in the Chivers et al. (2001) and Schalk et al. (2002) studies, leeches are believed to be egg predators and when presented eggs leeches in both studies consume them. But only the Chivers et al. study obtained the predicted response. Perhaps in Schalk et al. (2002) leeches will consume eggs when presented to them, but in the wild they may be stronger larval predators. Similarly, Laurila et al. (2002) may have populations of moor frogs that currently are experiencing stronger egg predation by leeches, dragonfly larvae, and fishes leading to induced hatching. Anderson and Petranka (2003) may have populations of wood frogs where dragonfly larvae are not strong predators leading to a minimal influence on frog embryos. Clearly future studies should provide some indication of natural predation levels to support conclusions or lack there of.

Predation is a strong selective force that shapes population and community structure, but the selection pressures on different populations may not be the same accounting for the differences in responses that we may see. Evolution of adaptive plasticity in amphibian life-cycles should provide a greater survival benefit and not a greater cost.

The responses that we see should be selected for and although they may not make intuitive sense to us at first, we must realize that some benefit must exist. However, we must be cautious in ruling out other factors such as pathogens, oxygen levels etc. that can also alter hatching characteristics before coming to this conclusion.

## 7. CONCLUSION

The sophistication of chemosensory assessment of predation risk seems remarkable. Even at the embryonic level we see the complexity and specificity of the chemical cues used in this assessment. Two patterns have emerged from the studies conducted to date; however, more studies are needed to verify these patterns. We also need to branch out and examine how life-history shifts at one point in the life-cycle influences the rest of the life cycle. Moreover, we need to consider the natural level of predation risk in each system we study because the selection pressures may differ from one population to the next. By examining each system more closely we start to understand how predation shapes population and community structure.

## 8. REFERENCES

Anderson, R. A., and Petranka, J. W., 2003, Odonate predator does not effect hatching time or morphology of two amphibians, *J. Herp.* **37**:65-71.

Chivers, D. P., and Smith, R. J. F., 1998, Chemical alarm signalling in aquatic predator/prey interactions: a review and prospectus, *Écoscience* **5**:338-352.

Chivers, D. P., and Mirza, R. S., 2001, Predator diet cues and the assessment of predation risk by aquatic vertebrates: a review and prospectus, in: *Chemical Signals in Vertebrates, Volume 9*, A. Marchlewska-Koj, J.J. Lepri, and D. Müller-Schwarze, eds, Plenum Press, New York, pp. 277-284.

Chivers, D. P., Kiesecker, J. M. Marco, A., Wildy, E. L., and Blaustein, A. R., 1999, Shifts in life history as a response to predation in western toads (*Bufo boreas*). *J. Chem. Ecol.* **25**:2455-2464.

Chivers, D. P., Kiesecker, J. M., Marco, A., DeVito, J., Anderson, M. T., and Blaustein, A. R., 2001, Predator-induced life-history changes in amphibians: egg predation induces hatching, *Oikos* **92**:135-142.

Duellman, W. E., and Trueb, L., 1994, *The Biology of Amphibians*, McGraw-Hill, New York, pp. 694.

Gallie, J. A., Mumme, R. L., and Wissinger, S. A., 2001, Experience has no effect on the development of chemosensory recognition of predators by tadpoles of the American toad, *Bufo americanus*, *Herpetologica*, **57**:376-383.

Kats, L. B., and Dill, L. M., 1998, The scent of death: chemosensory assessment of predation risk by prey animals, *Ecoscience* **5**:361-394.

Kats, L. B., Petranka, J. W., and Sih, A., 1988, Antipredator defenses and the persistence of amphibian larvae with fishes, *Ecology* **69**:1865-1870.

Kiesecker, J. M., Chivers, D. P., Anderson, M. T., and Blaustein, A. R., 2002, The effects of predator diet on life history shifts of red-legged frogs, *Rana aurora*, *J. Chem. Ecol.* **28**:1007-1015.

Laurila, A., Jutta, K., and Esa, R., 1998, Predator-induced changes in life history in two anuran tadpoles: Effects of predator diet, *Oikos* **83**:307-317.

Laurila, A., Crochet, P.-A., and Merilä, A., 2001, Predation-induced effects on hatchling morphology in the common frog (*Rana temporaria*), *Can. J. Zool.* **79**:26-30.

Laurila, A., Pakkasmaa, S., Crochet, P.-A., and Merilä, A., 2002, Predator-induced plasticity in early life history and morphology in two anuran amphibians, *Oecologia* **132**:524-530.

Mirza, R.S., and Chivers, D.P., 2001, Are chemical alarm signals conserved within salmonid fishes? *J. Chem. Ecol.* **27**:1641-1655.

Moore, R. D., Newton, B., and Sih, A., 1996, Delayed hatching as a response of streamside salamander eggs to chemical cues from predatory sunfish, *Oikos* **77**:331-335.

Petranka, J. W., Rushlowe, A. W., and Hopey, M. E., 1998, Predation by tadpoles of *Rana sylvatica* on embryos of *Ambystoma maculatum*: Implications of ecological role reversals by Rana (Predator) and Ambystoma (Prey), *Herpetologica* **54**:1-13.

Rowe, C. L., Sadinski, W. J., and Dunson, W. A., 1994, Predation on larval and embryonic amphibians by acid-tolerant caddisfly larvae (*Ptilostomis postica*), *J.Herp.* **28**:357-364.

Schalk, G., Forbes, M. R., and Weatherhead, P. J., 2002, Developmental plasticity and growth rates of green frog (*Rana clamitans*) embryos and tadpoles in relation to a leech (*Macrobdella decora*) predator, *Copeia* **2002**:445-449.

Sih, A., and Moore, R. D., 1993, Delayed hatching of salamander eggs in response to enhanced larval predation risk, *Am. Nat.* **142**:947-960.

Storfer, A., 1999, Gene flow and local adaptation in a sunfish-salamander system, *Behav. Ecol. Soc.* **46**:73-79.

Warkentin, K. M., 1995, Adaptive plasticity in hatching age: a response to predation risk trade-offs. *Proc. Natl Acad. Sci.*, **92**:3507-3510.

Warkentin, K. M., 2000, Wasp predation and wasp-induced hatching of red-eyed treefrog eggs, *Anim. Behav.* **60**:503-510.

Warkentin, K. M., Currie, C. C. and Rehner, S. A., 2001, Egg-killing fungus induces early hatching of red-eyed treefrog eggs. *Ecology* **82**:2860-2869.

# LATENT ALARM SIGNALS: ARE THEY PRESENT IN VERTEBRATES?

Ole B. Stabell[*]

## 1. INTRODUCTION

Chemical alarm signals are present in a number of aquatic animal species. In lower vertebrates in particular, alarm signals have been demonstrated in several fish and amphibian species (Pfeiffer, 1963, 1974). The chemical messengers in question are not only releasing behavioural reactions in conspecifics, but they also appear to affect secondary defence characteristics by inducing changes in morphology and life history shifts in the receivers. In that way, chemical alarm signals seem to fulfil the basic definition of being true pheromones (Karlson and Lüscher, 1959).

However, in several species of lower vertebrates, chemical alarm signals seem to be lacking (Pfeiffer, 1966, 1977). Still, many of these species are responding with behavioural, morphological, and life history changes to chemical cues released by predators, *i.e.* to presumed kairomones (Brown et al., 1970). In an evolutionarily context one may ask how this can be possible? A predator that is easily detected by its prey should be regarded as a looser in the long run. How then can complex physiological mechanisms be developed in a prey species in response to predator cues, when at the same time the predator is presumably being subjugated to a chemical 'arms race' with both competitors and its prey? This last question represents an obvious reason to assume that our current biological 'map' on predator-prey interactions is not giving the correct description of the 'landscape' to be covered.

The intention with this paper is to add a new functional 'air photo' to improve our current biological 'map', so that the area of chemical alarm signals and their functional properties may be covered in a better way. To do so, lessons learned from invertebrate studies will be focused on, and the term 'latent' alarm signal will be introduced. Of circumstantial evidence in favour of latent alarm signals in vertebrates, revealed from the

---

[*] Ole B. Stabell, Department of Natural Sciences, Agder University College, Serviceboks 422, N-4604 Kristiansand, Norway.

literature, will be focused on, and some preliminary data demonstrating the presence of such alarm signals in tadpoles of frogs will subsequently be presented.

## 2. CHEMICAL ALARM SIGNALS

### 2.1. Alarm Signals and Predator Labelling

A chemical alarm signal, as described by Karl von Frisch, is a substance (or mixture of substances) which is released involuntarily by an injured individual of a species, and which induces alarm behaviour in conspecifics (von Frisch, 1941; Pfeiffer, 1977). There seems to be similarities in alarm signal properties between closely related species, as demonstrated by the partial release of behavioural responses in cyprinid fishes to alarm signals from other species of cyprinids (Schutz, 1956). A predator, further, may eat a prey individual, and then subsequently over time release the alarm signals of that prey through the gastro-intestinal tract. Other individuals of the prey species, or closely related prey, may then detect the predator due to the alarm signals leaking from its gut, thereby responding indirectly to the signal. The disclosure of a predator to potential prey, due to chemical alarm signals from previously ingested prey, has been denoted 'predator labelling' (Mathis and Smith, 1993).

### 2.2. Alarm Signal Functions

Predator labelling may be the functional mechanism of release when a continuous exposure of prey to alarm signals, induces responses like morphology changes and life history modifications. In fact, morphology changes in fish resulting from long time exposure to alarm signals were discovered due to predator labelling (Brönmark and Miner, 1992; Stabell and Lwin, 1997). In that case, it was shown that a predator, the European pike (*Esox lucius*) could be labelled with alarm signals of a cyprinid, the crucian carp (*Carassius carassius*), subsequently inducing a body depth increase in the prey. Hitherto, life history modifications due to alarm signal exposure has not been described in fishes.

In amphibian larvae, several reports on behavioural responses to alarm signals in toads and salamanders have been given (reviewed by: Chivers and Smith, 1998). However, reports are scarce on morphology changes and life history shifts resulting from direct exposure to conspecific alarm signals in amphibians. The work by Chivers et al., (1999) on life history shift in tadpoles of the Western toad seems to be the only one within anurans. On the other hand, a number of reports on both behavioural, morphological and life history responses to predator cues can be found (Kats and Dill, 1998). While predator labelling may be suspected in those amphibian species that are possessing alarm signals (e.g., Petranka et al., 1987; Skelly and Werner, 1990; DeVito et al., 1998), the effects seems more inexplicable for responses obtained with predator cues in species where alarm signals are presumably lacking. Such responses are especially found in tadpoles of frogs within the *Hyla* and *Rana* genera (Table 1).

However, one striking feature revealed from most of the reports on species presumably devoid of alarm signals is that the predators have been fed the prey species of study

**Table 1.** List of work on tadpoles of frogs within the *Hyla* and *Rana* genera reporting from exposure to chemical cues released by predators.

| Prey Species | Predator | Effect | Reference |
|---|---|---|---|
| H. chrysoscelis | Green sunfish | Behaviour | Petranka et al. (1987) |
| R. lessonae / R. esculenta | Pike * | Behaviour | Stauffer and Semlitsch (1993) |
| R. aurora | Newts * | Behaviour | Wilson and Lefcort (1993) |
| R. temporaria | Three fish species * | Behaviour | Manteifel (1995) |
| H. chrysoscelis | Dragonfly larvae * | Morphology | McCollum and Leimberger (1997) |
| R. temporaria | Dragonfly larvae * | Behaviour | Laurila et al. (1997) |
| R. arvalis | Dytiscus larvae * | Morphology / Growth | Lardner (1998) |
| R. sylvatica | Dragonfly larvae * | Morpholgy | Van Buskirk and Relyea (1998) |
| R. temporaria | Dragonfly larvae * | Life history | Laurila et al. (1998) |
| H. regilla | Garter snakes * | Life history | DeVito et al. (1999) |
| R. temporaria | Perch / Dragonfly larvae * | Behaviour | Laurila (2000) |
| R. dalmatina / R. temporaria / R. arvalis / H. arborea | Dytiscus larvae * | Morphology / Life history | Lardner (2000) |
| R. temporaria | Dragonfly larvae / Notonecta / Newts* | Behaviour / Morphology | Van Buskirk (2001) |

\* The predators were fed tadpoles of the species studied.

(Table 1). This suggests that some type of predator labelling may be the functional mechanism after all, but that the chemical signals in question do not possess the instant releaser effects (Wilson and Bossert, 1963) normally associated with alarm pheromones. The signal type proposed here, denoted 'latent' alarm signals, has recently been described in aquatic invertebrates (Stabell et al., 2003). A short glance at the invertebrate story therefore seems appropriate.

## 2.3. Latent Alarm Signals

Water fleas (*Daphnia* spp.) respond with avoidance responses, and undergo phenotypic changes when exposed to chemical cues from predators. In particular, if parthenogenetic mother individuals are exposed to chemical cues from predators when carrying their eggs, the resulting offspring will develop elongated tail or neck spines, or 'helmets' (*i.e.* a morphological change in head form) compared to offspring from unexposed *Daphnia*. Morphological changes of this kind appear to reduce the chances of prey being captured by predators. Until recently, the chemical cues detected by *Daphnia* were assumed to be of predator origin (*i.e.* kairomones), since juices of crushed *Daphnia* had been found ineffective in behaviour experiments. Similar to the present case with lower vertebrates, a search in the literature revealed that in general, predators used as donors of the active chemical cues had been fed *Daphnia*. Accordingly, a hypothesis evolved suggesting that latent alarm signals could possibly be present in *Daphnia*, to be activated in predators following ingestion (Stabell et al., 2003).

To test this hypothesis, fish predators (threespine sticklebacks) were initially fed earthworms for ten weeks to remove *Daphnia* remains from their gastrointestinal tracts. When the fishes were subsequently fed either earthworms or alternatively *Daphnia* for another six days, morphological changes were induced in *D. galeata* with water conditioned by fish fed *Daphnia*, but not with water conditioned by fish fed earthworms. Further, extracts made from intestines of earthworm-fed fish, homogenised with earthworms, gave no morphological changes, but intestines of the same origin homogenised with *Daphnia* did. Similar results were found when earthworms and *Daphnia* were homogenised with fish liver. In both cases, the homogenates were left at room temperature for at least 2.5 h before preparation of extracts, allowing enzymes to be acting in the homogenates. When extracts made of freshly homogenised *Daphnia* were tested, no detectable changes were found in the first instar stage of test animals, whereas extracts made of homogenised *Daphnia* that had been kept at room temperature induced such changes. From the second instar stage, no differences were obtained with regard to previous treatment of homogenates, suggesting that an activation of the extract made from fresh *Daphnia* homogenates had later taken place in the water.

The result obtained with *Daphnia* revealed that the chemical cues in question do not stem from predators. The active components were suggested to be pheromones that have their origin in conspecific prey, but are present in the prey in an inactive form, to be activated during the passage through the intestine of a predator. Intestinal or bacterial enzymes in predators may be the probable agents responsible for signal activation, explaining why the process may also take place in the water. Accordingly, latent alarm signals (*i.e.* 'dormant' pheromones) seem to be a reality in aquatic invertebrates, and may well be present also in vertebrates.

## 3. WHAT IS REVEALED FROM THE LITERATURE?

Concerning rog larvae, Pfeiffer (1963, 1966) tested alarm responses in tadpoles of *Rana esculenta*, *R. temporaria*, *R. pipiens*, *Hyla arborea*, and *Xenopus laevis* to skin extracts of conspecifics, and concluded that fright reactions were not present in these species. Skin extract of *R. temporaria* was also negative in tests with toad (*Bufo bufo*) tadpoles, and *R. temporaria* tadpoles did not respond to skin extract of toad tadpoles. These results made him suggest that non-bufonids lack both the alarm substance and the ability to respond to the alarm substance.

However, some confusion has later developed as for the presence of alarm substances in frog larvae, due to the fact that an increased behavioural activity was reported in tadpoles of the Cascades frog (*R. cascadae*) resulting from exposure to extracts of conspecifics (Hews and Blaustein, 1985). Hokit and Blaustein (1995), however, later rejected the findings in *R. cascadae* and pointed out that performing tadpoles were even observed feeding on dead and injured conspecifics, suggesting that the increased activity previously measured may actually have been a feeding response. On the other hand, Wilson and Lefcort (1993) reported a reduced activity in tadpoles of the red-legged frog (*R. aurora*) resulting from exposure to macerated tadpoles. Early metamorphosis and small metamorphic size, resulting from long-term exposure to injured conspecifics, has also been reported in tadpoles of that same species (Kiesecker et al., 2002). Since bacteria in the water may activate possible 'latent' alarm signals present in homogenates, an influence of conspecific extracts on the long-term development in frog larvae is not surprising. An instant behavioural response in frog tadpoles to extract of conspecifics seems, however, more questionable unless tadpole homogenates have been subjected to incubation before use. Such occurrence could possibly be the case in the report by Wilson and Lefcort (1993), since an intermediate response was found to tadpole extract compared to water conditioned by tadpole-fed newts and water conditioned by insect-fed newts, being significantly different from both predator treatments.

Concerning tadpole responses to predator cues, the prey species in a study has mainly been used also as food for the predators, as previously mentioned (Table 1). The necessity for predators to feed on tadpoles to induce a behavioural response was suggested by Wilson and Lefcort (1993), who found that tadpoles of the red-legged frog (*R. aurora*) reduced their activity in response to predators feeding on conspecifics, but not when predators were feeding on insects. The need for feeding predators with the prey species of study has later been confirmed in studies of behaviour and life-history in tadpoles of the common frog, *R. temporaria* (Laurila et al., 1997, 1998), and studies of life-history in tadpoles of *R. aurora* (Kiesecker et al., 2002). Similar to *Daphnia*, a possibility also seems to exist for potential alarm signals in anuran tadpoles to adopt different signal properties following passage through the gut of the various predator species. Such a possibility is revealed from the report by Van Buskirk (2001), tentatively suggesting that tadpoles distinguished between different predators during development. Taken together, therefore, it seems that the literature gives a fair reason to propose that latent alarm signals are present also in vertebrates.

## 4. THE PRELIMINARY DATA

### 4.1. Morphology and Growth

Tadpoles of the red-legged frog were individually exposed to various tissue extracts and predator cues for a period of one month. Adult rough-skinned newts (*Tarica granulosa*) were used as predators, and both predator-conditioned waters and tissue extracts were introduced once a day to the boxes containing tadpoles of each series. No changes in morphology were obtained in this study, but changes in growth rates were found for several parameters. Data for percent increase in body weight and tail length are given in Table 2. Although a general trend emerged, no significant differences were found between waters from predators fed tadpoles and from predators fed control feed (*i.e.* earthworms). This may partly be due to the fact that predators had been fed tadpoles in the lab beforehand, but also that earthworm extract alone induced a diminished growth in tadpoles, suggesting earthworms as an unsuitable control feed in studies with tadpoles. However, another interesting feature was revealed. Water from newts fed tadpoles and extract of homogenised tadpoles both resulted in a growth significantly different from that of control treatment, but in opposite directions. Since the intestine of newts must be suspected of containing a different bacterial flora compared to the one present in water containing tadpoles, these results seem to fit in with a conclusion involving latent alarm signals.

**Table 2.** Changes in two growth parameters, measured for tadpoles of the red-legged frog (*R. aurora*) exposed individually to different treatments for one month[1]

| Treatment | % body weight increase (± SD) [2] | % tail length increase (± SD) [2] |
|---|---|---|
| Tadpole-fed newts | 312.6 ± 91.6 [a] | 101.9 ± 27.1 [a] |
| Earthworm-fed newts | 269.6 ± 139.6 [a,b] | 87.4 ± 33.9 [a,b] |
| Crushed tadpoles | 169.8 ± 90.6 [c] | 65.2 ± 22.9 [c] |
| Crushed earthworms | 187.6 ± 79.4 [c,d] | 67.5 ± 19.2 [b,c] |
| Untreated water | 236.4 ± 84.6 [b,d] | 80.2 ± 24.1 [b,c] |

[1] Data obtained from unpublished work by O.B. Stabell, E.L. Wildy, A.C. Hatch, and A.R. Blaustein.
[2] Means that were not significantly different at p<0.05 (Mann-Whitney U-test) are identified with similar superscript letters. (n = 20).

### 4.2. Behaviour

Tadpoles of the common frog have been individually tested in 35 x 27 cm polypropylene trays, containing 1 litre of water (1.5 cm depth). Six animals were tested in parallel, and the behaviour was monitored from above with a video camera. For each stimulus, 2 ml of aqueous solution was introduced to one end of each tray at start, and the behaviour was followed for 10 min. The test sequence was then repeated with six new tadpoles, with the inlet side of trays switched. Perch (*Perca fluviatilis*) were used as predators, and zebrafish (*Brachydanio rerio*) were used as control feed. The results for the two parameters, 'time spent on stimulus side' and 'time active', are given in Table 3.

Extracts were made from homogenates of tadpoles and zebrafish, aged at room temperature for 7-8 hrs before filtering and freezing. As shown in Table 3, aged extract of tadpoles released avoidance behaviour in conspecifics, similar to that found for water conditioned by a tadpole-fed predator. The activity, however, also decreased, but not at a significant level. Again, this may possibly result from differences in bacterial flora between predators and prey, with a resulting variation in the active compounds due to different chemical pathways followed during signal activation. In summary, however, these results clearly suggest that latent alarm signals, possessing behavioural releaser effects following activation, are present in tadpoles of the common frog.

### 4.3. Life-history

Tadpoles of the common frog have also been individually tested in work involving life-history strategies. Series of 20 tadpoles were exposed to various treatments until

**Table 3.** Behavioural responses of tadpoles of the common frog (*R. temporaria*) exposed individually to different treatments[1]

| Treatment | % of time (± SD) on stimulus side[2] | % of time (± SD) active[2] |
|---|---|---|
| Zebrafish-fed perch | 69.7 ± 17.1 [a] | 32.0 ± 13.9 [a,c] |
| Tadpole-fed perch | 54.7 ± 14.4 [b] | 21.5 ± 9.9 [b] |
| Crushed and aged tadpole extract | 57.5 ± 13.8 [b] | 30.8 ± 12.4 [b,c] |
| Crushed and aged zebrafish extract | 66.0 ± 12.1 [a] | 35.1 ± 12.6 [a,c] |

[1] Data obtained from unpublished work by O.B. Stabell.
[2] Means that were not significantly different at p<0.05 (Paited T-test, one-tailed) are identified with similar superscript letters. n=12.

Table 4. Life-history responses measured in tadpoles of the common frog (*R. temporaria*) exposed individually to different treatments[1]

| Treatment | Weigth in mg (± SD) as metamorphs[2] | Days (± SD) to reach metamorphosis[2] |
|---|---|---|
| Cichlid fed pellet sticks | 210 ± 25.3 [a] | 32 ± 2.7 [a] |
| Cichlid fed tadpoles | 166 ± 29.4 [b] | 34 ± 4.1 [b] |
| Extract of crushed tadpoles | 156 ± 25.5 [b] | 35 ± 3.5 [b] |
| Extract of crushed pellet sticks | 226 ± 29.7 [a] | 28 ± 2.1 [a] |

[1] Data obtained from unpublished work by O.B. Stabell.
[2] Means that were not significantly different at $p<0.05$ (Pared T-test, one-tailed) are identified with similar superscript letters. n=20.

reaching metamorphosis, and were then measured by weight. Texas cichlids (*Cichlastoma zonatum*), to be used as predators, were fed either tadpoles or pellets sticks, and extract of tadpoles and pellet sticks were used as treatments for comparison. Extracts and predator waters were added daily to the boxes containing tadpoles of each series. The results obtained are given in Table 4. Similar to findings reported for *R. aurora* by Kiesecker at al. (2002), tadpoles of *R. temporaria*, exposed to either water from tadpole-fed fish or extract of tadpoles, were found smaller in weight at metamorphosis compared to their respective controls. Tadpoles of *R. temporaria* exposed to conspecific cues, however, were using longer time to reach metamorphosis than those exposed to control treatments. This latter result is contrary to what was reported in *R. aurora* by Kiesecker et al. (2002), suggesting that species differences may be present, either in signal properties or in the bacterial flora of their guts.

## 5. CONCLUSIONS

Latent alarm signals seem to be present in vertebrates, especially in tadpoles of frogs, and the existence of such signals may explain some basic evolutionary questions in the relationships between predators and prey. The presumable inactive compounds in question may be activated by the bacterial flora within the gut of predators, explaining why taxonomically different predators may release diverse responses within one single prey species. Knowing the exact source of the alarm signals will be of utmost importance for future work within predator-prey interactions.

## 6. REFERENCES

Brown, W. L., Eisner, T., and Whittaker, R. H., 1970, Allomones and kairomones: transspecific chemical messengers, *Bioscience* **20**:21.

Brönmark, C., and Miner, J. G., 1992, Predator-induced phenotypic change in body morphology in Crucian carp, *Science* **258**:1348.

Chivers, D. P., and Smith, R. J. F., 1998, Chemical alarm signalling in aquatic predator-prey systems: A review and prospectus, *Écoscience* **5**:338.

Chivers, D. P., Kiesecker, J. M., Marco, A., Wildy, E. L., and Blaustein, A. R., 1999, Shifts in life history as a response to predation in western toads (*Bufo boreas*), *J. Chem. Ecol.* **25**:2455.

DeVito, J., Chivers, D., Kiesecker, J., Marco, A., Wildy, E., and Blaustein, A., 1998, The effects of snake predation on metamorphosis of western toads, *Bufo boreas* (Amphibia, Bufonidae), *Ethology* **104**:185.

DeVito, J., Chivers, D. P., Kiesecker, J. M., Belden, L. K., and Blaustein, A. R., 1999. Effects of snake predation on aggregation and metamorphosis of Pacific treefrog (*Hyla regilla*) larvae, *J. Herpetol.* **33**:504.
Hews, D. K., and Blaustein, A. R., 1985, An investigation of the alarm response in *Bufo boreas* and *Rana cascadae* tadpoles, *Behav. Neur. Biol.* **43**:47.
Hokit, D. G., and Blaustein, A. R., 1995, Predator avoidance and alarm-response behaviour in kin-discriminating tadpoles (*Rana cascadae*), *Ethology* **101**:280.
Karlson, P., and Lüscher, M., 1959, 'Pheromones': a new term for a class of biological active substances, *Nature* **183**:55.
Kats, L. B., and Dill, L. M., 1998, The scent of death: Chemosensory assessment of predation risk by prey animals, *Écoscience* **5**:361.
Kiesecker, J. M., Chivers, D. P., Anderson, M., and Blaustein, A. R., 2002, Effect of predator diet on life history shifts of red-legged frogs, *Rana aurora*, *J. Chem. Ecol.* **28**:1007.
Lardner, B., 1998, Plasticity or fixed adaptive traits? Strategies for predation avoidance in *Rana arvalis* tadpoles, *Oecologia* **117**:119.
Lardner, B., 2000, Morphological and life history responses to predators in larvae of seven anurans, *Oikos* **88**:169.
Laurila, A., 2000, Behavioural responses to predator chemical cues and local variation in antipredator performance in *Rana temporaria* tadpoles, *Oikos* **88**:159.
Laurila, A., Kujasalo, J., and Ranta, E., 1997, Different antipredator behaviour in two anuran tadpoles: Effects of predator diet, *Behav. Ecol. Sociobiol.* **40**:329.
Laurila, A., Kujasalo, J., and Ranta, E., 1998, Predator-induced changes in life history in two anuran tadpoles: effects of predator diet, *Oikos* **83**:307.
Manteifel, Y., 1995, Chemically-mediated avoidance of predators by *Rana temporaria* tadpoles, *J. Herpetol.* **29**:461.
Mathis, A., and Smith, R. J. F., 1993, Chemical labeling of northern pike (*Esox lucius*) by the alarm pheromone of fathead minnows (*Pimephales promelas*), *J. Chem. Ecol.* **19**:1967.
McCollum, S. A., and Leimberger, J. D., 1997, Predator-induced morphological changes in an amphibian: predation by dragonflies affects tadpole shape and color, *Oecologia* **109**:615.
Petranka, J. W., Kats, L. B., and Sih, A., 1987, Predator-prey interactions among fish and larval amphibians: use of chemical cues to detect predatory fish, *Anim. Behav.* **35**:420.
Pfeiffer, W., 1963, Alarm substances, *Experientia (Basel)* **19**:113.
Pfeiffer, W., 1966, Die Verbreitung der Schreckreaktion bei Kaulquappen und die Herkunft des Schreckstoffes, *Z. vergl. Physiol.* **52**:79.
Pfeiffer, W., 1974, Pheromones in Fish and Amphibia, in: *Pheromones*, M.C. Birch, ed., North Holland Publishing Company, Amsterdam, pp. 269-296.
Pfeiffer, W., 1977, The distribution of flight reaction and alarm substance cells in fishes, *Copeia* **1977(4)**:653.
Schutz, F., 1956, Vergleichende Untersuchungen über die Schreckreaction bei Fischen und deren Verbreitung, *Z. vergl. Physiol.* **38**:84.
Skelly, D., and Werner, E., 1990, Behavioral and life-historical responses of larval American toads to an odonate predator, *Ecology* **71**:2313.
Stabell, O. B., and Lwin, M. S., 1997, Predator-induced phenotypic changes in crucian carp are caused by chemical signals from conspecifics, *Env. Biol. Fishes* **49**:145.
Stabell, O. B., Ogbebo, F., and Primicerio, R., 2003, Inducible defences in *Daphnia* depend on latent alarm signals from conspecific prey activated in predators, *Chem. Senses* **28**:141.
Stauffer, H. P., and Semlitsch, R. D., 1993, Effects of visual, chemical and tactile cues of fish on the behavioural responses of tadpoles, *Anim. Behav.* **46**:355.
Van Buskirk, J., 2001, Specific induced responses to different predator species in anuran larvae, *J. Evol. Biol.* **14**:482.
Van Buskirk, J., and Relyea, R. A., 1998, Selection for phenotypic plasticity in *Rana sylvatica* tadpoles, *Biol. J. Linn. Soc.* **65**:301.
von Frisch, K., 1941, Über einen Schreckstoff der Fischhaut und seine biologishe Bedeutung, *Z. vergl. Physiol.* **29**:46.
Wilson, D. J., and Lefcort, H., 1993, The effect of predator diet on the alarm response of red-legged frog, *Rana aurora*, tadpoles, *Anim. Behav.* **46**:1017.
Wilson, E. O., and Bossert, W. H., 1963, Chemical communication among animals, *Rec. Prog. Horm. Res.* **19**:673.

# BLOOD IS NOT A CUE FOR POSTSTRIKE TRAILING IN RATTLESNAKES

Tamara L. Smith and Kenneth V. Kardong[*]

## 1. INTRODUCTION

Rattlesnakes in the wild usually strike, envenomate, and release rodents immediately (Klauber, 1956) avoiding potential injury from retaliation, but this may allow the struck rodent to scamper some distance from the site of initial envenomation while the cocktail of venom components immobilizes and eventually kills the rodent. Because the envenomated rodent often breaks visual contact, relocation of the dispatched prey may rely upon chemosensory cues emitted by the struck prey. On theoretical and on experimental grounds, these chemosensory cues may be carried in the blood. Theoretically, the strike includes deep penetration of the fangs (Kardong and Bels, 1998), which may carry away distinctive chemical cues used next to relocate the envenomated prey. Experimental trials with blood suggest that it prompts elevated interest in colubrid snakes (Chiszar et al., 1992a) and prompts high-rates of tongue flicking in rattlesnakes (Chiszar et al., 1993b).

Consequently, the purpose of our experiments was to see if blood *alone* carried chemical cues used during poststrike trailing by the rattlesnake to relocate its envenomated prey.

## 2. MATERIALS AND METHODS

Twenty-three individually housed northern Pacific rattlesnakes, *Crotalus viridis oreganus* (adult, long-term captives) collected locally in Whitman Co., WA, under State permits, were used in each of four experiments. Snakes were maintained on white laboratory mice (Balb/c or Swiss Webster), fed twice a month, and provided water *ad*

---

[*] Tamara L. Smith, School of Biological Sciences, and the Center for Teaching, Learning, and Technology, Washington State University, Pullman, WA 99164-4236. Kenneth V. Kardong, School of Biological Sciences, Washington State University, Pullman, WA 99164-4236.

*libitum*. All mice were fed Harlan Teclad 8640 Rodent Diet mouse chow, and kept on the same bedding of hard wood shavings. Safety procedures for snakes generally followed those of Gans and Taub (1964).

Trials were conducted in an arena (1.25 m each side) described elsewhere in detail (Alving and Kardong, 1996; Lavín-Murcio et al., 1993; Robinson and Kardong, 1991). Temperature in the test arena room was held between 25-30 C°. A Y-shaped outline made of black tape was placed on the floor of the arena and covered with a new piece of white butcher paper before each trial. Hatch-marks were placed perpendicular to the main Y-outline at 10 cm increments. The Y-outline, a 40 cm base and 40 cm each arm, could be seen through the white paper and was used to guide the placement of the paired scent trails. Before each trial, a snake was removed from its home cage and placed in a holding box stationed at the beginning of the Y-maze for a period of time no shorter than 12 hours, including overnight.

*Treatment 1* – (Struck). Each snake was presented a choice of two trails, one of a struck mouse, the other of a water trail. At the onset of all trials, the room lights were dimmed and a removable chute was placed in front of the holding box. A pre-weighed mouse, secured with fishing line by its tail, was introduced down the chute to the snake. After the strike, this mouse was removed and used to make the prey odor trail. Pairs of non-overlapping scent trails were made for each trial immediately after the strike. First, de-mineralized water was used to make the control trail applied to the maze with a cotton-tipped applicator along the base and out one arm. Next, while being held in forceps by the nape of the neck, the struck mouse, belly-side in contact with the paper, was moved along the main branch, taking care not to overlap the water trail, and out the other arm of the Y-maze.

*Treatment 2* – (Unstruck). The two trail choices consisted of an unstruck mouse versus water, using methods as in Treatment 1, except the mouse trail was made with an unstruck mouse. After making the trail with the unstruck mouse, the snake was only then allowed to strike this same mouse. (The strike is necessary to release poststrike trailing behavior (Chiszar et al., 1992b; Smith et al., 2000).)

*Treatment 3* – (Unstruck plus blood). The two trail choices consisted of unstruck mouse odor out both arms, but blood taken from the struck mouse (see next for technique) was stroked over one of the unstruck choice trails. To do this, the unstruck mouse was slid along the base of the Y-maze and out one arm; then this same mouse was slid along the base of the Y-maze and out the other arm. This mouse was next presented to the snake and struck. After death, blood was drawn from it by cardiac puncture, placed on a cotton-tipped swab, and stroked over one of the two previously laid unstruck odor trails.

*Treatment 4* – (Blood). The same group of snakes was presented with two trail choices, the first made with distilled water and the second made with blood drawn from the struck mouse. Blood was taken from the dead mouse just struck by the snake (about 2 min before) by cardiac puncture and then deposited onto a cotton-tipped applicator for making the trail. To prevent transfer of integumentary cues, the needle used in cardiac puncture was removed before loading collected blood from the syringe onto the cotton-tipped applicator.

Immediately (<2 min.) after laying the trails in all treatments, the door to the holding box was removed and all ensuing behaviors were recorded via a Panasonic camera capable of filming under low light. The experimenter stepped out of view and recorded tongue-flicks from a monitor. Additional data were collected later from replay of the videotape.

The snake was considered to be following a trail if its head stayed within the 10-cm guidelines placed on either side of the black tape Y-maze trail. If the snake went outside these guidelines for over 30 seconds or if it did not leave the holding box within the 20 minute trial period, the snake was scored as out of bounds (OB) or not trailing (NT), respectively. Three main variables were scored: CHOICE, whether or not the snake followed an arm of the Y-maze to completion; TRAIL DISTANCE, the distance traveled along the Y-maze expressed as a percentage of the number of hatch marks crossed during the trailing episode out of the total of nine, evenly spaced hatch marks; EMERGERTF, rate of tongue-flicking per minute immediately upon emergence from the holding box. Results were analyzed using Statmost (Datamost Corp.) statistical software package. Appropriate nonparametric statistical tests were used (McNemar, Wilcoxon signed-rank tests, Krustal-Wallis). These protocols met guidelines for animal care and were approved by the local IACUC.

## 3. RESULTS

For Treatment 1 (struck mouse versus water), all snakes (23/23) trailed and exhibited characteristic poststrike chemosensory trailing, showing sustained high rates of tongue-flicking (RTF) and movement consistent with strike-induced chemosensory searching (SICS) (Chiszar et al., 1977; Chiszar et al., 1983; Chiszar et al., 1992b); for a review, (Gillingham and Clark, 1981; Golan et al., 1982; Haverly and Kardong, 1996). Having trailed 100% of the time in all trials, TRAIL DISTANCE averaged 100% (9/9 hatch marks) (Figure 1). Trailing of the prey odor trail was close and continuous. The average EMERGRTF was 77.9 tf/min. Total time of trailing averaged 127.6 seconds for 23 trials.

For Treatment 2 (Unstruck versus water), snakes trailed the mouse odor 73.9% (17/23), none of the snakes trailed the water trail; 21.7% (5/23) snakes were scored as out-of-bounds (OB), and one (4.4%) snake failed to leave the holding box and was scored as not trailing (NT). For all trials including OB and NT, TRAIL DISTANCE averaged 88.4% (8.8 hatch-marks). EMERGERTF was 80.8 tf/min. For those trials that the snake completed trailing (17) the average time of trailing was 194.1 seconds.

For Treatment 3 (Unstruck versus unstruck plus blood), snakes trailed 65.2% (15/23), although 8 of these were out the unstruck mouse odor trail (no overlaying blood); 7 were of the unstruck mouse odor trail (overlaid with blood). For all trials, TRAIL DISTANCE averaged 74.4 (6.6 hatch-marks). EMERGERTF was 83.3 tf/min. For those trials that the snake completed trailing (15) the average time of trailing was 106.1 seconds.

For Treatment 4 (blood versus water), none out of the 23 (0%) snakes successfully trailed along the main branch and out one arm to completion. In six of the trials (26.1%), the snakes moved very little out of the holding box and were scored as not trailing (NT). In the other 17 trials (73.9%), the snakes left the holding box and imprecisely trailed along the main branch, but did not reach the intersection before veering completely off and abandoning trailing, scored as out-of-bounds (OB). For all trials of blood cues,

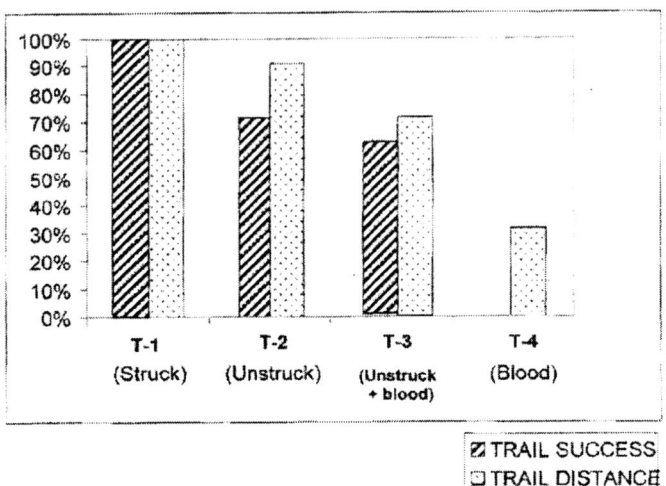

**Figure 1.** Trailing. Trail Success (CHOICE): whether or not the snake followed an arm of the Y-maze to completion, expressed as a percentage of average overall success for all trials. TRAIL DISTANCE: the distance traveled along the Y-maze expressed as percentage of the number of hatch marks crossed during the trailing episode out of the total hatch marks possible (9). Treatment 1 (T-1), Treatment 2 (T-2), Treatment 3 (T-3), Treatment 4 (T-4). Both trailing success and distance were significantly different in Treatment 1 compared to the other three treatments.

TRAIL DISTANCE averaged 30.9% (3.5 hatch-marks). EMERGERTF was 73.0 tf/min for those snakes where it was possible to record this rate (n= 20 trials). Since no snake completed trailing, there were no total times of trailing.

CHOICE was significantly different between Treatment 1 (Struck) and Treatment 2 (Unstruck) (100% vs. 73.9, $X^2 = 4.17$, P= 0.04) and between Treatment 1 (Struck) and Treatment 3 (Unstruck plus blood) (100% vs. 65.2%, $X^2 = 6.12$, P= 0.01). TRAILING DISTANCE was also significantly different between Treatment 1 (Struck) and both Treatment 2 (Unstruck) (Z= 2.2, P= 0.028) and Treatment 3 (Unstruck plus blood) (Z=2.5, P=0.01). Both CHOICE and TRAILING DISTANCE were significantly higher in all treatments compared to Treatment 4 (blood trail) (Z=3.97, P>0.001). However, there was no significant difference in CHOICE ($X^2$ =0.125, P=0.72) or in TRAILING DISTANCE (Z=1.24, P=0.21) between Treatment 2 (Unstruck) and Treatment 3 (Unstruck plus blood). The average rates of tongue flicking (RTF) upon first emergence from the holding box were significantly elevated above prestrike values (78.7 EMERGERTF up from 0.36 RTF prestrike) indicating that in all treatments the strike had released normal SICS behavior. Across the four treatments, the elevated EMERGERTF were not significantly different (H=5.10, P=0.16).

Overall, rattlesnakes did not follow the blood trail. In some trials with blood, snakes began to trail, exhibiting a predatory interest and SICS, but did not completely follow a blood (or water) trail to its end. TRAIL DISTANCE was used to express this level of trailing. Snakes completed significantly more trailing distance when the odor of a struck mouse was used than in any other treatment. This was especially true compared to the blood trail, wherein snakes, on average, ventured only about a third of the way along the trail. Note that adding blood to an unstruck mouse odor trail (Treatment 3) did not improve rattlesnake performance compared to the unstruck mouse odor trail only (Treatment 2). In fact, rattlesnakes in Treatment 3, when choosing an odor trail, made no significant discrimination between the two choices, (unstruck trail, n=8; unstruck plus blood, n=7). For total time of trailing, snakes trailed faster in Treatment 1 (Struck) trials than in Treatment 2 (unstruck) (127.6 sec. versus 194.1 sec., z-score = 2.44, p-value = 0.015, Wilcoxon sign-rank).

## 4. CONCLUSIONS

First, although the rattlesnakes in our study exhibited SICS, as defined by increased RTF and movement (Chiszar et al., 1977), they did not follow a poststrike odor trail made using only blood. This was not likely a consequence of low blood concentration along the trail. The blood trails presented to snakes were often still very wet when the snake first exited the holding box, whereby the snake body actually smeared the blood trail over which it had moved.

As in previous studies (Chiszar et al., 1983; Chiszar et al., 1992b; Smith and Kardong, 2000), we found a statistically higher EMERGERTF after the strike. However, in all four treatments, the elevated EMERGERTF's were statistically equivalent. Past studies showed increased RTF and other behaviors related to SICS when presented general biological materials (Chiszar et al., 1999a; Chiszar et al., 1992b; Chiszar et al., 1983; Chiszar et al., 1981; Chiszar et al., 1991; Chiszar et al., 1982). Other studies have also shown that snakes exhibit moderate rates of tongue-flicking, prestrike, when presented with biological odors of interest (Smith et al., 2000). Rattlesnakes relying on ambush for prey capture must first locate an area of high prey density to set up – sit-and-wait (Duvall et al., 1990; Duvall et al., 1985). Certainly in the wild, rattlesnakes use general chemosensory information prestrike to locate habitats occupied by rodents and there wait in ambush (Duvall et al., 1985) so that SICS in response to blood is not surprising. But it is important to note that SICS found in our study were elicited by the strike, and that blood added (or subtracted) nothing significant to (or from) the poststrike, RTF. As significant amounts of blood are very seldom lost by an envenomated mouse (Kardong, 1986), a blood trail in the wild would be an infrequent and quite unreliable consequence of envenomation. Therefore, it is not surprising that northern Pacific rattlesnakes, *Crotalus viridis oreganus*, do not use blood, alone, as a poststrike cue for successful chemosensory trailing.

Second, envenomation does produce a chemical cue that is formed quickly after the strike and immediately secreted from the rodent into the environment to produce an odor trail followed poststrike by the rattlesnake (Chiszar et al., 1992b; Kardong, 2001). Poststrike, rattlesnakes can follow an odor trail produced by an unstruck mouse (Treatment 2). However, the poststrike trailing success of rattlesnakes significantly improves if provided with the odor trail of a *struck* mouse (Treatment 1 compared to Treatment 2). Whatever this strike-induced odor cue might be, it is not carried in the blood to surface sites on the rodent for release to the environment.

Third, blood odor added to the odor of an unstruck mouse does not increase the poststrike trailing success of rattlesnakes (Treatment 3 compared to Treatment 2).

Essentially, poststrike trailing rattlesnakes ignored the blood cue, as they exhibited no preference for it over an unstruck mouse odor trail (Treatment 3). Therefore blood odor and mouse odor do not interact in a synergistic way to produce a scent trail of increased perceptibility for the rattlesnake.

Fourth, the chemical cue reaching the environment and producing the chemosensory image followed by the rattlesnake may be binary, composed of two or more necessary chemical principles presented simultaneously or in temporal sequence. For example, the binary cues used for poststrike trailing might involve integumentary (individual mouse odor) cues together with some other component related to envenomation. Successful poststrike trailing is a multiple step process entailing the delivery of an accurate strike, initially locating the beginning of the odor trail, selecting the trail of the struck mouse over other odors, and following the correct trail in the correct direction to completion. Thus multiple cues--multiple classes of cues--may be used at different stages of the entire trailing episode. Viewed as a binary system, our results could be interpreted as follows:

The increased EMERGERTF we observed in treatment 4 might be a normal response under poststrike conditions, picking up a general biological odor such as mouse scent. However an additional cue was unavailable—the odor of the individual struck mouse. Individual mouse odor, of this binary system, would be obligatory and the first chemosensory cue. The second chemosensory cue, induced by the strike/envenomation, serves to enhance or produce the odor for specific (Chiszar et al., 1992b; Chiszar et al., 1999b) discriminatory trailing. In our trials, rattlesnakes successfully followed an odor trail of the unstruck mouse (although at lower rate of success then of struck mice), suggesting that the cue provided by envenomation is not always necessary. The strike is necessary to release poststrike trailing, but so initiated, rattlesnakes can distinguish mouse odor trails even between littermates (Chiszar et al., 1992b). Cues provided by envenomation might only play a role when discriminating two closely related trails such as from the same mouse, before and after being struck. However, if individual mouse odor is absent, then an important part of the chemosensory binary image is absent.

In addition to odors picked up during contact with the rodent, the *specific* chemosensory cue(s), used for trailing, might also include a chemical principle generated by the strike. Envenomation may stimulate the rodent to release some type of alarm pheromone, secreted into the air and/or deposited on the substrate over which the struck rodent runs. Along with the odor of the mouse, rattlesnakes sample this cue via tongue-flicking and deploy this cue as a kairomone (chemicals used advantageously by the receiver [predator] than the emitter [rodent]) (Weldon, 1980) to successfully discriminate the particular trail of the struck rodent. The binary chemosensory cues would therefore include one part identifying the particular mouse (picked up during the strike) and a second part indicating envenomation (kairomone). Since the blood contained neither of these binary cues, this would account for why blood trails were not followed successfully. This also explains discrepancies between our results with blood (no trailing interest) and the results of others (tongue-flick interest) (Chiszar et al., 1993b). Different cues are emphasized during different predatory phases (Kardong, 1992; Kardong, 2001). Although blood may be biologically significant for snakes during other phases in their predatory behavior (Chiszar et al., 1993a), it may not have any biological significance during the poststrike trailing phase because it lacks chemical cues with information about the specific prey struck and about the success of envenomation.

In summary, although snakes showed SICS, poststrike, consistent with results from previous studies, this increased predatory interest did not lead to successful poststrike trailing of a blood trail. Therefore, whatever the chemical cue(s) used to trail poststrike, these are not carried in the blood to sites of surface release into the environment. Further, rattlesnakes do not use blood, alone, as a cue for poststrike trailing. We suggest that the chemosensory cue used in successful poststrike trailing is a binary chemosensory cue. This is consistent with the likely environmental chemosensory image produced by an envenomated rodent.

## 5. ACKNOWLEDGMENTS

We thank Paul A. Verrell for his generous help and loan of equipment; M. Rockwell Parker; and Doug Baker, who was especially helpful in supplying animal care.

## 6. REFERENCES

Alving, R. W., and Kardong, K. V., 1996, The role of the vomeronasal organ in rattlesnake (*Crotalus viridis oreganus*) predatory behavior, *Brain Behav. Evol.* 48:165-172.

Chiszar, D., Fox, K., and Smith, H. M., 1992a, Stimulus control of predatory behavior in the brown tree snake (*Boiga irregularis*), *Behav. Neural Biol.* 57:167-169.

Chiszar, D., Lee, R. K. K., Radcliffe, C. W., and Smith, H. M., 1992b, Searching behaviors by rattlesnakes following predatory strikes, in: *Biology of the Pit Vipers*, J. A. Campbell and E. D. Brodie, eds., Selva, Tyler, pp. 369-382.

Chiszar, D., Dunn, T., and Smith, H. M., 1993a, Response of brown tree snakes (*Boiga irregularis*) to human blood, *J. Chem. Ecol.* 19:91-96.

Chiszar, D., Hobika, G., and Smith, H. M., 1993b, Prairie rattlesnakes (*Crotalus viridis*) respond to rodent blood with chemosensory searching, *Brain Behav. Evol.* 41:229-233.

Chiszar, D., Radcliffe, C. W., Feiler, F., and Duvall, D., 1983, Strike-induced chemosensory searching by rattlesnakes: The role of envenomation-related chemical cues in the post-strike environment, in: *Chemical Signals in Vertebrates*, D. Müller-Schwarze and R. M. Silverstein, eds., Plenum Press, New York, pp. 1-24.

Chiszar, D., Radcliffe, C. W., O'Connell, B., and Smith, H. M., 1981, Strike-induced chemosensory searching in rattlesnakes (*Crotalus viridis*) as a function of disturbance prior to presentation of rodent prey, *Psychol. Rec.* 32:57-62.

Chiszar, D., Radcliffe, C. W., and Scudder, K., 1977, Analysis of the behavioral sequence emitted by rattlesnakes during feeding episodes. I. Striking and chemosensory searching. *Behav. Biol.* 21, 418-425.

Chiszar, D., Radcliffe, C. W., Smith, H. M., and Langer, P., 1991, Strike-induced chemosensory searching: Do rattlesnakes make one decision or two? *Bull.Maryland Herpet. Soc.* 27:90-94.

Chiszar, D., Smith, H. M., and Hoge, A. R., 1982, Post-strike trailing behavior in rattlesnakes, *Mem. I. Butantan* 46:195-206.

Chiszar, D., DeWelde, K., Garcia, M., Payne, D., and Smith, H. M., 1999a, Strike-induced chemosensory searching (SICS) in northern Pacific rattlesnakes (*Crotalus viridis oreganus*, Holbrook, 1840) rescued from substandard husbandry conditions. II. Complete recovery of function after two years, *Zoo Bio.* 18:141-146.

Chiszar, D., Walters, A, Urbaniak, J., Smith, H. M., and Mackessy, S. P., 1999b, Discrimination between envenomated and nonenvenomated prey by Western diamondback rattlesnakes (*Crotalus atrox*): Chemosensory consequences of venom, *Copeia* 1999:640-648.

Duvall, D., J. Goode, J., Hayes, W. K., Leonhardt, J. K., and Brown, D. G., 1990, Prairie rattlesnake vernal migration: Field experimental analysis and survival value, *Nat. Geographic Res.*, 6:457-469.

Duvall, D., King, M. B., and Gutzwiller, K. J., 1985, Behavioral ecology and ethology of the prairie rattlesnake, *Nat. Geographic Res.* 1:80-111.

Gans, C., and Taub, A., 1964, Precautions for keeping poisonous snakes in captivity, *Curator* 7:196-205.

Gillingham, J. C., and Clark, D. L., 1981, An analysis of prey searching behavior in the western diamondback rattlesnake (*Crotalus atrox*), *Behav. Neural Bio.* 32:235-240.

Golan, L., Radcliffe, C. W., Miller, T., O'Connell, B., and Chiszar, D., 1982, Trailing behavior in prairie rattlesnakes (*Crotalus viridis*), *J. Herpet.*, 16:287-293.

Haverly, J. E., and Kardong, K. V., 1996, Sensory deprivation effects on the predatory behavior of the rattlesnake, *Crotalus viridis oreganus*, *Copeia* **1996**:419-428.

Kardong, K. V., 1986, Predatory strike behavior of the rattlesnake, *Crotalus viridis oreganus*, *J. Comp. Psych.* **100**:304-313.

Kardong, K. V., 1992, Proximate factors affecting guidance of the rattlesnake strike, *Zool. J. Anat.* **122**:233-244.

Kardong, K. V., and Bels, V. L., 1998, Rattlesnake strike behavior: kinematics, *J. Exp. Biol.* **201**:837-850.

Kardong, K. V., and Bels, V. L. 2001, Functional morphology and evolution of the feeding apparatus in squamates, in: *Vertebrate Functional Morphology*, H. M. Dutta and J. S. D. Munshi, eds., Science Publishers, Inc., Enfield, pp. 173-219.

Klauber, L. M., 1956, *Rattlesnakes, their habits, life histories, and influence on mankind*, University of California Press, Berkeley, California.

Lavín-Murcio, P., Robinson, B. G., and Kardong, K. V., 1993, Cues involved in relocation of struck prey by rattlesnakes, *Crotalus viridis oreganus*, *Herpetologica* **49**:463-469.

Robinson, B. G., and K. V. Kardong, 1991, Relocation of struck prey by venomoid (venom-less) rattlesnakes *Crotalus viridis oreganus*, *Bull. Maryland Herpet. Soc.* **27**:23-30.

Smith, T. L., and Kardong, K. V., 2000, Cues used during post-strike trailing in rattlesnakes. *SICB* 2000, 384.

Smith, T. L., Kardong, K. V., and Lavín-Murcio, P. A., 2000, Persistence of trailing behavior: Cues involved in poststrike behavior by the rattlesnake (*Crotalus viridis oreganus*). *Behaviour* **137**:691-703.

Weldon, P. J., 1980, In defense of "kairomone" as a class of chemical releasing stimuli, *J. Chem. Ecol.* **6**:719-725.

# RATTLESNAKES CAN USE AIRBORNE CUES DURING POST-STRIKE PREY RELOCATION

M. Rockwell Parker and Kenneth V. Kardong[*]

## 1. INTRODUCTION

Rattlesnakes are a unique vertebrate system for studying chemosensory behaviors because unlike other venomous snakes such as elapids (Kardong, 1982), rattlesnakes break physical contact with the prey following an envenomating strike (Klauber, 1956) to reduce injury from retaliation by the prey (Chiszar et al., 1977; Estep et al., 1981; Golan, 1982; Cundall and Beaupre, 2001). Once released, the prey may then travel significant distances before succumbing to the effects of the venom (Estep et al., 1981; Kuhn et al., 1991). This presents an immediate problem to the snake: how to relocate the prey. Previous studies have shown that rattlesnakes are extremely efficient at relocating envenomated prey using substrate trails (see Kardong and Smith, 2002 for review). Rattlesnakes are capable of discriminating between unstruck and struck prey substrate trails (Furry et al., 1991), and they can successfully relocate struck prey when deprived of visual cues (Chiszar et al., 1977). However, cues such as urine (Chiszar et al., 1990) and blood (see Smith and Kardong, this volume) are of little importance to the snake during post-strike relocation.

Following the strike, rattlesnakes exhibit stereotyped innate behaviors termed strike-induced chemosensory searching (SICS) (Chiszar et al., 1977), including an increased tongue flick rate. This increase is indicative of subsequent increased stimulation of the vomeronasal organ (VNO) via cues received and transferred by the tongue (Burghardt and Pruitt, 1975). Previous work on rattlesnakes has suggested that the envenomating strike is the mechanism by which the snake acquires a partial chemosearching image (Melcer and Chiszar, 1989; Chiszar et al., 1991) and releases SICS behavior (Chiszar et al., 1977).

Relocation of the envenomated prey is crucial to the survival of the snake. Therefore, multiple cues could be used by the rattlesnake during relocation to insure

---

[*] M. Rockwell Parker and Kenneth V. Kardong, School of Biological Sciences, Washington State University, P.O. Box 644236, Pullman, WA, 99163.

recovery. Furthermore, substrate cues could be of primary utility to rattlesnakes, but it is doubtful that such cues are used exclusively in the process of prey relocation. To date, all previous studies investigating rattlesnake post-strike trailing behavior have only presented substrate-deposited chemical trails to the snakes. Although work has been conducted on rattlesnake olfaction (Cowles and Phelan, 1958), there was no distinction made between pre-strike and post-strike reactions of snakes to the odors presented. Therefore, our purpose was to examine the abilities of rattlesnakes (*Crotalus viridis oreganus*) to use airborne cues to relocate envenomated prey.

## 2. MATERIALS AND METHODS

Twenty-three northern Pacific rattlesnakes (*Crotalus viridis oreganus*) were used as the test animals. All individuals were housed in separate 10 gallon aquaria and kept on a 12h/12h l:d cycle at 30°C. Water was provided *ad libitum*, and the aquaria were lined with newspaper. Prey items used during all experiments were Swiss-Webster mice obtained the day of the trial from a large breeding colony.

The arena used for all trials has previously been described in detail (Lavín-Murcio et al., 1993). A Plexiglas Y-maze was created through which a moving airstream could be directed, and the maze fit directly over the Y-outline on the bottom of the arena (Figure 1). Small computer cooling fans (6 cm X 6 cm) were inserted into each arm of the Plexiglas Y-maze to provide and maintain air flow. Honeycomb flow blocks (1.5 cm-thick) were placed in front of the fans to reduce significant turbid air flow, and the holding box was fitted with a ¼" hardware cloth top to allow for unidirectional flow. The arena floor was covered with fresh butcher paper prior to each trial and then removed following the trial.

During a trial, the mouse's tail was tied with string, lowered down a vertical Plexiglas chute, struck by the snake, and then removed. Both fans were turned on immediately before removing the sliding door to the holding box, and both fans were on during all three treatments. All trials were recorded under low-light (Smith et al., 2000) using black and white security cameras and a VCR.

### 2.1. Treatments

Treatment 1 (substrate-only) was used as a baseline to test whether the snakes could use mouse-deposited substrate odors to relocate prey post-strike. To create substrate trails following the strike, the struck mouse was slid in one continuous motion, ventral surface down, along the base of the maze and out one of the two arms. The mouse was then removed from the maze so as not to present visual cues to the snake. Treatment 2 (airborne-only) presented snakes with airborne odors from struck mice to test whether snakes could relocate prey using only airborne information. To provide airborne cues, the struck mouse was placed in a wire-mesh basket, and that basket was placed on the intake side of one of the fans. The arm containing the odor was alternated from one trial to the next. Treatment 3 (substrate vs. airborne) presented snakes with both substrate and airborne cues from struck mice to determine if there was a preference for one cue over the other. To present both substrate and airborne cues in concert, the substrate trail was deposited along the base and out one arm as in Treatment 1, and then the same struck mouse was placed in a wire-mesh basket on the intake side of the fan of the opposite arm.

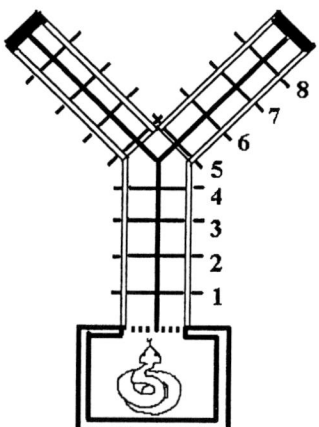

**Figure 1.** Generalized arena setup with Plexiglas Y-maze, holding box, and hatch-marks. The dashed line represents the location of the sliding door, and the black bars at the ends of the arms of the Y-maze represent the fans and honeycomb blocks.

We recorded several variables throughout a trailing episode: choice, tongue flicks, emergences, turnarounds, and temporal measures. Snakes were considered to have made a choice when the tip of the snake's rostrum passed the eighth hatch-mark (Figure 1), and choice was scored as either arm A, arm B, or no choice. Tongue flicks were observed as protrusions of the tongue, counted for every 10 cm section of the maze, and expressed as rate of tongue flick per minute. Tongue flicks were recorded per section of the maze until the snake's head passed the next sequential hatch-mark. An emergence was scored when the snake's head came out of the holding box during post-strike trailing. While trailing in the maze, snakes would often change their direction of travel, which provided another quantifiable character for analysis: a turnaround. A turnaround was defined as a deviation within the maze during a trailing episode where the snake moved past 90° in the left lateral or right lateral directions. Lastly, temporal measures were taken: time from last emergence to finish (total trailing time) and time spent per 10 cm section of the maze. Trials were considered finished when the snake's head passed the eighth hatch-mark in either arm (choice) or when 25 min. had elapsed (no trail).

All statistical analyses were done using SAS. Cochran's Q was used to analyze differences in choice between the three treatments. Friedman's signed-rank test was used to analyze differences between treatments in all other quantifiable measures.

## 3. RESULTS

In Treatment 1 (substrate-only), 22 snakes chose the "substrate" arm ($z = 4.38$, $p < 0.0001$) and one chose the "blank air" arm. In Treatment 2 (airborne-only), 19 snakes chose the "airborne" arm ($z = 3.13$, $p = 0.0017$) and four chose the "blank air" arm. In Treatment 3 (substrate vs. airborne), all 23 snakes chose the substrate arm ($z = 4.80$, $p <$

0.0001). The number of emergences was greater for Treatment 2 than in either Treatments 1 (p = 0.0017) or 3 (p = 0.0001). Also, the number of turnarounds was greater in Treatment 2 than in either Treatments 1 (p < 0.0001) or 3 (p = 0.0003) (Figure 2).

Rattlesnakes that trailed did so more quickly (mean = 61.61 sec) in the substrate-only treatment than in both the airborne-only (107.29 sec, p < 0.0001) and substrate-airborne treatments (88.0 sec, p = 0.0016). Tongue flick rates were lower at sections 1 (first 10 cm) and 2 (second 10 cm) in the airborne-only treatment than in both the substrate-only (section 1, p = 0.038; section 2, p = 0.0263) and substrate-airborne treatments (p = 0.0108; p = 0.0079).

## 4. CONCLUSIONS

Rattlesnakes can use airborne cues to relocate envenomated prey. Although mechanically deposited substrate trails are preferred, airborne cues provide sufficient information to the snake following the strike. Previous research has shown that rattlesnakes are capable of a suite of complex behaviors before (Duvall et al., 1985), during (Kardong and Bels, 1998), and after (Chiszar et al., 1977) an envenomating strike (see Kardong and Smith, 2002 for review). Our research extrapolates further on the complexity of rattlesnake chemosensory behavior following the strike.

Rattlesnakes are a unique system for studying predatory behaviors, but it should also be noted that several other species of reptiles serve as potential chemical models in terms of airborne cue use. Garter snakes are capable of using odorized air currents to locate prey (Waters, 1993) and respond to such odors by using their nasal chemical senses (Halpern et al., 1997). Geckos are known to use airborne odors to find fruit (Cooper and Perez- Mellado, 2001), and iguanas are well studied in their abilities to use airborne cues during mate searching (Alberts and Werner, 1993).

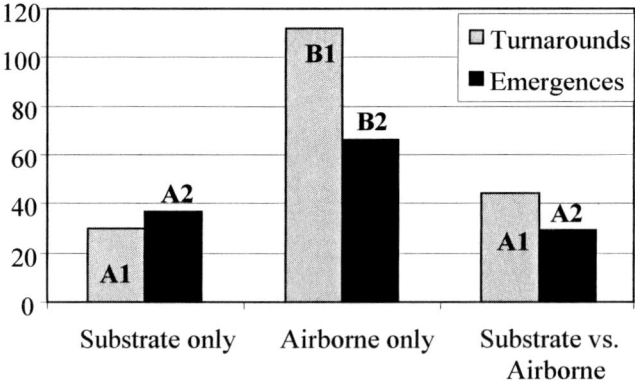

Figure 2. Numbers of turnarounds and emergences for all three treatments. Different letters with the same number (e.g. A1 and B1) represent statistically different groups.

Though rattlesnakes are capable of relocating prey using airborne cues, the process is very different than that occurring during substrate trailing. Differences observed between contexts may have several unique, plausible explanations. For example, tongue flick rates could have been lower in Treatment 2 (airborne-only) as a result of general ophidian neurological anatomy. Connections exist between the olfactory and vomeronasal systems at the hypoglossal nucleus which could result in residual signaling to the musculature of the tongue (Martinez-Marcos et al., 2002). Thus, tongue flicks seen during airborne cue presentation could be uninformative in that context. Conversely, there could be a difference in prey odor concentration from substrate trailing to airborne cue use, with airborne cues being more dilute. If such a concentration difference of the same cue were responsible for the differences in behavior seen, one would expect the snakes' behavior to also be notably different in Treatment 3 (airborne vs. substrate) where the snakes were presented with two informative cues. There was no such difference. Lastly, tongue flicks seen in the airborne-only treatment could be directed at cues in the moving air stream and/or at airborne cues that settled on the substrate in the maze. We did not attempt to distinguish between the tongue flick directions. Instead, the airborne-only treatment confirmed that cues carried in the air current could be used to successfully relocate prey following the strike.

The use of airborne cues post-strike is perplexing when we try to explain the origination of such a behavior. Selection should favor individuals capable of using all informative cues present to relocate prey. Therefore, could rattlesnakes be capable of airborne trailing because selection favored such a behavior secondarily, or were airborne cues of primary use at one time and substrate trailing arose in the more derived condition? Understanding how often and in what contexts such strategies have arisen is a difficult process. However, as these meetings highlight, two divergent research levels that characterize the study of chemosensory systems prove complementary: proximate and ultimate. As more species are studied and the proximate mechanisms responsible for prey relocation become better understood, comparative analyses will be possible and evolutionary events can be clarified.

## 5. ACKNOWLEDGEMENTS

Several people were responsible for the development and success of this project. S. J. Beaupre provided MRP with the opportunity to film post-strike trailing in the field, which provided the insight needed for this project. M. Halpern and T. Smith were helpful in their early critiques and advice on research methods. M. Evans helped greatly with statistical analyses, and R. Romjue assisted with technical aspects of the design.

## 6. REFERENCES

Alberts, A. C., and Werner, D. I., 1993, Chemical recognition of unfamiliar conspecifics by green iguanas: functional significance of different signal components, *Anim. Behav.* **46(1)**:197-199.
Burghardt, G. M., and Pruitt, C., 1975, Role of the tongue and senses in feeding of naïve and inexperienced garter snakes, *Phys. Behav.* **14**:185-194.
Chiszar, D., Radcliffe, C. W., and Scudder, K. M., 1977, Analysis of the behavioral sequence emitted by rattlesnakes during feeding episodes, *Behav. Biol.* **21**:418-425.

Chiszar, D., Melcer, T., Lee, R., Radcliffe, C. W., and Duvall, D., 1990, Chemical cues used by prairie rattlesnakes (*Crotalus viridis*) to follow prey trails of rodent prey, *J. Chem. Ecol.* **16(1)**:79-86.

Chiszar, D., Lee, R. K., Radcliffe, C. W., and Smith, H. M., 1991, Searching behaviors by rattlesnakes following predatory strikes, in: *Biology of the Pit Vipers*, J. A. Campbell and E. D. Brodie Jr., eds, Selva, Tyler, TX, pp. 369-382.

Cooper, W. E., and Pérez-Mellado, V., 2001, Location of fruit using only airborne odor cues by a lizard, *Physiol. Behav.* **74(3)**:339-342.

Cowles, R. B., and Phelan, R. L., 1958, Olfaction in rattlesnakes, *Copeia* **1958(2)**:77-83.

Cundall, D., and Beaupre, S. J., 2001, Field records of predatory strike kinematics in timber rattlesnakes, *Crotalus horridus*, *Amphibia-Reptilia* **22**:492-498.

Duvall, D., King, M. B., and Gutzwiller, K. J., 1985, Behavioral ecology and ethology of the prairie rattlesnake, *Nat. Geo. Res.* **1**:80-111.

Estep, K., Poole, T., Radcliffe, C. W., O'Connell, B., and Chiszar, D, 1981, Distance traveled by mice after envenomation by a rattlesnake (*C. viridis*), *Bull. Psych. Soc.* **18(3)**:108-110.

Furry, K., Swain, T., and Chiszar, D., 1991, Strike-induced chemosensory searching and trail following by prairie rattlesnakes (*Crotalus viridis*) preying upon deer mice (*Peromyscus maniculatus*): chemical discrimination among individual mice, *Herpetologica* **47(1)**:69-78.

Golan, L., Radcliffe, C., Miller, T., O'Connell, B., and Chiszar, D., 1982, Trailing behavior in prairie rattlesnakes (*Crotalus viridis*), *J. Herp.* **16(3)**:287-293.

Halpern, M., Halpern, J., Eichsen, E., and Borghjid, S., 1997, The role of nasal chemical senses in garter snake response to airborne odor cues from prey, *J. Comp. Psych.* **111(3)**:251-260.

Kardong, K. V., 1982, Comparative study of changes in prey capture behavior in the cottonmouth (*Agkistrodon piscivorus*) and Egyptian cobra (*Naja haje*), *Copeia* **1982**:337-343.

Kardong, K., 1986, Predatory strike of the rattlesnake, *Crotalus viridis oreganus*, *Comp. Psych.* **100**:304.

Kardong, K., and Bels, V. L., 1998, Rattlesnake strike behavior: kinematics, *J. Exp. Biol.* **201**:837-850.

Kardong, K., and Smith, T. L., 2002, Proximate factors involved in rattlesnake predatory behavior: a review, in: *Biology of the Vipers*, G. W. Schuett, ed., Eagle Mountain Publishing, Eagle Mountain, UT, pp. 253-266.

Klauber, L. M., 1956, *Rattlesnakes: their habits, life histories, and influence on mankind*, University of California Press, Berkeley, CA.

Kuhn, B. F., Rochelle, M. J., and Kardong, K. V., 1991, Effects of rattlesnake (*Crotalus viridis oreganus*) envenomation upon the mobility and death rate of laboratory mice (*Mus musculus*) and wild mice (*Peromyscus maniculatus*), *Bull. Md. Herp. Soc.* **27(4)**:189-194.

Lavín-Murcio, P. A., Robinson, B. G., and Kardong, K. V., 1993, Cues involved in relocation of struck prey by rattlesnakes, *Crotalus viridis oreganus*, *Herpetologica* **49(4)**:463-469.

Martinez-Marcos, A., Lanuza, E., and Halpern, M., 2002, Neural substrates for processing chemosensory information in snakes, *Brain Research Bull.* **57(3/4)**:543-546.

Melcer, T., and Chiszar, D., 1989, Striking prey creates a specific chemical search image in rattlesnakes, *Anim. Behav.* **37**:477-486.

Smith, T. L., Kardong, K. V., and Lavin-Murcio, P. A., 2000, Persistence of trailing behavior: cues involved in poststrike behavior by the rattlesnake (*Crotalus viridis oreganus*), *Behavior* **137**:691-703.

Waters, R. M., 1993, Odorized air current trailing by garter snakes, *Thamnophis sirtalis*, *Brain Beh. Evol.* **41**:219-223.

# THE SENSE OF SMELL IN PROCELLARIIFORMS
## An overview and new directions

Gregory B. Cunningham and Gabrielle A. Nevitt[*]

## 1. INTRODUCTION

Procellariiform seabirds are found in all oceans of the world, though both their diversity and numbers are highest in the southern hemisphere. Procellariiforms range in size from the least storm-petrel (*Halocyptena microsoma*), with an average weight of 19.5 g and a 32 cm wing span, to the wandering albatross (*Diomedea exulans*), with an average weight of 8.7 kg and a 3.2 m wing span. All members of this order are pelagic, and some navigate extreme distances to forage (for example, up to 15,000 km during an incubation shift for wandering albatrosses; Jouventin and Weimerskirch, 1990). These birds also share a complex life history, coming to land only to breed. Nesting behavior, however, varies with the species. For example, albatrosses and many larger petrel species nest above ground whereas other species such as shearwaters, storm-petrels, and smaller petrels nest in burrows which they construct themselves. These burrows can be over a meter deep. For both burrow- and surface-nesting species, parents take turns incubating a single egg, alternating foraging trips at sea with incubating shifts. Once the chick hatches, adults forage offshore often hundreds or thousands of kilometers from the breeding colony, and return periodically to provision the nestlings. Procellariiforms feed on a variety of prey items including squid, fish, krill, and other zooplankton (reviewed in Warham, 1990), and tend to aggregate in mixed species foraging flocks where food is abundant (Routh, 1949; Harrison et al., 1991).

Procellariiform seabirds share a keen sense of smell, and for the past forty years, research has focused on understanding how these birds use olfaction in central-place foraging, both with respect to locating foraging hotspots at sea, and in relocating the nest site once they return to the colony (Figure 1). With respect to foraging at sea, it has been shown that many species are attracted to odors related to food such as krill, fish, and squid (Grubb, 1972; Hutchison and Wenzel, 1980; Lequette et al., 1989; reviewed by

---

[*] Department of Neurobiology, Physiology and Behavior, University of California, Davis, California, 95616.

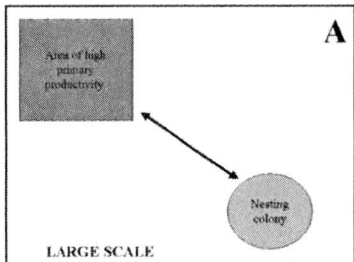 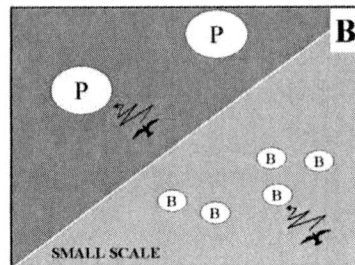

Figure 1. Olfactory behaviors. (A) Procellariiforms commute between nesting colonies and areas of high primary productivity where prey is likely to be found. (B) Procellariiforms locate prey patches (circled P's) and home burrows (circled B's) in part by tracking odors upwind. Diagrams not drawn to scale.

Nevitt, 2000). Recent work has shown that procellariiforms can also detect dimethyl sulfide (DMS; Nevitt et al., 1995; Nevitt, 2000; Nevitt and Haberman, 2003), an odor associated with primary productivity over the ocean (e.g., McTaggart and Burton, 1992). Nevitt has suggested that procellariiforms use odor cues at two spatial scales (see Nevitt, 2000; 2001). At large spatial scales (1000s of kilometers), procellariiforms may use olfactory information to recognize productive areas of the ocean where prey is likely to be found. Once the productive region is located, birds switch to a small-scale (10s or 100s of kilometers) area-restricted search mode to pinpoint prey patches directly. An area-restricted search is a multi-modal strategy to locate prey incorporating both olfactory and visual cues provided by specific ocean features, prey species, or the behavior of other predators (Haney, 1987; Brown, 1988; Haney et al., 1992; Silverman and Nevitt, 1995; reviewed by Nevitt and Viet, 1999).

While nearly every species of procellariiform that has been tested relies to some extent on olfactory cues to forage (see Nevitt, 2002), the best evidence to date suggests that species that depend on olfaction to relocate the nest tend to be nocturnal burrow-nesting species. These birds are able to relocate their burrows at night even under overgrown vegetation, or among thousands of conspecifics and heterospecifics in colonies that are heavily predated by other birds (for example, in the Southern hemisphere, skuas: *Catharacta skua lönnbergi*; Mougeot et al., 1998). Night vision seems to be poorly developed in the few species that have been studied (common diving petrels, *Pelecanoides urinatrix*, Brooke, 1989; Manx shearwater, *Puffinus puffinus*; Martin and Brooke, 1991), suggesting that other cues (auditory, spatial or olfactory) are likely more important than visual cues in relocating the nest sight.

With respect to olfaction, Grubb (1974) found that Leach's storm-petrels (*Oceanodroma leucorhoa*) consistently approached burrows from downwind, suggesting that olfactory cues guided this behavior. In Y-maze studies, Leach's storm-petrels tended to choose arms that were connected to their own nesting material over arms connected to forest-floor substrate (Grubb, 1974). Recently, Bonadonna et al. (2003a, 2003b) used Y-maze studies to demonstrate that several sub-Antarctic species (Antarctic prions, *Pachyptila desolata*; common diving petrels, *Pelecanoides urinatrix*; South-Georgian diving petrels, *Pelecanoides georgicus*) can identify their burrow by smell. A number of experiments have tested whether birds with an impaired sense of smell (i.e., by chemical ablation of the olfactory epithelium, obstructing the nostrils, or olfactory nerve transection) have the ability to relocate the burrow after release from distant locations (Grubb, 1974; Haftorn et al., 1988; Benvenuti et al., 1993; Bonadonna et al., 2001; Bonadonna and Bretagnolle, 2002). While results tend to suggest that burrowing species cannot relocate nest sites without a functional olfactory system, it is not easy to verify what the impact of these treatments are to survivorship at sea, since treated birds are often not seen again. It is possible, for example, that treated birds abandon the nest for the season (Blackmer et al., 2003) or that even temporary anosmia leads to increased morbidity. To date, no studies have approached this question in conjunction with year-to-year monitoring, so the effects of these treatments are not yet known.

## 2. OLFACTORY RESPONSES BY PROCELLARIIFORM CHICKS

While most work to date has focused on adults, the olfactory abilities of chicks have received little attention (but see Minguez, 1997). Most studies regarding the development of the olfactory system in birds have focused on domesticated species such as chickens (*Gallus domesticus*) and ducks (see Mabayo et al., 1996; Burne and Rogers, 1999 for behavioral studies; see Ayer-Le-Lievre et al., 1995 for a review of neuroanatomical studies). Unlike domesticated poultry, however, procellariiform chicks are abandoned by their parents prior to fledging. In fact, procellariiform fledglings typically need to learn to forage on their own for prey items that are hundreds of miles offshore (reviewed by Warham, 1990). How they accomplish this feat is a mystery.

To begin to address this question, we have recently modified the Porter method (see Porter et al., 1999) to test behavioral sensitivity of chicks to identified scented compounds in a controlled setting (Cunningham et al., 2003). The Porter method has been used successfully to assay olfactory sensitivity of chicken chicks in laboratory studies. This technique involves inducing a "sleep-like" state in a chick by warming it with a light bulb or a warm hand. Once asleep, odors (or controls) are puffed across the bird's bill, and its response is scored on a numerical scale. Working in the Kerguelen Archipelago, we have used this method to examine differences in olfactory ability among three species: the blue petrel (*Halobaena caerulea*), the thin-billed prion (*Pachyptila belcheri*), and the common diving petrel (*Pelecanoides urinatrix*). Results are summarized in Figure 2. We found that scores for blue petrel chicks were significantly higher for both a novel odor (phenyl ethyl alcohol, a rosy smell) and a food-related odor (DMS) compared to control scores. Thin-billed prion chicks scored significantly higher only to the novel odor as compared to control, but our investigations with thin-billed prions were limited by a small sample size and low statistical power. Results from both species are consistent with adult behavior, since both these species have been shown to associate with either experimentally or naturally elevated levels of DMS at sea (Nevitt et al., 1995; Nevitt, 2000). By contrast, common diving petrels gave no significant responses. These birds do not respond to DMS in experimental trials at sea (Nevitt et al., 1995; Nevitt, 2000).

One advantage of the Porter method is that it is robust enough to allow us to explore questions related to behavioral sensitivity. For example, working with the sub-Antarctic species assemblage near Elephant Island (61°05'S, 54°43'W) and South Georgia (54°17'S, 36°30'W), Nevitt has reported consistent species-related differences in

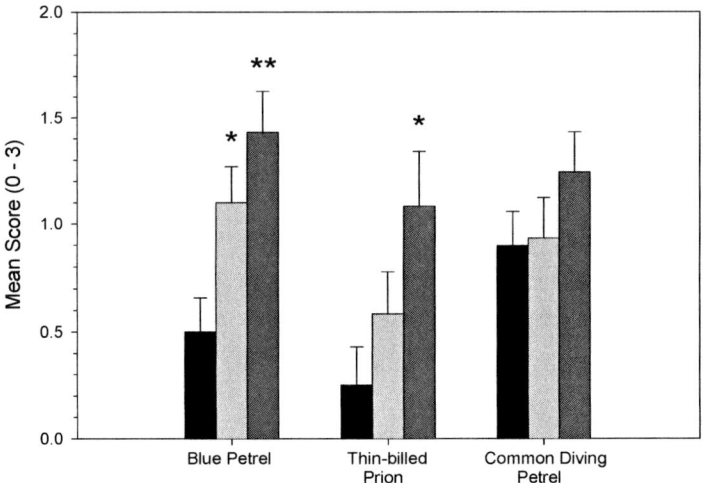

Figure 2. Average scores of blue petrels, thin-billed prions and common diving petrels to control (black), dimethyl sulfide (light gray) and phenyl ethyl alcohol (dark gray). For blue petrels, scores for PEA and DMS were significantly different from control (Wilcoxon signed rank test, **$P<0.01$, *$P<0.05$). For thin-billed prions, only scores for PEA were significantly different from control (Wilcoxon signed rank test, * $P<0.05$). Common diving petrels showed no significant differences in scores. (Adapted from Cunningham et al., 2003).

olfactory foraging behavior (Nevitt, 1999; reviewed in Nevitt, 2002). She has suggested that these differences in response are linked to life-history traits. Burrow nesters tend to forage opportunistically, and respond especially to odors associated with zooplankton grazing (i.e. DMS) whereas surface-nesters tend to recruit to odors more directly associated with macerated prey (Nevitt, 1999). Whether these differences in behavior are also linked to differences in olfactory sensitivity of chicks is unclear, but the Porter method will allow us to test these questions. In addition, we can examine whether behavioral sensitivity changes with a chick's age or satiation state, or by prior exposure to odors during normal growth and development (e.g., Sneddon et al., 1998). This technique may also prove useful in testing how chicks respond to social odors associated with their parents or con-specifics (e.g., Hagelin et al., 2003).

Ultimately, we hope to build a broader knowledge base of the olfactory abilities and sensitivities across many species of procellariiforms. Given another forty years, we may begin to understand how olfactory abilities may have influenced behavioral and life history changes, as well as macroevolutionary changes that lead to speciation and/or higher level taxonomic changes (i.e. at the genus and family level). This is an exciting area for future exploration.

## 3. ACKNOWLEDGEMENTS

This project was supported by NSF (IBN 0212467 and OPP 9814326) to GAN. We wish to sincerely thank Henri Weimerskirch and Francesco Bonadonna for the ongoing collaboration and the French Polar Institute (IPEV Program n° 109) for logistical support during the work at Kerguelen Island. The assistance of Christopher Burney, Vivien Chartendrault, and Nicolas Delelis in the field was invaluable. The comments of Dr. Richard Van Buskirk, Sean Lema, and Mark Hodges greatly improved this manuscript.

## 4. REFERENCES

Ayer-Le Lievre, C., Lapointe, F., and Leibovici, M., 1995, Avian olfactory neurogenesis, *Biol. Cell.* **84**:25-34.

Benvenuti, S., Ioale, P., and Massa, B., 1993, Olfactory experiments on Cory's shearwater (*Calonectris diomedea*): The effect of intranasal zinc sulphate treatment on short-range homing behaviour, *Boll. Zool.* **60**:207-210.

Brown, R. G. B., 1988, The influence of oceanographic anomalies on the distribution of storm-petrels (Hydrobatidae) in Nova Scotia waters, *Colon. Waterbirds* **11**:1-8.

Blackmer, A. L., Ackerman, J. T., and Nevitt, G. A., in press, Effects of investigator disturbance on hatching success and nest-site fidelity in a long-lived seabird, the Leach's Storm-petrel, *Biol. Conserv.*

Bonadonna, F., and Bretagnolle, V., 2002, Smelling home: A good solution for burrow-finding in nocturnal petrels? *J. Exp. Biol.* **205**:2519-2523.

Bonadonna, F., Cunningham, G. B., Jouventin, P., Hesters, F., and Nevitt, G. A., 2003a, Evidence for nest-odour recognition in two species of diving petrel, *J. Exp. Biol.* **206**:3719-3722.

Bonadonna, F., Hesters, F., and Jouventin, P., 2003b, Scent of a nest: discrimination of own-nest odours in Antarctic prions, *Pachyptila desolata*, *Behav. Ecol. Sociobiol.* **54**:174-178.

Bonadonna, F., Spaggiari, J., and Weimerskirch, H., 2001, Could osmotaxis explain the ability of blue petrels to return to their burrows at night? *J. Exp. Biol.* **204**:1485-1489

Brooke, M. D. L., 1989, Determination of the absolute visual threshold of a nocturnal seabird, the common diving petrel *Pelecanoides urinatrix*, *Ibis* **131**:290-300.

Burne, T. H. J., and Rogers, L. J., 1999, Changes in olfactory responsiveness by the domestic chick after early exposure to odorants, *Anim. Behav.* **58**:329-336.

Cunningham, G. B., Van Buskirk, R. W., Bonadonna, F., Weimerskirch, H., and Nevitt, G. A., 2003, A comparison of the olfactory abilities of three species of procellariiform chicks, *J. Exp. Biol.* **206**:1615-1620.

Grubb, T. C., 1972, Smell and foraging in shearwaters and petrels, *Nature* **237**:404-405.

Grubb, T. C., 1974, Olfactory navigation to the nesting burrow in Leach's petrel (*Oceanodroma leucrrhoa*), *Anim. Behav.* **22**:192-202.

Haftorn, S., Mehlum, F., and Bech, C., 1988, Navigation to nest site in the snow petrel (*Pagodroma nivea*), *Condor* **90**:484-486.

Hagelin, J. C., Jones, I. L., and Rasmussen, L. E. L., 2003, A tangerine-scented social odour in a monogamous seabird. *Proc. R. Soc. Lond. B. Biol. Sci.* **270**:1323-1329.

Haney, J. C., 1987, Ocean internal waves as sources of small-scale patchiness in seabird distribution on the Blake Plateau, *Auk* **104**:129-133.

Haney, J. C., Fristrup, K. M., and Lee, D. S., 1992, Geometry of visual recruitment by seabirds to femoral foraging flocks. *Ornis Scand.* **23**:49-62.

Harrison, N. M., Whitehouse, M. J., Heinemann, D., Prince, P. A., Hunt Jr., G. L., and Veit, R. R., 1991, Observations of multispecies seabird flocks around South Georgia, *Auk* **108**:801-810.

Hutchison, L. V., and Wenzel, B. M., 1980, Olfactory guidance in foraging by procellariformes, *Condor* **82**:314-319.

Jouventin, P., and Weimerskirch, H., 1990, Satellite tracking of Wandering albatrosses, *Nature* **343**: 746-748.

Lequette, B., Verheyden, C., and Jouventin, P., 1989, Olfaction in subantarctic seabirds: its phylogenetic and ecological significance. *Condor* **91**:732-735.

Mabayo, R. T., Okumura, J., Hirao, A., Sugita, S., Sugahara, K., and Furuse, M., 1996, The role of olfaction in oil preference in the chicken. *Physiol. Behav.* **59**:1185-1188.

Martin, G. R., and Brooke, M. D. L., 1991, The eye of a procellariiform seabird, the Manx shearwater, *Puffinus puffinus*: Visual fields and optical structure, *Brain Behav. Evol.* **37**:65-78.

McTaggart, A. R., and Burton, H., 1992, Dimethyl sulfide concentrations in the surface waters of the Australian Antarctic and Subantarctic oceans during an austral summer, *J. Geophys. Res.* **97**:14407-14412.

Minguez, E., 1997, Olfactory nest recognition by British storm-petrel chicks, *Anim. Behav.* **53**:701-707.

Mougeot, F., Genevois, F., and Bretagnolle, V., 1998, Predation on burrowing petrels by the brown skua (*Catharacta skua lonnbergi*) at Mayes Island, Kerguelen, *J. Zool. (Lond.)* **244**:429-438.

Nevitt, G. A., 1999, Olfactory foraging in Antarctic seabirds: a species-specific attraction to krill odors, *Mar. Ecol. Prog. Ser.* **177**:235-241.

Nevitt, G. A., 2000, Olfactory foraging by Antarctic procellariiform seabirds: Life at high Reynolds numbers, *Biol. Bull.* **196**:245-253.

Nevitt, G. A., 2001, Mechanisms of olfactory foraging in procellariiform seabirds, in: *Chemical Signals in Vertebrates, vol. 9*, A. Marchlewska-Koj, J. L. Lepri and D. Muller-Schwarze, eds., New York, Plenum Publishing Corp, pp. 27-33.

Nevitt, G. A., in press, Olfactory foraging strategies of procellariiform seabirds, in: *Proceedings of the 23$^{rd}$ International Ornithological Congress, Beijing, China*.

Nevitt, G. A., and Haberman, K., 2003, Behavioral attraction of Leach's storm-petrels (*Oceanodroma leucorhoa*) to dimethyl sulfide, *J. Exp. Biol.* **206**:1497-1501.

Nevitt, G. A., and Veit, R. R., 1999, Mechanisms of prey patch detection by foraging seabirds, in: *Proceedings of the 22$^{nd}$ International Ornithological Congress*, N. J. Adams and R. H. Slotow, eds., Johannesburg, South Africa, Birdlife, pp. 2072-2082.

Nevitt, G. A., Veit, R. R., and Kareiva, P., 1995, Dimethyl sulphide as a foraging cue for Antarctic procellariiform seabirds, *Nature* **376**:680-682.

Porter, R. H., Hepper, P. G., Bouchot, C., and Picard, M., 1999, A simple method for testing odor detection and discrimination in chicks, *Physiol. Behav.* **67**:459-462.

Routh, M., 1949, Ornithological observations in the Antarctic seas 1946-1947, *Ibis* **91**:577-606.

Silverman, E., and Nevitt, G. A., 1995, Evidence for network foraging by Antarctic seabirds, *Antarct. J. US Rev. Iss.* **30**:186-187.

Sneddon, H., Hadden, R., and Hepper, P. G., 1998, Chemosensory learning in the chicken embryo, *Physiol.Behav.* **64**:133-139.

Warham, J., 1990. The Petrels: Their Ecology and Breeding Systems, Academic Press, London.

# COTTONTAILS AND GOPHERWEED: ANTI-FEEDING COMPOUNDS FROM A SPURGE

Dietland Müller-Schwarze and José Giner[*]

## 1. ABSTRACT

Mammalian herbivores nearly never attack gopherweed, *Euphorbia lathyris* (Euphorbiaceae). To find the chemical principle responsible for the deterrent effect, we field-tested plant extracts in response-guided bioassays for antifeedant activity. We treated winter-dormant apple twigs with extract solutions and presented them to free-ranging eastern cottontails, *Sylvilagus floridanus*, at 3 different home ranges in Upstate New York during late winter. Extracts from all plant parts reduced feeding, measured as twig tips cut and amount of bark removed, but root extracts were most active. The bioassay led to isolation of putative antifeedant compounds from the root extract, a series of long-chain alkyl ferulates and gopherenediol, a new diterpenoid. Bioassay of docosanyl ferulate and gopherenediol demonstrated weak antifeedant activity. The bioassays are continuing.

## 2. INTRODUCTION

Wild mammals such as deer, lagomorphs, or rodents do not feed on Gopherweed (*Euphorbia lathyris*), and subterranean rodents feeding on the roots are said to be "killed because of its highly irritant gastric effect" (Hohmann et al., 1999). Gopherweed is also known as mole plant, caper spurge, or petroleum plant. It is a biennial spurge of Mediterranean and eastern European origin and grows up to 1 m tall and has white latex, typical of the spurges (*Euphorbiaceae*). During the 1970's this spurge had attracted interest as a potential, though expensive, source of hydrocarbons re-assembling gasoline (Calvin et al., 1982). Hence the name "petroleum plant" (e.g., Sachs, 1981). Gardeners

---

[*] Dietland Müller-Schwarze, Department of Environmental and Forest Biology; José Giner, Department of Chemistry; College of Environmental Science and Forestry, State University of New York, Syracuse, New York 13210 (dmullers@esf.edu; jlginer@syr.edu)

plant gopherweed as defense against moles, voles, and gophers. Gopherweed has escaped cultivation in some places. Although very toxic, the leaves, latex and seed oil have been used in folk medicine for a variety of ailments (Duke, 2003). In this project, we are searching for the active principle that repels mammalian herbivores by providing chemosignals in the absence of visual cues.

## 3. MATERIALS AND METHODS

### 3.1. Plant Extracts

The *Euphorbia* plants were grown outdoors from seed and harvested late in the growing season when they set seeds. In 1999, 16 first-year plants were harvested in late November, to be used in experiments in early 2000. They were 22 to 37 cm high, with roots 35 to 54 cm. They weighed 24.4 grams on average. We separated the roots from the plants, yielding 59.7 g roots and 272.3 g green tops. The *roots* were ground with 200 ml ethyl acetate (EtOAc) in a Waring blender. After decanting, this was repeated. Decanted again, the 2 extracts were combined and filtered. This yielded about 300 ml pale green cloudy extract. The roots were then again ground with 200 ml MeOH which yielded ca. 150 ml pale green extract. The plant *tops* were ground with 400 ml EtOAc. After decanting, they were extracted twice more with 300 ml. After filtering, 850 ml dark green-black extract remained. The plant tops were ground again with 400 ml MeOH. After filtering, ca. 500 ml green-black extract was obtained. In 2000, we harvested $2^{nd}$-year plants in mid-June. We divided them into roots (38 g), mid sections (226 g), and tops, including flowers and seeds (78 g). In 2001, 15 first-year plants were harvested in mid-October. After washing the soil off the roots, the total weight was 720 g (48 g per plant, on average).

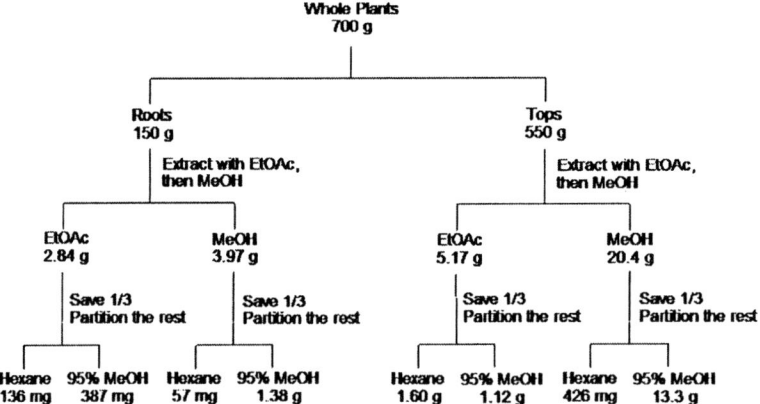

Figure 1. Fractionation Scheme.

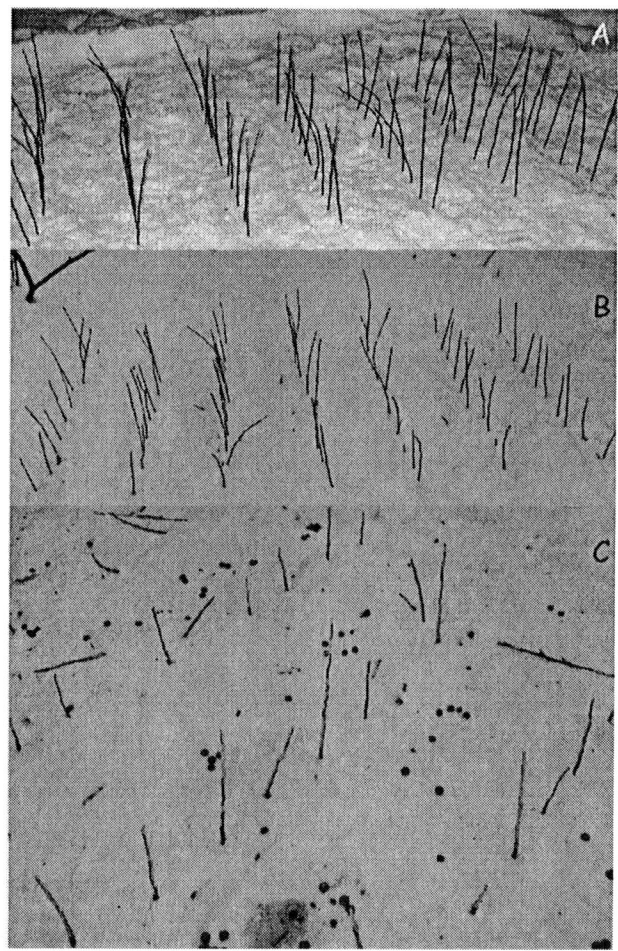

Figure 2. The experimental setup. A: Array of fresh apple twigs at beginning of test. B: After one night, some tips have been clipped. C: The end stage after several days. Tips and bark of many twigs have been totally consumed.

In 2000 and 2001 we tried to use the same proportions of solvents to biomass as were used in 1999. In 2002, extracts of roots and tops were partitioned between hexane and aqueous methanol, as shown in Figure 1. In 2003, we tested isolated and synthesized single compounds in different concentrations.

## 2.2. Field tests

The extracts were tested on free-ranging cottontails (*Sylvilagus floridanus*, Lagomorpha) in upstate New York during late winter (February/March) when deep snow (25-50 cm) covers alternate food sources. Three different home ranges were selected (cottontails have winter home ranges of about 4 ha). To apply the extracts to a food item, straight, unbranched first-year apple twigs (50 cm long) were dipped into graduated cylinders (100 ml) containing the extracts. Each twig took up 0.5 to 0.75 ml solution. Thus for the root extracts, each apple twig came to be coated with 0.04 (1/25) to 0.08 (1/12) of the contents of the root of one average plant. For the extracts from the plant tops, each apple twig received, on average, 0.014 (1/70) to 0.024 (1/42) of the contents of one plant.

The twigs were stuck 10 cm deep into the hard snow, arranged in rows spaced 30 cm apart. Each row had 10 replicates of the same treatment, 20 cm apart within the row (Figure 2). We used two controls: twigs dipped in solvent only, and untreated twigs without any solvent. The sequence of rows was random, and the field experimenter worked with coded samples. Since cottontails are nocturnal, the results of their feeding were checked in the morning. The experiment lasted 5 nights, or until the samples were consumed, whichever occurred first. The cottontails deplete the apple twigs in three stages. First, they bite off the tips. During the second night, they cut back the twigs some more, and start to chew off bark from the thicker, still standing parts of the twigs. Finally, the animals consume the remaining bark. Thus, we recorded the three variables numbers of tips cut off, length of twig remaining, and amount of bark peeled.

In 2000, 7 treatments were tested at 5 different sites: Root and tops extracted with MeOH or EtoAc; solvent only (MeOH or EtOAC), and blanks. At each site, each treatment was replicated by 10 twigs, resulting in a total of 350 treated twigs (7 x 10 x 5). In 2001 two trials were run, at 3 and 5 sites, respectively. The nine treatments included those from 2000, plus extracts from stem sections between tops and roots. With 10 replications for each treatment, a total of 720 twigs were used. In 2002, we tested various solvent fractions. Eighteen samples were tested at two sites, resulting in a total of 360 twigs. In 2003, 3 trials were run. In trial 1, 7 treatments were tested at 3 sites, or a total of 210 twigs. These included 3 concentrations of ferulate (Fig. 3), and 2 concentrations of gopherenediol (Figure 3; Giner et al., submitted), compounds that had been isolated from

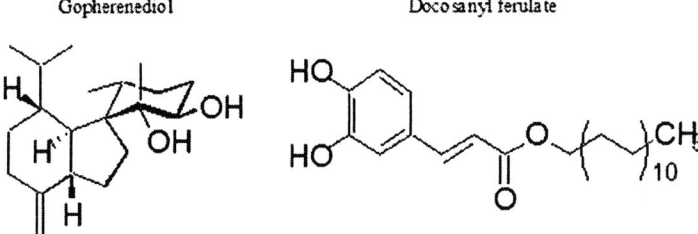

**Figure 3.** The two compounds tested in the present experiment: Gopherene diol, a novel diterpenoid compound, and docosanyl ferulate.

the fractions tested the year before; and solvent controls and blanks. The low concentration corresponded to the level found in the plant; a 3-fold increase was designated "high". Trial 2 repeated trial 1 exactly. Trial 3 followed up on the responses to the single compound in trial 1. Gopherenediol, solvent, and blank were tested with 30 twigs for each treatment (total: 90 twigs). Thus, a grand total of 1940 apple twigs were used in these bioassays during four winters.

## 3. RESULTS

### 3.1. Twig Cutting

In 2000 we compared responses to root extract with those to extract of the green tops, in EtOAc and MeOH. Measured as number of twigs cut off at the top, both root and top extracts in EtOAc (difference to solvent only: $p < 0.04$; Figure 4) were most repellent during the first night. However, after six nights, the cottontails had avoided the tops extract in EtOAc the most.

In 2001, in trial 1 the largest number of twigs surviving after 2 days was found in the stem EtOAc sample. In trial 2, the animals clipped the least number of twigs in the root MeOH and top EtOAc samples.

In 2002, most twigs of the sample "roots-EtOAc" survived even after 8 days. Similarly, the least bark was removed from roots-EtOAc (Figure 5).

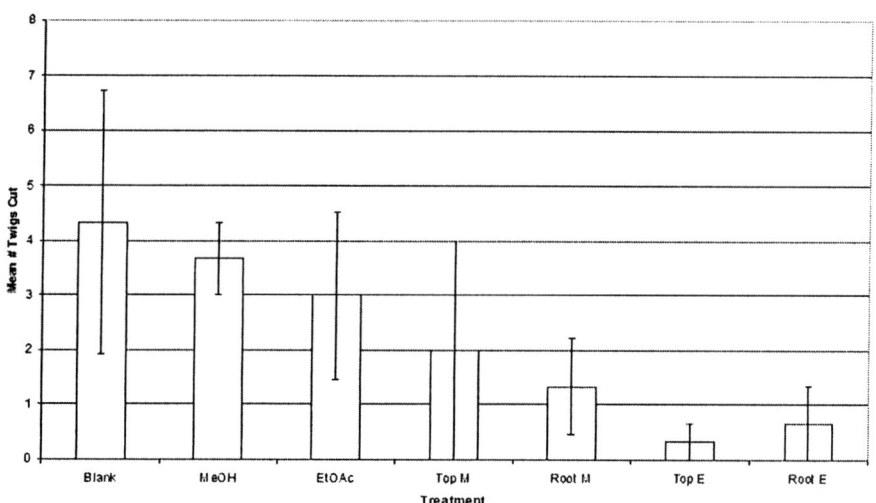

Figure 4. Mean numbers of twigs cut during first night at 3 sites in 2000. Root and tops extracts in methyl alcohol (M) or ethyl acetate (E). Solvent EtOAc vs. Tops E: $p < 0.04$.

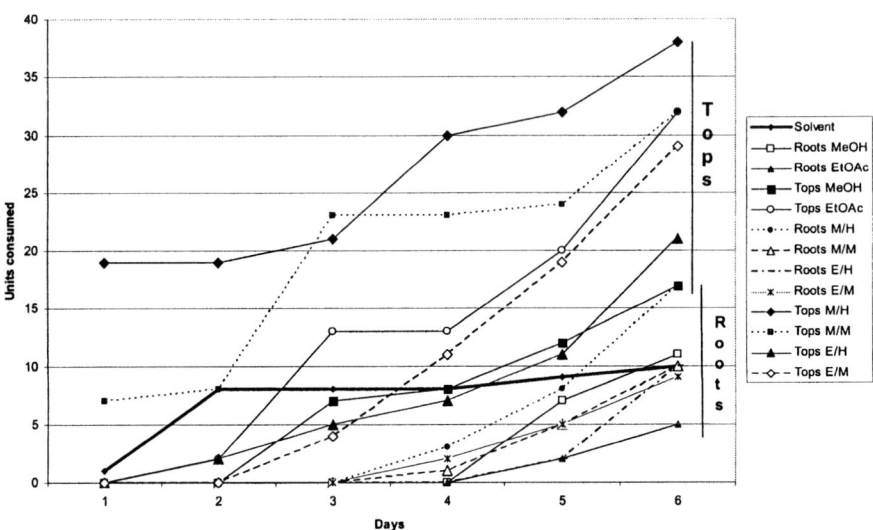

**Figure 5.** In 2002 tests, extracts from roots protected the experimental twigs better than extracts from tops. But solvent (acetone) alone (heavy line) was as active as root extracts. "Units removed": Total of tips, buds and 4-cm sections of bark.

In 2003, Trial 1, the results differed between the 3 sites. Tips cut off: At two sites (L and R), the cottontails cut off most tips during the first night, showing no discrimination between treatments. At the third site (B) twigs treated with solvent only were most avoided during the first night, and the high concentration of ferulate during the second night, with no additional feeding after the first night. By Day 3, all samples had at least 9 of 10 tips removed.

The cottontails also cut off sections of the twigs during the first night. At two sites little of each twig was cut off: at B from 14 to 26 cm of the original 40 cm were left standing after one night; and at R 29 – 37 cm. The amounts taken did not vary significantly between the treatments. By contrast, cottontails at site L cut the twigs much shorter (only 2 – 13 cm left standing) and discriminated more. Blank, solvent and low concentrations of ferulate were eaten most, while the high concentration of diterpene and ferulate and the medium concentration of ferulate were avoided most (Fig. 6). At site R, after 3 nights, blank, solvent, and "ferulate medium" were most depleted, while twigs with "diterpene low" had the longest sections left.

Trial 2 replicated Trial 1. Remaining *twig lengths* after Night 1 were greatest for "ferulate high", and least for "diterpene low", at all 3 sites combined.

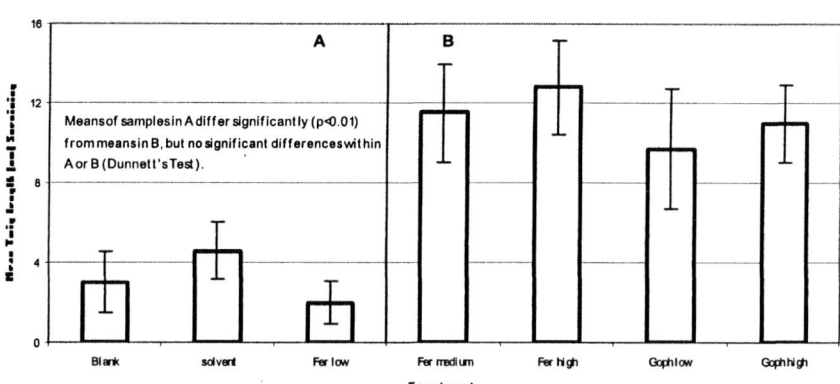

**Figure 6.** Surviving lengths of twigs treated with different concentrations of gopherene diol (Goph) and docosanyl ferulate (Fer). 2003 tests.

## 3.2. Bark Peeled

In 2001, in trial 1 the least bark was removed from twigs treated with stem extract in EtOAc, closely followed by root extract in EtoAc. In trial 2, root extract in MeOH was the most avoided sample in terms of bark eating, paralleling tip cutting. On day 3, the most bark remained on the twigs treated with root-MeOH extract.

In 2002, at 2 sites over 6 days, roots extract in EtOAc was most avoided, while tops extracted first in MeOH, then in hexane, were consumed most ($p < 0.01$). However, the difference between this root extract and the solvent (EtOAc) was not significant, leaving the results open to interpretation.

In 2003, during the first night at site B, the blank was peeled the most, and "ferulate high" the least. On Day 2, "ferulate low" was peeled the most, and "ferulate medium" and "diterpene low" the least. On Day 3, twigs at B and L were too depleted for measurements, while at R, "ferulate high" and "diterpene low" were the most consumed samples (37% of bark eaten), solvent was most avoided (9 % eaten).

Trial 2 replicated Trial 1. The cottontails removed the most bark from the blanks, and the least from "ferulate medium" during Night 1 at R. At B, after Night 2, most bark was still intact at "ferulate medium", and "ferulate low", "diterpene low", and blanks were completely depleted.

Trial 3 tested the high concentration of ferulate once more. The animals clipped very few tips during the first night and almost all of them the next, with little discrimination. However, the lengths of twigs left standing after the $3^{rd}$ night differed significantly: The cottontails avoided "ferulate high" more than solvent alone ($p<0.001$).

## 4. DISCUSSION AND SUMMARY

The experiments with extracts from various plant parts showed that the compounds responsible for inhibiting feeding on the plant occur practically in all parts of the plant, although the roots appear to be better protected against herbivory. In the course of this study, two natural products were identified as possible antifeedants in the crude root extracts, one of them being a novel compound, the diterpene gopherene diol.

The responses of the cottontails varied widely: During the same night, at some sites they fed on all experimental twigs, while they ignored all at another site. Also, their feeding over time varied: Sometimes they discriminated well on the first day; at other times, significant response differences were observed only later. The bioassay had to navigate through these seemingly inconsistent responses. Since the bioassay was run without replacements of twigs after the first night, the responses during the first night are the most meaningful. On the other hand, the cottontails often consumed very little over all during the first night.

We noted that the two solvents were not equivalent. EtOAc appeared to have a slight repellent effect by itself. This can be seen in Figure 4.

Lack of alternative food during late winter may have enhanced the acceptance of some or all samples. During the growing season, by contrast, the repellent effect may be considerably stronger because the herbivores have so many more choices.

These bioassays with wild cottontails have led to the discovery of new compounds first gopherweed compounds that may play a role in protecting gopherweed against mammalian herbivores. Compounds in leafy spurge, *E. esula*, deter cattle and rats from feeding on this plant (Halaweish et al. 2002, 2003; Hein and Miller 1992; Kronberg et al. 1993, 1995). This current project is continuing.

## 5. REFERENCES

Calvin, M., Nemethy, E.K., Redenbaugh, K., and Otvos, J. W., 1982, Plants as a direct source of fuel, *Experientia* **38**:18-22.

Duke, J. A., 1983/2003, *Euphorbia lathyris* L., Euphorbiaceae. Mole plant, petroleum plant, caper spurge, gopher plant. *Handbook of Energy Crops*. http://newcrop.hort.purdue.edu.

Giner, J.-L., Zhao, H., Schroeder, T., and Müller-Schwarze, D., *submitted*, Gopherenediol: A novel diterpenoid isolated as a putative rodent antifeedant from the roots of *Euphorbia lathyris*.

Halaweish, F., Kronberg, S., Hubert, M. B., and Rice, J. A., 2002, Toxic and aversive diterpenes of *Euphorbia esula*, *J. Chem. Ecol.* **28**:1599-1611.

Halaweish, F., Kronberg, S., and Rice, J. A., 2003, Rodent and ruminant ingestive response to flavanoids in *Euphorbia esula*, *J. Chem. Ecol.* **29**:1073-1082.

Hein, D. G., and Miller, S. D., 1992, Influence of leafy spurge on forage utilization by cattle, *J. Range Manage.* **45**:405-407.

Hohmann, J., Evanics, F., Vasas, A., Dombi, G., Jerkovich, G., and Máthé, I., 1999, A novel lathyrane diterpenoid from the roots of *Euphorbia lathyris*, *J. Nat. Prod.* **62**:176-178.

Kronberg, S. L., Lynch, W. C., Cheney, C. D., and Walker, J. W., 1995, Potential aversive compounds in leafy spurge for ruminants and rats, *J. Range Manage.* **21**:1387-1399.

Kronberg, S. L., Muntifering, R. B., Ayers, E. L., and Marlow, C. B., 1993, Cattle avoidance of leafy spurge: A case of conditioned aversion, *J. Range Manage.* **46**:364-366.

Sachs, R. M., Low, C. B., Macdonald, J. D., Awad, A. R., and Sully, M. J., 1981, *Euphorbia lathyris*: a potential source of petroleum-like products, *Cal. Agr.* **35**:29-32.

# INDEX

Accessory olfactory bulb
  electrophysiology of mitral cells, delivery mode of chemoattractant, 256–266
  garter snake, signal transduction in vomeronasal system, 242–254
  golden hamster, individual recognition, 278
  opossum, sexual dimorphism, 283–289
Acetaldehyde, African elephant females, 129, 134
Acetic acid, snake cloacal gland, 49
Acoustic signals/auditory signals/vocalizations
  elephants, *see* Elephants
  frogs, 24
  mice, 206
  toads, 78
  human neonates, 302
*Aethia cristatella* (crested auklet), 8
African civet (*Civetticus civetta*), 116–117
African elephant (*Loxodonta africana*)
  mate choice and mate guarding in, 119
  preovulatory urine analysis, 128–137
Age/maturity/life stage
  elephant
  response to frontalin, 140
  and variability of responses in bioassay, 143, 148, 149
  male mouse, prolactin effects, 171
  social recognition signals, 11
Age of scent signal
  house mouse, 84
  persistence of anti-predator cues in environment, 357–362
Aggression, *see also* Territory
  Asian elephant, testosterone and, 119
  conflict reduction in insect colonies, 12, 13, 14
  house and laboratory mice, 209–213
  lemur stinkfights, 160
  mouse urinary chemicals and, 81

Aggression (*cont.*)
  multicomponent pheromones, 9
  musth as honest signal, 119, 120
*Ailuropoda melanoleuca* (giant panda)
  behavioral responses to substrate odors of conspecifics of opposite sex, 101–109
  chemical signals in urine, 110–117
Alarm signals, 8
  fish, *see* Fish, predation/alarm cues/threat sensitivity in
  latent, 381–387
  snake cloacal gland secretion function, 49
Albatross (*Diomedea exulans*), 403
Albumin, frontalin and, 136
Alcohols and aldehydes, lemur scent marker chemistry, 162, 163
Alloparental care, 15–16
  parallels between social mammals and social insects, 16–17
  social mammals, 14
Allopatric salamander species, sex-specific and species-specific signal responses in, 32–40
Altruistic reproductive behavior, shared reproduction in social groups, 11
*Ambystoma barbouri* (streamside salamander), 375, 376
*Ambystoma maculatum* (spotted salamander), 375, 376, 377, 378
*Ambystoma mexicanum* (axolotl) vomeronasal system, 216–224
*Ambystoma talpoideum* (mole salamander), 220
*Ambystoma texanum* (small-mouthed salamander), 376
*Ambystoma tigrinum* (tiger salamander), 220, 222
American cockroach (*Periplaneta americana*), 9
Amino acid sequencing, 190–191

# INDEX

Amphibians, 3
  alarm signals, latent, 381–387
  *Ascaphus truei* (tailed frog), mate recognition in, 24–30
  chemically mediated life-history shifts in, 373–379
  garter snake-red-backed salamander interaction
    decline in avoidance of predator cues, 365–370
    diet cues and nocturnal shift in antipredator response, 349–355
    persistence of anti-predator cues in environment, 357–362
  red-spotted newts, repelling response, 42–47
  splendipherin, discovery and characterization of, 21–23
  tree frog (*Litoria splendida*), 8, 21–23
Amygdala, 218–219, 279, 280
Anal gland secretions
  beaver, 57–61
  rabbit, material costs of signal production, 79
Annual/seasonal cycles, panda urinary estrogen metabolites, 112, 113, 114, 115
Anogenital glands
  giant panda, 101–109
  lemur, 160, 161, 162, 162–163
  mara (*Dolichotis patagonum*), female enuration in, 89–92
Antarctic prions (*Pachyptila desolata*), 404
Antebrachial gland, lemur, 160, 161, 162, 164
Anthopleurine, 8
Anti-feeding compound, gopherweed, 409–416
Ants, 10
  queenless, 10
  shared reproduction in social groups, 11, 12
  snake cloacal gland secretions as repellent, 49
Anurans
  splendipherin, discovery and characterization of, 21–23
  vomeronasal-orbital communication, 232–233, 236
Aphids, 12
*Apis mellifera* (honeybee), 11, 13
*Aplysia* (sea slugs), 8
Applications, CSV, 4, 5
Aquatic invertebrates, 8
*Archocentrus nigraofasciatus* (convict cichlids), 317
*Ascaphus truei* (tailed frog), mate recognition in, 24–30
Asian elephant (*Elephas maximus*), 118–126, 128
Asian red-bellied newt (*Cynops pyrrhogaster*), 8
*Asterropteryx semipunctatus* (starry goby), 316
Audition, *see* Acoustic signals/auditory signals/vocalizations

Australian cashmere goats (*Capra hircus*), 70
Aversive substances
  farnesenes as, 81
  red-spotted newt, 42–47
Axolotl (*Ambystoma mexicanum*)
  mosaic signals, individual recognition, 269–280
  vomeronasal system, 216–224
    anatomy, 217–219
    chemosensory-guided behavior, 219–222
    electrophysiology, 222–224
    signal transduction in, 256–266

Badgers (*Meles meles*), 120
*Balantiopteryx plicata* (gray sac-winged bat), 93–95, 99
Bank vole (*Clethrionomys glareolus*), 56, 188
Bark beetles, 140
Barnacle egg hatching pheromone, 8
Basic Local Alignment Sequence Tool (BLAST), 191–192
Bass, largemouth (*Micropterus salmoides*), 317
Bats, emballonurid, 93–100
Beaver (*Castor canadensis*), 271
  social recognition signals, 11
  species and subspecies recognition in, 56–61
Bees, 11, 12, 13
Behavioral cues
  chemosensory bioassay, 141–142, 148
  suppression of breeding in subordinates, 16
Behavioral responses
  alarm signals and, 386
  predation and, 382, 383
  snake
    brown tree snake, 52
    strike-induced chemosensory searching (SICS), 397–401
Behavior assessment, 4
Birds, 3, 8
  sense of smell in procellariforms, 403–406
  signalling opportunities, 79
  vomeronasal-orbital communication, 235, 236
*Biziura lobata* (musk duck), 8
Black-tailed deer (*Odocoileus hemionus columbianus*), 56, 271
BLAST (Basic Local Alignment Sequence Tool), 191–192
Blindsnake (*Leptotyphlops dulcis*), 49
Blue petrel (*Halobaena caerulea*), 405–406
*Boiga irregularis* (brown tree snake), 49–54
Brachial gland, lemur, 160, 161, 162, 164
*Brachydanio rerio* (zebrafish), 386
Brazilian gray short-tailed opossum (*Monodelphis domestica*) vomeronasal organ sexual dimorphism, 283–289
Breeding, applications of CSV, 5

# INDEX

Brevicomin, 8, 81, 82
   male-male aggression, 210
   mouse protein characterization, 196
Brook charr (*Salvelinus fontinalis*), 330
Brook stickleback (*Culaea inconstans*), 321–326, 329, 330, 331
Brown tree snake (*Boiga irregularis*), 49–54
Bruce (pregnancy block) effect, 278, 283–284
*Bufo boreas* (western toad), 343, 374
*Bufo bufo* (common toad), 78, 374
*Bufo calamita* (Natterjack toad), 78
2-*sec*-Butyl-4,5,dihydrothiazole, 9

Caecilian vomeronasal-orbital communication, 228, 232, 236
*Caiman crocodilus* (cayman), 266
Calcium, vomeronasal system signal transduction, 244, 248–254
*Callithrix jaccus* (common marmoset), 15, 16–17, 270
Canids, breeding behavior, 14, 15
*Canis lupus* (grey wolf), 15
*Capra* (goat), 120, 254, 265
*Capra hircus* (Australian cashmere goat), 70
*Capreolus capreolus* (roe deer), 157
Capuchin, tufted (*Cebus apella*), 56
*Carassius auratus* (goldfish), 334–340
*Carassius carassius* (crucian carp), 342, 382
Carboxylic acids
   human male sweat, 308, 309, 311
   snake cloacal gland secretion composition, 49
Carp, crucian (*Carassius carassius*), 342, 382
Cascades frog (*Rana cascadae*), 374, 376
Caste
   insect queen pheromones, 12–13
   parallels between social mammals and social insects, 16–17
   social recognition signals, 11
   suppression of breeding in subordinates, 15
*Castor canadensis* (North American beaver), 56–61, 271
*Castor canadensis acadicus*, 58, 60, 61
*Castor canadensis leucodontus*, 58, 60, 61
Castoreum, 57–61
*Castor fiber* (Eurasian beaver), 57, 58, 60–61
*Catharacta skua lonnbergi* (skuas), 404
Cayman (*Caiman crocodilus*), 266
*Cebus apella* (tufted capuchins), 56
Central nervous system
   axolotl vomeronasal system projections, 218–219
   golden hamster
      individual recognition processes in, 278–280
      and odor-based sexual preference, 292, 294, 295–296

*Cephalophus* (duikers), 157
Cervid preorbital apparatus, 152–157
*Cervus elaphus* (red deer), 156, 157
CHARMM, 23
Charr, brook (*Salvelinus fontinalis*), 330
Chemical detection, 5
Chemical composition of pheromones
   discoveries, 7–8
   elephant females, 129, 134
   giant panda, 112, 113, 114, 115, 116
   human male sweat, 308, 309, 311
   multicomponent pheromones, 9
   snake cloacal glands, 49
Chin glands, rabbit, 79
Chromatography, 185–186, 187, 194–195
Cichlid
   convict (*Archocentrus nigrofasciatus*), 317
   Texas (*Cichlasoma zonatum*), 387
*cis*-4-Decenal, 8
*cis*-Civetone, giant panda, 114, 115, 116–117
*Civetticus civetta* (African Civet), 114, 116–117
Clan recognition, 10
*Clemmys japonica* (Japanese tortoise), 266
*Clethrionomys glareolus* (bank vole), 56, 188
Cloacal gland secretions, brown tree snake, 49–54
Cloning, cDNA, 192
Cockroach (*Periplaneta americana*), 9, 10
Colony member recognition, 10, 11
Competitive ability
   giant panda, 107–108
   mice, 209–210
   *Peromyscus maniculatus* (deer mouse) signaling, 77–86
Conflict, *see* Aggression; Territory
Conspecific recognition, *see* Species recognition
Contact chemoreception, 10, 22, 97
Convergence and contrast in insects and vertebrates, 7–17
Convict cichlids (*Archocentrus nigrofasciatus*), 317
Cooperative behaviors, 15–16
   queen pheromones, 13–14
   shared reproduction in social groups, 11–12
   social mammals, 14
Coquerel's sifaka (*Propithecus verreauxi coquereli*), 159–166
Coral reef shrimps, shared reproduction in social groups, 12
*Coregonus* (whitefish), 343
*Cormura brevirostris* (chestnut sac-winged bat), 93–95, 99
Corpora allata, 17
Cottontails (*Sylvilagus floridanus*), 409–416
Countermarking, 80, 84

Courting/mating site, red-spotted newt
  (*Notophthalmus viridescens*) repelling
  response, 42–47
Courtship
  mice, vomeronasal system and, 283
  snake cloacal gland secretion as inibitor, 50
Crested auklet (*Aethia cristatella*), 8
Crocodiles, 235, 236, 266
*Crotalus viridis* (rattlesnake) cloacal gland
  secretion composition, 49
*Crotalus viridis oreganus* (northern Pacific
  rattlesnake)
  poststrike prey location, 397–401
  poststrike trailing cues, 389–395
Crucian carp (*Carassius carassius*), 342, 382
*Culaea inconstans* (brook stickleback), 321–326,
  329, 330, 331
Cultural transmission in fish, 324
*Cynops* (newt), 9
*Cynops ensicauda* (Japanese sword-tailed newt), 22
*Cynops pyrrhogaster* (Japanese fire-belly newt), 8,
  22

Dace, finescale (*Phoxinus neogaeus*), 315
*Dama dama* (fallow deer), 157
Danaidone, 10
*Danaus gillippus* (Queen butterfly), 10
*Daphnia* (water flea), 383–384
Darter (*Etheostoma exile*), 330
Decenal, 8
Decanoic acid, 112, 113, 114, 115, 116
Decapeptides, 8
Deer
  black-tailed (*Odocoileus hemionus
    columbianus*), 56, 271
  fallow (*Dama dama*), 157
  mule (*Odocoileus hemionus hemionus*), 56, 157
  preorbital apparatus, sexual dimorphism in,
    156, 157
  red (*Cervus elaphus*), 156, 157
  roe (*Capreolus capreolus*), 157
  species and subspecies recognition in, 56
  white-tailed (*Odocoileus virginiaus*), 157
Deer mouse (*Peromyscus maniculatus*), 56, 77
Defensive response, cloacal secretion response, 54
Density effects on fish learning, 324
*Desmognathus ochrophaeus* (Alleghany mountain
  dusky salamander), 10
Dialysis, mouse protein characterization, 194–195
Diet
  meadow vole males, protein content and female
    receptivity, 70–75
  nocturnal shift in antipredator response of
    salamander, 349–355
  and pheromone content, 3

Dihydroedulans, 133, 134, 135, 137
*Diomedea exulans* (wandering albatross), 403
Distance for signals, 10
Diving petrel, common (*Pelecanoides urinatrix*),
  404, 405–406
Diving petrel, South Georgian (*Pelecanoides
  georgicus*), 404
Dodecanoic acid, 114
($Z$)-7-Dodecenyl acetate
  discovery of, 7–8
elephant
  African elephant (*Loxodonta africana*)
    females, 133, 136
  assessing chemical communications in, 140,
    141
*Dolichotis patagonum* (mara), female enuration in,
  89–92
Dominance, *see* Rank/status/dominance
*Drosophila melanogaster* (fruit fly), 10
Dwarf hamster (*Phodopus campbelli*), 271

Early research, 1–2
Eastern cottontails (*Sylvilagus floridanus*),
  gopherweed anti-feeding compound,
  409–416
Edulans, 133, 134, 135, 137
Electronic noses, 6
Electrophoresis, 184–185, 187, 188
Electrophysiology, vomeronasal system
  axolotl, 222–224
  garter snake mitral cells, delivery mode of
    chemoattractant and, 256–266
Electrospray ionization (ESI), 186, 187
Elephants, 118–126, 128–137
  African (*Loxodonta africana*), preovulatory
    urine analysis, 128–137
    collection of specimens, 129
    compounds identified, 134–136
    sampling and analysis, 129–132
  Asian (*Elephas maximus*), musth in, 118–126,
    128
    chemical senses, 120–121
    features of, 118–119
    responses, 122–125
    role of, 119–120
    urine bioassays, 121–122
  assessing chemical communication,
    140–149
    behaviors, 144–147
    factors affecting variation, modulation of
      receiver response, 147–149
    main behaviors of chemosensory bioassays,
      141–142
    receiver variation, 143–144
  $Z$-7 pheromone, 7–8

# INDEX

Emballonurid bat wing sacs, 93–100
  evolution, 99–100
  mating systems, 99
  morphology, 93–95, 99
  perfume blending, 96
  use for scent displays, 96–98
Enuration, female mara (*Dolichotis patagonum*), 89–92
Environment
  ecological relationships, 5
  and fish learning, 324, 326
  persistence of anti-predator cues in, 357–362
*Esox lucius* (northern pike), 315, 323, 324, 342, 382, 383
Esters, lemur scent mark chemistry, 162, 163
Estrogens, giant panda urinary metabolites, 110–117
Estrous cycle
  elephant
    assessing chemical communications in, 140, 144, 148
    preovulatory urine analysis, 128–137
    and variability of responses in bioassay, 148, 149
  female pheromone effects, 14
  humans, sweaty smell and, 308–311
  lemur scent marks, 160
  suppression of breeding in subordinates, 16
*Etheostoma exile* (Iowa darters), 315
*Euphorbia lathyris*, anti-feeding compounds from, 409–416
European/northern pike (*Esox lucius*), 315, 323, 324, 342, 382, 383
*Eurycea bislineata* (two-lined salamander), 350–355
Evacuated cannister capture followed by cryogenic trapping (ECC/CT), 129–130
Evolution of chemical cues into signals, 9; *see also* Insect-vertebrate convergence and contrasts
*Exo*-brevicomin, 9

Fallow deer (*Dama dama*), 157
Family membership/kin recognition, 10; *see also* Cooperative behaviors
  beaver anal gland secretions and castoreum, 57, 61
  elephant musth signal contents, 125
  shared reproduction in social groups, 11–12
  tailed frog, 25
Farnesenes, 81, 82, 210
Fathead minnows (*Pimephales promelas*), 314, 315, 329, 339
  learned recognition of heterospecific alarm cues, 321–326

Fathead minnows (*cont.*)
  predation effects on phenotypic and lifestyle variation, 342–347
Fatty acids
  giant panda, 112, 113, 114, 115, 116
  human male sweat, 308, 309, 311
  multicomponent pheromones, 9
  snake cloacal gland secretion composition, 49
Female-female interactions
  elephant, 143–144
  social mammals, 14
Field trials
  fish response to chemical cues, 328–333
  predator diet cues and nocturnal shift in antipredator response, 351–352, 353, 354–355
Fights, *see* Aggression
Finescale dace (*Phoxinus neogaeus*), 315
Fingerprint, olfactory, 200
Fish, 2, 3
  frog predation, 386
  major histocompatibility complex-associated odors, 173, 200
  multicomponent pheromones, 9
  predator labelling, 382, 383
  types of pheromones, 8
  vomeronasal system, 217
Fish, predation/alarm cues/threat sensitivity in, 2, 4, 313–319
  evidence for threat-sensitive behaviors, 316–318
    concentration-dependent shifts in response, 317–318
    increased vigilance toward secondary predator cues, 316–317
    threat-sensitive foraging tactics, 317
  juvenile goldfish, 334–340
  learned recognition of heterospecific alarm cues, 321–326
    ability to learn, 322
    cultural transmission, 324
    ecological factors affecting learning, 324–325
    factors affecting use of heterospecific cues, 325
    learning, 322–323
  overt anti-predator responses, 315–316
  overt responses and stimulus concentration, 314–315
  predation effects on phenotypic and lifestyle variation, 342–347
  recent field experiments, 328–333
    controls, distilled water, 329–330
    controls, unknown heterospecific, 330–331
    validation of ecological factors, 332

Fitness
  competitive ability, *see* Competitive ability
  shared reproduction in social groups, 11
Flehmen, 120, 121, 123, 124, 125, 133
Food location, tailed frog, 25
Frog, 8, 382
  Cascades (*Rana cascadae*), 374, 376, 384
  chemically mediated life-history shifts in embryonic amphibians, 373–379
  common (*Rana temporaria*), 346, 374, 375, 376, 383, 384, 385, 386, 387
  European tree (*Hyla arborea*), 383, 384
  green (*Rana clamitans*), 376, 378
  moor (*Rana arvalis*), 376, 378, 383
  red-legged (*Rana aurora*), 384, 385, 386, 387
  tailed (*Ascaphus truei*), 24–30
  tree, magnificent (*Litoria splendida*), 8, 21–23
  tree, Pacific (*Hyla regilla*), 374, 383
  wood (*Rana sylvatica*), 375, 376, 377, 383
Frontalin, 140, 143
Fruit fly (*Drosophila melanogaster*), 10

Gall thrips, 12
Garter snake (*Thamnophis sirtalis*)
  cloacal secretion composition, 49
  electrophysiology of mitral cells, delivery mode of chemoattractant and, 256–266
  honest signals, 10
  predation on frogs, latent alarm signals, 383
  predation on red-backed salamander
    decline in avoidance of predator cues, 365–370
    diet cues and nocturnal shift in antipredator response, 349–355
    persistence of anti-predator cues in environment, 357–362
  vomeronasal organ sexual dimorphism, 283
  vomeronasal system signal transduction, 242–254
    behavioral and anatomical studies, 243
    calcium imaging, 248–249, 250
    calcium role, 252–253, 254
    chemoattractant purification, 244, 245
    electrical potential generation, 246, 247
    G-protein coupled receptors, 245–246
    model of, 243–244, 253–254
    phosphorylation of membrane bound proteins, 251–252
    receptor binding and second messenger generation, 246, 247
Gas chromatography/mass spectrometry, *see also* Mass spectrometry
  African elephant (*Loxodonta africana*) female pheromones, 130, 133, 134, 135

Gas chromatography/mass spectrometry (*cont.*)
  lemur scent marker chemistry, 161
  mouse protein ligands, 194–195
*Gekko gecko* (Tokay gecko), 266
Gel electrophoresis, 184, 185, 188
Gender
  beaver anal gland secretions and castoreum, 57
  elephant
    response to frontalin, 140
    and variability of responses in bioassay, 143, 148, 149
  golden hamster, 272
  sex-specific and species-specific signal responses in allopatric and sympatric salamander species, 32–40
  social mammals, 14
  social recognition signals, 11
General arousal hypothesis, 67
Genes/genetics
  immediate-early, 279
  laboratory mice, 212, 213
  major histocompatibility complex-associated odors, 173–180, 200
Genetic conflicts of interest, 12
*Geoclemys reevesii* (Reeves' turtle) 266
Geographic isolation, salamander species, 32–40
Giant panda (*Ailuropoda melanoleuca*)
  behavioral responses to substrate odors of conspecifics of opposite sex, 101–109
  chemical signals in urine, 110–117
Glowlight tetra (*Hemigrammus erythrozonus*), 316
Glycopeptides, contact pheromones, 10
Glycoproteins, snake cloacal gland secretions, 49
Goat (*Capra*), 120
  Australian cashmere (*Capra hircus*), 70
  nonvolatile substances, 254, 265
Goby, starry (*Asterropteryx semipunctatus*), 316
Golden hamster (*Mesocricetus auratus*)
  mosaic signals, individual recognition, 269–280
  neurobiology of odor-based sexual preference, 291–297
Golden lion tamarin (*Leontopithecus rosalia*), 16
Goldfish (*Carassius auratus*), 9, 334–340
Gonadal steroids
  African elephant (*Loxodonta africana*) females, 129
  giant panda urinary metabolites, 110–117
  neurobiology of odor-based sexual preference in golden hamster, 291–292
  pheromones based on, 8
  testosterone, *see* Testosterone
Gonadotropins, *see* Luteinizing hormone
Gopherweed anti-feeding compound, 409–416

# INDEX

G-protein coupled receptors, vomeronasal system signal transduction, 243–244, 245–246, 247, 253–254
Grassland mouse (*Mus spretus*), 193–194
Gray short-tailed opossum (*Monodelphis domestica*), 283–289
Gray's waterbuck (*Kobus megaceros*), 120
Greater sac-winged bat (*Saccopteryx bilineata*), 93–100
Green frog (*Rana clamitans*), 376, 378
Green sunfish (*Lepomis cyanellus*), 317–318, 383
Grey wolf (*Canis lupus*), 15
Growth, alarm signals and, 385–386
Guinea pig, 254, 265

Habitat effects on fish learning, 324–325
*Halobaena caerulea* (blue petrel), 405–406
*Halocyptena microsoma* (least storm petrel), 403
Hamster, 11
  dwarf (*Phodopus campbelli*), 271
  golden (*Mesocricetus auratus*), *see* Golden hamster
  Turkish (*Mesocricetus brandtii*), 273
Harderian gland
  muntjac, 152, 153, 154, 155, 157
  vomeronasal-orbital communication, 228–237
Harris's sparrows (*Zonotrichia querula*), 79
Health, colony queen, 13
Hearing, in human neonates, 302
Helper roles, shared reproduction in social groups, 11, 14
*Hemigrammus erythrozonus* (glowlight tetra), 316
Herbivory, 4, 409–416
*Heterocephalus glaber* (naked mole rats), 11, 12, 15
Hippocampus, 279, 280
Historical overview of chemosensory research, 1–6
  early developments, 1–2
  recent trends, 2–6
Homing, tailed frog, 25
Honest signals, 10, 13, 119, 120
Honeybees (*Apis mellifera*), 11, 13
Hormones
  giant panda, urinary steroid metabolites, 110–117
  neurobiology of odor-based sexual preference in golden hamster, 291–292
  and pheromone production, 10
  prenatal endocrine factors, 17
  role of birth process in early olfactory learning, 302–304
  soiled bedding from group-housed females, effects on male reproductive condition, 168–171

House mice (*Mus domestica*), *see* Mice
HPLC, splendipherin, 21–22
Humans, 4
  major histocompatibility complex-associated odors, 173, 200
  menstrual cycle, sweaty smell and, 308–311
  retention of olfactory memories in infants, 300–305
Hyena, spotted, 11–12
*Hyla arborea* (European tree frog), 383, 384
*Hyla chrysoscelis*, (Cope's gray treefrog)382
*Hyla regilla* (Pacific tree frog), 374, 383
Hymenoptera, 12
Hypothalamus, 17
Hypoxanthine-3-$N$-oxide, 314, 329, 330

Imaging studies, CNS activity in individual recognition, 279
Immediate-early genes, 279
Inbreeding, laboratory mice, 212, 213
Individuality/individuals/individual recognition, 8
  beaver anal gland secretions and castoreum, 57
  elephant variability of responses in bioassay, 143–144
  laboratory mice, inbreeding and, 213
  major histocompatibility complex-associated odors, 173–180
  ownership, 199–206
    assessment of signatures, 201–203
    individuality and, 200–201
    individual recognition, 204–206
    mechanism of signal, 203–206
  recognition in golden hamsters, underlying neural mechanisms, 269–280
  social recognition signals, 11
Infant care, 15
Insect predation, latent alarm signals, *see* Latent alarm signals
Insect-vertebrate convergence and contrasts, 7–17
  distance for signals, 10
  evolution of chemical cues into signals, 9
  honest signals, 10
  inhibition or suppression of subordinate reproduction, 16
  molecules used as pheromones, 7–8
  parallels between social mammals and social insects, 16–17
  pheromones and reproduction in, 11–12
  primer pheromones in social animals, 14
  signature odors, 10–11
  in singular cooperatively breeding mammals with high reproductive skew, 15–16
  social insect queens, 12–14
  social recognition, 10–11
  species specificity, 9

Interspecific interactions, recent trends, 4, 5
Invertebrates, frontalin, 140
Invertebtrates, aquatic, 8
Ion exchange chromatography, 186, 187
Iowa darter (*Etheostoma exile*), 315
Isoelectric focusing (IEF), 184, 185

Kairomones, 383–384
Keratin, material costs of signal production, 78–79
King snake (*Lampropeltis getulus*), 54
Kin recognition, *see* Family membership/kin recognition
*Kobus megaceros* (Gray's waterbuck), 120

*Lampropeltis getulus* (common king-snake), 54
Land pheromones, 8
Largemouth bass (*Micropterus salmoides*), 317
Latent alarm signals, 381–387
　evidence for in literature, 384–385
　function of alarm signals, 382–383
　origin of, 383–384
　predator labelling, 382, 383
　preliminary studies, 385–387
　　behavior, 386
　　life-history, 386–387
　　morphology and growth, 385–386
Lateral entorhinal cortex, 278
Learning, 278
　and chemical communication, 3
　human neonate, 300–305
　individual recognition, 280
　mice, MUP signals, 205, 206
　recognition of heterospecific alarm cues in fish
　　ability to learn, demonstrating, 322
　　cultural transmission, 324
　　ecological factors affecting learning, 324–325
　　factors affecting use of heterospecific cues, 325
　　learning, 322–323
Least storm petrel (*Halocyptena microsoma*), 403
Lemur
　mouse (*Microcebus murinus*), 266
　scent marking chemistry in *Lemur catta* and *Propithecus verreauxi coquereli*, 159–166
*Leontopithecus rosalia* (golden lion tamarin), 16
*Lepomis cyanellus* (green sunfish), 317–318, 383
*Lepomis gibbosus* (pumpkinseed sunfish), 318
*Leptotyphlops dulcis* (blindsnake), 49
Life-history
　alarm signals and, 386–387
　chemically mediated shifts in embryonic amphibians, 373–379
　predation effects, 342–347, 382, 383
Life stage, *see* Age/maturity/life stage
Ligands, 8

Light intensity, fish response to chemical cues, 332
Lipids, *see also* Fatty acids
　preorbital apparatus of muntjac, 154, 155,
　snake cloacal gland secretion composition, 49
Lipocalins
　African elephant (*Loxodonta africana*) females, 136
　individuality and ownership, 200–201
*Litoria splendida* (magnificent tree frog), 8
Lizards
　major histocompatibility complex-associated odors, 173
　nonvolatile substances, 264
*Loxondonta africana* (African elephant)
　mate choice and mate guarding in, 119
　preovulatory urine analysis, 128–137
Luteinizing hormone
　elephant (*Loxodonta africana*) females, 129, 133, 134, 136, 137
　elephant male gonadotropin surge, 9
　male mice, female odor effects, 171

Magnificent tree frog (*Litoria splendida*), 8, 21–23
Main olfactory organ
　electrophysiology of mitral cells, 256–266
　elephants, 120–121
　golden hamster, individual recognition in, 278
Major histocompatibility complex (MHC), 200
　experimental paradigm, 174–178
　　habituation/dishabituation, 176–177
　　mate choice and kin recognition, 174–175
　　pregnancy block, 177–178
　　trained discrimination, 175–176
　function of male scent signals in competition, 210–211
　golden hamster, 272
　investigation and preference, 178–179
　mice, 173–180
　significance of MHC-associated odors, 179–180
Mandibular pheromone, queen, 12
Manx shearwater (*Puffinus puffinus*), 404
Mara (*Dolichotis patagonum*), female enuration, 89–92
Marine animals, shared reproduction in social groups, 12
Marmoset, common (*Callithrix jacchus*)
　cooperative breeding, 15
　individual recognition, 270
　parallels between social mammals and social insects, 16–17
　suppression of breeding in subordinates, 16
Mass spectrometry, *see also* Gas chromatography/mass spectrometry
　African elephant female pheromones, 130, 131, 132, 133, 134, 135, 136

Mass spectrometry (cont.)
  mouse protein characterization, 186, 187, 189–195
  polymorphisms, 193–194
  previously characterized proteins/genomes, 189–190
  unidentified proteins/genomes, 190–192
Mate recognition, tailed frog, 24–30
Mate selection
  African elephants, 119
  golden hamsters, neurobiology of odor-based sexual preference, 291–297
Matrix associated laser desorption ionisation (MALDI), 186, 190, 193–194
Maturity, see Age/maturity/life stage
Meadow voles (*Microtus pennsylvanicus*)
  protein content of male diet, 70–75
  self-grooming, 64–68
*Meles meles* (badger), 120
Memory
  and individual recognition, 273
  individual recognition, mosaic signals, 280
  retention of olfactory memories in human infants, 300–305
*Mesocricetus auratus* (golden hamster)
  individual recognition, mosaic signals, 269–280
  neurobiology of odor-based sexual preference, 291–297
*Mesocricetus brandtii* (Turkish hamster), 273
Methyl hexenoic acid, 308–311
Mice
  brevicomin, 8
  honest signals, 10
  major histocompatibility complex-associated odors, 173–180
  male-male aggression, 209–213
    aggression in laboratory mice, 211–213
    competitive ability signaling, 209–210
    function of male scent signals in competition, 210–211
  multicomponent pheromones, 9
  nonvolatile substances, 254, 265
  ownership, 199–206
    assessment of signatures, 201–203
    individuality and, 200–201
    individual recognition, 204–206
    mechanism of signal, 203–206
  signaling of competitive ability by male house mice, 77–86
    costs of marking, 80
    material costs of signal production, 78–79, 81–82
    opportunities for spatial deposition of scent marks, 82–83
    signal age, 84

Mice (cont.)
  signaling of competitive ability (cont.)
    socially related signal costs, 85
    socially related signaling opportunities, 79–80
    spatially related signaling opportunities, 79
    temporally related signaling opportunities, 80
  soiled bedding from group-housed females, effects on males, 168–171
  urinary proteins, see Mouse urinary proteins
  vomeronasal system, 283
Microbial degradation of scent signals, 357–362
*Microcebus murinus* (mouse lemur), 266
*Micropterus salmoides* (largemouth bass), 317
*Microtus ochrogaster* (prairie vole), 15–16, 65
*Microtus pennsylvanicus* (meadow voles)
  protein content of male diet, 70–75
  self-grooming in, 64–68
*Microtus pinetorum* (pine vole), 16
Mitral cells neurophysiology, delivery mode of chemoattractant and, 256–266
Molecular modeling, splendipherin structure, 23
Mole rats
  naked (*Heterocephalus glaber*), 11, 12, 15
  subterranean (*Spalax ehrenbergi*), 56
*Monodelphis domestica* (gray short-tailed opossum), 283–289
Moor frog (*Rana arvalis*), 376, 378
Morphology
  alarm signals and, 385–386
  predation and, 382, 383
  queens, 12
Mosaic signals, 269–280
  individual recognition in social context, 275–277
  methods used to assess, 272–273
  neural mechanisms of individual recognition, 278–280
    central nervous system, 278–280
    main olfactory system and accessory olfactory system, 278
  representation of individuals, 273–275
  sources of individually distinct odors, 271–272
Moths, 9
  honest signals, 10
  Z-7 pheromone, 7–8
Mouse lemur (*Microcebus murinus*), 266
Mouse urinary proteins (MUP), 8, 278
  analytical methods, see Protein chemistry
  costs of producing, 83
  function of male scent signals in competition, 210–211
  male-male aggression, 210
  ownership, 200–201

Mule deer (*Odocoileus hemionus hemionus*), 56, 157
Multicomponent pheromones, 3, 9
Multidimensional scaling ordination (MDS), 161–162
Multisensory context, 3
*Muntiacus muntjak* (barking deer), 153, 156
Muntjac (*Muntiacus reevesi*), 152–157
　harderian gland, 152, 153, 154, 155, 157
　preorbital sac, 153–154, 155, 156
Muscone, 116
*Mus domestica* (house mice), *see* Mice
Musk duck (*Biziura lobata*), 8
*Mus spretus* (grassland mouse), 193–194
Musth, 118–126, 128
　elephant temporal gland secretions, 136
　frontalin and, 140

Naked mole rats (*Heterocephalus glaber*), 11, 12, 15
Neonates, human, retention of olfactory memories in, 300–305
　role of birth process in early olfactory learning in human infants, 302–304
　senses
　　audition, 301–302
　　hearing, 302
　　vision, 300–301
Neural mechanisms
　individual recogntion in golden hamster, 278–280
　central nervous system, 278–280
　main olfactory system and accessory olfactory system, 278
　role of birth process in early olfactory learning in human infants, 302–304
　vomeronasal system electrophysiology
　　axolotl, 222–224
　　garter snake mitral cells, delivery mode of chemoattractant and, 256–266
　vomeronasal system signal transduction, garter snake, 242–254
Newt, 22
　Asian red-bellied (*Cynops pyrrhogaster*), 8, 22
　frog predation, 382
　red-spotted (*Notophthalmus viridescens*), 42–47, 220
　rough-skinned (*Taricha granulosa*), 385
　species specificity of pheromones, 9
Nocturnal shift in antipredator response to diet cues, 349–355
Nonvolatile components, 5, 8
　proteins, *see* Protein chemistry; Proteins
　snake cloacal gland secretion composition, 49

North American beaver (*Castor canadensis*), 11, 56–61, 271
Northern pike (*Esox lucius*), 315, 323, 324, 342, 382, 383
*Notonecta* (backswimmer), 383
*Notophthalmus viridescens* (red-spotted newt), 42–47, 220
Nuclear magnetic resonance (NMR) spectroscopy, splendipherin structure, 22–23

*Oceanodroma leucorhoa* (Leach's storm petrel), 404
Octanal, 8
Octanoic acid, 116
*Odocoileus*, species and subspecies recognition in, 56
*Odocoileus hemionus hemionus* (mule deer), 157
*Odocoileus hemionus columbianus* (black-tailed deer), 56, 271
*Odocoileus virginiaus* (white-tailed deer), 157
Odor-based sexual preference in golden hamster, 291–297
Odor detection, 4
Odor mixtures, social recognition, 10–11
Oestrus, *see* Estrous cycle
Oleic acid, snake cloacal gland secretions, 49
Olfaction in human neonates, 300–301
Olfactory fingerprints, 200
Olfactory signals
　evolution of chemical cues into signals, 9
　suppression of breeding in subordinates, 16
*Oncorhynchus mykiss* (rainbow trout), 314, 323, 339
Opossum
　nonvolatile substances, 254, 265
　vomeronasal organ sexual dimorphism, 283–289
Orbital gland, 157
Ovarian status, *see* Estrous cycle; Reproductive state
Overmarking, 80
Ownership, mouse studies, 199–206
　assessment of signatures, 201–203
　individuality and, 200–201
　individual recognition, 204–206
　mechanism of signal, 203–206

*Pachyptila belcheri* (thin-billed petrel), 405–406
*Pachyptila desolata* (Antarctic prion), 404
Pacific tree frog (*Hyla regilla*), 374, 383
Palmitic acid, 49
Parahippocampal area, 278

*Pelecanoides georgicus* (south Georgian diving petrel), 404
*Pelecanoides urinatrix* (common diving petrel), 404, 405–406
Peptide mass fingerprinting, 189–190
Peptides, 8
 specificity, mechanisms of achieving, 9
 splendipherin, discovery and characterization of, 21–23
*Perca flavescens* (yellow perch), 314, 323
*Perca fluviatus* (perch), 382, 386
*Periplaneta americana* (American cockroach), 9, 10
Periplanone-B, 9
Peri-rhinal cortex, 278
*Peromyscus maniculatus* (deer mouse), 56, 77
*Peromyscus polionotus* (Oldfield mouse), 56
*Peropteryx macrotis* (Peter's sac-winged bat), 93–95, 99
Pest control, 5
Petrel
 blue (*Halobaena caerulea*), 405–406
 common diving (*Pelecanoides urinatrix*), 404, 405–406
 Leach's storm (*Oceanodroma leucorhoa*), 404
 least storm (*Halocyptena microsoma*), 403
 South Georgian diving (*Pelecanoides georgicus*), 404
Phenotypic variation, predation effects in fish, 342–347
Pheromones
 primer, and reproduction in social animals, 14
 and reproduction in social groups, 11–12
 social insect queens, 12–14
 terminology, 3
Philopatry, 15, 17
*Phodopus campbelli* (dwarf hamster), 271
*Phoxinus neogaeus* (finescale dace), 315
Phylogeny, Emballonuridae, 94
Physiological condition, *see* Reproductive state/condition
Pike (*Esox lucius*), 315, 323, 324, 342, 382, 383
*Pimephales promelas* (fathead minnow), 314, 315, 329, 339
 learned recognition of heterospecific alarm cues, 321–326
 predation effects on phenotypic and lifestyle variation, 342–347
Pine vole (*Microtus pinetorum*), 16
Plankton, 8
Plant compounds
 gopherweed anti-feeding compound, 409–416
 pheromones derived from, 10

*Plethodon cinereus* (red-backed salamander)
 predation by garter snakes
  decline in avoidance of predator cues, 365–370
  diet cues and nocturnal shift in antipredator response, 349–355
  persistence of anti-predator cues in environment, 357–362
 vomeronasal organ sexual dimorphism, 283
*Plethodon jordani* (Jordan's salamander), 10, 22
*Plethodon* (*P. shermani*, *P. montanus*, and *P. teyahalee*), sex- and species-specific signal responses in allopatric and sympatric species, 32–40
Poisons, plant, 10
Polar molecules, 8
Policing, insect social controls, 14
Polyacrylamide gel electrophoresis, 184, 185, 190
Polypeptides, 8
Population density, social mammals, 14
Prairie voles (*Microtus ochrogaster*), 15–16, 65
Predation, 4
 chemically mediated life-history shifts in embryonic amphibians, 373–379
 differential firing of mitral cells, delivery mode of chemoattractant, 256–266
 early developments and recent trends, 2, 4
 fish, *see* Fish, predation/alarm cues/threat sensitivity in
 garter snake (*Thamnophis sirtalis*), signal transduction in, 242–254
 garter snake–red-backed salamander
  decline in avoidance of predator cues, 365–370
  diet cues and nocturnal shift in antipredator response, 349–355
  persistence of anti-predator cues in environment, 357–362
 herbivory, gopherweed anti-feeding compounds, 409–416
 latent alarm signals, 381–387
 life history and phenotypic variation in aquatic vertebrates, 342–347
 rattlesnakes
  poststrike prey relocation, 397–401
  poststrike trailing cues, 389–395
 snake cloacal gland secretion function, 49
 tailed frog recognition, 25
Preening, material costs of signal production, 78–79
Preganediol, and civet, 114
Pregnancy block effect, 278, 283–284
Prenatal hormonal environment, 17
Preorbital apparatus, muntjac, 152–157

Preputial gland
  house mouse, 79
  mice, 210
Prion, Antarctic (*Pachyptila desolata*), 404
Prion, thin-billed (*Pachyptila belcheri*), 405–406
Procellariforms, sense of smell in, 403–406
Proceptivity, 71, 111
Prolactin, 163, 164, 165
Propanoic acid, snake cloacal gland secretions, 49
*Propithecus verreauxi coquereli* (Coquerel's sifaka), 159–166
Prosimians, lemur scent marker chemistry, 159–166
Prostaglandin pheromones, 9
Protein chemistry, 183–196
  analysis of protein ligands, 194–196
    protein-associated molecules, 194–195
    protein-ligand interactions, 196
  identification, 188–192
    previously characterized proteins/genomes, 189–190
    strategies for, 188
    unidentified proteins/genomes, 190–192
  polymorphisms, characterization of, 193–194
  scent mark protein characterization, 183–196
  techniques, 183, 184–187
    chromatography, 185–186, 187
    combined methods, 187
    electrophoresis, 184–185
    mass spectrometry, 186, 187
Proteins, 8
  elephant pheromones, 136, 140
  major histocompatibility complex-associated odors, 173–180
  major urinary protein (MUP) complex, 200–201
  snake cloacal gland secretion composition, 49
*Pseudemys scripta* (slider), 266
*Puffinus puffinus* (Manx shearwater), 404
Pumpkinseed sunfish (*Lepomis gibbosus*), 318
Pyridine-*N*-oxide, 329, 330

Queen butterflies (*Danaus gillipus*), 10
Queenless ants, 10
Queens, insect, 13
  parallels between social mammals and social insects, 16–17
  shared reproduction in social groups, 12

Rabbits
  gopherweed anti-feeding compound, 409–416
  material costs of signal production, 79
Rainbow trout (*Oncorhynchus mykiss*), 314, 323, 339
*Rana arvalis* (moor frog), 376, 378, 383
*Rana aurora* (red-legged frog), 384, 385, 386, 387

*Rana cascadae* (Cascades frog), 374, 376, 384
*Rana clamitans* (green frog), 376, 378
*Rana dalmatina* (agile frog), 383
*Rana esculenta* (edible frog), 383, 384
*Rana lessonae* (pool frog), 383
*Rana sylvatica* (wood frog), 374, 376, 377, 383
*Rana temporaria* (common frog), 346, 374, 375, 376, 383, 384, 385, 386, 387
*Rangifer tarandus* (reindeer), 156, 157
Rank/status/dominance
  elephant
    musth signal contents, 125
    response to frontalin, 140
    and variability of responses in bioassay, 143, 147, 148
  farnesenes and, 81
  honest signals, 10
  lemur antebrachial gland marking rate, 160
  mice
    and density of scent marks of male house mouse, 82
    male-male aggression, 209–210
    mouse scent mark signals, 201
  signaling opportunities, 79
  social mammals versus social insects, 14, 16–17
  suppression of breeding in subordinates, 15
Rats
  major histocompatibility complex-associated odors, 272
  nonvolatile substances, 254, 265
Rattlesnake (*Crotalis viridis*), cloacal gland secretion composition, 49
Rattlesnake (*Crotalus viridis oreganus*)
  poststrike prey relocation, 397–401
  poststrike trailing cues, 389–395
Recent research trends and developments, 2–6
Reception studies, 4
Red-backed salamander, *see Plethodon cinereus*
Red deer (*Cervus elaphus*), 156, 157
Red-legged frog (*Rana aurora*), 384, 385, 386, 387
Red-spotted newt (*Notophthalmus viridescens*), 42–47, 220
Redirected behavior hypothesis, 66
Reindeer (*Rangifer tarandus*), 120, 156, 157
Repelling response/aversive substances
  farnesenes as, 81
  red-spotted newts, 42–47
Reproduction/reproductive behavior, 2
  amphibian
    axolotl courtship, 219–222
    red-spotted newt (*Notophthalmus viridescens*) repelling response, 42–47
    splendipherin, discovery and characterization of, 21–23

# INDEX

Reproduction/reproductive behavior (*cont.*)
  axolotl courtship (*cont.*)
    tailed frog (*Ascaphus truei*), mate recognition in, 24–30
  elephant
    African elephant (*Loxodonta africana*), preovulatory urine analysis, 128–137
    musth, 118–126, 128, 136, 140
  emballonurid bat wing sac role, 93–100
  giant panda, *see* Giant panda
  golden hamster, neurobiology of odor-based sexual preference, 291–297
  insect-vertebrate convergence and contrasts
    inhibition or suppression of subordinate reproduction, 16
    insect queen pheromones, 12–13
    parallels between social mammals and social insects, 16–17
    pheromones and reproduction in social groups, 11–12
    primer pheromones in social animals, 14
    in singular cooperatively breeding mammals with high reproductive skew, 15–16
  mara (*Dolichotis patagonum*), female enuration, 89–92
  meadow voles (*Microtus pennsylvanicus*), protein content of male diet, 70–75
  mice, vomeronasal system and, 283–284
  multicomponent pheromones, 9
  odor-based sexual preference in golden hamster, 291–297
  shared reproduction in social groups, 11–12
  snake cloacal gland secretion and, 49, 50
Reproductive skew
  shared reproduction in social groups, 11–12
  social mammals, 14
Reproductive state/condition, *see also* Estrous cycle
  elephant
    musth, 118–126, 128, 136, 140
    and variability of responses in bioassay, 143, 144
  giant panda, urinary steroid metabolites, 110–117
  golden hamster, 272
  insect social controls, 14
  mice, effects of soiled bedding from group-housed females, 168–171
  social recognition signals, 11
Reptiles, 3
Ring-tailed lemur (*Lemur catta*), 159–166
Roe deer (*Capreolus capreolus*), 157
Rough-skinned newt (*Taricha granulosa*), 385
Ryanodine, 248

*Saccopteryx bilineata* (greater sac-winged bat), 93–100
*Saccopteryx leptura* (lesser white-lined bat), 93–95, 99
Saddle-back tamarin (*Saguinus fuscicollis*), 11
Salamander, two-lined (*Eurycea bislineata*), 350–355
Salamander (*Ambystoma* species)
  axolotl (*Ambystoma mexicanum*) vomeronasal system, 216–224
  mole (*Ambystoma talpoideum*), 220
  small-mouthed (*Ambystoma texanum*), 376
  spotted (*Ambystoma maculatum*), 375, 376, 377, 378
  streamside (*Ambystoma barbouri*), 375, 376
  tiger (*Ambystoma tigrinum*), 220, 222
Salamander (*Plethodon* species)
  contact and nonvolatile pheromones, 10, 22, 264
  garter snake-red-backed salamander interaction
    decline in avoidance of predator cues, 365–370
    diet cues and nocturnal shift in antipredator response, 349–355
    persistence of anti-predator cues in environment, 357–362
  sex- and species-specific signal responses in allopatric and sympatric species, 32–40
    collection and maintenance of specimens, 34
    data analysis, 35
    female behavior, 36–37
    female odor discrimination, 35
    male behavior, 37–38
    male odor discrimination, 35–36
    terrestrial species, 33–34
    test protocol, 34–35
  vomeronasal organ sexual dimorphism, 283
*Salvelinus fontinalis* (brook charr), 330
SDS-PAGE, 184, 185, 190
Sea anemones, 8
Seabirds, sense of smell in procellariforms, 403–406
Sea slug (*Aplysia*), 8
Sex, *see* Gender
Sex pheromones, multicomponent, 9
Sexual dimorphism
  elephant temporal glands, 119–120
  neurobiology of odor-based sexual preference in golden hamster, 291–297
  preorbital apparatus of muntjac, 152–157
  vomeronasal system
    axolotl, 216–224
    golden hamster, 278
    gray short-tailed opossum, 283–289
    salamander, 37
Sexual isolation, salamander species, 32–40

Sexual preference, golden hamster (*Mesocricetus auratus*), 291–297
  central olfactory system structures, 294
  vomeronasal system and basal diencephalic structures, 295–296
  vomeronasal versus olfactory system control, 292–294
  anatomy, 292
  female preference, 293–294
  male preference, 292–293
Shearwater, Manx (*Puffinus puffinus*), 404
Shoals, fish response to chemical cues in presence of, 332
Short-tailed opossum, gray (*Monodelphis domestica*), 283–289
Sifaka, Coquerel's (*Propithecus verreauxi coquereli*), 159–166
Signal transduction, garter snake vomeronasal system, 242–254
Signature odors, 10–11
Silefrin, 22
Singular breeders, mammals, 14
Size exclusion chromatography, 186, 194–195
Skew, reproductive, 11–12
Skin gland anatomy, 2
Skin pheromones
  snake, 10, 49
  splendipherin, discovery and characterization of, 21–23
Skuas (*Catharacta skua lonnbergi*), 404
Small-mouthed salamander (*Ambystoma texanum*), 376
Snakes
  brown tree snake (*Boiga irregularis*) cloacal secretions, effects of, 49–54
  garter (*Thamnophis sirtalis*), *see* Garter snake
  honest signals, 10
  rattlesnakes
    poststrike prey relocation, 397–401
    poststrike trailing cues, 389–395
    vomeronasal-orbital communication, 228, 230–232
Social groups/social communications
  insect-vertebrate convergence and contrasts
    parallels between social mammals and social insects, 16–17
    pheromones and reproduction, 11–12
    primer pheromones in social animals, 14
    recognition/signature odors, 10–11
    social insect queens, 12–14
  major histocompatibility complex-associated odors, 173–180
  mice, 79–80
  pheromones and reproduction, 11–12
Socially related signaling opportunities, 79–80

Social recognition, 10–11
Social status, *see* Rank/status/dominance
Sodefrin, 8, 22
Solid-phase microextraction
  African elephant (*Loxodonta africana*) female pheromones, 131, 132, 133, 134, 135, 136
  mouse protein characterization, 195
  panda urinary steroid metabolites, 112, 113
Solubility, 8
South Georgian diving petrels (*Pelecanoides georgicus*), 404
*Spalax ehrenbergi* (subterranean mole rat), 56
Spatial deposition of scent marks, mice, 83–84
Spatially related signalling opportunities, 79
Spatial orientation, tailed frog, 25
Species recognition, 9, 11
  barnacle larvae, 8
  *Castor canadensis* (North American beaver), 56–61
  elephant musth signal contents, 125
  salamanders, 32–40
  snake cloacal gland secretion effects, 49
  splendipherin, 22
  tailed frog, 25
Species specificity of newt pheromones, 9
Sperm, contact pheromones, 10
*Sphenodon* (tuatara), 233–234, 236
Spiders, 12
Splendipherin, 21–23
Spotted hyena (*Crocuta crocuta*), 11–12
Spotted newt (*Notophthalmus viridescens*), 42–47, 220
Spotted salamander (*Ambystoma maculatum*), 375, 376, 377, 378
Squamates, *see* Snakes
Starry gobies (*Asterropteryx semipunctatus*), 316
Status, *see* Rank/status/dominance
Steroid hormones, *see* Gonadal steroids
Stickleback, brook (*Culaea inconstans*), 321–326, 329, 330, 331
Stickleback, threespine (*Gasterosteus aculeatus*), 384
Stinkfights, lemur, 160
Storm petrel (*Halocyptena microsoma*), 403
Storm petrel, Leach's (*Oceanodroma leucorhoa*), 404
Streamside salamander (*Ambystoma barbouri*), 375, 376
Stress, maintenance of dominance, 17
Strike-induced chemosensory searching (SICS), 397–401
Subiculum, 278, 280
Subordinates, *see* Rank/status/dominance

Subspecies recognition
  *Castor canadensis* (North American beaver), 56–61
  tamarin, 11
Subterranean mole rat (*Spalax ehrenbergi*), 56
Sunfish, green (*Lepomis cyanellus*), 317–318, 383
Sunfish, pumpkinseed (*Lepomis gibbosus*), 318
Swordtails (*Xiphophorus helleri*), 323, 324, 325, 330
*Sylvilagus floridanus* (eastern cottontail), 409–416

Tailed frog (*Ascapus truei*), mate recognition in, 24–30
*Taricha granulosa* (rough-skinned newt), 385
Tarsal gland secretion, deer, 56
Temporal cortex, 278
Temporal gland secretions
  African elephant females, 136
  Asian elephant, 119–120, 140
Temporally related signalling opportunities, 80
Terminology, 3
Termites, 11, 12
Territory
  beaver anal gland secretions and castoreum, 57
  emballonurid bat signals, 97
  giant panda, 107
  honest signals, 10
  mice
    countermarking, 84
    scent mark signals, 201
Testosterone
  elephant
    aggression in Asian elephant, 119
    musth state parameters, 122
  mice, female bedding effects on, 170, 171
Tetra, glowlight (*Hemigrammus erythrozonus*), 316
Texas cichlid (*Cichlastoma zonatum*), 387
*Thamnophis sirtalis*, see Garter snake
Thapsigargin, 248
Theaspiranes, 133, 134, 135, 137
Thiazole, 82, 196, 210
Thigmotaxis, brown tree snake, 51
Thin-billed petrel (*Pachyptila belcheri*), 405–406
Threat sensitivity, fish, see Fish, predation/alarm cues/threat sensitivity in
Threespine sticklebacks (*Gasterosteus aculeatus*), 384
Throat glands, lemur, 162, 163, 164
Tiger moth (*Utetheisa ornatrix*), 10
Tiger salamander (*Ambystoma tigrinum*), 220, 222
Toad
  common (*Bufo bufo*), 374
  signaling competitive ability, 78
  western (*Bufo boreas*), 343, 374

Tree-frog
  *Hyla arborea* (European treefrog), 383, 384
  magnificent (*Litoria splendida*), 8, 21–23
  Pacific (*Hyla regilla*), 374, 383
Trimethylamine
  African elephant (*Loxodonta africana*) females, 129
  snake cloacal gland secretion composition, 49
*Triturus alpestris* (alpine newt), 342
*Triturus helveticus* (palmate newt), 342
Trout, rainbow (*Oncorhynchus mykiss*), 323, 339
Tufted capuchins (*Cebus apella*), 56
Turkish hamster (*Mesocritus brandtii*), 273
Turtle, 235, 236, 266
Two-component pheromones, 9
Two-lined salamanders (*Eurycea bislineata*), 350–355

Urine
  African elephants, female, 136
  steroid metabolites, giant panda, 110–117
  mara (*Dolichotis patagonium*), female enuration, 89–92
  mouse, 8; see also Mouse urinary proteins
  musth, 120
Urodeles, vomeronasal-orbital communication, 233, 236
*Utetheisa ornatrix* (tiger moth), 10

Vertebrates
  convergence and contrast in insects and vertebrates, 7–17
  ethology, early developments and recent trends, 2
Vision, in human neonates, 301–302
Visual signals
  fish learning, 325, 332
  frogs, 25, 26, 27
  material costs of signal production, 78–79
  predator response in fish, 316
  suppression of breeding in subordinates, 16
Vocalizations, see Acoustic signals/auditory signals/vocalizations
Volatile components
  mouse urine, 81
    major urinary protein (MUP) complex, 203, 204, 205, 206
    protein characterization, 194–195, 196
  snake cloacal gland secretion composition, 49
Volatility, 8
Vole
  bank (*Clethrionomys glareolus*), 56, 188
  meadow (*Microtus pennsylvanicus*), 64–68
  nonvolatile substances, 254, 265

Vole (*cont.*)
   parallels between social mammals and social insects, 16–17
   prairie (*Microtus ochrogaster*), 65
   suppression of breeding in subordinates, 16
Vomeronasal system, 4
   axolotls, 216–224
   differential firing of mitral cells, delivery mode of chemoattractant and, 256–266
   elephants
      chemosensory bioassay, 142, 146
      function of, 120–121
   golden hamsters, individual recognition, 278
   mice, MUP ownership signals, 203–204, 206
   opossum, 283–289
   orbital connections, 228–237
      anurans, 232–233, 236
      caecilians, 228, 232, 236
      mammals, 234–235, 236
      *Sphenodon*, 233–234, 236
      squamates, 228, 230–232, 236
      turtles, crocodiles, and birds, 235, 236
      urodeles, 233, 236
   salamander, sexual dimorphism, 37

Wandering albatross (*Diomedea exulans*), 403
Wasps, shared reproduction in social groups, 11, 12
Water fleas (*Daphnia*), 383–384
Water solubility, 8
Weaver ant colony, 13
Western toad (*Bufo boreas*), 343, 374
Whitefish (*Coregonus*), 343
White-tailed deer (*Odocoileus virginiaus*), 157
Wildlife management, 5
Wood frog (*Rana sylvatica*), 376, 378
Worker caste
   fertility suppression in social mammals, 15
   insect queen pheromones, 12–13
   policing, 14

*Xenopus laevis*, 384
*Xiphophorus helleri* (swordtails), 323, 324, 325, 330

Yellow perch (*Perca flavescens*), 314, 323

Z-7-dodecen-1-yl acetate, 7–8, 133, 136, 140, 141
Zebrafish (*Brachydanio rerio*), 386